# Neuroprotective Effects of Phytochemicals in Brain Ageing

Surajit Pathak • Antara Banerjee
Editors

# Neuroprotective Effects of Phytochemicals in Brain Ageing

Springer

*Editors*
Surajit Pathak
Faculty of Allied Health Sciences
Chettinad Academy of Research and Education (CARE), Chettinad Hospital and Research Institute (CHRI)
Chennai, Tamil Nadu, India

Antara Banerjee
Faculty of Allied Health Sciences
Chettinad Academy of Research and Education (CARE), Chettinad Hospital and Research Institute (CHRI)
Chennai, Tamil Nadu, India

ISBN 978-981-99-7268-5 ISBN 978-981-99-7269-2 (eBook)
https://doi.org/10.1007/978-981-99-7269-2



This Springer imprint is published by the registered company Springer Nature Singapore Pte Ltd.
The registered company address is: 152 Beach Road, #21-01/04 Gateway East, Singapore 189721, Singapore

*Dedicated to our beloved mother the late Mrs. Bela Rani Pathak.*

# Preface

This book, ***Neuroprotective Effects of Phytochemicals in Brain Ageing***, deals with the mechanism of brain ageing and the prospective therapeutic approaches rendered by phytochemicals for the treatment.

The effects of ageing on the brain and cognition are widespread and have multiple aetiologies. This book will discuss how natural compounds can modulate brain ageing and repair function of degenerated neurons. This book will also address some recent crucial topics on cognitive impairment, hormetic phytochemicals in brain ageing, and molecular mechanisms of action of phytochemicals to delay neurological disorders and brain ageing.

As we age, our bodies undergo natural changes, and the brain is no exception. It is particularly vulnerable to the impact of senescence, making it even more crucial to explore the connections between ageing and neurological health. Through advanced neuroimaging techniques, scientists have discovered brain age as a biomarker that can predict our chronological age. This discrepancy between predicted and actual age, termed brain ageing, provides valuable insights into age-related brain alterations.

This book explores the details of brain ageing and its potential links to neurodegenerative diseases. This book will explore the effects of ageing on the DNA methylation process and the aggregation of specific proteins like phosphorylated tau, α-synuclein, and Aβ. Yet, it will also uncover the gaps in our knowledge, as we still seek to understand whether these age-related changes lead to cognitive impairment.

Currently, treatments for neurodegenerative diseases are limited, often providing only temporary relief with unwanted side effects. This brings us to an exciting area of exploration—the potential role of dietary interventions and nutraceuticals in mitigating neurological impairment. Phytochemicals hold promise as supportive therapeutic agents. This book will examine the neuroprotective properties of various phytochemicals and their potential to counteract brain ageing.

We genuinely hope that basic scientists, nutritionists, and doctoral students, working on any neurological disorders, or related research, or research linked to alternative medicine for neurological disorders will find this book to be useful.

We sincerely thank Springer Nature for their help in publishing this book.

Chennai, Tamil Nadu, India Surajit Pathak
Chennai, Tamil Nadu, India Antara Banerjee

# Contents

# About the Editors

**Surajit Pathak, Ph.D** is currently working as a Professor at Chettinad Academy of Research and Education, Kelambakkam, Chennai, since 2015. He worked as a Principal Research Scientist in Linkoping University, Linkoping, Sweden (2022 June–Dec 2022). He has received his Ph.D. degree from India in 2007 and completed his Post-Doctoral training (2007–2015) at the University of Alabama, USA, University of Padova, Italy, and University of Linkoping, Sweden.

**His current research focuses on the role of intracellular and extracellular microRNAs for the early detection of colorectal cancer.** He has published more than **125 research and review articles** in peer-reviewed international journals of repute with 3238 citations and 32 h-index. He received various research grants from the national and international funding agencies. He has active research collaboration with top research laboratories in Sweden, Italy, Norway, Turkey, Mexico, and India.

He is the editorial board member of various renowned international high-impact journals. Prof. Pathak has edited several books for **Springer Nature:** ***Cancer Stem Cells: New Horizons in Cancer Therapies*****; MRW book** ***Handbook of Animal Models and its Uses in Cancer Research*****; and** ***Evidence-Based Functional Foods for Prevention of Age-related Diseases***. He is a member of various professional research bodies of India, including **Indian Association of Cancer Research, Indian Science Congress.**

**Antara Banerjee, Ph.D** is presently working as an Associate Professor at Chettinad Academy of Research and Education, Chennai, since 2016. She completed her Ph.D. in 2008 and pursued her Post-Doctoral research at the University of Padova, Italy (2009–2012), and later she joined as Senior Research Associate at the University of Linkoping, Sweden (2013–2015).

She has published over **120 research and review articles** in peer-reviewed international and national journals with a citation index of **2403, an i10-index of 56, and an h-index of 25**. She serves as an editorial board member and reviewer of international and national journals. She serves as the editor of books published by Springer, namely ***Cancer Stem Cells: New Horizons in Cancer Therapies***; ***Handbook of Animal Models and its Uses in Cancer Research***; **and** ***Evidence-Based Functional Foods for Prevention of Age-related Diseases***. She is a lifetime member of Indian Association of Cancer Research.

# Phytochemicals as Protective Agents for Brain Aging

# 1

Mohamed Y. Zaky, Hadeer M. Morsy, and Osama M. Ahmed

**Abstract**

Aging is the major danger agent for many age-related disorders, including neurological diseases. Knowing the biological mechanisms of aging is critical to achieve brain health. In this regard, brain aging is characterized by neurophysiological functions gradually decline, impaired adaptive neuroplasticity, dysregulation of neuronal calcium homeostasis, neuroinflammation, and oxidatively modified molecules and organelles. Several molecular and cellular indicators of brain aging include mitochondrial dysfunction, accumulation of oxidatively damaged molecules, impaired lysosome and proteasome function, dysregulation of neuronal calcium homeostasis, inflammation, and impaired neurogenesis. Thus, therapeutic approaches target the beneficial effects of phytochemicals on aging-associated impairment of brain function. In cellular models of neurological diseases, phytochemicals, or food's second metabolites, preserve neuronal cells against cell death, and the neuroprotective effect is attributed to antioxidant, antiinflammatory, and antiaging properties. Phytochemicals activate the neuronal antioxidant defense and survival. Mechanisms associated with neuroprotective effects of phytochemicals on the brain such as cellular defense against oxidative stress (the Nrf2-ARE antioxidant system), cell survival system (the TrkB signaling pathway), and crosstalk between TrkB signaling pathway and Nrf2-ARE antioxidant system were outlined. The recent evidence in this chapter describes the hallmarks of brain aging which affect impairment in brain function, acknowledgment of antiaging,

Mohamed Y. Zaky and Hadeer M. Morsy contributed equally with all other contributors.

M. Y. Zaky · H. M. Morsy · O. M. Ahmed (✉)
Physiology Division, Zoology Department, Faculty of Science, Beni-Suef University, Beni-Suef, Egypt
e-mail: osama.ahmed@science.bsu.edu.eg



S. Pathak, A. Banerjee (eds.), *Neuroprotective Effects of Phytochemicals in Brain Ageing*, https://doi.org/10.1007/978-981-99-7269-2_1

antioxidant, and antiinflammatory effects of phytochemicals against oxidative stress and emphasizes their importance as a protective agent for brain aging and neurodegenerative disease resistance.

**Keywords**

Brain aging · Phytochemicals · Trkb signaling pathway · Nrf2-ARE antioxidant system

## 1.1 Introduction

Brain aging is the main risk factor for neurodegenerative diseases, which is defined by continuous declines in memory and cognition. Senescent cells proliferate with age and contribute to brain dysfunction (Rossiello et al. 2022). It is characterized by a progressive deterioration of physiological processes of neurons, diminished neuroplasticity adaptation, uncontrolled calcium homeostasis of neurons, inflammation of neurons, and modified oxidatively molecules and organelles. Lots of methods contribute to brain aging, including oxidative stress (OS) elevation, inflammation, and disruptions in energy utilization, which contribute to neurodegenerative disorders (Zia et al. 2021).

Moreover, the impacts of aging on the brain and cognition are prevalent and have a variety of reasons. Aging influences molecules, cells, vasculature, gross morphology, and cognitive ability. Our brains shrink as a result of age, especially in the frontal cortex. The threat of stroke and ischemia raises as the blood vessels shrink leading to hypertension, and the white matter begins to develop lesions. Impairment of memory occurs due to age, and stimulation of the brain for memory exercises becomes more bilateral. This could be an effort for restitution and acquirement of new networks, or it might be due to specific areas that are not any more readily accessible. Brain aging is affected by neurotransmitters, genetics, experience, and hormones (Peters 2006).

Several of neuron's functions such as chemical, biological, and physical are changed as we age, which causes memory failure, alterations in behavior, dementia, cognitive function loss, and immune system dysfunction. Furthermore, aging is a major danger agent for prevalent neurodegenerative diseases and disorders (NDDs) (Zia et al. 2021).

Rather than just high brain aging, rates of neuron loss have also been attributed to modest modifications in the neural function and structure in specific neural circuits. Neuronal degeneration in the aging brain is compensated by increasing dendritic arbors and synaptic connections. In the case of neurodegenerative disorders associated with age, the brain loses dendritic arbors and synaptic contacts. As a result, it is unable to adjust for neuronal loss (Samarghandian et al. 2019; Huffman 2012). There is no efficient treatment for aging-related neurological disorders.

Nutritional interventions may be regarded as a viable strategy among therapeutic interventions aimed at preventing or delaying the normal and pathological aging processes (Bastianetto and Quirion 2002).

Due to the brain being the greatest complicated body organ, not every medication is approved for the therapy of age-related neurodegenerative diseases and disorders (NDDs). The present treatments are inefficient in treating NDDs with multifunctional pathogenic disorders, including aging (Vijayakumar et al. 2016). In recent years, there has been a rise in scientific study on medicinal plants, and their active ingredients may aid in the discovery of new multifunctional therapeutic agents. In the previous 30 years, the Drug Administration (FDA) and US Food has approved numerous plant-based natural and synthetic neuroprotective medicines (Ramalingam et al. 2018).

Recent research strongly suggests that phytochemicals may have significant protective, immunomodulatory, antioxidant, and/or antiinflammatory effects in the background of brain aging (Corbi et al. 2016; Forni et al. 2019). Phytochemicals are plant nonnutrient active components present in almost all plants. They are linked to a lower chance of chronic and age-related deteriorating diseases (Qaragholi et al. 2022). They have protected neuronal cells from apoptosis in neurodegenerative disease, have neuroprotective activity, and inhibit mitochondrial membrane permeabilization (Wu et al. 2017).

Moreover, the target of this chapter is to acknowledge brain aging's cellular and molecular markers and imply the mechanisms of phytochemicals in protecting brain cells from aging.

## 1.2 Cellular and Molecular Hallmarks of Brain Aging

Brain aging is marked by a complex interplay of cellular and molecular changes. These include DNA damage and mutations, shortened telomeres (protective caps on chromosomes), alterations in gene expression, disrupted protein balance, mitochondrial decline, the rise of senescent cells, changes in nutrient sensing, and impaired communication between cells. These interconnected hallmarks contribute to the decline in brain function and increased vulnerability to neurodegenerative diseases that often accompany the aging process (Machiela and Southwell 2020).

### 1.2.1 Mitochondrial Dysfunction

Mitochondria are the "energy powerhouse of the cell" due to cell's allowance to generate ATP production more than that of oxidative phosphorylation alone. They are also essential to our survival because they offer cellular function by supporting biochemical energy (Loh and Reiter 2022). Mitochondria are required for energy production, calcium homeostasis maintenance, and the regulation of different intracellular signaling pathways (Li et al. 2023). Mitochondria can expand, merge with others, and split. Moreover, lysosomes can clear the impairment of mitochondria

(Mattson and Arumugam 2018). Furthermore, it has been noticed that mitochondria play a nonenergetic function in aging, death, and metabolic process regulation (He et al. 2020). Several technological methods had previously been utilized to establish the effect of aging on mitochondria in brain cells (Grimm and Eckert 2017). The majority of mitochondrial genes are transferred to nuclear DNA (nDNA). Mitochondrial DNA (mtDNA) encodes for 13 protein subunits involved in oxidative phosphorylation. The other 76 subunits have been encoded by cytosolic nDNA that is then transported to the mitochondria, where both genomes interact functionally (Lu et al. 2023). Nuclear and mitochondrial DNA functional communication is required for mitochondrial activity and health. The absence of such an association leads to mitochondrial dysfunction and a decrease in ATP synthesis (Zia et al. 2021). Comparisons between mitochondrial structures extracted from animal brain tissue reveal a variety of age changes, including expansion of mitochondria, enhanced oxidative damage, impaired electron transport chain (ETC) function, elevation of depolarized membranes of mitochondria, $Ca^{2+}$ handling impairment, and a lower threshold for triggering mitochondrial membrane permeability transition pores formation (Mattson and Arumugam 2018). The decline in the mitochondrial performance linked to brain aging requires a drop in cellular $NAD^+$ levels and the NAD: NADH ratio, which is expected to impair the activities of $NAD^+$-dependent enzymes important for the function and viability of neurons, as protein deacetylases of the sirtuin family (Braidy et al. 2014; Fang et al. 2017). Because of the brain's susceptibility to mitochondrial dysfunction, aging caused a significant drop in mitochondrial performance (Yankner et al. 2008) (Fig. 1.1).

### 1.2.2 Accumulation of Oxidatively Damaged Molecules

All of the primary characteristics of aging are linked to changes in genetic material and its primary metabolic products, proteins, DNA damage accumulation caused by endogenous risks like an error of DNA replication and reactive oxygen species (ROS) generation, both of which elevate with age and help to the resurgence of age-related disorders (Manni et al. 2023). For the following reasons, the brain is especially susceptible to oxidative stress. The brain, which accounts for about 2% of total body mass, expends 20% of total oxygen usage. Furthermore, it contains polyunsaturated fatty acids that are easily peroxidizable. Furthermore, the brain lacks antioxidant defenses (Lau et al. 2005).

Furthermore, an unbalanced oxidative response causes an increase in ROS production and/or decrease in antioxidant defenses, malfunctioning neuron accumulation, proteins that have aggregated, and mitochondria during aging. The most common ROS in neurons is superoxide anion radicals ($^{-}O_2$), and by different oxidases, hydrogen peroxide process produces hydroxyl radicals that react with $Fe^{2+}$ or $Cu^+$ and nitric oxide (NO) produced as a result of increased intracellular $Ca^{2+}$ levels (Mattson and Arumugam 2018). Subcellular structures and membranes can be destroyed as a result of neuronal degradation. Indeed, besides too many intrinsic elements malfunctioning, different key groups of large molecules are

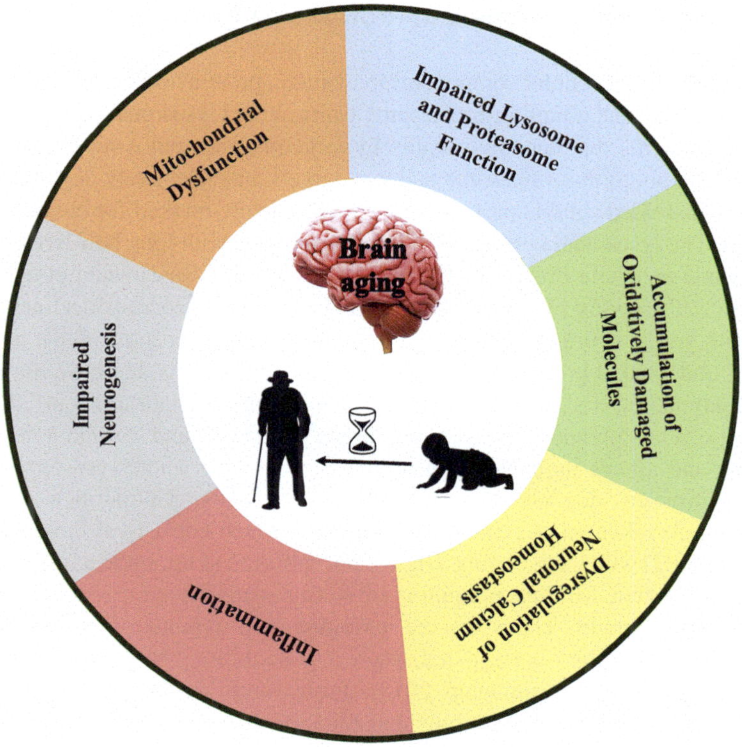


**Fig. 1.1** Cellular and molecular hallmarks of brain aging (Mattson and Arumugam 2018)

especially vulnerable to the negative implications of oxidative reaction during brain aging. A number of investigations have found that active oxidative damage to nucleic acids adds to the genesis, occurrence, and progression of NDDs (Davinelli et al. 2016). Furthermore, several groups of neurons, like cerebellar granule cells, retinal ganglion, amacrine, and horizontal cells, showed age-related accumulation of oxidative DNA damage, such as the presence of the modified base 8-hydroxydeoxyguanosine (8-OHdG) (Klein and Ackerman 2003). There is proof that lowered antioxidant defenses, and a reduced capacity to eliminate oxidatively damaged molecules is enough to speed up aging. Ubiquitination frequently targets oxidatively changed proteins for proteasomal degradation, whereas autophagy targets oxidatively damaged membranes and mitochondria to lysosomes. Excessive oxidative stress, on the other hand, can hinder the operation of proteasomes and lysosomes (Mattson and Arumugam 2018).

### 1.2.3 Impaired Lysosome and Proteasome Function

Removing damaged cellular components to maintain protein homeostasis is essential for cell survival. The ubiquitin-proteasome pathway and lysosomal pathway are the two main systems that accomplish this by targeting and removing damaged cell proteins. The ubiquitin-proteasome pathway targets and selectively degrades short-lived and misfolded soluble proteins that are specifically marked for elimination by ubiquitin. The lysosomal system can eliminate proteins with long half-lives, insoluble proteins, or whole organelles by cellular autophagy (Bustamante et al. 2018). With aging, the ability of neurons to maintain protein homeostasis declines. Compared with young neurons, aged neurons demonstrate greater intracellular accumulation of undegraded proteins and greater amounts of proteins that have undergone polyubiquitination, a sign of proteasome dysfunction (Graham and Liu 2017). In autophagy, a membranous phagophore holds the cargo and combing it with a lysosome and allows its contents to be released into the acidic lysosomal lumen where hydrolases degrade the phagophore's contents. Ubiquitination directs proteasomal breakdown of protein, a mechanism which consists of three enzymes (E1, E2, and E3) with the E3 being a ligase that shifts the ubiquitin to a lysine residue of the target protein. Multiple ubiquitins are subsequently linked together to produce the detected ubiquitin chain by the unfolded 19S regulatory subunit of the proteasome, which then is degraded when it goes through the barrel of the 20S subunit (Mattson and Arumugam 2018). Proteasome activity was measured in numerous brain regions of aging rats and found to be significantly lower in some areas unlike others, implying that neuronal populations are vulnerable to malfunction of proteasome differently during aging (Keller et al. 2000).

### 1.2.4 Dysregulation of Neuronal Calcium Homeostasis

Evaluations of proteasome action in aged rats' brain areas demonstrated notable reductions in a number of regions, but not others, implying that neuronal populations are vulnerable to malfunction of proteasome differently during the age (Burgoyne et al. 2019). The stimulating neurotransmitter glutamate induces $Na^{+}$-fluxing AMPA receptors on the postsynaptic dendrite upon discharge from presynaptic axon terminals, leading to membrane depolarization and $Ca^{2+}$ influx via the NMDA glutamate receptor channel and voltage-dependent $Ca^{2+}$ channels (Mattson and Arumugam 2018). This triggers a transient elevation in cytoplasmic $Ca^{2+}$ concentration as $K^{+}$ channels and $Na^{+}$ "pumps" are caused to maintain membrane potential, and $Ca^{2+}$ is eliminated via $Ca^{2+}$ ATPases located in the plasma membrane and endoplasmic reticulum membrane. Cytosolic kinases and phosphatases are elevated due to temporary $Ca^{2+}$ release, which can affect the phosphorylation of dendritic proteins involved in glutamate receptor trafficking to and from the membrane, cytoskeletal reorganization, and local protein synthesis (Mattson and Arumugam 2018). Dysregulated calcium flux may be a potential parallel between ionizing radiation exposure and brain aging, as ionizing radiation can alter calcium flux in both target and bystander cells (Lyng et al. 2006; Shao et al. 2006). Furthermore, proteins related to calcium transport and signaling were upregulated after exposure to low doses of $^{56}$Fe irradiation (Britten et al. 2017).

### 1.2.5 Inflammation

In human brains, aging is linked with abnormal inflammatory responses. Proinflammatory cytokines, in particular, increase with age, while antiinflammatory molecules decrease (Corbi et al. 2016). Inflammation is closely related to increased ROS production. Neurodegenerative diseases have also been linked to oxidative stress-mediated inflammation. Neuroinflammation is linked to increased signal transmission, which results in the release of genes like inducible nitric oxide synthase (iNOS), interleukin-1 (IL-1), tumor necrosis factor-α (TNF-α), and nuclear factor kappa B (NF-κB). Inflammation has been shown to activate neural-protecting microglial cells, causing them to generate high levels of ROS and cytokines (Lau et al. 2005). Oxidative stress caused by inflammatory pathways enhancement, which results in mitochondrial damage and dysfunction, looks to be playing a crucial function in neuronal degeneration and in the development of NDDs (Godos 2022). In addition, when neurons are damaged due to aging or degeneration of neurons, the release of neurotransmitters, ATP, cytokines, and ion changes triggers microglia. Neuroinflammatory activation of microglia has been linked to neuronal loss, reduced behavioral function of neurons, and cognition malfunction. Inflammation of neurons creates complex interactions with oxidizing agents through redox sensors that have been detected in receptors, transcription factors, and enzymes. These variables influence neuronal connectivity and function, culminating in neurodegenerative changes. Furthermore, following being activated, microglial cells generate a form of NOS that can be stimulated and produce large amounts of NO, which promotes oxidative neuronal damage (Zia et al. 2021). Furthermore, the neuropathological changes caused by oxidative stress and a pro-inflammatory state involve the accumulation of insoluble materials like amyloid-β (Aβ) plaques and neurofibrillary tangles (NFTs), which are the primary causes of cell apoptosis and dementia (Davinelli et al. 2016).

In various human and mouse tissues, the NF-κB transactivation pattern found to be most strongly associated with aging (Adler et al. 2007). Furthermore, the most popular antiinflammatory factors that improve lifespan are AMP-activated protein kinase (AMPK) and silent information regulator 1 (SIRT1) (Salminen et al. 2012). It was revealed that several signatures of aging are most notably involving an activation of inflammation/immune response genes (de Magalhães et al. 2009).

### 1.2.6 Impaired Neurogenesis

Aging is associated with impaired neurogenesis, specifically decreased self-renewal, proliferation, and capacity for regenerating neurons to integrate into the central

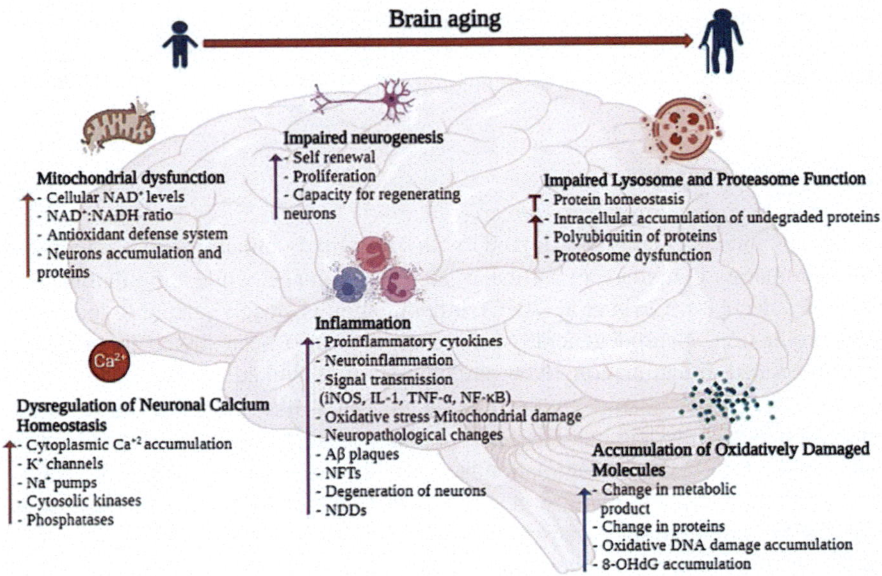


**Fig. 1.2** Cellular and molecular hallmarks of brain aging (Mattson and Arumugam 2018)

nervous system network (Cavallucci et al. 2016). In age-related neurodegenerative diseases, individuals have markedly lower numbers of new neurons compared with age-matched healthy controls (Moreno-Jiménez et al. 2019). Evidence suggests that radiation-induced cognition malfunction causes changes in blood vessels of the brain and glial cell groups. In general, brain radiation injury causes a long-term change in the brain's environment, with inflammation playing a critical part (Pazzaglia et al. 2020). Many other aging characteristics may add to neurogenesis impairment throughout aging. Diminishing of mitochondrial oxidative metabolism and genetic compromise of mitochondrial electron transport chain (ETC) function as a result of aging neural progenitor cells. Furthermore, age-related neurogenesis decreases may be exacerbated by oxidative stress, poor DNA repair, and inflammation (Mattson and Arumugam 2018) (Fig. 1.2).

## 1.3 Phytochemicals

Phytochemicals are a potent class of compounds that are secondary metabolites of plants and include a wide variety of chemical entities (Tousson and El-Gharbawy 2023). Phytochemicals enhance health by way of their numerous antioxidant, antimicrobial, anticarcinogenic, antiinflammatory, neuroprotective, and antiaging properties (Okoro et al. 2021). Several lines of bioactive phytochemicals have been established as antioxidants in in vivo and in vitro studies by eliminating free radicals and ROS, modulating cell aging, besides, neurons preserving from cell apoptosis. However, neuroprotection is now thought to be dependent not only on antioxidant mechanism, but also on the ion and $Ca^{2+}$ chelation, anti-inflammation, modulation of signaling in cells, and the stimulation of pro-survival antiapoptotic Bcl-2 proteins and neurotrophic compounds (Wu et al. 2017). As a result, phytochemicals function as moderate stressors, triggering stress-protective genes that are expressed adaptively in neural cells and enhancing opposition to brain aging processes (Davinelli et al. 2016).

Phytochemicals represent a major light for the creation of novel drug classes for treating aging-associated conditions. These compounds may trigger aging-related pathways such as autophagy and DNA repair, as well as combat aging-related systemic inflammation as well as oxidative stress (Vaiserman et al. 2021). Similarly, numerous pharmacological studies have proven the benefits and improvements in the therapy of phytochemicals in dementia cure, depression, and NDDs, which offer nutrients to normal cells combat agents of disease, and boost immunity (Balakrishnan et al. 2021).

Several studies have also found that some phytochemicals' antioxidant activity can trigger nuclear factor E2-related factor 2 (Nrf2)/antioxidant response element (ARE) signaling pathways. Furthermore, phytochemicals activate the phosphatidylinositol 3-kinase/protein kinase B (PI3K/Akt) and extracellular signal-regulated kinase (ERK) pathways which inhibit NF-κB pathways (Mandel and Youdim 2004; Firuzi et al. 2015). Phytochemicals have also been shown to decrease mitochondrial dysfunction and prevent the formation of inflammatory responses and α-synuclein accumulation-induced oxidative stress (Balakrishnan et al. 2021).

### 1.3.1 Activating of Neuronal Antioxidant Defense by Phytochemicals

Various plants with biologically active compounds are said to have ROS neutralization and boost the antioxidant system (Hannan et al. 2020). Neuroprotective effects of phytochemicals are associated with reduced levels of oxidative stress due to possessing antioxidant properties (Mattson and Cheng 2006).

Furthermore, phytochemicals may enhance the survival of cell pathways by stimulating many growth signaling pathways, as well as nontoxic aggregate formation and proteolytic system activation to alleviate A-induced neuronal mitochondrial dysfunction. Furthermore, phytochemicals stimulate α-secretase or suppress β-secretase activity, besides, hindering fibril and toxic oligomer production (Rahman et al. 2020). Furthermore, phytochemicals have been proven to safeguard neurons from oxidative stress by triggering TrkB signaling pathways and the Nrf2-ARE system, which are considered therapeutic targets for NDD treatments (Hannan et al. 2020). This chapter examines the research on the phytochemicals that have neuroprotective properties that can activate these neural defense systems.

## 1.4 Mechanisms Associated with Neuroprotective Effects of Phytochemicals on Brain

### 1.4.1 Effect of Cellular Defense Against Oxidative Stress: The Nrf2-ARE Antioxidant System

Antioxidant defense mechanisms exist in cells including antioxidant enzymes and other molecules that protect against OS-mediated injury. The main cellular mechanism that controls the mechanism of antioxidant defense is the Keap1-Nrf2 pathway. Nrf2 is a primary redox homeostasis regulator in cells, and its stimulation is controlled in a variety of strategies. Under normal circumstances, Keap1 sequesters Nrf2 in the cytoplasm, where it is polyubiquitinated and triggers a proteasomal breakdown. The Nrf2 Keap1 complex has been disturbed, resulting in Nrf2 release and nuclear translocation (Hannan et al. 2020). When Nrf2 enters the nucleus, it creates heterodimers with small musculoaponeurotic fibrosarcoma (MAF) proteins, and these Nrf2-MAF heterodimers recognize an enhancer sequence known as ARE, which is found in over 250 genes in the regulatory regions (Ma 2013). Stimulation of the Nrf2-ARE pathway via Keap1-Nrf2 impairment generates an increase in the expression of several genes encoding a network of collaborating enzymes that compose an antioxidant defense system. This antioxidant system's functions include redox homeostasis, involving SOD, CAT, sulfaredoxin (Srx), thioredoxin (Trx), and peroxiredoxin (Prdx); GSH synthesis and metabolism, involving Gpx, GR, g-glutamine cysteine ligase (GCL) and synthase (GCS); quinone recycling, involving NAD(P)H quinone oxidoreductase (Nqo1) and iron homeostasis-heme oxygenase 1 (HO-1) (Hannan et al. 2020).

Nrf2 is mainly found in the cytoplasm of the degenerative brain, implying that Nrf2 activity is inadequate to facilitate effective antioxidant reactivity (Ramsey et al. 2007). Proof suggests that knocking out Nrf2 makes neurons more vulnerable to oxidative stress, but Nrf2 elevation reverses the impact (Lee and Johnson 2004). As a result, it is thought that stimulating the antioxidant defense system through the Nrf2 signaling targeting provides a possible treatment for NDDs, brain injury, and brain aging (Hannan et al. 2020).

### 1.4.2 Cell Survival System: The TrkB Signaling Pathway

Neurotrophin signaling is critical for the survival of neurons and synaptic plasticity, besides memory and learning (Costa et al. 2022). Among the neurotrophins found in the mature central nervous system, brain-derived neurotrophic factor (BDNF) is the most abundant. BDNF binds TrkB and promotes the survival of neurons and plasticity of synapses by activating conventional signaling pathways such as the mitogen-activated protein kinase (MAPK) route, the PI3K/Akt system, and the phospholipase C-γ (PLCγ) pathway (Kowiański et al. 2018). The main TrkB-mediated survival strategy that enhances the survival of neurons while also protecting against death is the PI3K/Akt signaling pathway. Activated Akt additionally regulates cell viability by balancing proteins that promote and inhibit apoptosis. The MAPK and PLCγ pathways control neuronal growth and survival by expressing numerous genes in a CREB (cAMP response element binding protein)-dependent pathway, which encodes BDNF and other proteins linked with synaptic plasticity (Hannan et al. 2020).

Meis et al. (2018) reported that Fear learning impairments are observed in BDNF heterozygous knockout mice. Also, Jiang et al. (2019) there is a strong link between low BDNF levels and degeneration of neurons in NDDs. These findings indicate that the BDNF/TrkB signaling pathway is critical for survival, plasticity of synapses, and development of neurons. Various natural components have proven to trigger the TrkB signaling pathway and support the longevity of cells, making them potential treatments for NDDs, brain injury, and brain aging (Hannan et al. 2020).

### 1.4.3 Cross-Talk Between TrkB Signaling Pathway and Nrf2-ARE Antioxidant System

Networks of neurotrophin signaling such as PI3K/Akt, MAPK (Ras/ref./Erk), and PLCg (PKC or CaMK) keep neurons alive and connected. According to the evidence, the TrkB signaling pathway, which is the primary neurotrophin route in neurons that have matured, might function as an elevation of the Nrf2-ARE system. Phosphorylation of Akt, by way of example, facilitates Nrf2 nuclear translocation and triggers the Nrf2-ARE system, which leads to the transcription of many antioxidant enzyme-encoding genes (Qi et al. 2017). The PI3K/Akt pathway controls hemeoxygenase-1 (HO-1) which is important in cellular homeostasis preservation (Pischke et al. 2005). Furthermore, data indicate that the Nrf2-ARE antioxidant system and the TrkB signaling pathway may interact. Another low signaling mechanism of the TrkB pathway, the MAPK (ERK/p38 MAPK) pathway, has been demonstrated to control Nrf2 transcriptional activity (Singh et al. 2010; Tufekci et al. 2011). The TrkB signaling pathway and the Nrf2-ARE antioxidant system are thus complementary, and concurrent stimulation of these pathways demonstrated that conferring neuroprotection toward oxidative stress besides mitigating cognition impairments and memory among individuals suffering from NDDs such as aging or brain injury (Hannan et al. 2020). TrkB is linked with $p75^{NTR}$ signaling dependent on BDNF, which may also stimulate the Nrf2 pathway. TrkB is the main BDNF receptor, with two functional isoforms: TrkB.FL (full-length) and TrkB.T1 (truncated) receptors. TrkB.FL suppresses ceramide production via its tyrosine kinase activity at its low-affinity receptor, $p75^{NTR}$. In contrast, TrkB.T1, which lacks an intracellular tyrosine kinase domain, promotes ceramide synthesis after BDNF activation. The $p75^{NTR}$ signaling pathway has the potential to be a sword with two edges (Ishii and Mann 2018).

The overproduction of $p75^{NTR}$ leads to excess ceramide accumulation and is referred to as being linked to death (Kropf and Fahnestock 2021). Activity of TrkB.FL receptor tyrosine kinase, on the other hand, restricts sphingomyelinase, securing cells from ceramide toxicity and inhibiting the $p75^{NTR}$-mediated prodeath signaling

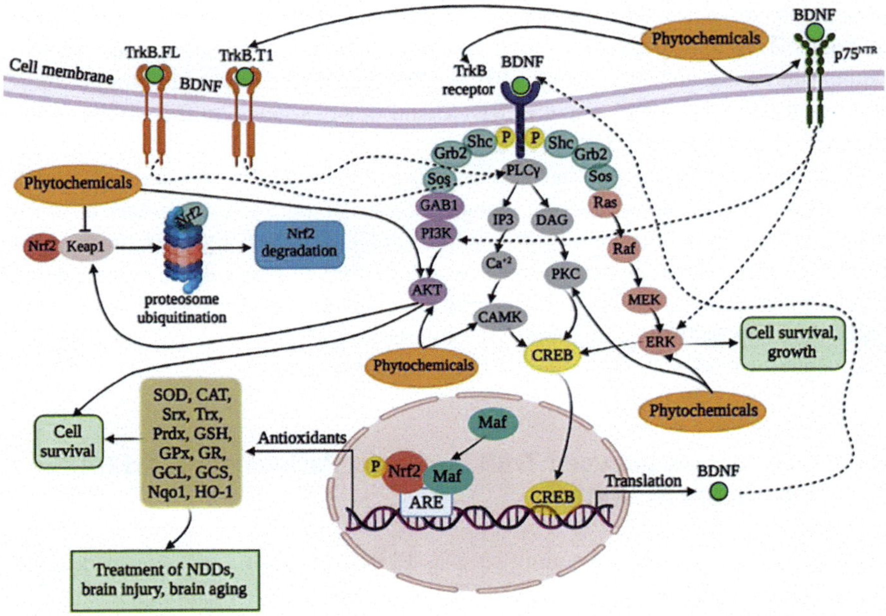


**Fig. 1.3** Mechanisms associated with neuroprotective effects of phytochemicals on the brain (Alam et al. 2016; Guerzoni et al. 2017; Hannan et al. 2020)

cascade (Ishii et al. 2018). Besides, the BDNF TrkB.T1-p75$^{NTR}$ signaling affects physiological accumulations of ceramide to stimulate protein kinase Cς (PKCς) resulting in stimulation of casein kinase 2 (CK2) and Nrf2; consequently, the antioxidant capacity of cells is regulated (Ishii et al. 2018). TrkB.FL and TrkB.T1 receptors are expressed by neurons, but the receptor ratio amounts vary depending on neuronal activation (Gomes et al. 2012). Excitotoxic stimulation of rat hippocampus neurons in culture associated with TrkB.FL downregulation and TrkB.T1 expression upregulation altered the frequency considerably different between the two receptors (Gomes et al. 2012), which permits BDNF to control Nrf2 stimulation when there is p75$^{NTR}$ (Ishii and Mann 2018). This action shields neurons from oxidative damage during excitotoxic glutamate activation, which is a common event in neural NDDs, injury, and stroke (Ishii and Mann 2018). Furthermore, BDNF-TrkB.T1-p75$^{NTR}$Nrf2-induced neuroprotection is context-dependent, meaning that it is based on the level of activation (Hannan et al. 2020) (Fig. 1.3).

## 1.5 Conclusion

Brain aging is the main risk factor for a variety of neurodegenerative diseases. Brain diseases associated with aging have lately emerged as the leading causes of disability and death. It caused brain dysfunctions on the cellular and molecular level like mitochondrial dysfunction, accumulation of oxidatively damaged molecules, impaired lysosome and proteasome function, dysregulation of neuronal calcium homeostasis, inflammation, and impaired neurogenesis. Moreover, phytochemicals are secondary metabolites of plants and have antiaging, antioxidant, and antiinflammatory properties. In this chapter, we briefed the possible mechanisms of phytochemicals for protecting from aging. In conclusion, phytochemicals are expected to be a promising approach as a protective agent for brain aging. This suggests that while both treatment approaches offer benefits, a combination tailored to the patient's specific needs might be most effective. Further research is needed to explore how factors such as age, severity of the condition, and individual response affect the optimal treatment plan.

## References


Adler AS, Sinha S, Kawahara TL, Zhang JY, Segal E, Chang HY (2007) Motif module map reveals enforcement of aging by continual NF-κB activity. Genes Dev 21(24):3244–3257

Alam MA, Rallabandi VP, Roy PK (2016) Systems biology of immunomodulation for post-stroke neuroplasticity: multimodal implications of pharmacotherapy and neurorehabilitation. Front Neurol 7:1–16

Balakrishnan R, Azam S, Cho DY, Su-Kim I, Choi DK (2021) Natural phytochemicals as novel therapeutic strategies to prevent and treat Parkinson's disease: current knowledge and future perspectives. Oxidative Med Cell Longev 2021:1–32

Bastianetto S, Quirion R (2002) Natural extracts as possible protective agents of brain aging. Neurobiol Aging 23(5):891–897

Braidy N, Poljak A, Grant R, Jayasena T, Mansour H, Chan-Ling T, Guillemin GJ, Smythe G, Sachdev P (2014) Mapping $NAD^+$ metabolism in the brain of ageing Wistar rats: potential targets for influencing brain senescence. Biogerontology 15:177–198

Britten RA, Jewell JS, Davis LK, Miller VD, Hadley MM, Semmes OJ, Lonart G, Dutta SM (2017) Changes in the hippocampal proteome associated with spatial memory impairment after exposure to low (20 cGy) doses of 1 GeV/n 56Fe radiation. Radiat Res 187(3):287–297

Burgoyne RD, Helassa N, McCue HV, Haynes LP (2019) Calcium sensors in neuronal function and dysfunction. Cold Spring Harb Perspect Biol 11(5):a035154

Bustamante HA, González AE, Cerda-Troncoso C, Shaughnessy R, Otth C, Soza A, Burgos PV (2018) Interplay between the autophagy-lysosomal pathway and the ubiquitin-proteasome system: a target for therapeutic development in Alzheimer's disease. Front Cell Neurosci 12:126

Cavallucci V, Fidaleo M, Pani G (2016) Neural stem cells and nutrients: poised between quiescence and exhaustion. Trends Endocrinol Metab 27(11):756–769

Corbi G, Conti V, Davinelli S, Scapagnini G, Filippelli A, Ferrara N (2016) Dietary phytochemicals in neuroimmunoaging: a new therapeutic possibility for humans? Front Pharmacol 7:364

Costa RO, Martins LF, Tahiri E, Duarte CB (2022) Brain-derived neurotrophic factor-induced regulation of RNA metabolism in neuronal development and synaptic plasticity. Wiley Interdiscip Rev RNA 13(5):e1713

Davinelli S, Maes M, Corbi G, Zarrelli A, Willcox DC, Scapagnini G (2016) Dietary phytochemicals and neuro-inflammaging: from mechanistic insights to translational challenges. Immun Ageing 13(16):1–17

de Magalhães JP, Curado J, Church GM (2009) Meta-analysis of age-related gene expression profiles identifies common signatures of aging. Bioinformatics 25(7):875–881

Fang EF, Lautrup S, Hou Y, Demarest TG, Croteau DL, Mattson MP, Bohr VA (2017) NAD+ in aging: molecular mechanisms and translational implications. Trends Mol Med 23:899–916

Firuzi O, Moosavi F, Hosseini R, Saso L (2015) Modulation of neurotrophic signaling pathways by polyphenols. Drug Des Devel Ther 10:23–42

Forni C, Facchiano F, Bartoli M, Pieretti S, Facchiano A, D'Arcangelo D, Norelli S, Valle G, Nisini R, Beninati S, Tabolacci C, Jadeja RN (2019) Beneficial role of phytochemicals on oxidative stress and age-related diseases. Biomed Res Int 2019:1–16

Godos J (2022) Do antioxidant phytochemicals play a role in neurodegenerative disorders? The case of polyphenols. Nutrients 14(22):4826

Gomes JR, Costa JT, Melo CV, Felizzi F, Monteiro P, Pinto MJ, Inácio AR, Wieloch T, Almeida RD, Grãos M, Duarte CB (2012) Excitotoxicity downregulates TrkB. FL signaling and upregulates the neuroprotective truncated TrkB receptors in cultured hippocampal and striatal neurons. J Neurosci 32(13):4610–4622

Graham SH, Liu H (2017) Life and death in the trash heap: the ubiquitin proteasome pathway and UCHL1 in brain aging, neurodegenerative disease and cerebral Ischemia. Ageing Res Rev 34: 30–38

Grimm A, Eckert A (2017) Brain aging and neurodegeneration: from a mitochondrial point of view. J Neurochem 143(4):418–431

Guerzoni LP, Nicolas V, Angelova A (2017) In *vitro* modulation of TrkB receptor signaling upon sequential delivery of curcumin-DHA loaded carriers towards promoting neuronal survival. Pharm Res 34:492–505

Hannan MA, Dash R, Sohag AA, Haque MN, Moon IS (2020) Neuroprotection against oxidative stress: phytochemicals targeting TrkB signaling and the Nrf2-ARE antioxidant system. Front Mol Neurosci 13:116

He Y, Makarczyk MJ, Lin H (2020) Role of mitochondria in mediating chondrocyte response to mechanical stimuli. Life Sci 263:118602

Huffman KJ (2012) The developing, aging neocortex: how genetics and epigenetics influence early developmental patterning and age-related change. Front Genet 3:212

Ishii T, Mann GE (2018) When and how does brain-derived neurotrophic factor activate Nrf2 in astrocytes and neurons? Neural Regen Res 13(5):803

Ishii T, Warabi E, Mann GE (2018) Circadian control of p75 neurotrophin receptor leads to alternate activation of Nrf2 and c-Rel to reset energy metabolism in astrocytes via brain-derived neurotrophic factor. Free Radic Biol Med 119:34–44

Jiang L, Zhang H, Wang C, Ming F, Shi X, Yang M (2019) Serum level of brain-derived neurotrophic factor in Parkinson's disease: a meta-analysis. Prog Neuropsychopharmacol Biol Psychiatry 88:168–174

Keller JN, Hanni KB, Markesbery WR (2000) Possible involvement of proteasome inhibition in aging: implications for oxidative stress. Mech Ageing Dev 113(1):61–70

Klein JA, Ackerman SL (2003) Oxidative stress, cell cycle, and neurodegeneration. J Clin Investig 111(6):785–793

Kowiański P, Lietzau G, Czuba E, Waśkow M, Steliga A, Moryś J (2018) BDNF: a key factor with multipotent impact on brain signaling and synaptic plasticity. Cell Mol Neurobiol 38:579–593

Kropf E, Fahnestock M (2021) Effects of reactive oxygen and nitrogen species on TrkA expression and signalling: implications for proNGF in aging and Alzheimer's disease. Cell 10(8):1983

Lau FC, Shukitt-Hale B, Joseph JA (2005) The beneficial effects of fruit polyphenols on brain aging. Neurobiol Aging 26(1):128–132

Lee JM, Johnson JA (2004) An important role of Nrf2-ARE pathway in the cellular defense mechanism. BMB Rep 37(2):139–143

Li L, Li X, Yang C, Li L (2023) Peanut AhmTERF1 regulates root growth by modulating mitochondrial abundance. Genes 14(1):209

Loh D, Reiter RJ (2022) Melatonin: regulation of viral phase separation and epitranscriptomics in post-acute sequelae of COVID-19. Int J Mol Sci 23(15):8122

Lu H, Wang X, Li M, Ji D, Liang D, Liang C, Liu Y, Zhang Z, Cao Y, Zou W (2023) Mitochondrial unfolded protein response and integrated stress response as promising therapeutic targets for mitochondrial diseases. Cell 12(1):20

Lyng FM, Maguire P, McClean B, Seymour C, Mothersill C (2006) The involvement of calcium and MAP kinase signaling pathways in the production of radiation-induced bystander effects. Radiat Res 165(4):400–409

Ma Q (2013) Role of nrf2 in oxidative stress and toxicity. Annu Rev Pharmacol Toxicol 53:401–426

Machiela E, Southwell AL (2020) Biological Aging and the Cellular Pathogenesis of Huntington's Disease. J Huntingtons Dis 9(2):115–128. https://doi.org/10.3233/JHD-200395

Mandel S, Youdim MB (2004) Catechin polyphenols: neurodegeneration and neuroprotection in neurodegenerative diseases. Free Radic Biol Med 37(3):304–317

Manni G, Buratta S, Pallotta MT, Chiasserini D, Di Michele A, Emiliani C, Giovagnoli S, Pascucci L, Romani R, Bellezza I, Urbanelli L, Fallarino F (2023) Extracellular vesicles in aging: an emerging hallmark? Cell 12(4):527

Mattson MP, Arumugam TV (2018) Hallmarks of brain aging: adaptive and pathological modification by metabolic states. Cell Metab 27(6):1176–1199

Mattson MP, Cheng A (2006) Neurohormetic phytochemicals: low-dose toxins that induce adaptive neuronal stress responses. Trends Neurosci 29(11):632–639

Meis S, Endres T, Munsch T, Lessmann V (2018) The relation between long-term synaptic plasticity at glutamatergic synapses in the amygdala and fear learning in adult heterozygous BDNF-knockout mice. Cereb Cortex 28(4):1195–1208

Moreno-Jiménez EP, Flor-García M, Terreros-Roncal J, Rábano A, Cafini F, Pallas-Bazarra N, Ávila J, Llorens-Martín M (2019) Adult hippocampal neurogenesis is abundant in neurologically healthy subjects and drops sharply in patients with Alzheimer's disease. Nat Med 25(4): 554–560

Okoro NO, Odiba AS, Osadebe PO, Omeje EO, Liao G, Fang W, Jin C, Wang B (2021) Bioactive phytochemicals with anti-aging and lifespan extending potentials in Caenorhabditis elegans. Molecules 26(23):7323

Pazzaglia S, Briganti G, Mancuso M, Saran A (2020) Neurocognitive decline following radiotherapy: mechanisms and therapeutic implications. Cancers 12(1):146

Peters R (2006) Ageing and the brain. Postgrad Med J 82(964):84–88

Pischke SE, Zhou Z, Song R, Ning W, Alam J, Ryter SW, Choi AM (2005) Phosphatidylinositol 3-kinase/Akt pathway mediates heme oxygenase-1 regulation by lipopolysaccharide. Cell Mol Biol 51(5):461–470

Qaragholi ZM, Abu-Raghif AR, Kadhim EJ (2022) Phytochemical screening of anti-aging medicinal plants of celery (Apium graveolens), safflower (Carthamus tinctorious) and pomegranate (Punica granatum) aqueous extracts. BNIHS 140:1161–1174

Qi G, Mi Y, Wang Y, Li R, Huang S, Li X, Liu X (2017) Neuroprotective action of tea polyphenols on oxidative stress-induced apoptosis through the activation of the TrkB/CREB/BDNF pathway and Keap1/Nrf2 signaling pathway in SH-SY5Y cells and mice brain. Food Funct 8(12): 4421–4432

Rahman MA, Rahman MH, Biswas P, Hossain MS, Islam R, Hannan MA, Hannan A, Uddin J, Rhim H (2020) Potential therapeutic role of phytochemicals to mitigate mitochondrial dysfunctions in Alzheimer's disease. Antioxidants 10(1):23

Ramalingam M, Kim H, Lee Y, Lee YI (2018) Phytochemical and pharmacological role of liquiritigenin and isoliquiritigenin from radix glycyrrhizae in human health and disease models. Front Aging Neurosci 10:348

Ramsey CP, Glass CA, Montgomery MB, Lindl KA, Ritson GP, Chia LA, Hamilton RL, Chu CT, Jordan-Sciutto KL (2007) Expression of Nrf2 in neurodegenerative diseases. J Neuropathol Exp Neurol 66(1):75–85

Rossiello F, Jurk D, Passos JF, d'Adda di Fagagna F (2022) Telomere dysfunction in ageing and age-related diseases. Nat Cell Biol 24(2):135–147

Salminen A, Kauppinen A, Kaarniranta K (2012) Phytochemicals suppress nuclear factor-κB signaling: impact on health span and the aging process. Curr Opin Clin Nutr Metab Care 15(1):23–28

Samarghandian S, Azimi-Nezhad M, Farkhondeh T (2019) Thymoquinone-induced antitumor and apoptosis in human lung adenocarcinoma cells. J Cell Physiol 234(7):10421–10431

Shao C, Lyng FM, Folkard M, Prise KM (2006) Calcium fluxes modulate the radiation-induced bystander responses in targeted glioma and fibroblast cells. Radiat Res 166(3):479–487

Singh S, Vrishni S, Singh BK, Rahman I, Kakkar P (2010) Nrf2-ARE stress response mechanism: a control point in oxidative stress-mediated dysfunctions and chronic inflammatory diseases. Free Radic Res 44(11):1267–1288

Tousson E, El-Gharbawy DM (2023) Impact of Saussurea lappa root extract against copper oxide nanoparticles induced oxidative stress and toxicity in rat cardiac tissues. Environ Toxicol 38(2): 415–421

Tufekci KU, Civi Bayin E, Genc S, Genc K (2011) The Nrf2/ARE pathway: a promising target to counteract mitochondrial dysfunction in Parkinson's disease. Parkinsons Dis 2011:1–14

Vaiserman A, Koliada A, Lushchak O (2021) Phyto-nanotechnology in anti-aging medicine. Aging 13(8):10818

Vijayakumar S, Prabhu S, Rajalakhsmi S, Manogar P (2016) Review on potential phytocompounds in drug development for Parkinson disease: a pharmacoinformatic approach. Inform Med Unlocked 5:15–25

Wu Y, Shamoto-Nagai M, Maruyama W, Osawa T, Naoi M (2017) Phytochemicals prevent mitochondrial membrane permeabilization and protect SH-SY5Y cells against apoptosis induced by PK11195, a ligand for outer membrane translocator protein. J Neural Transm 124: 89–98

Yankner BA, Lu T, Loerch P (2008) The aging brain. Annu Rev Pathol 3:41–66

Zia A, Pourbagher-Shahri AM, Farkhondeh T, Samarghandian S (2021) Molecular and cellular pathways contributing to brain aging. Behav Brain Funct 17(1):6

# Phytonutraceuticals Modulate Cell Survival Signaling and Regulate Sympathetic Innervation in Aging and Disease

# 2

Barathi Seetharaman, Ramasamy Vasantharekha, UshaRani Balu, and S. Winkins Santosh

**Abstract**

As individual's age, they experience an upsurge in cellular oxidative stress levels, autophagy, and chronic inflammation, resulting in several physiological and metabolic changes. Phytochemicals are secondary metabolites produced in plants that boost survival mechanisms and can modulate the psychological parameters of various life forms inhabiting specific environments. Polyphenols, a phytochemical derived from fruits, nuts, and vegetation, possess antioxidant properties that counteract oxidative stress and metabolic dysregulation. Preventing age-related diseases and cognitive deterioration and promoting mental agility, particularly among older people, has become a matter of vital public interest. Aged individuals face an elevated risk of contracting diseases due to a decline in immune functions and defense mechanisms and a chance of an increase in autoimmune diseases. Due to defective T cell receptor signaling, chronic low-grade inflammation attenuates disease progression with advancing age. Abnormal signaling cascades ERK, CREB, PI3K/Akt/GSK-3β, and NF-kB pathways were observed in aging and neurodegeneration. To enhance the immune and metabolic functions and improve cognition levels imposed by old age, one should consider adopting dietary practices that center around consuming fruits, nuts, and vegetables. Combining evidence-based pharmacological therapies with naturally derived compounds may provide a practical approach to slow down age-related pathological progression while minimizing side effects.

B. Seetharaman · R. Vasantharekha · U. Balu
EDART Laboratory, Department of Biotechnology, School of Bioengineering, SRM Institute of Science and Technology, Kattankulathur, Chennai, India

S. W. Santosh (✉)
PG & Research Department of Advanced Zoology and Biotechnology, Government Arts College for Men, Nandanam, Chennai, Tamil Nadu, India



S. Pathak, A. Banerjee (eds.), *Neuroprotective Effects of Phytochemicals in Brain Ageing*, https://doi.org/10.1007/978-981-99-7269-2_2

Targeting the sympathetic nervous system's involvement would help minimize age-related immune dysfunction and diseases. Resveratrol, quercetin, epicatechin, and curcumin are plant-derived nutraceutical components with antioxidant activities that can improve cognitive functionality across different age groups and mitigate age-related functional losses of synaptic and neuroplasticity.

**Keywords**

Aging · Phytochemicals · Resveratrol · Epicatechin · Quercetin · Curcumin

## 2.1 Introduction

Within stationary autotrophs, each plant genus or species has developed the ability to produce a unique combination of "phytochemicals." These compounds, known as "secondary metabolites," do not contribute to the plant's primary metabolism or provide essential macronutrients. However, their synthesis enhances the plant's capacity to adapt to local challenges by facilitating environmental interactions. The roles of these phytochemicals encompass various aspects such as general protection (e.g., antioxidant, ultraviolet light absorption, and antiproliferative properties), regulation of the plant's interactions with pathogenic and symbiotic microorganisms above and below the ground, defense against neighboring competitor plants, and management of the plant's interactions with more complex organisms (Kennedy 2014a, b; Kennedy and Wightman 2011).

Increasing evidence shows that aging is highly associated with a chronic increase in reactive oxygen species (ROS), a low-grade pro-inflammatory phenotype accumulation, and a reduction in age-related autophagy. This suggests that these factors may play essential roles in promoting aging. Indeed, reducing ROS and low-grade inflammation and promoting autophagy by calorie restriction or other dietary manipulation can extend lifespan in a broad spectrum of model organisms. The class of phytochemicals known as polyphenols mainly consists of phenolic rings. These are plant products found in a variety of fruits and vegetables. These secondary metabolites are a buffer when plants experience environmental or biological stress. These are also recognized to protect certain disorders linked to aging when included in the human diet. The antioxidant qualities of polyphenols are well recognized and protect against oxidative stress (Deepika and Maurya 2022). Several plant foods high in polyphenols, anti-inflammatory dietary patterns, and plant-based dietary patterns like the Mediterranean diet, which contains a range of fruits, vegetables, legumes, nuts, and whole grains, all appear to benefit brain health. These dietary patterns contain bioactive substances such as polyphenols, unsaturated fatty acids, antioxidant vitamins, and other phytochemicals (Rajaram et al. 2019). Aging is associated with weakened immune reactivity increasing vulnerability to infections, neurodegeneration, and cancer. Cognition involves intricate mental processes for information processing and navigation. Human studies utilize tools like the Mini-Mental State Examination (MMSE) to assess various functions, including orientation, recall, language, praxis, attention, and calculation (Vasantharekha et al. 2017). Elderly individuals commonly experience cognitive decline, highlighting the need to address it in services for older people. Dementia, which affects 37 million individuals in Europe, is expected to increase due to the aging population (Park et al. 2003).

Researchers are investigating the role of the sympathetic nervous system in age-related immune dysfunction and neurodegenerative diseases (Madden et al. 1997). Eating more fruits, nuts, and vegetables improves cognitive function and can delay cognitive decline with age (Miller et al. 2017). Phytochemicals hold potential against oxidative stress-related diseases, especially when combined with drugs like DMARDs. Repurposing existing drugs, including those from traditional medicine, is gaining importance due to socioeconomic factors. The future will see increased interest in phytochemical-based drugs for inflammation and oxidative stress-related diseases (Forni et al. 2019). The roles of phytochemicals such as Resveratrol, Epicatechin, Quercetin, Curcumin, Green tea extract (EGCG), and a few others affecting aging and cognition have been discussed.

## 2.2 Aging Process

Age-related cognitive decline increases the likelihood of developing dementia and other brain illnesses connected to it. The outcome of cognitive aging and its progression to pathological states are determined by genetic predisposition and environmental exposures (Singh et al. 2021). In the general population, some cognitive abilities degrade with age, along with some forgetfulness, a lower capacity for maintaining attention, and a decreased capacity for problem-solving (Howes et al. 2020). The biological activities of an organism gradually deteriorate as it ages due to a range of morphological and functional changes. For many chronic illnesses, aging is a significant risk factor.

Age-associated cognitive decline and mild cognitive impairment are frequently observed both in men and women of the elderly population. To illustrate the challenges involved in this endeavor, the accumulation of β-amyloid (Aβ) proteins, neurofibrillary tangles (NFT) formation, and deposition of Lewy bodies in the brain have long been regarded as the primary pathological mechanisms in neurodegenerative diseases like Alzheimer's disease (AD), which is the primary contributor to cognitive deterioration in the elderly (Herrup 2015). The information storage process, a fundamental aspect of normal cognition, is not yet comprehensively understood. However, several mechanisms have been implicated, including excitatory and inhibitory neurotransmitters such as acetylcholine, GABA, glycine, and glutamate, the biosynthesis of biogenic amines dopamine, epinephrine, norepinephrine serotonin and cortisol, synaptic structures like dendritic spines, and neurophysiological responses such as short and long-term potentiation (LTP) (Howes et al. 2020). As individuals age, certain cognitive functions may decline, accompanied by varying degrees of forgetfulness and reduced ability to maintain focus and problem-solving skills, commonly observed in the general population. Biomarkers associated with cognitive decline in the normal aging process of the brain, mild cognitive impairment, and dementia have not been extensively explored. However, oxidative stress (caused by factors like free radicals), mitochondrial dysfunction, apoptosis, telomeric DNA depletion, and other factors implicated in general aging are believed to play a significant role.

Additionally, there is an increase in white matter lesions, Aβ plaques, neurofibrillary tangles, tau pathology, Lewy bodies, α-synucleinopathy, as well as a decline in neurotransmitter synthesis and transport in the neurons, including the cholinergic system. Vascular lesions and cortical shrinkage are also observed, particularly in the hippocampus. These changes become more pronounced in neurodegenerative diseases, each of which is associated with specific cognitive disorders (Howes et al. 2020; van Ginneken 2017). Various brain disorders, both early and late-onset, contribute to cognitive decline during the aging process. Each disorder has its specific biomarkers associated with it. For instance, early and late-onset Alzheimer's disease (AD) is linked to biomarkers such as Aβ42, total tau (T-tau), and phosphorylated tau (P-tau). Lewy body dementia and Parkinson's disease (PD) are associated with synuclein and phosphorylated synuclein biomarkers. Vascular dementia, other vascular conditions, and stroke do not have specific biomarkers identified yet. In addition to disease-specific protein abnormalities, other biomarkers are relevant to pharmacological treatment. These include indicators of neurodegeneration in cholinergic, dopaminergic, and serotonergic systems (such as presynaptic and postsynaptic uptake or receptor sites, as well as metabolic enzymes). Abnormal neurotrophic function, impaired neurogenesis, and various proteins involved in immune responses are also considered significant biomarkers. Extensive research is currently focused on investigating a broad range of protein abnormalities through neuroimaging, cerebrospinal fluid (CSF), and blood studies (Howes et al. 2020; van Ginneken 2017; Maclin et al. 2019).

The studies consistently revealed a near-universal cognitive decline among older people, albeit slight in individuals with initially good cognitive functioning. This highlights the significance of cognitive decline as a crucial factor that demands careful attention when designing services for older individuals. The likelihood of cognitive decline progressing to dementia indicates that it will emerge as a highly significant medical condition in the twenty-first century. With dementia already impacting the lives of 37 million individuals in Europe, its prevalence is expected to experience a substantial surge due to the aging population (Park et al. 2003). Recent findings have shed light on the potential for aerobic and cognitive training to mitigate or reverse age-related changes in brain structure and function. However, individuals who are aging with a disability face a significant dilemma. Conditions that directly impair brain function or the large muscle groups required for fitness enhancement and prevention of age-related declines may hinder their ability to participate in aerobic conditioning, cognitive training, or both. Numerous disabling conditions, such as multiple sclerosis (MS), cerebral palsy, and traumatic brain injuries, affect both the brain and neuromuscular control mechanisms. Other conditions, like post-polio syndrome or permanent amnesia in survivors of herpes encephalitis, lack recognized cognitive or neuromotor components. While this chapter does not provide an exhaustive review of the emerging evidence on cognitive and physical training for individuals aging with disabilities, MS, with its high prevalence and extensive literature on the heterogeneous cognitive and motor aspects of the disease, serves as an illustrative example of how principles of aging can be applied to a specific condition (Borson 2010).

The aging process affects the mind in diverse ways, with cognitive abilities showing divergent trajectories. Cognitive aging research suggests that mental processes, particularly those requiring effort and conscious deliberation, decline universally as individuals age. However, specific automatic processes and expertise-related functions remain intact. Notably, there is evidence of remarkable plasticity in cognitive function even in old age. Older adults can improve their performance through practice, and positive beliefs about aging and memory can also enhance their performance. This malleability suggests that motivation and emotion play crucial roles in the cognitive lives of older individuals. Therefore, when studying cognitive aging, it is essential to consider the changes in emotional experiences that older adults undergo.

In contrast to cognitive deterioration, emotional experiences and emotion regulation tend to improve throughout the lifespan. Most emotional processes remain intact, and any changes observed, such as decreases in negative affect or improvements in emotional control, indicate enhancements in the emotional lives of older adults. To truly understand the aging mind, it is essential to explore the intersection between cognitive decline and emotional enhancement (Carstensen et al. 2006).

## 2.3 Sympathetic Innervation in Aging and Disease

The aging process is linked to a decline in immune reactivity, leading to higher susceptibility to infectious diseases and cancer among the elderly. To understand age-related immune dysfunction, researchers have initiated evaluations of the potential involvement of the sympathetic nervous system. Specifically, they examine the characteristics of sympathetic noradrenergic (NA) innervation in the lymphoid organs of aged animals. Changes that occur with age, specifically in the sympathetic noradrenergic (NA) innervation and norepinephrine (NE) levels in the thymus and spleen of BALB/c mice, could potentially contribute to immune dysfunction associated with aging (Madden et al. 1997). It has previously been reported that there has been a remarkable rise in sympathetic noradrenergic (NA) innervation within the involuting thymus of F344 rats (Bellinger et al. 1988). This increase is characterized by heightened density of NA nerves surrounding septa and blood vessels and increased infiltration into parenchymal regions, noticeable as early as 12 months of age and becoming more extensive in the 24-month-old thymus. Interestingly, although the concentration of norepinephrine (NE) increased with age, the total NE content in the thymus remained unaltered, indicating that the maintenance of NA innervation is maintained as the thymus undergoes involution (Madden et al. 1997).

The sympathetic noradrenergic (NA) nerve fibers establish communication with target cells that express cell surface adrenergic receptors, also known as adrenoceptors. This communication is facilitated through the neurotransmitter norepinephrine (NE). Various immune system cells, including mature T and B lymphocytes, macrophages, and others, possess adrenoceptors of the h- and i-adrenergic types (Spengler et al. 1990; Abrass et al. 1985; Genaro and Borda 1989; Fuchs et al. 1988; Marchetti et al. 1990; Radojcic et al. 1991). Thymocytes, in their unfractionated state, exhibit minimal levels of i-adrenoceptor expression, and the expression of i-adrenoceptors may increase as thymocyte maturation progresses (Fuchs et al. 1988; Radojcic et al. 1991) The age-related augmentation of sympathetic noradrenergic (NA) innervation observed in the thymus and spleen of mice implies that norepinephrine (NE) may increasingly contribute to signaling potential targets within these organs as the animal ages. According to this hypothesis, pharmacological modulation of the sympathetic nervous system (SNS) may exert a more pronounced influence on immune reactivity in older animals when compared to younger animals (Madden et al. 1997).

Numerous experiments involving chemical sympathectomy of secondary lymphoid organs conducted on various strains of mice and rats consistently indicate that sympathectomy substantially reduces cell-mediated immune responses. This reduction is particularly evident in older animals with age-related deterioration of NA nerve fibers. Conversely, sympathectomy enhances humoral immune responses, particularly IgM and IgG, especially in animals with a Th1 dominance, and this effect is most noticeable in older animals. The significant immunological consequences of sympathectomy in aged animals suggest that the sympathetic nervous system (SNS) plays a crucial role as an external regulator of the immune system, significantly when immune function is compromised. Furthermore, these pronounced immunological effects in older animals might indicate a heightened sensitivity to changes in sympathetic tone and activation in intact old animals, mainly when compensatory mechanisms are already functioning at their maximum capacity (Madden et al. 1995).

## 2.4 Phytonutraceuticals in Neurological Impairments

According to epidemiological research, consuming fruits, nuts, and vegetables is favorably related to cognitive function; nevertheless, these foods, which contain a range of neuroprotective phytochemicals, are typically underutilized. It may be possible to prevent or delay the start of cognitive deterioration as we age by increasing our consumption of fruit, nuts, and vegetables (Miller et al. 2017). A review has identified several compounds showing protective effects against neurological impairments. These include polyphenolics from sources like strawberry, spinach, blueberry, tea, red bell pepper, and aged garlic extract, as well as herbal remedies like ding lang root extract, excluding ginkgo and ginseng (Youdim and Joseph 2001).

Several phytochemicals possess cognitive and alertness/arousal effects that could contribute to enhanced athletic performance. These effects can be linked to the ecological functions associated with specific groups of phytochemicals. Among these groups, phenolics and their polyphenol subgroup, currently the subject of extensive research, are essential for maintaining good health and cardiovascular function when included in a regular diet. The promising evidence of terpenes' beneficial cognitive and mood effects is particularly notable in several edible herbs that express monoterpene compounds, such as sage, rosemary, and peppermint. The diterpene-rich *G. biloba* and various triterpene-rich extracts, including ginseng and bacopa, also show potential. Alkaloids, known as the classic compounds for plant defense, frequently strongly affect neurotransmission in herbivores and human consumers. However, due to their potency, toxicity, and legal considerations, only nicotine and caffeine currently exhibit evident potential for cognitive enhancement in a sporting context.

Nonetheless, it is essential to note that nicotine is addictive, and factors such as habituation and withdrawal effects, as well as dosage patterns, must be carefully considered. Additionally, intriguing data is emerging regarding potential additive effects and synergies when phytochemicals are consumed alongside low doses of caffeine. In this context, the rational approach to enhancing cognitive function may involve the consumption of phytochemical or nutrient-rich extracts that contain lower doses of caffeine (Kennedy 2019).

The study of dietary interventions on cognitive function in older individuals encounters various challenges, including diagnosis, dosage determination, standardization, placebo control, blinding (for the subjects), and accounting for genetic and life experience variations among participants. Furthermore, the complexity is heightened by the potential impact of phyto-nutraceuticals on brain functions beyond cognition, such as sleep or mood, which, in turn, can influence cognitive abilities. These compounds may also affect other bodily systems, such as cardiovascular health or blood glucose regulation, which can indirectly influence brain function (Howes et al. 2020).

There is a significant interest in exploring the potential use of phytochemicals against human diseases related to oxidative stress, particularly those where inflamed cells play a critical role in pathogenic mechanisms. Utilizing phytochemicals in combination with drugs such as DMARDs can significantly reduce side effects and improve cost-effectiveness. Moreover, socioeconomic factors increasingly influence new drug development, leading to a rising focus on repurposing and repositioning existing drugs, including those commonly used in traditional medicine. Consequently, phytochemical-based drugs are anticipated to garner growing interest in treating inflammation and oxidative stress-related diseases (Forni et al. 2019). To enhance human health and prevent age-related degenerative disorders, it is imperative to conduct extensive research on natural dietary compounds in the future. Various phytochemicals have demonstrated antioxidant and antiaging properties through different models, whether in vitro, in vivo, or ex vivo. These chemical components exhibit such activities due to their pharmacophoric structure. Polyphenols and bio-flavonoids, in particular, exhibit remarkable antiaging properties, and plant extracts rich in polyphenols possess significant antiaging capabilities. However, further scientific studies are required to establish the safety and efficacy of natural therapies in both reawakening and detoxification procedures in a consistent manner (Mondal et al. 2014).

Increasing evidence supports that specific phytochemicals in foods, such as resveratrol, green tea extract, EGCG, epicatechin, quercetin, and curcumin, possess antiaging properties. Dietary consumption of these compounds has been shown to promote health and extend lifespan in various animal models, operating through multiple mechanisms. These mechanisms include reducing oxidative stress, suppressing chronic inflammation, inducing autophagy, and regulating molecules crucial for mitochondrial function and energy balance. Notably, these mechanisms align with those associated with calorie restriction, indicating that incorporating these phytochemicals into the diet may offer a viable and practical approach to promoting healthy human aging (Si and Liu 2014).

### 2.4.1 Resveratrol

Resveratrol, also known as trans-3.4′,5-trihydroxystilbene (RES), is a polyphenolic phytoalexin compound that is small in size and not classified as a flavonoid. It is found in wine (especially grape skin), fruits, peanuts, and certain plants. It demonstrates effects on modulating cell growth and possesses anticarcinogenic properties (Martin et al. 2016). The initial investigations into the field of epidemiology linked the consumption of wine to a decreased risk of developing dementia. However, there have been proposals to integrate this data by examining additional factors, such as smoking habits and the presence of ApoE4, as it has been advised that this evidence should be considered in conjunction (Perry and Howes 2011). While the impact of wine and grapes on the risk of dementia is inconclusive, the potential advantages have primarily been associated with a specific compound called resveratrol (Howes et al. 2020). While grapes and wine account for a significant portion of dietary resveratrol, this compound can also be found in other food sources such as peanuts, pistachios, and blueberries (Dominguez and Barbagallo 2018; Howes and Perry 2011).

Regarding the pathology associated with dementia, resveratrol exhibits anti-Aβ properties through various mechanisms. Compared to specific antioxidants derived from plants, resveratrol emerges as a more potent inhibitor of Aβ42 aggregation, effectively reducing the formation of Aβ plaques in a transgenic Alzheimer's disease model (Williams et al. 2011). Additionally, resveratrol diminishes the expression of transthyretin, a carrier protein known to decrease the aggregation and harmful effects of amyloidogenic proteins (Costa et al. 2017). Furthermore, in a mouse model of Alzheimer's disease, resveratrol exhibited an extension of lifespan, decreased cognitive impairments, neuroprotective effects, and reductions in both Aβ accumulation and tau hyperphosphorylation (Howes and Perry 2011). More investigation is required to understand the implications and potential synergistic effects of the interaction between resveratrol and other polyphenols, vitamins, amino acids, micronutrients, and commonly used drugs. Resveratrol's multiple intracellular targets make it necessary to gather additional data to determine the outcomes of these interactions (Malaguarnera 2019). The cognitive deficit improvement observed in an animal model of Alzheimer's disease was attributed to the impact of grape supplementation, which contains a combination of resveratrol, catechins, flavonols, and various phenolic acids. This was determined based on its ability to modulate altered amyloid precursor protein processing and its anti-inflammatory and antioxidant effects (Sheng et al. 2018). The neuroprotective action of resveratrol, which is believed to be associated with its impact on cognition, stems from its antioxidant effects. In experiments conducted on rat hippocampal slices, resveratrol demonstrated the ability to safeguard astrocytes against oxidative stress induced by hydrogen peroxide ($H_2O_2$) by elevating the levels of glutathione (GSH) (Howes et al. 2020; Williams et al. 2011). In mouse studies, administering resveratrol at a daily dose of 40 mg per kilogram for 1 month stimulated neurogenesis. Enhanced levels of hippocampal BDNF mRNA accompanied this improvement compared to the control group. Grapeseed extract has also been shown to enhance adult neurogenesis in vivo (Poulose et al. 2017), making it a potential nutraceutical of interest for maintaining cognitive functions. The potential neuroprotective effects of resveratrol might arise from its ability to activate the sirtuin 1 miR-134 pathway and increase the expression of CREB/BDNF, thereby contributing to its neuroprotective mechanisms (Shen et al. 2018). An additional investigation demonstrated that resveratrol positively affected the proliferative states of neural stem cells in the hippocampus of rats. Restoring BDNF levels mitigated stress-induced learning deficits and minimized hippocampal degeneration in vivo (Phillips 2017).

In a study spanning 26 weeks, the combination of 200 mg of daily resveratrol with flavonol quercetin demonstrated the ability to enhance memory retention in healthy individuals who were overweight (Zaplatic et al. 2019). Moreover, under this discovery, consuming grape juice over 12 weeks positively affected verbal learning among older individuals experiencing cognitive decline. However, no such improvement was observed in dementia patients (Phillips 2017). An alternative study indicated that among participants with mild cognitive decline, the administration of a grape-based formulation demonstrated potential in safeguarding against metabolic decline in brain regions typically affected during the early stages of Alzheimer's disease, in comparison to a placebo (Lee et al. 2017). Within a group of individuals diagnosed with Alzheimer's disease ($n = 119$), the consumption of resveratrol (starting at 500 mg daily and gradually escalating every 13 weeks for a total of 52 weeks, reaching 1000 mg twice daily) resulted in a reduction of Aβ40 levels in both plasma and cerebrospinal fluid (CSF). Additionally, it was determined that the tested doses of resveratrol were safe and well tolerated by the participants (Cicero et al. 2018).

### 2.4.2 Epicatechin

(−)-Epicatechin (EC) is a secondary metabolite found in various plants and is one of the most abundant plant polyphenols consumed in the human diet (Aron and Kennedy 2008). Epicatechin is found in apples, cocoa, pear, peach, apricot, cherry (sweet and sour), raspberry, blackberry, strawberry, etc. (Tsanova-Savova et al. 2005). EC is renowned for its remarkable antioxidant activity. Metabolites derived from EC can be detected in the plasma and accumulate in several organs, including the brain, liver, heart, intestine, kidney, and others. This suggests that EC could benefit various tissues and organs throughout the human body (Serra et al. 2011; Abd El Mohsen et al. 2002). In recent years, scientists have investigated natural products' therapeutic and preventive properties for treating chronic diseases, with EC being one of the most extensively studied. EC has been shown to downregulate the expression of nuclear factor-kappa B (NF-kB) and tumor necrosis factor-alpha (TNF-$\alpha$), effectively reducing the incidence of chronic inflammation associated with oxidative stress (Vazquez-Prieto et al. 2012).

Consuming cacao products containing a high concentration of (−)-epicatechin is associated with a decreased risk of cardiovascular issues and enhanced vascular function (Ramirez-Sanchez et al. 2018). Epidemiological studies establish a connection between the consumption of cocoa products and a decreased risk of cardiometabolic issues (Buijsse et al. 2010; McCullough et al. 2012). Furthermore, studies have indicated that cocoa consumption can stimulate the mobilization of angiogenic cells, enhance memory, and increase brain blood volume in normal 60-year-old individuals after 3 months of supplementation (Heiss et al. 2010; Brickman et al. 2014). The study's findings provide evidence for the neuroprotective potential of low-dose EC in countering age-related decline in brain function. This potential can be attributed to EC's ability to effectively mitigate oxidative stress, inflammation, and factors contributing to brain degeneration while promoting neuron growth associated with functional improvements. Notably, these effects may also extend to young animals, suggesting a potential role in preserving brain health. Observations have demonstrated that flavonoids can hinder oxidative stress-induced neuronal harm by inhibiting the activation of caspase-3. This finding provides evidence supporting their anti-apoptotic solid action. Specifically, EC and 3-O-methyl-epicatechin have been found to safeguard neurons against oxidative damage by suppressing JNK, c-jun, and pro-caspase-3, which are downstream partners involved in this mechanism (Navarrete-Yañez et al. 2020). Epicatechin has been found to elevate the GSH (glutathione) levels in astrocytes, leading to improved memory. In rats with A$\beta$-induced hippocampal toxicity, epicatechin can mitigate lipid peroxidation and reduce ROS (reactive oxygen species), positively impacting memory function (Williams et al. 2011). EC exhibits inhibitory effects on Ar (androgen receptor) activity and regulates the availability of l-arginine.

Additionally, EC and Norv (a drug) enhance endothelial and vascular function in aging by modulating eNOS (endothelial nitric oxide synthase). As a result, EC shows potential as a therapeutic candidate for conditions characterized by endothelial dysfunction, such as aging, cardiometabolic diseases, and hypertension. These findings provide a basis for considering and implementing clinical trials involving EC (Garate-Carrillo et al. 2020). A study has shown that EC can counteract the detrimental effects of aging on organ function in mice. EC achieves these effects even at low doses, distinguishing it from other polyphenols (Moreno-Ulloa et al. 2015). The study revealed that after a 12-week supplementation period with 900 mg of CF (containing 138 mg of epicatechin), participants in the high CF group exhibited a response time that was 630 ms faster compared to those in the low CF control group. This improvement was observed through a Modified-Benton test designed to assess the function of the dentate gyrus region in the hippocampus (Brickman et al. 2014).

### 2.4.3 Quercetin

Quercetin, a flavonoid primarily derived from edible plants, is recognized as one of the most potent antioxidants of plant origin (Brüll et al. 2015). Certain food items have a notably high concentration of quercetin in their edible portions. In a study conducted by Hertog et al. (1992), hydrolyzed samples from various fruits and vegetables were examined to determine the total aglycone content. Onions were found to have the highest concentration of quercetin (284–486 mg/kg), followed by kale (110 mg/kg), French beans (32–35 mg/kg), broccoli (30 mg/kg), lettuce (14 mg/kg), and tomatoes (8 mg/kg). Among the fruits examined, the average quercetin concentration was 15 mg/kg, with apples displaying the highest concentration (21–72 mg/kg) (Formica and Regelson 1995).

It has been reported that dietary supplementation of quercetin at 200 mM extended the lifespan of *C. elegans* by 20%. Quercetin is one of the most commonly found phytochemicals. It is commonly found in many plant-based products. The largest concentration of quercetin is often found in onions. Antioxidant qualities are present in this plant component, which is thought to have antiaging benefits. Both the free and conjugated versions of it are known to exist. Quercetin possesses antioxidative, anti-inflammatory, antiproliferative, anticarcinogenic, antidiabetic, and antiviral qualities. Due to the molecule's lipophilicity and ease of blood-brain barrier crossing, it offers protection from neurodegenerative disorders. Quercetin is also found to treat neurodegeneration, diabetes, cancer, and inflammation. Several medicinal products already use the powerful antioxidant quercetin (Deepika and Maurya 2022).

QUER and its derivative, QU-CAP, have been identified as potent proteasome activators with antioxidant properties. These properties significantly impact cellular lifespan, survival, and viability. Research has revealed that these compounds rejuvenate primary fibroblasts in the middle-aged and terminally senescent stages (Chondrogianni et al. 2010). Extensive research has been conducted on the neuroprotective effects of quercetin. Quercetin can counteract cell toxicity caused by oxidative stress in neurons at low micromolar concentrations. It also exhibits anti-inflammatory properties by reducing the expression of pro-inflammatory cytokines like NF-kB and iNOS while promoting neuronal regeneration. Following absorption, quercetin undergoes metabolization through glucuronidation, methylation, or sulfation, and all its metabolites have shown neuroprotective effects. The neuroprotective efficacy of quercetin has been investigated in both in vitro and in vivo models (Costa et al. 2016; Caruana et al. 2016; Jantan et al. 2015). Multiple studies have reported the neuroprotective effects of quercetin when administered in vivo. It can shield neuronal cells from oxidative stress caused by different chemicals, thereby preventing hippocampal apoptosis (Costa et al. 2016; Dong et al. 2014). The antioxidant properties of quercetin have been associated with improved memory, learning, and cognitive functions. Numerous studies have demonstrated these effects, highlighting quercetin's ability to enhance cognitive performance (Singh et al. 2016). In a mouse model of hypoxic-ischemic brain injury, researchers examined the neuroprotective and anti-inflammatory effects of a fraction enriched with flavonoids, including quercetin and its glucosides. The study demonstrated that repeated administration of the flavonoid-enriched fraction before experimental stroke induced by hypoxic ischemia prevented neuronal loss in the striatum and dorsal hippocampus.

Due to the low bioavailability of quercetin and its glucosides, the neuroprotective effects required injection via the intraperitoneal or intravenous route (Keddy et al. 2012). Quercetin, a flavonoid widely present in various plants and commonly found in our daily diet, holds great promise as a pharmacological compound with therapeutic potential. Its remarkable effects have been observed in both in vitro and in vivo models, demonstrating its ability to counteract the underlying mechanisms associated with Alzheimer's disease (AD). Quercetin exhibits neuroprotective properties by mitigating oxidative stress and neuroinflammation, effectively safeguarding neuronal cells. Its anti-AD properties encompass inhibiting the aggregation of Aβ and phosphorylation of tau proteins.

Furthermore, it restores acetylcholine levels by inhibiting the hydrolysis of acetylcholine through the AChE enzyme. Future clinical trials focusing on quercetin and its analogs as neuroprotective agents should prioritize improving its bioavailability. Developing related compounds with enhanced gut and brain penetrability can enhance the clinical efficacy of these compounds in treating neurodegenerative disorders (Khan et al. 2019).

### 2.4.4 Curcumin

Reports suggest that populations who regularly consume curry exhibit a lower occurrence of Alzheimer's disease, and older individuals within these populations tend to demonstrate superior cognitive functions (Dominguez and Barbagallo 2018). Based on a study involving individuals aged 60 to 90 ($n = 1010$), it was determined that those who consumed curry frequently or occasionally displayed notably higher MMSE scores compared to individuals who were non-consumers or consumed curry rarely (Perry and Howes 2011). As a result, abundant research has been dedicated to unraveling the mechanical impacts of turmeric rhizome and its components commonly found in curry. Most studies investigating cognitive decline in dementia and normal aging have primarily focused on curcuminoids, with particular attention given to curcumin (Howes and Perry 2011). The association between oxidative stress and aging and cognitive decline has been established, prompting significant interest in the antioxidant properties of curcumin (Nelson and Tabet 2015). The curcumin molecule's enol structure and intramolecular hydrogen bond are crucial in its ability to scavenge free radicals. The phenolic hydrogens within curcumin are essential for its antioxidant activity and interactions with free radicals.

Additionally, curcumin can form complexes with metal ions through its diketone and pairs of phenol and methoxy groups, thereby mitigating metal-induced amyloid aggregation and oxidative neurotoxicity (Williams et al. 2011). Numerous investigations have explored the effects of curcumin in diverse animal models of cognitive decline, including those representing accelerated and normal aging as well as dementia. These studies, employing various research designs, have consistently reported positive outcomes, demonstrating curcumin's beneficial effects in most cases (Williams et al. 2011; Sanei and Saberi-Demneh 2019; Sarker and Franks 2018). The capacity of curcumin to enhance memory in numerous studies can be attributed to its antioxidant mechanisms, which involve reducing oxidative damage by enhancing glutathione (GSH) levels and decreasing lipid peroxide levels in brain tissue (Williams et al. 2011; Perry and Howes 2011; Sarker and Franks 2018). Additionally, animal studies have demonstrated that curcumin enhances memory performance and can reverse stress-induced reductions in neurogenesis, which are associated with brain-derived neurotrophic factors (Poulose et al. 2017). Considering that the aging process and neurodegenerative diseases have been described as inflammatory conditions, the anti-inflammatory mechanisms of curcumin may play a significant role in its potential to alleviate cognitive decline (Poulose et al. 2017; Nelson and Tabet 2015).

Furthermore, curcumin preserves the functionality of protein disulfide isomerase, an enzyme that facilitates the maturation of proteins containing disulfide bonds. By doing so, curcumin helps prevent the accumulation of misfolded protein variants resistant to protein disulfide isomerase and tends to accumulate under oxidative stress conditions. These misfolded proteins are implicated in the pathogenesis of Alzheimer's disease (Howes and Perry 2011). About both cognitive functions in aging and dementia, curcumin exhibited the ability to rescue the age-related decline in the specific input specificity of long-term potentiation (LTP) in the hippocampus by promoting the activity of NMDA receptors (Phillips 2017). Regarding the proposed association between high cholesterol levels and the risk of Alzheimer's disease, the cholesterol-lowering effects of curcumin, as evidenced by various animal studies, have also been suggested as an additional factor in reducing the risk of dementia and cognitive decline associated with aging (Goozee et al. 2016). According to a meta-analysis of trials investigating the effects of curcumin interventions across different groups of individuals, the findings suggested that curcumin demonstrated greater efficacy in enhancing cognitive functions among the elderly compared to individuals diagnosed with Alzheimer's disease (Zhu et al. 2019). The increasing preclinical evidence regarding the mechanical impacts of curcumin instills optimism that through rigorous studies assessing its effects on cognitive functions in the general aging population, as well as its relation to the risk of Alzheimer's disease, curcumin may potentially play a role in preserving cognitive abilities as individuals age. This may be further enhanced by the development of formulations with improved bioavailability (Howes et al. 2020).

### 2.4.5 Green Tea Extract and EGCG

Tea, derived from the leaves and buds of the *Camellia sinensis* plant, is widely recognized as the second most popular beverage globally, surpassing coffee, beer, wine, and carbonated soft drinks in consumption (Costa et al. 2002; Cabrera et al. 2006). For centuries, green tea has been consumed in China and other Asian countries to maintain and enhance overall health. In the present day, green tea has gained significant recognition as a dietary agent with promising potential for the prevention and treatment of various diseases, leading to extensive research worldwide. Numerous studies conducted on different animal models have demonstrated that the aqueous extract of green tea, rich in significant catechins (EGCG, EGC, ECG, and EC), exhibits a wide range of beneficial properties. These include antioxidant, antimutagenic, antidiabetic, anti-inflammatory, antibacterial, antiviral, and, notably, cancer-preventive effects.

Additionally, epidemiological studies have suggested that regular consumption of green tea may offer protective benefits against the development of several types of cancers. Furthermore, emerging research indicates that green tea consumption may have a positive impact on neurodegenerative diseases such as Parkinson's and Alzheimer's disease (Lee et al. 2000). The antiaging effect of green tea extract and its major bioactive component, EGCG, was not reported until recently. Still, all of these studies were performed with rodents or lower model organisms. Green tea has been used to regulate digestion and brain activity. Epigallocatechin gallate (EGCG) is the primary active compound in green tea. Although it has been hypothesized that both EGCG and epigallocatechin (EGC) can penetrate the blood-brain barrier and reach the brain parenchyma, EGCG is more effective than EGC at promoting neuronal differentiation. Additionally, it has been proposed that EGCG and its breakdown by the gut microbiota have a delayed effect on nerve cells and that both EGCG and its degradation products encourage the differentiation of nerve cells (Unno and Nakamura 2021).

A study showed that EGCG exhibited a neuroprotective effect against hippocampal neuronal damage in a gerbil model of global ischemia. Importantly, this neuroprotective effect was dose-dependent, indicating that higher doses of EGCG resulted in more excellent protection. The neuroprotective mechanism of EGCG was attributed to its potent antioxidant properties, which effectively countered the damaging effects of oxidative stress in the hippocampus. This suggests that the antioxidant activity of EGCG played a significant role in its ability to protect neurons from ischemic injury in the brain (Lee et al. 2000). The diverse range of actions exhibited by green tea catechins suggests their potential significance in providing neuroprotection. As potent radical scavengers, these polyphenols possess multiple modes of action, engaging various cellular mechanisms involved in neuroprotection and neurorescue. These mechanisms include iron chelation, scavenging of oxygen and nitrogen radicals, activation of the PKC signaling pathway, and stimulation of prosurvival genes. By acting on these various pathways, green tea catechins demonstrate their ability to provide comprehensive cellular neuroprotection and promote neuronal survival (Mandel et al. 2008). L-theanine is a constituent of Green Tea. When administered alone, it has been shown to have various effects on the central nervous system (CNS), blood flow, and heart rate.

Additionally, when combined with caffeine, L-theanine exhibits an antagonistic effect by counteracting the arousing effects of caffeine and reducing blood pressure. L-theanine has also been found to impact mood and cognitive functions, including improved reaction time, long- and short-term memory, and the ability to suppress distractions. EGCG has been primarily observed to mitigate elevated oxidative stress, regulate psychological stress responses, inhibit the progressive rise in systolic blood pressure, and modulate GABAergic systems (Dietz and Dekker 2017). The metabolism of green tea catechins has been examined in animal models and human subjects. It has been observed that tea polyphenols can accumulate in the brain and exert neuroprotective effects through their ingestion alone (Hong et al. 2000). Upon oral administration, green tea catechins are absorbed and retained in the brain, allowing their antioxidative and neuroprotective properties to positively impact age-related decline in spatial cognition.

Furthermore, the use of green tea catechins may serve as a preventive measure against neurodegenerative disorders like Alzheimer's disease, which is linked to oxidative damage and neurotoxicity (Haque et al. 2006). The current study provides evidence that the extended pre-administration of PE effectively mitigates spatial cognitive learning deficits induced by Aβ1–40 in rats with an Alzheimer's disease (AD) model. PE pre-administration consistently attenuated the elevated levels of lipid peroxidation (LPO) and reactive oxygen species (ROS) in both the brain and plasma, indicating that the antioxidative properties of PE may play a crucial role in preventing cognitive impairment in rats infused with Aβ (Haque et al. 2008).

### 2.4.6 Other Phytochemicals and Disease

Berries and citrus fruits encompass various phytochemicals, including flavonoids, anthocyanidins/anthocyanins, tannins, phenolic acids, coumarins, and terpenoids. These compounds have been extensively studied due to their potential impacts on health and cognition (Howes and Simmonds 2014; Jimenez-Garcia et al. 2013; Smeriglio et al. 2017). An instance of this is protocatechuic acid, which has been shown to diminish cell death, excitotoxicity, endoplasmic reticulum stress, intracellular calcium release, and reactive oxygen species (ROS) production induced by Aβ, glutamate, and $H_2O_2$ in cultured neurons. Similarly, the anthocyanidin cyanidin exhibits comparable effects (Hornedo-Ortega et al. 2016; Thummayot et al. 2016; Winter et al. 2017). Research has examined the impact of citrus flavonoids on cognitive functions in various animal models. When rats were orally administered the citrus flavanone naringenin for 14 days, it effectively enhanced spatial memory deficits and reduced lipid peroxidation and oxidative stress caused by okadaic acid administered intracerebroventricularly.

Moreover, it also decreased the activation of TNF-α, TNF-β, IL-1, NF-κB, and caspase-3 (Sachdeva and Chopra 2015). After being administered intraperitoneally for 40 days, the hexamethoxyflavone nobiletin demonstrated memory-enhancing effects and effectively reduced tau hyperphosphorylation and oxidative stress in the hippocampal tissue of senescence-accelerated mice (SAM) (Nakajima et al. 2013). In vivo, graphene, a coumarin compound distinct from citrus flavonoids, exhibits neuroprotective properties and influences inflammatory pathways. It accomplishes this by inhibiting microglial activation and COX-2 expression by astrocytes within the central nervous system (CNS) (Okuyama et al. 2013). Coffee, derived from Coffea species, contains significant amounts of caffeine, which acts as an antagonist for the adenosine A2A receptor. By inhibiting glutamate release, caffeine contributes to neuroprotection.

Additionally, caffeine reduces Aβ levels, indicating its involvement in preserving or enhancing cognitive functions through distinct mechanisms (Howes 2013). Various studies indicate that caffeine exhibits multiple mechanical effects that are relevant to cognitive functions and Alzheimer's disease (AD). In coffee, a pyridine alkaloid called trigonelline exists, which has been found to promote neurite outgrowth and inhibit acetylcholinesterase (AChE) in vitro. Moreover, trigonelline can reduce oxidative stress, suppress astrocyte activity, and alleviate inflammation. These effects contribute to its protective actions against Aβ and its potential to enhance memory in animal models of Alzheimer's disease (AD) (Howes 2013; Fahanik-Babaei et al. 2019). Randomized controlled trials (RCTs) have demonstrated that administering saffron extracts at a daily dosage of 30 mg for either 16 or 22 weeks can enhance cognitive functions in patients with Alzheimer's (AD). Notably, these improvements were observed when comparing saffron extracts to the AChE inhibitor drug, donepezil, while exhibiting fewer adverse effects.

Furthermore, when saffron extracts were taken at the exact dosage for 1 year, they were comparable to memantine, an NMDA receptor antagonist, to reduce cognitive decline in AD patients (Howes et al. 2017). Saffron extracts and their components have demonstrated mechanical effects pertinent to glucose control, diabetes, and ischemic heart disease. Since these conditions are recognized as risk factors for cognitive decline and dementia, saffron might indirectly impact cognitive functions during aging through its mechanisms relevant to the abovementioned diseases (Christodoulou et al. 2015). Studies conducted on a prospective basis have revealed that consuming high amounts of polyunsaturated fatty acids (PUFAs) and monounsaturated fatty acids (MUFAs) can help prevent cognitive decline in older individuals. Additionally, cohort studies have indicated that the consumption of omega-3 PUFAs is associated with a reduced risk of developing Alzheimer's disease (AD) (Gillette-Guyonnet et al. 2013). Clinical investigations have primarily examined the impacts of polyunsaturated fatty acids (PUFAs) from fish oils or docosahexaenoic acid (DHA). Several studies have reported cognitive advantages in various populations, including healthy adults, healthy older adults, and individuals with mild cognitive dysfunction associated with Alzheimer's disease (AD) (Freund-Levi et al. 2006; Witte et al. 2014). The production of pro-inflammatory cytokines contributes to disease severity. Differential expression of signaling molecules initiates Wnt, JNK, and apoptotic pathways and progress the disease (Vasantharekha et al. 2023). The JAK/STAT pathways differentiate naïve T cells into subsets of T cells and cytokine signals. The cross-talk and dysregulation of JAK/STAT, MAPK, PI-3 K, and Akt modulate signaling cascades at the molecular level and influence the disease progression through neuroendocrine dysfunction (Pratap et al. 2020). Aging results in an imbalance in pro- and anti-apoptotic signals and an age-related reduction in the expression of intracellular signaling pathways like m-TOR, p-ERK/ERK, p-CREB/CREB, and Akt. These signaling molecules and their pathways associated with the immune cells lead to immune dysregulation (Hima et al. 2020).

## 2.5 Conclusion

As discussed, cognition and aging are two of humans' most essential functions. Over the years, research in this field has unveiled many of the secrets underlying these processes. The discovery of these functions and pathways has and will lead to better drug delivery systems. It is thus pivotal to understand these processes. On the other hand, the aging process is often attributed to oxidative stress, and it has been concluded that as individuals grow old, most mental processes decline. The pathways involved in these processes are potent targets for drugs that can slow down or inhibit them. The aging process is also linked to a decline in immune reactivity, which consequently leads to higher susceptibility to infectious diseases and cancer among the elderly. Thus, evaluations of the potential involvement of the sympathetic nervous system have been done to understand age-related immune disorders.

That is where phytochemicals have proven vital. Phytochemicals are found in almost every fruit, vegetable, berries, nut, etc. These chemicals include resveratrol, epicatechin, quercetin, EGCG, curcumin, and many more. As discussed, resveratrol exhibits an extension of lifespan, decreased cognitive impairments, neuroprotective effects, and reductions in Aβ accumulation and tau hyperphosphorylation. The potential neuroprotective effects of resveratrol might be attributed to their ability to activate the sirtuin 1 miR-134 pathway and increase the expression of CREB/BDNF, consequently contributing to its neuroprotective mechanisms. Epicatechin is renowned for its antioxidant activity, an essential mechanism for arresting aging and cognitive decline. EC has been shown to downregulate the expression of nuclear factor-kappa B (NF-kB) and tumor necrosis factor-alpha (TNF-α), thereby reducing chronic inflammation associated with oxidative stress. Potent proteasome activators with antioxidant properties, the compound QUER and its derivative QU-CAP, have been discovered to significantly affect cellular lifespan, survival, and viability. Research findings demonstrate that these compounds rejuvenate primary fibroblasts at middle-aged and terminally senescent stages. Curcumin has shown greater efficacy in enhancing cognitive functions among the elderly population. Green tea phytochemicals demonstrate their ability to provide comprehensive cellular neuroprotection and promote neuronal survival through iron chelation, scavenging of oxygen and nitrogen radicals, activation of the PKC signaling pathway, and stimulation of prosurvival genes. Other plant extracts like flavonoids, anthocyanidins/anthocyanins, tannins, phenolic acids, coumarins, and caffeine have shown similar neuroprotective and cognition-enhancing effects. Saffron extracts and their components have also shown effects pertinent to glucose control, diabetes, and ischemic heart disease. Because these conditions are recognized as risk factors for cognitive decline and dementia, saffron might indirectly impact cognitive functions during aging through its mechanistic effects relevant to these diseases. Phytochemicals offers immense potential for developing personalized therapeutic strategies to combat age-related cognitive decline and neurodegeneration. Thus, it is clear that many phytochemicals function in neuroprotection, enhancing cognition and lowering aging rates. Overcoming challenges associated with bioavailability and developing effective delivery systems for phytochemicals will be crucial for their successful clinical translation. Additionally, establishing clear biomarkers for monitoring the efficacy of these interventions will be vital for personalized treatment plans. The future of neurodegenerative disease prevention and treatment holds immense promise with the exploration of phytochemicals. By harnessing the power of these natural compounds through personalized medicine, combination therapies, and overcoming translational hurdles, we can pave the way for a healthier and more cognitively vibrant future for aging populations.

## References


Abd El Mohsen MM, Kuhnle G, Rechner AR, Schroeter H, Rose S, Jenner P, Rice-Evans CA (2002) Uptake and metabolism of epicatechin and its access to the brain after oral ingestion. Free Radic Biol Med 33(12):1693–1702

Abrass CK, O'connor SW, Scarpace PJ, Abrass IB (1985) Characterization of the beta-adrenergic receptor of the rat peritoneal macrophage. J Immunol 135(2):1338–1341

Aron PM, Kennedy JA (2008) Flavan-3-ols: nature, occurrence and biological activity. Mol Nutr Food Res 52(1):79–104

Bellinger DL, Felten SY, Felten DL (1988) Maintenance of noradrenergic sympathetic innervation in the involuted thymus of the aged Fischer 344 rat. Brain Behav Immun 2(2):133–150

Borson S (2010) Cognition, aging, and disabilities: conceptual issues. Phys Med Rehabil Clin N Am 21(2):375–382

Brickman AM, Khan UA, Provenzano FA, Yeung LK, Suzuki W, Schroeter H, Small SA (2014) Enhancing dentate gyrus function with dietary flavanols improves cognition in older adults. Nat Neurosci 17(12):1798

Brüll V, Burak C, Stoffel-Wagner B, Wolffram S, Nickenig G, Müller C, Egert S (2015) Effects of a quercetin-rich onion skin extract on 24 h ambulatory blood pressure and endothelial function in overweight-to-obese patients with (pre-) hypertension: a randomised double-blinded placebo-controlled cross-over trial. Br J Nutr 114(8):1263–1277

Buijsse B, Weikert C, Drogan D, Bergmann M, Boeing H (2010) Chocolate consumption in relation to blood pressure and risk of cardiovascular disease in German adults. Eur Heart J 31(13): 1616–1623

Cabrera C, Artacho R, Giménez R (2006) Beneficial effects of green tea—a review. J Am Coll Nutr 25(2):79–99

Carstensen LL, Mikels JA, Mather M (2006) Aging and the intersection of cognition, motivation, and emotion. In: Handbook of the psychology of aging. Academic Press, London, pp 343–362

Caruana M, Cauchi R, Vassallo N (2016) Putative role of red wine polyphenols against brain pathology in Alzheimer's and Parkinson's disease. Front Nutr 3:31

Chondrogianni N, Kapeta S, Chinou I, Vassilatou K, Papassideri I, Gonos ES (2010) Anti-ageing and rejuvenating effects of quercetin. Exp Gerontol 45(10):763–771

Christodoulou E, Kadoglou NP, Kostomitsopoulos N, Valsami G (2015) Saffron: a natural product with potential pharmaceutical applications. J Pharm Pharmacol 67(12):1634–1649

Cicero AF, Fogacci F, Banach M (2018) Botanicals and phytochemicals active on cognitive decline: the clinical evidence. Pharmacol Res 130:204–212

Costa LM, Gouveia ST, Nóbrega JA (2002) Comparison of heating extraction procedures for Al, Ca, Mg, and Mn in tea samples. Anal Sci 18(3):313–318

Costa LG, Garrick JM, Roquè PJ, Pellacani C (2016) Mechanisms of neuroprotection by quercetin: counteracting oxidative stress and more. Oxidative Med Cell Longev 2016:1

Costa C, Tsatsakis A, Mamoulakis C, Teodoro M, Briguglio G, Caruso E, Fenga C (2017) Current evidence on the effect of dietary polyphenols intake on chronic diseases. Food Chem Toxicol 110:286–299

Deepika, Maurya PK (2022) Health benefits of quercetin in age-related diseases. Molecules 27(8): 2498

Dietz C, Dekker M (2017) Effect of green tea phytochemicals on mood and cognition. Curr Pharm Des 23(19):2876–2905

Dominguez LJ, Barbagallo M (2018) Nutritional prevention of cognitive decline and dementia. Acta Biomed 89(2):276

Dong YS, Wang JL, Feng DY, Qin HZ, Wen H, Yin ZM, Li C (2014) Protective effect of quercetin against oxidative stress and brain edema in an experimental rat model of subarachnoid hemorrhage. Int J Med Sci 11(3):282

Fahanik-Babaei J, Baluchnejadmojarad T, Nikbakht F, Roghani M (2019) Trigonelline protects hippocampus against intracerebral Aβ (1–40) as a model of Alzheimer's disease in the rat: insights into underlying mechanisms. Metab Brain Dis 34:191–201

Formica JV, Regelson W (1995) Review of the biology of quercetin and related bioflavonoids. Food Chem Toxicol 33(12):1061–1080

Forni C, Facchiano F, Bartoli M, Pieretti S, Facchiano A, D'Arcangelo D, Jadeja RN (2019) Beneficial role of phytochemicals on oxidative stress and age-related diseases. BioMed Res Int 2019:1

Freund-Levi Y, Eriksdotter-Jönhagen M, Cederholm T, Basun H, Faxen-Irving G, Garlind A, Palmblad J (2006) ω-3 fatty acid treatment in 174 patients with mild to moderate Alzheimer disease: OmegAD study: a randomized double-blind trial. Arch Neurol 63(10):1402–1408

Fuchs BA, Albright JW, Albright JF (1988) β-Adrenergic receptors on murine lymphocytes: density varies with cell maturity and lymphocyte subtype and is decreased after antigen administration. Cell Immunol 114(2):231–245

Garate-Carrillo A, Navarrete-Yañez V, Ortiz-Vilchis P, Guevara G, Castillo C, Mendoza-Lorenzo P, Ramirez-Sanchez I (2020) Arginase inhibition by (−)-Epicatechin reverses endothelial cell aging. Eur J Pharmacol 885:173442

Genaro AM, Borda E (1989) Alloimmunization-induced changes in β-adrenoceptor expression and cAMP on B lymphocytes. Immunopharmacology 18(1):63–70

Gillette-Guyonnet S, Secher M, Vellas B (2013) Nutrition and neurodegeneration: epidemiological evidence and challenges for future research. Br J Clin Pharmacol 75(3):738–755

Goozee KG, Shah TM, Sohrabi HR, Rainey-Smith SR, Brown B, Verdile G, Martins RN (2016) Examining the potential clinical value of curcumin in the prevention and diagnosis of Alzheimer's disease. Br J Nutr 115(3):449–465

Haque AM, Hashimoto M, Katakura M, Tanabe Y, Hara Y, Shido O (2006) Long-term administration of green tea catechins improves spatial cognition learning ability in rats. J Nutr 136(4): 1043–1047

Haque AM, Hashimoto M, Katakura M, Hara Y, Shido O (2008) Green tea catechins prevent cognitive deficits caused by Aβ1–40 in rats. J Nutr Biochem 19(9):619–626

Heiss C, Jahn S, Taylor M, Real WM, Angeli FS, Wong ML, Yeghiazarians Y (2010) Improvement of endothelial function with dietary flavanols is associated with mobilization of circulating angiogenic cells in patients with coronary artery disease. J Am Coll Cardiol 56(3):218–224

Herrup K (2015) The case for rejecting the amyloid cascade hypothesis. Nat Neurosci 18(6): 794–799

Hima L, Patel MN, Kannan T, Gour S, Pratap UP, Priyanka HP, Vasantharekha R, ThyagaRajan S (2020) Age-associated decline in neural, endocrine, and immune responses in men and women: involvement of intracellular signaling pathways. J Neuroimmunol 345(2020):577290. https://doi.org/10.1016/j.jneuroim.2020.577290

Hong JT, Ryu SR, Kim HJ, Lee JK, Lee SH, Kim DB, Kim PY (2000) Neuroprotective effect of green tea extract in experimental ischemia-reperfusion brain injury. Brain Res Bull 53(6): 743–749

Hornedo-Ortega R, Alvarez-Fernández MA, Cerezo AB, Richard T, Troncoso AM, Garcia-Parrilla MC (2016) Protocatechuic acid: inhibition of fibril formation, destabilization of preformed fibrils of amyloid-β and α-synuclein, and neuroprotection. J Agric Food Chem 64:7722–7732

Howes MJR (2013) Alkaloids and drug discovery for neurodegenerative diseases

Howes MJR, Perry E (2011) The role of phytochemicals in the treatment and prevention of dementia. Drugs Aging 28:439–468

Howes MJR, Simmonds MS (2014) The role of phytochemicals as micronutrients in health and disease. Curr Opin Clin Nutr Metab Care 17(6):558–566

Howes MJR, Fang R, Houghton PJ (2017) Effect of Chinese herbal medicine on Alzheimer's disease. Int Rev Neurobiol 135:29–56

Howes MJR, Perry NS, Vásquez-Londoño C, Perry EK (2020) Role of phytochemicals as nutraceuticals for cognitive functions affected in ageing. Br J Pharmacol 177(6):1294–1315

Jantan I, Ahmad W, Bukhari SNA (2015) Plant-derived immunomodulators: an insight on their preclinical evaluation and clinical trials. Front Plant Sci 6:655

Jimenez-Garcia SN, Guevara-Gonzalez RG, Miranda-Lopez R, Feregrino-Perez AA, Torres-Pacheco I, Vazquez-Cruz MA (2013) Functional properties and quality characteristics of bioactive compounds in berries: biochemistry, biotechnology, and genomics. Food Res Int 54(1):1195–1207

Keddy PG, Dunlop K, Warford J, Samson ML, Jones QR, Rupasinghe HV, Robertson GS (2012) Neuroprotective and anti-inflammatory effects of the flavonoid-enriched fraction AF4 in a mouse model of hypoxic-ischemic brain injury. PLoS One 7(12):e51324

Kennedy DO (2014a) Polyphenols and the human brain: plant "secondary metabolite" ecologic roles and endogenous signaling functions drive benefits. Adv Nutr 5(5):515–533

Kennedy DO (2014b) Plants and the human brain. Oxford University Press, New York, NY

Kennedy DO (2019) Phytochemicals for improving aspects of cognitive function and psychological state potentially relevant to sports performance. Sports Med 49:39–58

Kennedy DO, Wightman EL (2011) Herbal extracts and phytochemicals: plant secondary metabolites and the enhancement of human brain function. Adv Nutr 2(1):32–50

Khan H, Ullah H, Aschner M, Cheang WS, Akkol EK (2019) Neuroprotective effects of quercetin in Alzheimer's disease. Biomol Ther 10(1):59

Lee SR, Suh SI, Kim SP (2000) Protective effects of the green tea polyphenol (−)-epigallocatechin gallate against hippocampal neuronal damage after transient global ischemia in gerbils. Neurosci Lett 287(3):191–194

Lee J, Torosyan N, Silverman DH (2017) Examining the impact of grape consumption on brain metabolism and cognitive function in patients with mild decline in cognition: a double-blinded placebo controlled pilot study. Exp Gerontol 87:121–128

Maclin JMA, Wang T, Xiao S (2019) Biomarkers for the diagnosis of Alzheimer's disease, dementia Lewy body, frontotemporal dementia and vascular dementia. Gen Psychiatr 32(1): e100054

Madden KS, Felten SY, Felten DL, Bellinger DL (1995) Sympathetic nervous system-immune system interactions in young and old Fischer 344 rats. Ann N Y Acad Sci 771(1):523–534

Madden KS, Bellinger DL, Felten SY, Snyder E, Maida ME, Felten DL (1997) Alterations in sympathetic innervation of thymus and spleen in aged mice. Mech Ageing Dev 94(1–3): 165–175

Malaguarnera L (2019) Influence of resveratrol on the immune response. Nutrients 11(5):946

Mandel SA, Amit T, Weinreb O, Reznichenko L, Youdim MB (2008) Simultaneous manipulation of multiple brain targets by green tea catechins: a potential neuroprotective strategy for Alzheimer and Parkinson diseases. CNS Neurosci Ther 14(4):352–365

Marchetti B, Morale MC, Pelletier G (1990) Sympathetic nervous system control of thymus gland maturation: autoradiographic characterization and localization of the beta2-adrenergic receptor in the rat thymus gland and presence of a sexual dimorphism during ontogenic development. Prog Neuroimmunol 3:103–115; Thymus gland maturation: autoradiographic characterization and localization of the beta2-adrenergic receptor in the rat thymus gland and presence of a sexual dimorphism during ontogenic development. Prog Neuroendocr, 3, 103–115.

Martin LM, Johnson PJ, Amorim JR, Honaker AR, Donaldson RS, DeClue AE (2016) Investigation of the potential immunomodulatory effects of resveratrol on equine whole blood: an in vitro investigation. Res Vet Sci 106:97–99

McCullough ML, Peterson JJ, Patel R, Jacques PF, Shah R, Dwyer JT (2012) Flavonoid intake and cardiovascular disease mortality in a prospective cohort of US adults. Am J Clin Nutr 95(2): 454–464

Miller MG, Thangthaeng N, Poulose SM, Shukitt-Hale B (2017) Role of fruits, nuts, and vegetables in maintaining cognitive health. Exp Gerontol 94:24–28

Mondal SC, Singh P, Kumar B, Singh SK, Gupta SK, Verma A (2014) Ageing and potential anti-aging phytochemicals: an overview. World J Pharm Pharm Sci 4(1):426–454

Moreno-Ulloa A, Nogueira L, Rodriguez A, Barboza J, Hogan MC, Ceballos G, Ramirez-Sanchez I (2015) Recovery of indicators of mitochondrial biogenesis, oxidative stress, and aging with (−)-epicatechin in senile mice. J Gerontol A Biol Sci Med Sci 70(11):1370–1378

Nakajima A, Aoyama Y, Nguyen TTL, Shin EJ, Kim HC, Yamada S, Yamada K (2013) Nobiletin, a citrus flavonoid, ameliorates cognitive impairment, oxidative burden, and hyperphosphorylation of tau in senescence-accelerated mouse. Behav Brain Res 250:351–360

Navarrete-Yañez V, Garate-Carrillo A, Rodriguez A, Mendoza-Lorenzo P, Ceballos G, Calzada-Mendoza C, Ramirez-Sanchez I (2020) Effects of (−)-epicatechin on neuroinflammation and hyperphosphorylation of tau in the hippocampus of aged mice. Food Funct 11(12): 10351–10361

Nelson L, Tabet N (2015) Slowing the progression of Alzheimer's disease; what works? Ageing Res Rev 23:193–209

Okuyama S, Minami S, Shimada N, Makihata N, Nakajima M, Furukawa Y (2013) Anti-inflammatory and neuroprotective effects of auraptene, a citrus coumarin, following cerebral global ischemia in mice. Eur J Pharmacol 699(1–3):118–123

Park HL, O'Connell JE, Thomson RG (2003) A systematic review of cognitive decline in the general elderly population. Int J Geriatr Psychiatry 18(12):1121–1134

Perry E, Howes MJR (2011) Medicinal plants and dementia therapy: herbal hopes for brain aging? CNS Neurosci Ther 17(6):683–698

Phillips C (2017) Lifestyle modulators of neuroplasticity: how physical activity, mental engagement, and diet promote cognitive health during aging. Neural Plast 2017:1

Poulose SM, Miller MG, Scott T, Shukitt-Hale B (2017) Nutritional factors affecting adult neurogenesis and cognitive function. Adv Nutr 8(6):804–811

Pratap UP, Hima L, Kannan T, Thyagarajan C, Priyanka HP, Vasantharekha R, Pushparani A, Thyagarajan S (2020) Sex-based differences in the cytokine production and intracellular

signaling pathways in patients with rheumatoid arthritis. Arch Rheumatol 35(4):545–557. https://doi.org/10.46497/ArchRheumatol.2020.7481
Radojcic T, Baird S, Darko D, Smith D, Bulloch K (1991) Changes in β-adrenergic receptor distribution on immunocytes during differentiation: an analysis of T cells and macrophages. J Neurosci Res 30(2):328–335
Rajaram S, Jones J, Lee GJ (2019) Plant-based dietary patterns, plant foods, and age-related cognitive decline. Adv Nutr 10(Supplement_4):S422–S436
Ramirez-Sanchez I, Mansour C, Navarrete-Yañez V, Ayala-Hernandez M, Guevara G, Castillo C, Villarreal FJ (2018) (−)-Epicatechin induced reversal of endothelial cell aging and improved vascular function: underlying mechanisms. Food Funct 9(9):4802–4813
Sachdeva AK, Chopra K (2015) Naringin mitigate okadaic acid-induced cognitive impairment in an experimental paradigm of Alzheimer's disease. J Funct Foods 19:110–125
Sanei M, Saberi-Demneh A (2019) Effect of curcumin on memory impairment: a systematic review. Phytomedicine 52:98–106
Sarker MR, Franks SF (2018) Efficacy of curcumin for age-associated cognitive decline: a narrative review of preclinical and clinical studies. Geroscience 40:73–95
Serra A, Macià A, Romero MP, Anglès N, Morelló JR, Motilva MJ (2011) Distribution of procyanidins and their metabolites in rat plasma and tissues after an acute intake of hazelnut extract. Food Funct 2(9):562–568
Shen J, Xu L, Qu C, Sun H, Zhang J (2018) Resveratrol prevents cognitive deficits induced by chronic unpredictable mild stress: Sirt1/miR-134 signalling pathway regulates CREB/BDNF expression in hippocampus in vivo and in vitro. Behav Brain Res 349:1–7
Sheng K, Shui S, Yan L, Yu J, Hao G, Qu H, Zheng L (2018) The beneficial effects of dietary grape supplementation on improving cognitive deficits in APP/PS1 double transgenic mice. J Funct Foods 49:224–234
Si H, Liu D (2014) Dietary anti-aging phytochemicals and mechanisms associated with prolonged survival. J Nutr Biochem 25(6):581–591
Singh SK, Srivastav S, Yadav AK, Srikrishna S, Perry G (2016) Overview of Alzheimer's disease and some therapeutic approaches targeting Aβ by using several synthetic and herbal compounds. Oxidative Med Cell Longev 2016:1
Singh P, Sivanandam TM, Konar A, Thakur MK (2021) Role of nutraceuticals in cognition during aging and related disorders. Neurochem Int 143:104928
Smeriglio A, Barreca D, Bellocco E, Trombetta D (2017) Proanthocyanidins and hydrolysable tannins: occurrence, dietary intake and pharmacological effects. Br J Pharmacol 174(11): 1244–1262
Spengler RN, Allen RM, Remick DG, Strieter RM, Kunkel SL (1990) Stimulation of alpha-adrenergic receptor augments the production of macrophage-derived tumor necrosis factor. J Immunol 145(5):1430–1434
Thummayot S, Tocharus C, Suksamrarn A, Tocharus J (2016) Neuroprotective effects of cyanidin against Aβ-induced oxidative and ER stress in SK-N-SH cells. Neurochem Int 101:15–21
Tsanova-Savova S, Ribarova F, Gerova M (2005) (+)-Catechin and (−)-epicatechin in Bulgarian fruits. J Food Compos Anal 18(7):691–698
Unno K, Nakamura Y (2021) Green tea suppresses brain aging. Molecules 26(16):4897
van Ginneken V (2017) Are there any biomarkers of aging? Biomarkers of the brain. Biomed J Sci Tech Res 1(1):2017
Vasantharekha R, Priyanka HP, Swarnalingam T, Srinivasan AV, ThyagaRajan S (2017) Interrelationship between mini-mental state examination scores and biochemical parameters in patients with mild cognitive impairment and Alzheimer's disease. Geriatr Gerontol Int 17:1737–1745
Vasantharekha R, Priyanka HP, Nair RS, Hima L, Pratap UP, Srinivasan AV, ThyagaRajan S (2023) Alterations in immune responses are associated with dysfunctional intracellular signaling in peripheral blood mononuclear cells of men and women with mild cognitive impairment and Alzheimer's disease. Mol Neurobiol. https://doi.org/10.1007/s12035-023-03764-3

Vazquez-Prieto MA, Bettaieb A, Lanzi CR, Haj FG, Fraga CG, Oteiza PI (2012) (−)-Epicatechin prevents adipose tissue inflammation in animal and cell models. Free Radic Biol Med 53:S8
Williams P, Sorribas A, Howes MJR (2011) Natural products as a source of Alzheimer's drug leads. Nat Prod Rep 28(1):48–77
Winter AN, Brenner MC, Punessen N, Snodgrass M, Byars C, Arora Y, Linseman DA (2017) Comparison of the neuroprotective and anti-inflammatory effects of the anthocyanin metabolites, protocatechuic acid and 4-hydroxybenzoic acid. Oxidative Med Cell Longev 2017:1
Witte AV, Kerti L, Hermannstädter HM, Fiebach JB, Schreiber SJ, Schuchardt JP, Flöel A (2014) Long-chain omega-3 fatty acids improve brain function and structure in older adults. Cereb Cortex 24(11):3059–3068
Youdim KA, Joseph JA (2001) A possible emerging role of phytochemicals in improving age-related neurological dysfunctions: a multiplicity of effects. Free Radic Biol Med 30(6): 583–594
Zaplatic E, Bule M, Shah SZA, Uddin MS, Niaz K (2019) Molecular mechanisms underlying protective role of quercetin in attenuating Alzheimer's disease. Life Sci 224:109–119
Zhu LN, Mei X, Zhang ZG, Xie YP, Lang F (2019) Curcumin intervention for cognitive function in different types of people: a systematic review and meta-analysis. Phytother Res 33(3):524–533

# 3 Molecular Mechanism of Brain Aging: Protective Effects of Phytochemicals

Sathya Narayanan Govindarajulu, Dheepthi Jayamurali, Nivedita Manoharan, Nivetha Ravishankar, and Padmini Sateesha Acharya


**Abstract**

Brain aging is an irretrievable process that occurs in the elderly population and inescapable phase of the entity. Aging is associated with complex and multifactorial portent with tedious decrement in behavioral and physiological actions. The decline in memory and cognition corresponds with the anatomical changes in neuronal plasticity with dendritic spines. This augments with neurochemical and neurophysiological alterations. Brain aging is unified by anatomical and physiological changes that include the brain weight and volume, synaptic density, thickness of cortex, and dysregulation of energy metabolism. Many features of aging are meticulous molecular mechanisms that are still unclear. The signaling pathway of brain aging embraces autophagy, oxidative stress, mitochondrial dysfunction, cell proliferating capacity, senescent associated secretions, and epigenetical alterations. This chapter also discusses the novel phytochemical that acts as the antiaging agents. Phytochemicals have a profound effect on brain aging in many aspects which maintain chemical balances. Phytochemicals are secondary metabolites of the plants. The phytochemicals affect the modulation of cellular signaling pathways such as disruption in mitochondrial function, neuroinflammation, neurotropic factor deficiency, apoptosis, oxidative stress, and abnormal protein aggregation to conflict against age-related neurodegeneration. Phytochemicals like berberine, phenolic compound, resveratrol, omega 3 fatty



S. N. Govindarajulu (✉) · D. Jayamurali · N. Manoharan · N. Ravishankar
Department of Physiology, Dr. ALM Post Graduate Institute of Basic Medical Sciences, The University of Madras, Chennai, Tamil Nadu, India

P. S. Acharya
Department of Anatomy, Dr. ALM Post Graduate Institute of Basic Medical Sciences, The University of Madras, Chennai, Tamil Nadu, India







S. Pathak, A. Banerjee (eds.), *Neuroprotective Effects of Phytochemicals in Brain Ageing*, https://doi.org/10.1007/978-981-99-7269-2_3

acids, and cocoa and their actions on cellular signaling are widely discussed in this chapter.

**Keyword**

Aging · Cellular signaling · Mitochondrial dysfunction · Epigenetical alteration · Phytochemicals · Neurodegeneration

## 3.1 Introduction

Aging is an inevitable phase in all living organisms. Generally, aging is associated with intricate and multifactorial phenomena with plodding decrement in behavioral and physiological actions (Kiss et al. 2008). Brain aging is an irretrievable process that occurs in the elderly population (Hou et al. 2019). Aging can be defined as manisfestation of decline in memory and cognition along with the changes in the structural plasticity of the dendritic spines (Sikora et al. 2021). Aging is not only accompanied by anatomical deterioration but also by neurochemical and neurophysiological alterations. The anatomical alteration concerns cellular and molecular changes. Brain aging is interconnected to the anatomical and physiological changes that include the brain weight and volume, synaptic density, thickness of cortex, and dysregulation of energy metabolism (Fox and Schott 2004; Salat et al. 2004; Goyal et al. 2014). In previous studies, aging concerns the neuronal loss in the hippocampus and the prefrontal cortex (PFC) (Burke and Barnes 2006). The reduced dendritic branching were seen in PFC and hippocampal region but there was no marked reduction observed in subregions of the human brain (Flood et al. 1987; Flood 1991; Grill and Riddle 2002). At the cellular levels, there was a loss of myelinated fibers in cerebral hemispheres and the reduction is comparatively high in corpus callosum and PFC (Heden and Gabrieli 2004; Bartzokis et al. 2003).

Aging causes reduction in the neurochemical system like cholinergic, dopaminergic and noradrenergic pathways which results in cognitive impairments (Eriksdotter-Nilsson et al. 1989). The inhibitory neurotransmitter GABA diminishes during aging, and consequently, balance differs between the excitatory and inhibitory neurons (Loerch et al. 2008). In human PET studies of aged brains, there was reduced serotonin receptors in basal ganglia and frontal cortex (Wong et al. 1984). Thus characterized by physical, chemical, and biological alteration in neurons which increases the risk of cardiovascular disease, cancer, and neurodegenerative diseases and leads to death (Mariani et al. 2005; Lopez-Otin et al. 2013). Aging causes slow deterioration in brain's physiological capacity due to its effect on cells, molecules, gross morphology, and vasculature (Peters 2006). In spite of many features of aging in the neuronal network of aging, the meticulous molecular mechanisms are still unclear.

Several studies related to aging have shown the signaling pathway and the molecular targets comprising autophagy, oxidative stress, mitochondrial dysfunction, cell proliferating capacity, senescent-associated secretions, and epigenetic alterations. The neuroanatomical and physiological alterations play an integral role in aging and age-related disorders.

## 3.2 Molecular Mechanism of Aging

### 3.2.1 Gene Expression Profiling

Gene expression profiling use advanced technology such as microarrays and RNA sequencing to evaluate the activity of many genes concurrently. Through the comparison of gene expression patterns in aged tissues, scientists are able to identify genes that undergo alterations in activity as a result of aging.

#### 3.2.1.1 Mitochondrial Dysregulations

Mitochondria are the powerhouse of the cell as it produces more ATP via oxidative phosphorylation. Mitochondria also plays a major role in homeostasis within the cells and modulates the signal transcription of nuclear genes. It has been also reported that mitochondria have other roles in aging and metabolism (Gonzalez-Freire et al. 2015; Held and Houtkooper 2015). As the organism's ages, the efficacy of the electron transport chain declines and the production of ATP declines and disturbs axonal transport in the neurons (Green et al. 2011). In the brain, dysfunctional mitochondrion gradually accumulates and surges the ROS production which in turn causes degradation and cell damage (Harman 1965). The molecular mechanism underlying the age-related dysfunction remains unclear. In gene profiling studies, the degeneration of mitochondria could contribute to aging particularly brain and muscle as they are more susceptible to mitochondrial dysfunction (Tripathi 2012). The mitochondria extracted from animals have shown mitochondrial fragmentation, enhanced oxidative damage to mitochondrial DNA, and impairment in respiratory chain electrons and $Ca^{2+}$ which diminishes the formation of mitochondrial membrane permeability transition pores (mPTP) (Zia et al. 2021). The studies suggest that epigenetic changes can also contribute to mitochondrial dysfunction in which the age-related upregulation of BAZ2B (bromodomain adjacent to zinc finger domain 2B and EHMT1 (euchromatic histone lysine methyltransferase1) involved in the downregulation of genes in metabolism has been reported in Alzheimer's disease and mild cognitive impairments (Yuan et al. 2020).

#### 3.2.1.2 Oxidative Stress

Reactive oxygen species cause damage to the cell including proteins, DNA, and lipids which leads to a decline in physiological functions and biological chemicals through aging (Harraan 1956). Oxidative damage is associated with neurodegenerative diseases where it depends on the chemical characteristics of oxidative moiety, the level between the antioxidant enzymes and the reactive oxidative species (Kumar et al. 2018). In aging, the escalating redox signaling can also alter $Ca^{2+}$ regulation which can also be an underlying mechanism. The redox signaling influences the cysteine disulfide bond of the ryanodine receptor for intracellular $Ca^{2+}$ and determines its release after hyperpolarization. The disulfide bonds are restricted to NMDA receptor subunits and modify the function of NMDA receptors. In aging, the reducing agent dithiothreitol augments the NMDA receptor in the prefrontal cortex and hippocampus (Barter et al. 2019). This might lead to other age-related disorders.

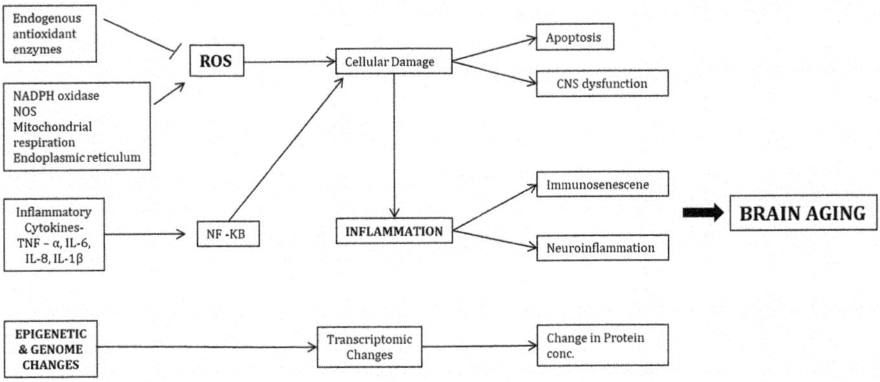


**Fig. 3.1** Molecular mechanism of brain aging

Mitochondria are considered to be the key origin of reactive oxidative species in aging where the impairment leads to a high generation of mitochondrial ROS which is associated with respiratory chain deficiency and mitochondrial DNA mutation (Harman 1972). In a study, the genetic modification has been conducted to elevate the ROS in mice model and it has been reported that increased ROS in the mitochondria has increased the lifespan of worms and while increased ROS in the cytoplasm has decreased it (Schaar et al. 2015; Veal et al. 2007) (Fig. 3.1).

### 3.2.2 Cell Proliferation Capacity

In aging, the shortening of telomere which plays a crucial role in maintaining the genome and promoting stability in replication thus preventing chromosomal fusion and recombination (Muraki et al. 2012; Blackburn et al. 2015). The role of telomere shortening and aging were not completely illustrated. The age-related alterations specifically in neurons are not studied as neurons are postmitotic; in neurons after terminal replication, the cell division is absent (Mosch et al. 2007). In addition, studies have also shown that in somatic cells the telomerase-induced activity has restored the aging features (Bar et al. 2016). In telomerase knockout mice model, the telomerase reverse transcriptase protein (TERT) in hippocampal region has shown to modulate the mood and behaviour by regulating the neural progenitor cells (NPCs) with reduced in the hippocampal dependent learning and memory (Rolyan et al. 2011). In another study, the olfactory dysfunctions are improved in telomerase reactivated in knockout mice (Jaskelioff et al. 2011). For age-related neurodegenerative disease associated with neuronal loss, the neural stem cell transplantation is considered to be the potential treatment (Lunn et al. 2011).The structural brain indices the link between the brain parenchymal fraction (BPF) and leukocyte telomere length (LTL) in relation to cognition function (Gampawar et al. 2020).

### 3.2.3 Epigenetic Alteration

Epigenetics is the alteration in gene expression that is not due to the alteration in DNA sequences. The epigenetic informatin is coded in the post-translational modifications on the N-terminal tails of histones, noncoding RNAs. The cognitive decline is associated with the neural and synaptic activity genes. The change in gene expression can also be due to lifestyle modification where they induce changes in the histones through which the cognition might decline (Barter et al. 2019). 353 CpG site has been proposed as the epigenetic clock with an alteration in DNA methylation (Horvath 2013). The H4K16ac plays a vital role in aging process and neurodegenerative disease, and marked loss of H4K16ac is seen in AD patients (Nativio et al. 2018). A recent study in AD patient has shown that BAZ2B and EHMT1 expression escalates with aging, and it has been positively correlated in the frontal cortex of the human brain. In aged mice, affecting the H3K9 methylation in the promoter region of a gene associated with metabolism in the ablation of BAZ2B improved cognitive dysfunction (Yuan et al. 2020).

Another epigenetic mechanism for regulating aging through microRNA (miRNA) inhibits translation by binding to the posttranscriptional target mRNA. The dysfunction of miRNA affects the age-related disorders and neuroinflammations (Danka et al. 2017). An astrocyte apolipoprotein E3 (APOE3) vesicle carries the miscellaneous regulatory miRNA which silences the genes tangled in the synthesis of cholesterol and promotes the histone acetylation and activates the expression in the neurons in contradictory to APOE4 which carries minimal miRNA in AD patients and thus downregulates histone acetylation and leads to cognitive decline (Li et al. 2021) (Fig. 3.1).

### 3.2.4 Inflammation and Senescence-Associated Secretory Phenotype (SASP)

Senescence-associated secretory phenotype is the cell that stops the cell division and changes its activity to reinforce the secretory function. The brain has its separate subset of immunological specific cells and microglia (Sikora et al. 2021). The abnormal activation of microglial cell escalates the level of inflammatory cystokines and chemokines (Pluvinage et al. 2020). Other than microglia, astrocytes also promote inflammation. A1 reactive astrocytes are secreted in response to the liposaccharides in the aged mouse brain where microglia secretes the cytokines highlighting the communication of propagation of inflammation (Clarke et al. 2018; Liddelow et al. 2017). The studies have confirmed that long-term cultured cortical and hippocampal neurons correlated with the SASP gene expressions and increased IL-6 and Purkinje neurons in the aged mice brain (Jurk et al. 2012). In vitro studies, astrocyte-induced estradiol treatment has elevated levels of TNF-α, IL-1β, IL-6, and IL-8 in the hypothalamus of aged mice (Dai et al. 2020). A new insight of study related to DNA damage, brain function, telomeres and inflammation, reduced Ft1 causes DNA instability, aberration of telomeres, and cell senescence. In Ft1 knockout mice, there was a significant increase in TGF-β and IL-6 (Burla et al. 2018). In another study, nuclear protein complexes (NPC) link between the nucleus and cytoplasm as a key target to senescence, and they are expected to function in brain senescing (Moreno-Blas et al. 2019; Schlachetzki et al. 2020) (Fig. 3.1).

### 3.2.5 Autophagy

Autophagy is an intracellular process where it disposes the unnecessary and damaged molecules and organelles. Autophagy regulates the cellular homeostasis (Sikora et al. 2021). Aging, neurodegenerative diseases, and age-related disorders are said to be majorly regulated by autophagy. As in autophagy, the protein aggregates and the unnecessary cell organelles accumulate which affects the metabolism and cellular homeostasis and impairs the autophagy-related dysfunction leading to cell death (Wong et al. 2020). In age dependent, there was a decline in autophagy and protein metabolism which leads to the accumulation of protein toxicity in the cell. In neurons, during postmitotic, the protein toxicity does not eliminate and lead to age-related disorders (Menzies et al. 2017). During aging, the morphological change in the neurons is that accumulation of lipofuscin is mostly seen in cardiac muscles, nerve cells, and skin (Moreno-Garcia et al. 2018). A study demonstrated downregulation of autophagy genes like BECN1, ATG5, and ATG7 is seen in aging (Lipinski et al. 2010). A study has also shown that dysfunction of lysosomal membrane permeabilization (LMP) leads to lysosomal leakage and impairment in autophagy in the cytosol (Gomez-Sintes et al. 2016) and increases the cathepsin in aged rat brain (Stoka et al. 2016). Several studies and investigation have shown the link between autophagy and senescence in rat brains where in the cortical neural cells lysosomal–autophgosome fusion impairs and inhibits the autophagic influx, and treatment with rapamycin reactivates the autophagy by unblocking (Ishikawa and Ishikawa 2020). Another study that links autophagy with senescent phenotype is increased GATA4 (transcriptional factor) which is stabilized by autophagy and also activates NF-kB as a result of SASP (Moreno-Blas et al. 2019; Kang et al. 2015).

## 3.3 Protective Effects of Phytochemicals

### 3.3.1 Berberine

Berberine, a novel drug-relevant alkaloid, is a natural molecule extracted from plant families which include Annonaceae, Papaveraceae, Berberidaceae, Menispermaceae, and Rutaceae (Neag et al. 2018). Berberine has been in the spotlight in recent decades because of its multitargeted pharmacological effects like modulating the—(1) oxidation-related pathway, (2) gut microbiota axis, (3) cell viability, and (4) inflammatory pathway (Shou et al. 2022). The various aspects of the candidate molecule berberine make it a potent supplement to slow aging. In D-galactose-induced aging rat models, berberine treatment restored the antioxidant system, the activity of $Na^+/K^+$ ATPase and $Ca^{2+}$ ATPase in the RBCs that concluded the drug provided redox homeostasis during aging (Yadawa et al. 2023). Also, the transgenic APP/PS1 mice when treated with berberine on a dosage of 50 and 100 mg/kg for 14 days limited the pathology of Alzheimer's disease by inhibiting the formation of Amyloid β via enhancing the α-Secretases (Cai et al. 2018).

Berberine has also been reported to be a nascent drug to combat aging and its related disorders as well as to attenuate cellular senescence in rodent models via targeting mitochondria by AMPK activation (Xu et al. 2017). The mechanism behind the berberine and the mitochondrial biogenesis was reported via reducing the oxidative stress and inflammation by stimulation of AMPK-dependent uncoupling protein 2 expression in vivo (Wang et al. 2011). Berberine has been discovered to improve aging-related cognitive dysfunction in mice. The drug attenuated the spatial learning and memory impairment by suppressing the expression of aging biomarkers in D-Gal-induced senescence mice by regulating Akt1, Ctnnb1, Tp53, and Jun (Yao et al. 2023). Insights from recent studies have unleashed the crosstalk between age-related cognitive deficits and muscular dysfunction as well as its amelioration by berberine (100 mg/kg/6 months) in Wistar rats. AMPK/SIRT1/PGC-1a pathway activation may be the underlying mechanism by which berberine mitigated cognitive decline and improved muscular function (Yu et al. 2018). Berberine has also been highlighted for its neuroprotective effects against doxorubicin-induced cognitive impairment in rats via modulating the brain growth factors like brain-derived neurotrophic factor (BDNF), cAMP response element binding protein (CREB), and sirtuin 1 expression. The drug also elicited antiinflammatory, antiapoptotic, and antioxidative effects (Shaker et al. 2021).

### 3.3.2 Phenolic Compounds

Flavonoids and other polyphenols are widespread substances in plants that play an important role in managing pathogenic and symbiotic soil bacteria and fungus as well as safeguarding the surrounding species from a variety of biotic and abiotic stressors. They are fundamental to the human diet, and research indicates that consuming them may help to regulate a wide range of illnesses (Kennedy et al. 2014). In a study conducted by Chacar et al. (2019), animal groups treated with phenolic compounds showed a decrease in plasma inflammatory and fibrotic markers as well as increased extracellular matrix like remodeling and downregulation of p38 pathway. Phenolic compound consumption has preserved the ill effects of aging. The phenolic compound was found to easily cross the blood–brain barrier (BBB) which reached the brain cells improving brain health by reducing the occurrence of neurodegenerative diseases (Figueira et al. 2017). Consuming meals packed with polyphenols throughout the lifespan may prevent or reverse age-related declines in cognitive function and minimize neurodegeneration (Vauzour 2017).

A review by Ammar et al. (2020a, b) stated that more positive effects of phenol-rich compounds seem to result from early acute and/or chronic administration of low- to high-dose phenolic components with high absorption rates (30%). A diet high in polyphenols may help older persons' brain and cognitive abilities to improve in several ways. The amount consumed and bioavailability of polyphenols appear to possess their enhancing effects. The findings indicate that to cross the brain–blood barrier and significantly impact cognitive health, polyphenols must be consumed at least at an intermediate dose (500 mg). And intermediate (9%) to high (43%) bioavailability rates are required (Ammar et al. 2020a, b).

### 3.3.3 Resveratrol

Resveratrol is found in grape skin and has antioxidants as well as antitumor properties. The grapes are the major source of wine which has a wide variety of antioxidants such as catechin, epicatechin, and proanthocyanidins including resveratrol. Bertelli et al. (2009) have stated that wine and grapes are very effective against many cardiovascular complications, especially resveratrol present in the wine and grapes found to have cardioprotective abilities. The resveratrol present in the wine is the major dietary component of the Mediterranean lifestyle which is the reference diet prescribed for most of the lifestyle diseases including cardiovascular diseases (Guerreroa et al. 2009). It also improves cognition and mood and mitigates the hippocampal plasticity while enhancing the adult hippocampal neurogenesis (AHN). However, some studies have shown the opposite effects of resveratrol against the AHN (Dias et al. 2016). The enzyme deacetylase Sir2/SIRT1 was resistant to stress and aging but the expression of this enzyme was increased by resveratrol. The increase in SIRT1 and related enzymes by chronic treatment of resveratrol results in a change of neuronal transcription that aids an improvement in antiapoptotic activity.

Animal studies have observed that chronic treatment of resveratrol in old rats enhanced memory performance and the prevention of cognitive impairment. In Alzheimer's disease-induced rats, the resveratrol treatment has prevented amyloid-β toxicity and hippocampal neuronal degeneration and alleviated neurogenesis as well as neuronal survival via brain-derived neurotrophic factor (BDNF) expression. Thus, resveratrol was found to exert a potent effect on neuronal plasticity in the brains of aged rats (Flores et al. 2016). In vivo studies conducted by Pal et al. (2022) observed that the animals when treated orally with R-Cu (resveratrol-copper) for about 12 months led to a decrease in several hallmarks of brain aging and resulted in the healthy aging of brain cells. The antioxidant properties of resveratrol play a vital role in delaying cognitive brain aging, and it also reduces the markers of aging in the brain (Terracina et al. 2022). Zhou et al. (2021) have reported that resveratrol could be a safe and novel drug for the treatment and prevention of brain aging and aging-related diseases. Though resveratrol possesses a wide range of properties, work by Oomen et al. (2009) found that resveratrol has no effects on cholinergic cell number and fiber density.

### 3.3.4 Cocoa

"At no other time has Nature concentrated such a wealth of valuable nourishment into such a small space as in the cocoa bean"—Alexander Von Humboldt (Zeli et al. 2022). Cocoa also known as chocolate is an amazing popular phytochemical with versatile usage throughout the globe. It is widespread throughout the globe and is consumed by all age groups. The popularity made the scientific analysis of the compound largely successful and interesting. All of the celebrations are marked by the baked goods and products made by the chocolate. Cocoa is obtained from the *Theobroma cocoa* beans which are rich in flavonoids and antioxidants (Zeli et al. 2022). The process of obtaining the edible aspect of cocoa is through a careful controlled fermentation process combined with drying and crushing it to produce a smooth, rich, flavourful, aromatic liquid which then is shaped and marketed in a variety of shapes and forms with lots of added flavors (fruit bits, nuts, sugar, and alcohol). The activity of the antioxidants is higher in cocoa compared to blueberries, green tea, and red wine (Septianti and Langkong 2020). Theobromine present in the cocoa protects the brain cells by different mechanisms such as by reducing lipid peroxidation and glutamate excitotoxicity (Bhat et al. 2021). The mechanism of action of the cocoa depends on various phytoactive molecules. In a study, the aqueous extract of cocoa was administered to the SH-SY5Y differentiated human neuroblastoma cells which showed reduced glutathione and significant neuroprotection against the ROS species (Carballeda et al. 2021). The percentage of phenolic compounds and flavonoids present in dark chocolate with a higher amount of cocoa is five times higher in milk chocolates. Flavonoids are the source of the antioxidant property of cocoa as they quench the ROS through a hydroxyl group. Epicatechin molecule has been stated to be responsible for the cellular antiinflammatory response which ultimately led to the neuroprotective behavior of the cocoa. Certain compounds of cocoa were reported to regulate the gut microbiota which on a further aspect helped in the prevention of neurodegeneration (Fanton et al. 2021). Alzheimer's disease, Parkinson's disease, and Amyotrophic lateral sclerosis like neurodegenerative disorders exhibit severe degeneration due to plaque formation and apoptosis. The procyanidin component of the cocoa helps in preventing apoptosis of the neurons (Sutcliffe et al. 2017). The active compounds of the cocoa were found to be psycho-physiologically active molecules boosting cellular protection (neuroprotection) as well as a great mood enhancer. It aids in synchronization of the circadian rhythm, acting as a neuromodulator that impacts directly neurogenesis, brain function, and mood. Hence, it is also referred to as a chrono nutrient (Garbarino et al. 2022). Theobromine acts as a long-term neuroprotectant exhibiting significant improvement in the working memory and an excellent cognitive enhancer by neuromodulation especially in elders to prevent cognitive decline. Hence, to conclude in terms of Alzheimer's disease and other dementia-related disorders, it can be reduced by regular intake of cocoa products from an early period of time (Cova et al. 2019). In conclusion, the long-term benefits of cocoa consumption can be neuroprotective, decadent neuromodulator, imparting stabilized cognitive patterns.

### 3.3.5 Omega 3-Polyunsaturated Fatty Acid (ω-3 PUFA)

There have been emerging trends on the protective role of ω-3 PUFA on brain aging. ω-3 PUFA are essential fatty acids that have a crucial role in maintaining good health and brain health in recent decades. The neuronal membranes and glia's are made of phospholipids especially ω-3 PUFA which are necessary for various significant brain functions. Denis et al. (2015) analyzed the data and concluded that ω-3 PUFA has a possible role in neuroprotection, neurotransmission, as well as neurogenesis preventing brain aging via reducing lipid peroxidation, neuronal inflammation, and excitotoxicity. The optimal concentration of docosahexaenoic acid (DHA)—the novel ω-3 PUFA is essential for brain repair mechanisms limiting age-related brain aging via influencing the neuronal and glial functions—optimizes the synaptic homeostasis. The DHA targets the hippocampal glutamatergic transmission and neurogenesis thus improving the memory as well as cognitive function respectively (Denis et al. 2013). Emerging trends exhibited a positive association between the long ω-3 PUFA with both language and memory scores but not with motor coordination. The compound has been underlined with evidence for blood–brain barrier integrity and cognitive enhancement thus promoting brain health but not ω-6 FA (Barnes et al. 2021). The long-chain ω-3 PUFA such as DHA and eicosapentaenoic acid (EPA), on supplementation in the older adults have proven to have promising potential in maintaining neuronal health and in healthy brain aging (Mora et al. 2022). A recent review underlined the association between the brain morphology, grey matter, brain volume, and white matter lesions especially in the hippocampal region which revealed—higher ω-3 PUFA and higher brain volume as well as higher volume, larger grey matter and lower white matter lesion with higher ω-3 PUFA. The author recommended natural intake of ω-3 PUFA is profoundly important to enhance cognitive function and maintain brain health (Macaron et al. 2021). There is a tremendous amount of evidence that has shown the beneficial effects of ω-3 PUFA in all aspects of brain health and also in the potential treatment of various neurodegenerative diseases thus preventing brain aging (Dyall 2015). It is highly suggestive for personalized dietary intake of ω-3 PUFA to older adults to mitigate age-related cognitive as well as neuronal decline.

### 3.3.6 Epigallocatechin-3-Gallate

Green tea polyphenols especially epigallocatechin-3-gallate (EGCG), an established nutraceutical, have numerous medicinal properties such as antioxidant, antiinflammatory, and antiaging properties. Aging is found to be a key factor in the development of neurodegenerative diseases like AD and PD in association with the immunosenescence of microglia. EGCG supplementation possesses antiinflammatory property by inhibiting microglial activation (Payne et al. 2022). Hyperhomocysteinemia (HHcy) induced inflammation, oxidative stress, apoptosis, DNA damage, and unbalanced neurotransmitter along with consecutive cognitive and memory impairment transmission in the hippocampus of BALB-c mice was reversed by the treatment with EGCG (Mostafa et al. 2022). EGCG inhibited the astrocyte-released neuroinflammatory cytokines in the AD mouse model and enhanced memory by preventing amyloidogenesis (Young-Jung Lee et al. 2013). EGCG at a dosage of 2 mg/kg or 6 mg/kg when given orally to mice prevented D-galactose-induced apoptosis exhibiting its neuroprotective effect on the aging mice. The neuroenhancer EGCG reversed the D-gal-induced learning and memory impairment and elevated the antioxidant system (He et al. 2009).

## 3.4 Conclusion

In this chapter we have discussed the molecular mechanisms of brain aging which includes gene expression profiling—mitochondrial dysregulations and oxidative stress, cell proliferation capacity, epigenetic alterations, inflammation and senescence-associated secretory phenotype, and autophagy. Mitochondria have been identified as the target organelle of various underlying processes of brain aging via mitochondrial fragmentation, oxidative damage to the mitochondria, and further DNA damage. Whereas a part of natural aging includes telomere shortening its molecular aspects need further investigation. miRNA has also been the culprit of the brain aging process by epigenetic modifications causing neuroinflammation. Glial cells also contribute to the aging by secreting inflammatory cytokines and chemokines. The normal autophagy mechanism seems disrupted in old age leading to protein toxicity in the neuronal cell leading to neuronal damage. Modern medicine recognizes natural compounds to be of profound potential in curing various neurological disorders. Although the phytochemical approach toward therapeutic usage is blooming, the bioactive compounds need scientific evidence for acceptance and practice. Here in this chapter, we have discussed scientific evidence of various phytochemicals that prevent brain aging. Phytochemicals provide neuroprotection via stimulation of AMPK-dependent mitochondrial biogenesis, by increasing BDNF activity, by downregulating the p38 pathway thus preventing apoptosis of neuronal cells and so on. Finally, these phytochemicals, their respective receptors, the signaling cascade which they activate, and the ability of these phytochemicals to cross the blood–brain barrier should be carefully monitored and studied further.

## References

Ammar A, Trabelsi K, Müller P, Bouaziz B, Boukhris O, Glenn JM, Bott N, Driss T, Chtourou H, Müller N, Hökelmann A (2020a) The effect of (poly)phenol-rich interventions on cognitive functions and neuroprotective measures in healthy aging adults: a systematic review and meta-analysis. J Clin Med 9(3):835. https://doi.org/10.3390/jcm9030835

Ammar A, Trabelsi K, Boukhris O, Bouaziz B, Müller P, Glenn JM, Hökelmann A (2020b) Effects of polyphenol-rich interventions on cognition and brain health in healthy young and middle-aged adults: systematic review and meta-analysis. J Clin Med 9(5):1598. https://doi.org/10.3390/jcm9051598

Bär C, Blasco MA (2016) Telomeres and telomerase as therapeutic targets to prevent and treat age-related diseases. F1000Res 5:78

Barnes S, Chowdhury S, Gatto NM, Fraser GE, Lee GJ (2021) Omega-3 fatty acids are associated with blood–brain barrier integrity in a healthy aging population. Brain Behav 11(8):e2273. https://doi.org/10.1002/brb3.2273

Barter J, Foster T (2019) Cellular and molecular mechanisms for age-related cognitive decline. In: Heilman K, Nadeau S (eds) Cognitive changes and the aging brain. Cambridge University Press, Cambridge, pp 16–27. https://doi.org/10.1017/9781108554350.003

Bartzokis G, Cummings JL, Sultzer D et al (2003) White matter structural integrity in healthy ageing adults and patients with Alzheimer disease: a magnetic resonance imaging study. Arch Neurol 60:393–938

Bertelli AA, Das DK (2009) Grapes, wines, resveratrol, and heart health. J Cardiovasc Pharmacol 54(6):468–476. https://doi.org/10.1097/FJC.0b013e3181bfaff3

Bhat JA, Gupta S, Kumar M (2021) Neuroprotective effects of theobromine in transient global cerebral ischemia-reperfusion rat model. Biochem Biophys Res Commun 571:74–80. https://doi.org/10.1016/j.bbrc.2021.07.051

Blackburn EH, Epel ES, Lin J (2015) Human telomere biology: a contributory and interactive factor in aging, disease risks, and protection. Science 350(6265):1193–1198

Burke SN, Barnes CA (2006) Neural plasticity in the ageing brain. Nat Rev Neurosci 7:30–40

Burla R, La Torre M, Zanetti G, Bastianelli A, Merigliano C, Del Giudice S et al (2018) p53-sensitive epileptic behavior and inflammation in Ft1 hypomorphic mice. Front Genet 9:581. https://doi.org/10.3389/fgene.2018.00581

Cai Z, Wang C, He W, Chen Y (2018) Berberine alleviates amyloid-beta pathology in the brain of APP/PS1 transgenic mice via inhibiting β/γ-secretases activity and enhancing α-secretases. Curr Alzheimer Res 15(11):1045–1052. https://doi.org/10.2174/1567205015666180702105740

Carballeda Sangiao N, Chamorro S, de Pascual-Teresa S, Goya L (2021) Aqueous extract of cocoa phenolic compounds protects differentiated Neuroblastoma SH-SY5Y cells from oxidative stress. Biomol Ther 11:1266. https://doi.org/10.3390/biom11091266

Chacar S, Hajal J, Saliba Y, Bois P, Louka N, Maroun RG, Faivre J-F, Fares N (2019) Long-term intake of phenolic compounds attenuates age-related cardiac remodeling. Aging Cell 18:e12894. https://doi.org/10.1111/acel.12894

Clarke LE, Liddelow SA, Chakraborty C, Munch AE, Heiman M, Barres BA (2018) Normal aging induces A1-like astrocyte reactivity. Proc Natl Acad Sci U S A 115:E1896–E1905. https://doi.org/10.1073/pnas.1800165115

Cova I, Leta V, Mariani C, Pantoni L, Pomati S (2019) Exploring cocoa properties: is theobromine a cognitive modulator? Psychopharmacology 236(2):561–572. https://doi.org/10.1007/s00213-019-5172-0

Dai X, Hong L, Shen H, Du Q, Ye Q, Chen X, Zhang J (2020) Estradiol-induced senescence of hypothalamic astrocytes contributes to aging-related reproductive function declines in female mice. Aging (Albany NY) 12(7):6089–6108. https://doi.org/10.18632/aging.103008. Epub 2020 Apr 7. PMID: 32259796; PMCID: PMC7185128

Danka Mohammed CP, Park JS, Nam HG, Kim K (2017) microRNAs in brain aging. Mech Ageing Dev 168:3–9

Denis I, Potier B, Vancassel S, Heberden C, Lavialle M (2013) Omega-3 fatty acids and brain resistance to ageing and stress: body of evidence and possible mechanisms. Ageing Res Rev 12(2):579–594. https://doi.org/10.1016/j.arr.2013.01.007

Denis I, Potier B, Heberden C, Vancassel S (2015) Omega-3 polyunsaturated fatty acids and brain aging. Curr Opin Clin Nutr Metab Care 18(2):139–146. https://doi.org/10.1097/mco.0000000000000141

Dias GP, Cocks G, do Nascimento Bevilaqua MC, Nardi AE, Thuret S (2016) Resveratrol: a potential hippocampal plasticity enhancer. Oxidative Med Cell Longev 2016:1–14. https://doi.org/10.1155/2016/9651236

Dyall SC (2015) Long-chain omega-3 fatty acids and the brain: a review of the independent and shared effects of EPA, DPA and DHA. Front Aging Neurosci 7:52. https://doi.org/10.3389/fnagi.2015.00052

Eriksdotter-Nilsson M, Garhardt G, Granholm AC et al (1989) Multiple changes in noradrenergic mechanisms in the coeruleo—hippocampal pathway during ageing. Structural and functional correlates in intraocular double grafts. Neurobiol Ageing 10:117–124

Fanton S, Cardozo LFMF, Combet E, Shiels PG, Stenvinkel P, Vieira IO, Narciso HR, Schmitz J, Mafra D (2021) The sweet side of dark chocolate for chronic kidney disease patients. Clin Nutr 40(1):15–26. https://doi.org/10.1016/j.clnu.2020.06.039

Figueira I, Garcia G, Pimpão RC et al (2017) Polyphenols journey through blood-brain barrier towards neuronal protection. Sci Rep 7:11456. https://doi.org/10.1038/s41598-01711512-6

Flood DG (1991) Region specific stability of dendritic extent in normal human ageing and regression in alzheimer's disease. II subiculum. Brain Res 540:83–95

Flood DG, Guarnaccia M, Caleman PD (1987) Dendritic extent in human CA2–3 hippocampal pyramidal neurons in normal ageing and senile dementia. Brain Res 409:88–96

Flores G, Vázquez-Roque RA, Diaz A (2016) Resveratrol effects on neural connectivity during aging. Neural Regen Res 11(7):1067–1068. https://doi.org/10.4103/1673-5374.187029

Fox NC, Schott JM (2004) Imaging cerebral atrophy: normal ageing to Alzheimer's disease. Lancet 363:392–394

Gampawar P, Schmidt R, Schmidt H (2020) Leukocyte telomere length is related to brain parenchymal fraction and attention/speed in the elderly: results of the Austrian stroke prevention study. Front Psych 11:100

Garbarino S, Garbarino E, Lanteri P (2022) Cyrcadian rhythm, mood, and temporal patterns of eating chocolate: a scoping review of physiology, findings, and future directions. Nutrients 14:3113. https://doi.org/10.3390/nu14153113

Gomez-Sintes R, Ledesma MD, Boya P (2016) Lysosomal cell death mechanisms in aging. Ageing Res Rev 32:150–168. https://doi.org/10.1016/j.arr.2016.02.009

Gonzalez-Freire M, De Cabo R, Bernier M, Sollott SJ, Fabbri E, Navas P et al (2015) Reconsidering the role of mitochondria in aging. J Gerontol A Biol Sci Med Sci 70(11):1334–1342

Goyal MS, Hawrylycz M, Miller JA, Snyder AZ, Raichle ME (2014) Aerobic glycolysis in the human brain is associated with development and neotenous gene expression. Cell Metab 19:49–57

Green DR, Galluzzi L, Kroemer G (2011) Mitochondria and the autophagy-inflammation-cell death axis in organismal aging. Science 333:1109–1112

Grill JD, Riddle DR (2002) Age related and laminar specific dendritic changes in the medial frontal cortex of the rat. Brain Res 937:8–21

Guerreroa RF, García-Parrillab MC, Puertasa B, Cantos-Villara E (2009) Wine resveratrol and health: a review. Nat Prod Commun 4(5):635–658

Harman D (1965) The free radical theory of aging: effect of age on serum copper levels. J Gerontol 20:151–153

Harman D (1972) The biologic clock: the mitochondria? J Am Geriatr Soc 20(4):145–147

Harraan D (1956) Aging: a theory based on free radical and radiation chemistry. J Gerontol 11(3):298–300

He M, Zhao L, Wei M-J, Yao W-F, Zhao H-S, Chen F-J (2009) Neuroprotective effects of (−)-Epigallocatechin-3-gallate on aging mice induced by D-Galactose. Biol Pharm Bull 32(1):55–60. https://doi.org/10.1248/bpb.32.55

Heden T, Gabrieli JD (2004) Insights into the ageing mind: a view from cognitive neuroscience. Nat Rev Neurosci 5:87–96

Held NM, Houtkooper RH (2015) Mitochondrial quality control pathways as determinants of metabolic health. BioEssays 37(8):867–876

Horvath S (2013) DNA methylation age of human tissues and cell types. Genome Biol 14:R115

Hou Y, Dan X, Babbar M, Wei Y, Hasselbalch SG, Croteau DL et al (2019) Ageing as a risk factor for neurodegenerative disease. Nat Rev Neurol 15:565–581

Ishikawa S, Ishikawa F (2020) Proteostasis failure and cellular senescence in long-term cultured postmitotic rat neurons. Aging Cell 19:e13071. https://doi.org/10.1111/acel.13071

Jaskelioff M, Muller FL, Paik J-H, Thomas E, Jiang S, Adams AC et al (2011) Telomerase reactivation reverses tissue degeneration in aged telomerase deficient mice. Nature 469(7328): 102–106

Jurk D, Wang C, Miwa S, Maddick M, Korolchuk V, Tsolou A et al (2012) Postmitotic neurons develop a p21-dependent senescence-like phenotype driven by a DNA damage response. Aging Cell 11:996–1004. https://doi.org/10.1111/j.1474-9726.2012.00870.x

Kang C, Xu Q, Martin TD, Li MZ, Demaria M, Aron L et al (2015) The DNA damage response induces inflammation and senescence by inhibiting autophagy of GATA4. Science 349: aaa5612. https://doi.org/10.1126/science.aaa5612

Kennedy DO (2014) Polyphenols and the human brain: plant "secondary metabolite" ecologic roles and endogenous signaling functions drive benefits. Adv Nutr 5(5):515–533. https://doi.org/10.3945/an.114.006320

Kiss H, Mihalik A, Nanasi T, Őry B, Spiro Z, Sőti C et al (2008) Ageing as a price of cooperation and complexity: self-organization of complex systems causes the ageing of constituent networks. Nat Prec 21:1–16

Kumar A, Yegla B, Foster TC (2018) Redox signaling in neurotransmission and cognition during aging. Antioxid Redox Signal 28:1724–1745

Lee Y-J, Choi D-Y, Yun Y-P, Han SB, Oh K-W, Hong JT (2013) Epigallocatechin-3-gallate prevents systemic inflammation-induced memory deficiency and amyloidogenesis via its anti-neuroinflammatory properties. J Nutr Biochem 24(1):298–310. https://doi.org/10.1016/j.jnutbio.2012.06.011

Li X, Zhang J, Li D, He C, He K, Xue T et al (2021) Astrocytic ApoE reprograms neuronal cholesterol metabolism and histone-acetylation-mediated memory. Neuron 109:957–970.e8

Liddelow SA, Guttenplan KA, Clarke LE, Bennett FC, Bohlen CJ, Schirmer L et al (2017) Neurotoxic reactive astrocytes are induced by activated microglia. Nature 541:481–487. https://doi.org/10.1038/nature21029

Lipinski MM, Zheng B, Lu T, Yan Z, Py BF, Ng A et al (2010) Genomewide analysis reveals mechanisms modulating autophagy in normal brain aging and in Alzheimer's disease. Proc Natl Acad Sci U S A 107:14164–14169. https://doi.org/10.1073/pnas.1009485107

Loerch PM, Jianbin W, Kelly AD et al (2008) Evaluation of the ageing brain transcripteome and synaptic regulation. PLoS One 3:e3329

Lopez-Otin C, Blasco MA, Partridge L, Serrano M, Kroemer G (2013) The hallmarks of aging. Cell 153(6):1194–1217

Lunn JS, Sakowski SA, Hur J, Feldman EL (2011) Stem cell technology for neurodegenerative diseases. Ann Neurol 70:353–361

Macaron T, Giudici KV, Bowman GL, Sinclair A, Stephan E, Vellas B, de Souto Barreto P (2021) Associations of Omega-3 fatty acids with brain morphology and volume in cognitively healthy older adults: a narrative review. Ageing Res Rev 67:101300. https://doi.org/10.1016/j.arr.2021.101300

Mariani E, Polidori MC, Cherubini A, Mecocci P (2005) Oxidative stress in brain aging, neurodegenerative and vascular diseases: an overview. J Chromatogr B 827(1):65–75

Menzies FM, Fleming A, Caricasole A, Bento CF, Andrews SP, Ashkenazi A, Fullgrabe J, Jackson A, Jimenez Sanchez M, Karabiyik C et al (2017) Autophagy and neurodegeneration: pathogenic mechanisms and therapeutic opportunities. Neuron 93:1015–1034

Mora I, Arola L, Caimari A, Escoté X, Puiggròs F (2022) Structured long-chain Omega-3 fatty acids for improvement of cognitive function during aging. Int J Mol Sci 23(7):3472. https://doi.org/10.3390/ijms23073472

Moreno-Blas D, Gorostieta-Salas E, Pommer-Alba A, Mucino-Hernandez G, Geronimo-Olvera C, Maciel-Baron LA et al (2019) Cortical neurons develop a senescence-like phenotype promoted by dysfunctional autophagy. Aging 11:6175–6198. https://doi.org/10.18632/aging.102181

Moreno-Garcia A, Kun A, Calero O, Medina M, Calero M (2018) An overview of the role of lipofuscin in age-related neurodegeneration. Front Neurosci 12:464. https://doi.org/10.3389/fnins.2018.00464

Mosch B, Morawski M, Mittag A, Lenz D, Tarnok A, Arendt T (2007) Aneuploidy and DNA replication in the normal human brain and Alzheimer's disease. J Neurosci 27(26):6859–6867

Mostafa MD, ElKomy MA, Othman AI, Amer ME, El-Missiry MA (2022) Epigallocatechin-3-gallate enhances cognitive and memory performance and protects against brain injury in methionine-induced Hyperhomocysteinemia through interdependent molecular pathways. Neurotox Res 40(6):2103–2116. https://doi.org/10.1007/s12640-022-00605-4

Muraki K, Nyhan K, Han L, Murnane JP (2012) Mechanisms of telomere loss and their consequences for chromosome instability. Front Oncol 2:135

Nativio R, Donahue G, Berson A, Lan Y, Amlie-Wolf A, Tuzer F et al (2018) Dysregulation of the epigenetic landscape of normal aging in Alzheimer's disease. Nat Neurosci 21:497–505

Neag MA, Mocan A, Echeverría J, Pop RM, Bocsan CI, Crişan G, Buzoianu AD (2018) Berberine: botanical occurrence, traditional uses, extraction methods, and relevance in cardiovascular, metabolic, hepatic, and renal disorders. Front Pharmacol 9:557. https://doi.org/10.3389/fphar.2018.00557

Oomen C (2009) Resveratrol preserves cerebrovascular density and cognitive function in aging mice. Front Aging Neurosci 1:4. https://doi.org/10.3389/neuro.24.004.2009

Pal K, Raghuram GV, Dsouza J, Shinde S, Jadhav V, Shaikh A, Rane B, Tandel H, Kondhalkar D, Chaudhary S, Mittra I (2022) A pro-oxidant combination of resveratrol and copper down-regulates multiple biological hallmarks of ageing and neurodegeneration in mice. Sci Rep 12(1):17209. https://doi.org/10.1038/s41598-022-21388-w. PMID: 36241685; PMCID: PMC9568542

Payne A, Nahashon S, Taka E, Adinew GM, Soliman KFA (2022) Epigallocatechin-3-Gallate (EGCG): new therapeutic perspectives for Neuroprotection, aging, and Neuroinflammation for the modern age. Biomol Ther 12(3):371. https://doi.org/10.3390/biom12030371

Peters R (2006) Ageing and the brain. Postgrad Med J 82(964):84–88. https://doi.org/10.1136/pgmj.2005.036665

Pluvinage JV, Wyss-Coray T (2020) Systemic factors as mediators of brain homeostasis, ageing and neurodegeneration. Nat Rev Neurosci 21:93–102

Rolyan H, Scheffold A, Heinrich A, Begus-Nahrmann Y, Langkopf BH, Hölter SM et al (2011) Telomere shortening reduces Alzheimer's disease amyloid pathology in mice. Brain 134(7):2044–2056

Salat DH, Buckner RL, Snyder AZ, Greve DN, Desikan RS, Busa E, Morris JC, Dale AM, Fischl B (2004) Thinning of the cerebral cortex in aging. Cereb Cortex 14:721–730

Schaar CE, Dues DJ, Spielbauer KK, Machiela E, Cooper JF, Senchuk M et al (2015) Mitochondrial and cytoplasmic ROS have opposing effects on lifespan. PLoS Genet 11(2):e1004972

Schlachetzki JCM, Toda T, Mertens J (2020) When function follows form: nuclear compartment structure and the epigenetic landscape of the aging neuron. Exp Gerontol 133:110876. https://doi.org/10.1016/j.exger.2020.110876

Septianti E, Langkong J (2020) Profile of bioactive compounds, antioxidant and aromatic component from several clones of cocoa beans during fermentation. IOP Conf Ser Earth Environ Sci 575:012009

Shaker FH, El-Derany MO, Wahdan SA, El-Demerdash E, El-Mesallamy HO (2021) Berberine ameliorates doxorubicin-induced cognitive impairment (chemobrain) in rats. Life Sci 269:119078. https://doi.org/10.1016/j.lfs.2021.119078

Shou J-W, Shaw P-C (2022) Therapeutic efficacies of Berberine against neurological disorders: an update of pharmacological effects and mechanisms. Cell 11(5):796. https://doi.org/10.3390/cells11050796

Sikora E, Bielak-Zmijewska A, Dudkowska M, Krzystyniak A, Mosieniak G, Wesierska M, Wlodarczyk J (2021) Cellular senescence in brain aging. Front Aging Neurosci 13:646924. https://doi.org/10.3389/fnagi.2021.646924

Stoka V, Turk V, Turk B (2016) Lysosomal cathepsins and their regulation in aging and neurodegeneration. Ageing Res Rev 32:22–37. https://doi.org/10.1016/j.arr.2016.04.010

Sutcliffe T, Winter A, Punessen N, Linseman D (2017) Procyanidin B2 protects neurons from oxidative, nitrosative, and excitotoxic stress. Antioxidants 6(4):77. https://doi.org/10.3390/antiox6040077

Terracina S, Petrella C, Francati S, Lucarelli M, Barbato C, Minni A, Ralli M, Greco A, Tarani L, Fiore M, Ferraguti G (2022) Antioxidant intervention to improve cognition in the aging brain: the example of Hydroxytyrosol and resveratrol. Int J Mol Sci 23:15674. https://doi.org/10.3390/ijms232415674

Tripathi A (2012) New cellular and molecular approaches to ageing brain. Ann Neurosci 19(4): 177–182. https://doi.org/10.5214/ans.0972.7531

Vauzour D (2017) Polyphenols and brain health. OCL 24(2):A202. https://doi.org/10.1051/ocl/2017008

Veal EA, Day AM, Morgan BA (2007) Hydrogen peroxide sensing and signaling. Mol Cell 26(1): 1–14

Wang Q, Zhang M, Liang B, Shirwany N, Zhu Y, Zou MH (2011) Activation of AMP-activated protein kinase is required for berberine-induced reduction of atherosclerosis in mice: the role of uncoupling protein 2. PLoS One 6(9):e25436. https://doi.org/10.1371/journal.pone.0025436

Wong DF, Wanger HN, Douglass KH et al (1984) Effects of age on dopamine and serotonin receptors measured by PET in the living human brain. Science 266:1393–1396

Wong SQ, Kumar AV, Mills J, Lapierre LR (2020) Autophagy in aging and longevity. Hum Genet 139:277–290

Xu Z, Feng W, Shen Q, Yu N, Yu K, Wang S, Chen Z, Shioda S, Guo Y (2017) Rhizoma coptidis and berberine as a natural drug to combat aging and aging-related diseases via anti-oxidation and AMPK activation. Aging Dis 8(6):760–777. https://doi.org/10.14336/AD.2016.0620

Yadawa A, Srivastava P, Singh A, Kesherwani R, Bhoumik S, Kumar R, Arya J, Rizvi S (2023) Berberine may provide redox homeostasis during aging in rats. Z Naturforsch C J Biosci 78(7–8):307–315. https://doi.org/10.1515/znc-2022-0213

Yao J, Wei W, Wen J, Cao Y, Li H (2023) The efficacy and mechanism of berberine in improving aging-related cognitive dysfunction: A study based on network pharmacology. Front Neurosci 17:1093180. https://doi.org/10.3389/fnins.2023.1093180

Yu Y, Zhao Y, Teng F et al (2018) Berberine improves cognitive deficiency and muscular dysfunction via activation of the AMPK/SIRT1/PGC-1a pathway in skeletal muscle from naturally aging rats. J Nutr Health Aging 22:710–717. https://doi.org/10.1007/s12603-018-1015-7

Yuan J, Chang SY, Yin SG, Liu ZY, Cheng X, Liu XJ et al (2020) Two conserved epigenetic regulators prevent healthy ageing. Nature 579:118–122

Zeli C, Lombardo M, Storz MA, Ottaviani M, Rizzo G (2022) Chocolate and cocoa-derived biomolecules for brain cognition during ageing. Antioxidants 11:1353. https://doi.org/10.3390/antiox11071353

Zhou DD, Luo M, Huang SY, Saimaiti A, Shang A, Gan RY, Li HB (2021) Effects and mechanisms of resveratrol on aging and age-related diseases. Oxidative Med Cell Longev 2021:9932218. https://doi.org/10.1155/2021/9932218

Zia A, Pourbagher-Shahri AM, Farkhondeh T, Samarghandian S (2021) Molecular and cellular pathways contributing to brain aging. Behav Brain Funct 17(1):6. https://doi.org/10.1186/s12993-021-00179-9

# Response of Cellular Stress Toward the Hormetic Phytochemicals in Brain Aging

4

Aloke Saha, Tanmay Sanyal, Pronoy Mukherjee, Koushik Sen, and Nithar Ranjan Madhu

**Abstract**

As people age, their brain cells undergo various types of stress that can lead to neurodegeneration and cognitive decline. One potential strategy for preventing or mitigating this process is hormesis, a concept that involves exposure to low doses of stressors that can elicit adaptive responses and protect against subsequent exposure to higher doses of the same stressor. This chapter summarizes the evidence suggesting that phytochemicals found in various plants and vegetables can elicit hormetic effects and protect against oxidative stress, inflammation, and other cellular stressors that contribute to brain aging. This chapter discusses several phytochemicals that have been shown to elicit hormetic effects in brain cells, including resveratrol, curcumin, sulforaphane, and epigallocatechin-3-gallate (EGCG). The chapter also investigates the molecular mechanisms behind these phytochemicals' hormetic actions, such as activation of the Keap1-Nrf2 pathway and induction of autophagy. The chapter discusses the role of phytochemicals in supporting good aging and reducing age-related neurological disorders. Overall, this chapter gives an overview of the involvement of hormesis

A. Saha
Department of Zoology, University of Kalyani, Kalyani, West Bengal, India

T. Sanyal (✉)
Department of Zoology, Krishnagar Government College, Krishnagar, West Bengal, India

P. Mukherjee
Department of Zoology, Rishi Bankim Chandra College, Naihati, West Bengal, India

K. Sen
Department of Zoology, Jhargram Raj College, Jhargram, West Bengal, India

N. R. Madhu
Department of Zoology, Acharya Prafulla Chandra College, New Barrackpore, Kolkata, West Bengal, India





S. Pathak, A. Banerjee (eds.), *Neuroprotective Effects of Phytochemicals in Brain Ageing*, https://doi.org/10.1007/978-981-99-7269-2_4

and phytochemicals in response to cellular stress and brain aging. It implies that including phytochemical-rich foods in one's diet might be a realistic method to promote healthy aging and lower the risk of age-related cognitive impairment.

**Keywords**

Neuroprotection · Oxidative stress · Hormesis · Phytochemicals · Brain aging

## 4.1 Introduction

Our bodies change as we age, and our brain is no exception. The continuous deterioration of cognitive and motor skills and the structure and function of brain cells characterize brain aging (Peters 2006). Several variables contribute to brain aging, including oxidative stress, inflammation, and mitochondrial dysfunction (Grimm and Eckert 2017).

Cellular stress is an important factor in brain aging because it contributes to the build-up of damage to brain cells. When cells are stressed, they initiate a series of adaptive responses that help them cope and avoid additional harm (Fulda et al. 2010). However, excessive or persistent stress can exceed the cell's adaptive mechanisms, resulting in damage and malfunction. Cellular stress is a broad term that includes many different types of stressors, including oxidative stress, endoplasmic reticulum (ER) stress, heat shock, and DNA damage (Mensch and Zierz 2020). Many sources, including environmental pollutants, infections, physical or emotional stress, and aging, can induce these stressors.

One of the most well-studied kinds of cellular stress is oxidative stress, which is a key contributor to brain aging. It happens when there is a mismatch between the number of reactive oxygen species (ROS) produced and the ability of cells to neutralize them using antioxidants. ROS may cause damage to cellular components such as lipids, proteins, and DNA, resulting in cellular malfunction and death (Pizzino et al. 2017). ER stress is another form of cellular stress that might contribute to brain aging. The ER is in charge of protein folding and quality control, and any disturbance in these activities can result in the buildup of misfolded proteins and ER stress (Rao and Bredesen 2004). When cells are subjected to extreme temperatures, they experience heat shock, a sort of cellular stress (Somero 2020). Another sort of cellular stress that might contribute to brain aging is DNA damage. Cells have a variety of methods for repairing DNA damage, but if the damage is substantial or protracted, it can cause cellular malfunction and contribute to neurodegenerative disorders (Madabhushi et al. 2014). Understanding the processes of cellular stress and the adaptive responses that protect cells from stress can lead to new treatment approaches for brain aging and neurodegenerative illnesses.

Hormesis is a phenomenon that occurs when modest dosages of a stressor have positive effects on an organism, but excessive amounts are damaging (Mattson 2008). This notion has gained popularity in recent years since it has been demonstrated to have significant consequences for human health and illness.

Phytochemicals, which are naturally occurring molecules found in plants, have been demonstrated to have hormetic effects and are being researched for their potential advantages in a variety of medical diseases. They have been proven to have several biological actions, including antioxidant, antiinflammatory, and anticancer properties (Son et al. 2008). Phytochemicals have also been shown to interact with various cellular pathways, including those implicated in stress reactions (Lee et al. 2014). Resveratrol, a polyphenol present in grapes and red wine, has been proven to have antioxidative and antiinflammatory effects on brain cells (Salehi et al. 2018). Curcumin, a polyphenol present in turmeric, has been demonstrated to have hormetic properties as well as to protect against oxidative damage, inflammation, and ER stress (Concetta Scuto et al. 2019). Other phytochemicals examined for their hormetic effects on brain cells include green tea's epigallocatechin gallate (EGCG), broccoli's sulforaphane, and onions' and apples' quercetin (Si and Liu 2014).

Ayurveda, or ancient Indian medicine, has a long history of employing medicinal plants and herbs. Herbs and plants have been used in Ayurveda for over 5000 years and are still commonly used in India today. Ayurvedic medicine emphasizes using natural treatments to enhance health and well-being, and many of these therapies contain phytochemicals (Jaiswal and Williams 2017). Brahmi, also known as *Bacopa monnieri*, is an Ayurvedic herb that improves memory and cognitive function (Aguiar and Borowski 2013). Ashwagandha, commonly known as *Withaniasomnifera*, is an adaptogenic plant that helps the body cope with stress (Singh et al. 2011a, b). Shankhapushpi, also known as *Convolvulus pluricaulis*, is another Ayurvedic plant believed to improve memory and cognition (Balkrishna et al. 2020).

The importance of studying cellular stress responses to hormetic phytochemicals in brain aging stems from the possibility of developing innovative treatment techniques for brain aging and neurodegenerative illnesses (Calabrese et al. 2010). The discovery of phytochemicals with hormetic effects on brain cells and the understanding of the processes behind these effects can lead to the creation of novel medications for brain health (Sahebnasagh et al. 2022). Biomarkers of cellular stress and hormetic responses can be used to evaluate the efficiency of treatment interventions in brain aging and neurodegenerative disorders (Young et al. 2020). Furthermore, the use of phytochemicals and natural treatments in traditional medical systems such as Ayurveda gives a chance to investigate their potential advantages in brain aging and neurodegenerative illnesses (Kumar and Khanum 2012).

## 4.2 Cellular Stress and Its Role in Brain Aging

### 4.2.1 Overview of Cellular Stress and Its Effects on Brain Cells

The disturbance of cellular homeostasis produced by numerous internal and external causes is referred to as cellular stress. Cells are continually confronted with challenges and stimuli that might disrupt their regular functioning (Chovatiya and

Medzhitov 2014). These stressors can be classified as cellular stressors, such as oxidative, endoplasmic reticulum, heat shock, and metabolic stress.

An imbalance between the generation of reactive oxygen species (ROS) and the cellular antioxidant defense system causes oxidative stress. ROS (reactive oxygen species) are extremely reactive molecules created during regular cellular metabolism (Schieber and Chandel 2014). However, excessive ROS production or insufficient antioxidant defenses can result in oxidative stress, which can cause cellular damage. Oxidative stress has been linked to the etiology of several neurodegenerative illnesses and brain aging (Uttara et al. 2009).

ER stress arises when excessively unfolded or misfolded proteins overburden the endoplasmic reticulum. ER stress that is prolonged or unresolved can cause cellular malfunction and contribute to neurodegenerative disorders (Doyle et al. 2011). Heat shock stress develops when cells are subjected to high temperatures or other environmental stimuli. Heat shock proteins (HSPs) are important in cellular protection and are linked to neuroprotection and brain aging (Yenari 2002). Metabolic stress is caused by cellular metabolic imbalances such as nutrition scarcity, energy depletion, or dysregulated metabolism (Wellen and Thompson 2010). These disruptions can cause cellular malfunction and accelerate brain aging.

Because of their high energy needs and limited regeneration ability, brain cells, particularly neurons, are extremely sensitive to cellular stress. Neurons perform complicated and highly specialized jobs, and any disruption in their cellular homeostasis can have far-reaching consequences for brain function (Martino et al. 2011). Cellular stress can cause neuronal injury, synaptic dysfunction, decreased neurotransmission, and the activation of cell death pathways in brain cells. Neuronal injury can appear as a change in shape, decreased connection, and decreased neuronal survival (Gorman 2008). Synapses are essential in the brain for neural transmission and information processing. Cellular stress can affect synapse function, decreasing neurotransmission and neuronal communication (Lovinger 2008). Cognitive decline and other neurological symptoms linked with brain aging might be exacerbated by synaptic disruption (Bishop et al. 2010). Neurotransmission is required for signal transmission between neurons. Cellular stress can disrupt neurotransmitter production, release, and receptor function, disrupting neural signaling, and impairing brain function (Teleanu et al. 2022). Cell death mechanisms such as apoptosis or necrosis can be activated by prolonged or severe cellular stress, resulting in neuronal loss and neurodegeneration. Cell death pathways can be triggered by cellular damage buildup or the failure of cellular repair processes (Chi et al. 2018). To summarize, cellular stress substantially impacts brain cells and is implicated in brain aging and the development of neurodegenerative disorders.

### 4.2.2 Oxidative Stress and Its Implications in Brain Aging

Oxidative stress is a major factor in the brain's aging and the development of neurodegenerative disorders. It refers to a mismatch between the generation of reactive oxygen species (ROS) and the antioxidant defense systems of the cell.

ROS is naturally occurring by-products of cellular metabolism that play critical roles in cell signaling and physiological processes (Singh et al. 2019). Excessive ROS generation or a weakened antioxidant defense system, on the other hand, can cause oxidative stress, resulting in cellular damage and malfunction (Birben et al. 2012). Oxidative stress has been widely investigated in the context of brain aging and is thought to be a key factor in age-related cognitive decline and neurodegenerative diseases (Kandlur et al. 2020). This section provides an overview of oxidative stress and its role in brain aging.

#### 4.2.2.1 Sources of Oxidative Stress in the Brain

The brain is extremely vulnerable to oxidative stress because of its high metabolic activity, oxygen consumption, and relatively poor antioxidant capacity. ROS production in the brain is caused by various factors, including mitochondrial malfunction, inflammation, and environmental pollutants (Lushchak et al. 2021). The principal generator of ROS in cells is the mitochondria. Electrons can leak from the electron transport chain during the oxidative phosphorylation process, generating superoxide radicals. Mitochondrial malfunction, which happens with aging, can enhance ROS generation even more (Zorov et al. 2014).

As a defensive mechanism against infections or damage, neuroinflammation, which is characterized by the activation of immune cells in the brain, can create ROS (Mittal et al. 2014). Chronic inflammation, on the other hand, can cause long-term oxidative stress, which contributes to neuronal damage and brain aging (Liguori et al. 2018). Toxins in the environment, such as heavy metals, pesticides, and air pollution, can cause an increase in ROS generation in the brain. These poisons can cause oxidative stress and disturb cellular homeostasis, speeding up the brain's aging process (Mitra et al. 2022).

#### 4.2.2.2 Effects of Oxidative Stress on Brain Aging

Oxidative stress significantly impacts brain cells and contributes to the etiology of neurodegenerative disorders. The brain is vulnerable to oxidative stress due to its high polyunsaturated fatty acid content, high oxygen consumption, and inadequate antioxidant defenses (Chen et al. 2012). Protein degradation and aggregation, lipid peroxidation, DNA damage, and a weakened antioxidant defense system are some of the primary consequences of oxidative stress in brain aging (Guo et al. 2013).

Oxidative stress can cause protein oxidation and alterations in the brain. This can cause protein structure and function to be disrupted, affecting cellular activities (Salim 2017). Furthermore, oxidatively damaged proteins can combine, resulting in hazardous protein aggregates seen in neurodegenerative disorders such as Alzheimer's and Parkinson's. ROS can cause lipid peroxidation by attacking polyunsaturated fatty acids in cell membranes. This process creates lipid radicals, which spread oxidative damage further. The integrity and fluidity of cell membranes can be compromised by lipid peroxidation, compromising cellular communication and neural function (Petrovic et al. 2020). DNA damage caused by oxidative stress might include strand breakage, base changes, and oxidative lesions. DNA damage can affect genomic stability, alter gene expression, and impede cellular function.

Age-related cognitive decline and increased sensitivity to neurodegenerative illnesses are linked to DNA damage in neurons (Ayala et al. 2014).

The efficacy of the cellular antioxidant defense system decreases with age. Superoxide dismutase (SOD), catalase, and glutathione peroxidase are antioxidant enzymes that play critical roles in neutralizing ROS and maintaining redox equilibrium (Ighodaro and Akinloye 2018). Age-related oxidative stress, on the other hand, might overwhelm the antioxidant defense mechanism, resulting in an imbalance and increased cellular damage.

### 4.2.3 Other Forms of Cellular Stress and Their Potential Effects on Brain Aging

Other types of cellular stress, in addition to oxidative stress, play an important role in brain aging and the development of neurodegenerative disorders. Endoplasmic reticulum (ER) stress, heat shock stress, and metabolic stress can all affect cellular equilibrium and lead to cellular failure (Sazonova et al. 2021). Understanding the consequences of these stresses on brain aging is critical for discovering new treatment targets. This section will look at several types of cellular stress and how they could affect brain aging.

#### 4.2.3.1 Endoplasmic Reticulum (ER) Stress

The endoplasmic reticulum (ER) is a vital organelle that participates in protein synthesis, folding, and quality control. When misfolded or unfolded proteins accumulate in the ER lumen, this causes ER stress. ER stress is caused by various factors, including genetic alterations, calcium imbalance, oxidative damage, and altered protein glycosylation (Lin et al. 2008). The unfolded protein response (UPR) is a biological process that restores ER equilibrium and promotes cell survival in response to ER stress. Prolonged or severe ER stress, on the other hand, can cause neuronal malfunction and contribute to brain aging and neurodegenerative disorders (Remondelli and Renna 2017).

ER stress affects protein folding, resulting in a buildup of misfolded or unfolded proteins. Protein aggregation and the creation of harmful protein aggregates, which are characteristic of neurodegenerative disorders, can ensue. As a protective reaction to ER stress, the UPR is activated (Lindholm et al. 2006). Reducing the load of protein folding requires the overexpression of chaperone proteins and the suppression of global protein translation. Prolonged UPR activation, on the other hand, might result in neuronal cell death (Lindholm et al. 2017). Calcium homeostasis is disrupted by ER stress, resulting in aberrant calcium signaling. Excitotoxicity, a condition in which high calcium influx causes neuronal injury, can be exacerbated by calcium dysregulation (Cascella and Cecchi 2021).

A number of hormetic phytochemicals have been proven to reduce ER stress and enhance ER equilibrium. Curcumin, for example, has been shown to lower ER stress-induced apoptosis and increase the expression of ER chaperone proteins (Zhang et al. 2018). Resveratrol has also been demonstrated to protect against ER stress by lowering misfolded protein accumulation and boosting UPR activation (Lee et al. 2019). These phytochemicals can exert their effects by activating cellular stress response pathways and increasing antioxidant defenses, thereby decreasing ER stress-induced brain damage.

#### 4.2.3.2 Heat Shock Stress

Heat shock stress develops when cells are subjected to high temperatures or other environmental stimuli. It triggers the heat shock response (HSR), a cellular defensive mechanism that involves the activation of heat shock proteins (HSPs) (Hasday and Singh 2000). Under stressful situations, HSPs operate as molecular chaperones, facilitating normal protein folding and preventing protein aggregation. HSPs and heat shock stress have been linked to brain aging and neuroprotection (Webster et al. 2019).

Heat shock stress increases the production of HSPs, aiding protein stabilization and preventing aggregation. This can help maintain protein homeostasis and protect against protein damage (Almalki et al. 2021). HSPs have cytoprotective properties and can aid in the repair of damaged proteins as well as the maintenance of cellular integrity. They can also regulate cellular signaling pathways and protect antioxidants (Moura et al. 2018). In aging and neurodegenerative disorders, heat shock stress and HSPs have been linked to neuroprotection and better cognitive performance (Leak 2014).

Hormetic compounds like resveratrol and curcumin have been demonstrated to increase HSP expression and trigger the heat shock response. These phytochemicals can potentially improve cellular stress response and protein stability, eventually contributing to brain health and perhaps decreasing the consequences of brain aging (Dattilo et al. 2015).

#### 4.2.3.3 Metabolic Stress

Metabolic stress refers to cellular metabolic abnormalities such as nutrition scarcity, energy depletion, and dysregulated metabolism. To operate normally, brain cells require a steady supply of nutrition and energy (Wellen and Thompson 2010). Nutrient imbalances, mitochondrial malfunction, and dysregulated energy metabolism can all cause metabolic stress. It has the potential to have far-reaching effects on brain aging and neurodegenerative illnesses (Misrani et al. 2021).

Metabolic stress can cause energy depletion, impairing cellular processes and reducing neuronal survival. Mitochondrial dysfunction is frequently linked to metabolic stress (Picard and McEwen 2018). Excessive ROS generation, impaired ATP synthesis, and compromised cellular energy metabolism can all result from dysfunctional mitochondria. Metabolic stress can interfere with neurotransmitter production, release, and receptor function, resulting in disrupted neural signaling and reduced brain function (Misrani et al. 2021).

Several hormetic substances, such as resveratrol, have been found to increase energy metabolism and mitochondrial function. These phytochemicals have the ability to activate cellular stress response pathways such as AMP-activated protein kinase (AMPK), which can enhance energy generation and ameliorate the negative effects of metabolic stress on brain cells (Murugaiyah and Mattson 2015).

## 4.3 Hormesis and Phytochemicals

In recent years, there has been a surge of interest in the possible advantages of hormetic phytochemicals in the field of aging research. Hormesis is a toxicological term that refers to a biphasic dose–response relationship in which modest concentrations of stressors induce favorable adaptation responses in organisms (Calabrese 2014). Plant-derived phytochemicals, or bioactive substances, have been intensively researched for their potential health-promoting benefits.

### 4.3.1 Definition and Mechanisms of Hormesis

Hormesis is a dose-response phenomenon in which low amounts of stress stimulate biological systems, fostering resilience, and improved functioning (Schirrmacher 2021). Put another way, what doesn't kill you strengthens you. Hormesis can be traced back to ancient Greek physicians who discovered that small quantities of poison might have medicinal benefits (Calabrese et al. 2013). Hormesis is now recognized as a fundamental biological phenomenon with ramifications for a wide range of professions, including toxicology, medicine, and aging research (Bhakta-Guha and Efferth 2015).

Hormesis can occur via a variety of cellular pathways and molecular processes. One notable method is activating cellular stress response pathways such as the heat shock response (HSR) and the unfolded protein response (UPR). These pathways are responsible for protein homeostasis and cellular viability (Stetler et al. 2010). Stressors at low levels can activate the upregulation of molecular chaperones and antioxidant enzymes, increasing cellular defense systems (Voth and Jakob 2017). Hormetic phytochemicals can also stimulate signaling pathways such as the nuclear factor erythroid 2-related factor 2 (Nrf2). Nrf2 is a transcription factor that controls the expression of several antioxidant and detoxification genes. Hormetic phytochemicals activate Nrf2, resulting in increased synthesis of endogenous antioxidants, which reduces oxidative stress and inflammation in cells (Ma 2013). Hormetin-induced mitochondrial biogenesis is another mechanism driving hormesis. Mitochondria, or cell powerhouses, play an important role in aging and neurodegenerative disorders. Low dosages of stressors, such as phytochemicals, can induce the formation of new mitochondria, thereby increasing energy output and cellular resilience (Ristow and Schmeisser 2014).

### 4.3.2 Overview of Phytochemicals and Their Potential Hormetic Effects

In the field of aging research, the notion of hormesis, which defines the biphasic dose–response relationship in which modest doses of stressors induce favorable adaptive responses in animals, has received a lot of attention (Calabrese and Mattson 2017). Phytochemicals are a wide range of substances found in plants, such as fruits, vegetables, herbs, and seeds. These chemicals' bioactive characteristics and possible health benefits have been widely researched. Several phytochemicals have demonstrated potential benefits in modifying cellular stress responses and enhancing brain health in the context of hormesis and brain aging (Kumar et al. 2023).

#### 4.3.2.1 Polyphenols

Polyphenols are a kind of phytochemical that is distinguished by its phenolic structure. They may be found in a variety of plant-based meals and beverages, including fruits, vegetables, tea, and wine. Polyphenols have a variety of biological actions, such as antioxidant, antiinflammatory, and neuroprotective properties (Pandey and Rizvi 2009).

Resveratrol, a polyphenol found in grapes, berries, and red wine, has gotten a lot of interest because of its possible hormetic effects on the brain. It has been demonstrated to have antioxidant and antiinflammatory effects (Vauzour 2012). According to research, resveratrol can activate sirtuins, a protein family involved in cellular metabolism and stress tolerance (Grabowska et al. 2017). In animal models of brain aging, activating sirtuins, notably SIRT1, has been related to increased mitochondrial activity, decreased oxidative stress, and improved cognitive performance (Jęśko et al. 2017).

Green tea contains the catechin epigallocatechin gallate (EGCG). It is high in antioxidants and has anti-inflammatory effects. EGCG has the ability to scavenge reactive oxygen species (ROS), prevent lipid peroxidation, and alter cellular signaling pathways involved in oxidative stress. Several studies have suggested that it may have neuroprotective properties in several brain aging models and neurodegenerative disorders (Singh et al. 2011a, b; Mokra et al. 2022).

Curcumin, the major bioactive ingredient in turmeric, is another polyphenol with possible hormetic effects. Curcumin has been proven to have antiinflammatory, antioxidant, and neuroprotective effects (Hewlings and Kalman 2017). Curcumin has the ability to scavenge ROS, reduce lipid peroxidation, and boost the activity of antioxidant enzymes. It has demonstrated encouraging results in preclinical models of neurodegenerative disorders, indicating its promise as an antioxidative stress agent (Silvestro et al. 2021; Maiti and Dunbar 2018). It has the ability to affect several signaling pathways involved in cellular stress response, including the nuclear factor erythroid-related factor 2 (Nrf2). When Nrf2 is activated, the expression of antioxidant enzymes increases, improving cellular defense against oxidative stress and inflammation (Shahcheraghi et al. 2021).

#### 4.3.2.2 Flavonoids

Flavonoids are phytochemicals found in fruits, vegetables, herbs, and drinks such as tea and chocolate. They have a flavone backbone and a variety of biological actions, including antioxidant, antiinflammatory, and neuroprotective properties (Ullah et al. 2020).

Quercetin, a flavonoid present in apples, onions, and berries, has shown potential hormetic effects in the brain. Quercetin can activate the Nrf2 pathway, leading to the upregulation of antioxidant enzymes and the reduction of oxidative stress-induced neuronal damage (Salehi et al. 2020). Additionally, quercetin has been reported to modulate multiple signaling pathways associated with neuroinflammation and neuronal survival (Benameur et al. 2021).

Epicatechin, another flavonoid found in cocoa and dark chocolate, is plentiful. It has been linked to enhanced cerebral blood flow and better cognitive performance (Haskell-Ramsay et al. 2018). Epicatechin has antioxidant and antiinflammatory characteristics, which may contribute to its hormetic effects on brain aging. Furthermore, it has been demonstrated to trigger neuroprotection and synaptic plasticity cellular pathways (Navarrete-Yañez et al. 2020).

#### 4.3.2.3 Other Phytochemicals

Other phytochemicals, including polyphenols and flavonoids, have indicated potential hormetic effects in the context of brain aging. Curcuminoids, which are generated from turmeric, are antioxidative, antiinflammatory, and neuroprotective. These phytochemicals have been demonstrated to activate Nrf2 and improve cellular defense systems against oxidative stress and inflammation (Boyanapalli et al. 2014; Concetta Scuto et al. 2019). Anthocyanins, pigments found in berries, grapes, and other fruits, exhibit potent antioxidant and antiinflammatory effects. Anthocyanins have been associated with improved cognitive function and neuronal protection in preclinical studies (Medina Dos Santos et al. 2019). Sulforaphane, a cruciferous vegetable component present in broccoli and kale, stimulates the Nrf2 pathway and increases the production of detoxification enzymes. It has been shown in animal models of brain aging to have neuroprotective properties and to improve cognitive performance (Santín-Márquez et al. 2019).

Phytochemicals are an abundant supply of potential hormetic agents that can alter cellular stress responses and boost brain health as we age. Polyphenols, flavonoids, and other phytochemicals have antioxidant, antiinflammatory, and neuroprotective properties, which contribute to their potential hormetic properties (Kumar and Khanum 2012). Understanding the mechanisms behind these phytochemicals' hormetic activities is critical for designing tailored therapies to combat age-related cognitive decline and neurodegenerative disorders. More study is needed to determine the appropriate dose, bioavailability, and particular molecular targets of phytochemicals in order to fully realize their promise for supporting healthy brain aging (Fig. 4.1 and Table 4.1).

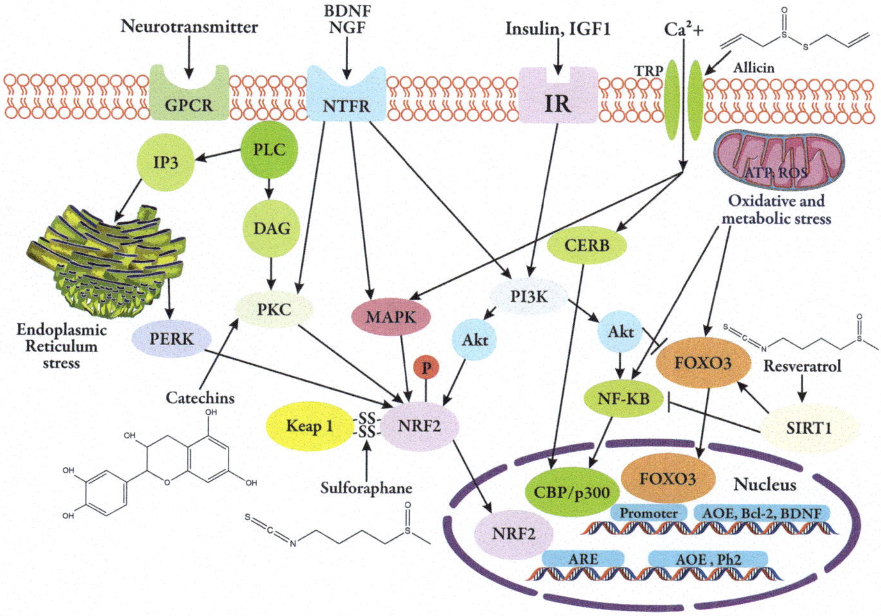


**Fig. 4.1** Molecular mechanisms by which phytochemicals activate hormetic stress response pathways in combatting cellular stress in brain aging

### 4.3.3 Bioavailability and Metabolism of Hormetic Phytochemicals

Hormesis, which is defined by the positive effects of low-dose stressors, and phytochemicals, which are bioactive molecules produced from plants, both have substantial potential for enhancing brain health throughout aging (Martel et al. 2019). Phytochemical bioavailability and metabolism are important factors in determining their effectiveness and therapeutic potential. Understanding hormetic phytochemical absorption and metabolism is critical for optimizing their therapeutic efficacy in fighting age-related cognitive decline and neurodegenerative disorders (Vaiserman et al. 2020). Several variables alter these chemicals' absorption, distribution, metabolism, and excretion, affecting their availability and action in the brain.

Their absorption from the gastrointestinal system determines hormetic phytochemicals' bioavailability. Chemical structure, solubility, molecular weight, and the presence of other food components all impact their absorption (Speciale et al. 2014). Due to their limited water solubility and quick metabolism, phytochemicals such as curcumin and resveratrol are known to have low bioavailability (Sohn et al. 2021). To boost absorption and bioavailability, options such as the use of nanotechnology-based delivery systems, coadministration with absorption enhancers, and formulation with lipid carriers have been investigated (Aqil et al. 2013).

Once ingested, hormonal phytochemicals must be transported throughout the body, including the brain. The blood–brain barrier (BBB), which divides the

**Table 4.1** Hormetic phytochemicals and their mechanism of action in combatting cellular stress in brain aging

| Sl No. | Phytocompound | Chemical structure | Source plant | Plant part | Extract type | Assay | Mechanism of action | References |
|---|---|---|---|---|---|---|---|---|
| 1 | Resveratrol | 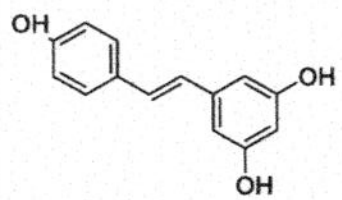 | *Polygonum cuspidatum* | Root | Ethanol extract | In Vivo | Activation of sirtuins and AMPK, antioxidant, anti-inflammatory | Navarro-Cruz et al. (2017) |
| 2 | Curcumin | 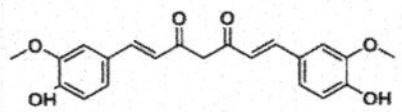 | *Curcuma longa* | Rhizome | Ethanol extract | In Vivo | Modulation of NF-κB and Nrf2 pathways, antioxidant, antiamyloidogenic | Cianciulli et al. (2022) |
| 3 | Epigallocatechin-3-gallate (EGCG) | 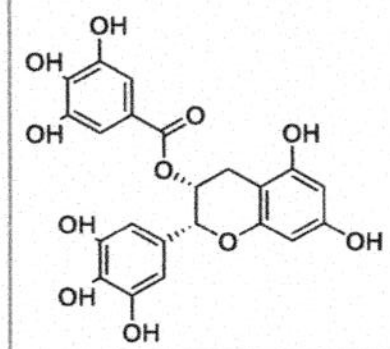 | *Camellia sinensis* | Leaves | Aqueous extract | In vitro | Activation of Nrf2, inhibition of oxidative stress and neuroinflammation | Payne et al. (2022) |
| 4 | Quercetin | 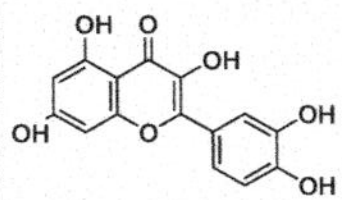 | *Allium cepa* | Bulb | Ethanol extract | In vitro | Activation of Nrf2 and SIRT1, inhibition of apoptosis, and neuroinflammation | Nishihira et al. (2021) |
| 5 | Caffeic acid | 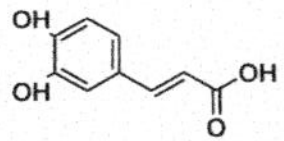 | *Coffea arabica* | Seeds | Ethanol extract | In vivo | Antioxidant, antiinflammatory, induction of phase II enzymes | Socała et al. (2020) |
| 6 | Lycopene |  | *Solanum lycopersicum* | Fruit | Oil extract | In vivo | Antioxidant, antiinflammatory, neuroprotective | Imran et al. (2020) |

| | | | | | | | | |
|---|---|---|---|---|---|---|---|---|
| 7 | Apigenin | 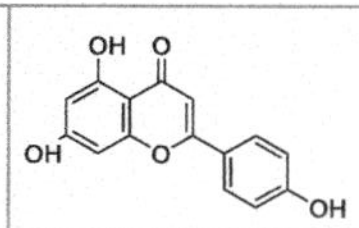 | *Matricaria chamomilla* | Flowers | Ethanol extract | In vitro | Activation of Nrf2, antiinflammatory, antioxidant | Salehi et al. (2019) |
| 8 | Genistein | 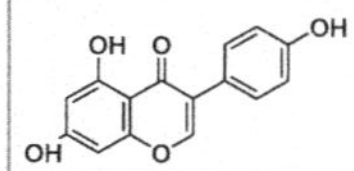 | *Glycine max* | Seeds | Ethanol extract | In vivo | Estrogenic effects, antioxidant, antiinflammatory | Duan et al. (2021) |
| 9 | Gingerol | 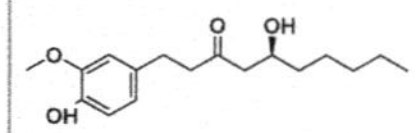 | *Zingiber officinale* | Rhizome | Ethanol extract | In vitro | Antiinflammatory, antioxidant, neuroprotective | Mohd Sahardi and Makpol (2019) |
| 10 | Sulforaphane | 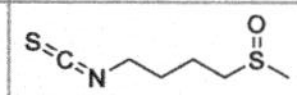 | *Brassica oleracea* | Broccoli Sprouts | Aqueous extract | In vitro | Activation of Nrf2, antioxidant, detoxification | Santín-Márquez et al. (2019) |
| 11 | Kaempferol | 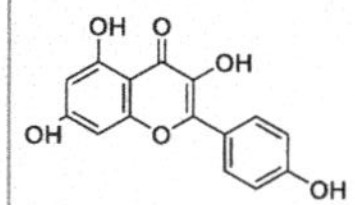 | *Kaempferia galanga* | Rhizome | Ethanol extract | In vitro | Antioxidant, antiinflammatory, neuroprotective | Silva dos Santos et al. (2021) |
| 12 | Allicin | 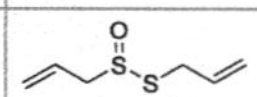 | *Allium sativum* | Bulb | Aqueous extract | In vitro | Antioxidant, antiinflammatory, neuroprotective | Tedeschi et al. (2022) |
| 13 | Berberine | 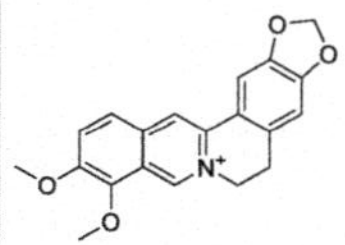 | *Coptis chinensis* | Rhizome | Ethanol extract | In vivo | Antiinflammatory, antioxidant, regulation of autophagy | Xu et al. (2017) |
| 14 | Ellagic acid | 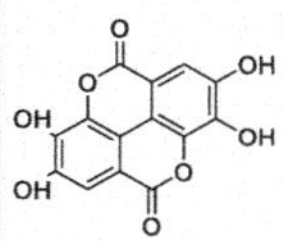 | *Punica granatum* | Fruit | Ethanol extract | In vitro | Antioxidant, antiinflammatory, antiapoptotic | Gupta et al. (2021) |

(continued)

**Table 4.1** (continued)

| Sl No. | Phytocompound | Chemical structure | Source plant | Plant part | Extract type | Assay | Mechanism of action | References |
|---|---|---|---|---|---|---|---|---|
| 15 | Capsaicin | 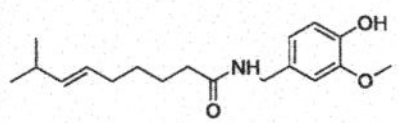 | *Capsicum annuum* | Fruit | Ethanol extract | In vivo | Antiinflammatory, antioxidant, neuroprotective | Thornton et al. (2023) |
| 16 | Rutin | 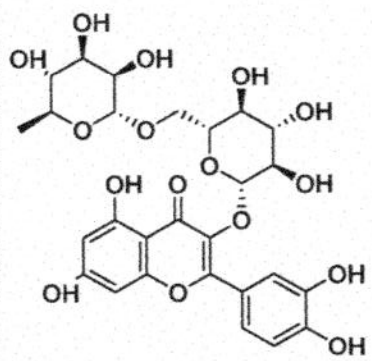 | *Dimorphandra mollis* | Leaves | Ethanol extract | In vitro | Antioxidant, antiinflammatory, neuroprotective | da Silva et al. (2020) |
| 17 | Betulinic acid | 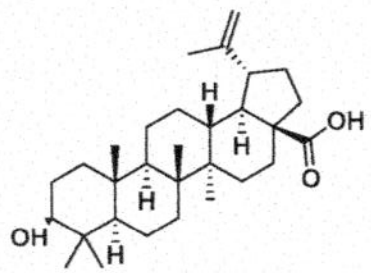 | *Betula* spp. | Bark | Ethanol extract | In vivo | Antiinflammatory, antioxidant, induction of neurotrophic factors | Hordyjewska et al. (2019) |
| 18 | Echinacoside | 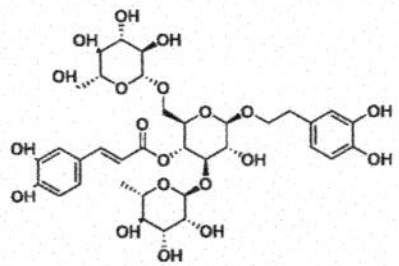 | *Cistanche deserticola* | Herb | Ethanol extract | In vitro | Antioxidant, antiinflammatory, neuroprotective | Gu et al. (2016) |
| 19 | Hesperidin | 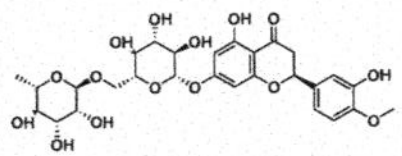 | *Citrus* spp. | Fruit | Aqueous extract | In vivo | Antioxidant, antiinflammatory, regulation of synaptic plasticity | Hajialyani et al. (2019) |
| 20 | Silymarin | 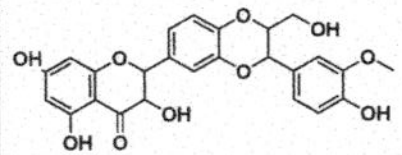 | *Silybum marianum* | Seeds | Ethanol extract | In vitro | Antioxidant, antiinflammatory, modulation of cellular signaling | Guo et al. (2019) |

| | | | | | | | | |
|---|---|---|---|---|---|---|---|---|
| 21 | Ursolic acid | 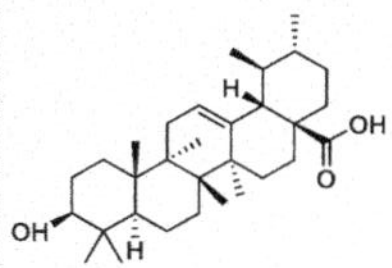 | *Ocimum sanctum* | Leaves | Ethanol extract | In vivo | Antioxidant, antiinflammatory, neuroprotective | Ahmad et al. (2018) |
| 22 | Piperine | 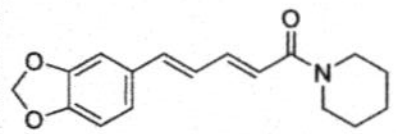 | *Piper nigrum* | Fruits | Ethanol extract | In vivo | Antioxidant, antiinflammatory, modulation of neurotransmitters | Ademuyiwa et al. (2021) |
| 23 | Salidroside | 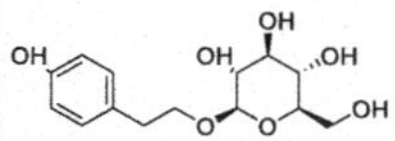 | *Rhodiola rosea* | Rhizome | Ethanol extract | In vivo | Antiinflammatory, antioxidant, neuroprotective | Zhuang et al. (2019) |
| 24 | Icariin | 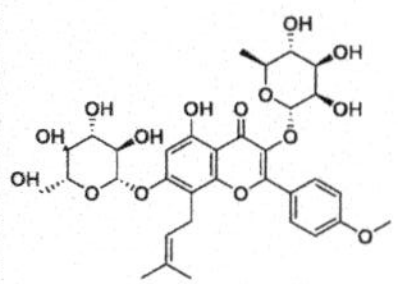 | *Epimedium* spp. | Aerial parts | Ethanol extract | In vitro | Neuroprotective, antiinflammatory, antioxidant | Li et al. (2022a, b) |

circulation from the brain, is critical in determining the entrance of phytochemicals into the brain (Liu et al. 2022). Hormetic phytochemicals' capacity to cross the BBB is determined by their physicochemical characteristics (Sánchez-Martínez et al. 2022). Certain lipophilic substances, like polyphenols, have been demonstrated to improve BBB permeability. Furthermore, active transport mechanisms may promote phytochemical entry into the brain (Figueira et al. 2017).

Metabolism is a critical step that influences hormetic phytochemical bioavailability and action. Metabolism can occur in various organs, including the liver, gut, and brain. The basic purpose of metabolism is to enhance detoxification and enable the removal of foreign substances (García-García et al. 2020). Enzymatic events such as oxidation, reduction, and hydrolysis occur during Phase I metabolism and can create more polar metabolites. Conjugation processes occur during Phase II metabolism, in which phytochemicals are conjugated with endogenous compounds such as glucuronic acid, sulfate, or glutathione. These conjugated metabolites are more water-soluble and simpler to defecate in general (Cooper and Hanigan 2018). Hormetic phytochemical metabolism can vary widely depending on the substance and individual circumstances (Franco et al. 2019). Curcumin, for example, undergoes significant metabolism, mostly by glucuronidation and sulfation, resulting in curcumin glucuronide and curcumin sulfate (Dei Cas and Ghidoni 2019). Resveratrol, too, travels through phase II metabolism, generating glucuronide and sulfate conjugates (Springer and Moco 2019).

Hormetic phytochemicals and their metabolites are predominantly eliminated from the body by renal and biliary excretion. Renal excretion involves processes in the kidneys such as filtration, reabsorption, and secretion, whereas biliary excretion involves the excretion of chemicals into the bile, which finally enters the gut and can be expelled in feces (De Vries et al. 1988). The excretion efficiency varies based on the chemical and its metabolites. Some phytochemicals, like quercetin, have high renal excretion, while others, like curcumin, have low urine excretion (Batiha et al. 2020). Biliary excretion is especially crucial for the clearance of phytochemical metabolites that move via enterohepatic circulation and are reabsorbed into circulation from the colon (Yang and Tsai 2019).

Hormetic phytochemical absorption and metabolism are critical elements influencing their efficacy in improving brain health throughout aging. Increasing the availability and activity of phytochemicals in the brain through optimization of these pathways might maximize their therapeutic potential (Sahebnasagh et al. 2022). Improved formulation and delivery technologies and better knowledge of individual differences in metabolism can help increase the bioavailability and efficacy of hormetic phytochemicals (Vesely et al. 2021). Realizing these compounds' full potential for reducing brain aging and neurodegenerative diseases, further study is needed to explain the intricate connections between phytochemicals, cellular processes, and individual variability.

## 4.4 Hormetic Phytochemicals and Their Effects on Cellular Stress in Brain Aging

This section seeks to give an in-depth examination of animal and human research data on the impact of hormetic phytochemicals on cellular stress in brain aging. It will go through various study methods, findings, and putative mechanisms of action involved in response to cellular stress in particular.

### 4.4.1 Evidence from Animal Studies and Human Studies

Animal studies have been critical in determining the effects of hormetic phytochemicals on cellular stress in the setting of brain aging. To investigate the influence of phytochemical therapies on cellular stress indicators and related consequences, this research uses various experimental techniques, including in vivo and in vitro models (Calabrese et al. 2012). While animal research has offered useful insights, human studies must be examined to evaluate the effects of hormetic phytochemicals on cellular stress in brain aging (Sahebnasagh et al. 2022). This section seeks to give a complete review of animal and human research on the effects of hormetic phytochemicals on cellular stress, including study methods, findings, and probable mechanisms of action.

#### 4.4.1.1 Study Design and Outcomes

Animal research on the effects of hormetic phytochemicals on cellular stress in brain aging uses a variety of experimental techniques to explain the processes involved and assess their functional relevance (Gessner et al. 2017). Phytochemicals are commonly administered through oral gavage, dietary supplementation, or intraperitoneal injection in these investigations, and cellular stress indicators are assessed using a mix of biochemical, molecular, and behavioral assays (Turner et al. 2011).

Markers of cellular stress commonly investigated in animal studies include oxidative stress, inflammation, protein misfolding, mitochondrial dysfunction, and neuronal damage (Picca et al. 2020). Oxidative stress markers, such as reactive oxygen species (ROS) levels, antioxidant enzyme activities, and lipid peroxidation, are frequently measured to evaluate the antioxidant potential of phytochemicals (Marrocco et al. 2017). Inflammatory markers, such as cytokines and chemokines, are assessed to determine the anti-inflammatory effects of phytochemical interventions (Chen et al. 2018). Additionally, measures of protein misfolding, mitochondrial function, and neuronal integrity provide insights into the potential neuroprotective properties of hormetic phytochemicals (Hannan et al. 2020). Animal research on the impact of hormetic phytochemicals on cellular stress in brain aging has yielded encouraging results. The effects include reduced oxidative stress, decreased inflammation, increased mitochondrial function, improved protein homeostasis, and retained neuronal structure and function (Limanaqi et al. 2020). Furthermore, cognitive performance tests, such as spatial learning and memory tasks, have shown that phytochemical therapies increase cognitive function.

Human research on the effects of hormetic phytochemicals on cellular stress in brain aging uses a variety of study approaches, such as randomized controlled trials (RCTs), observational studies, and intervention studies. These studies use a mix of clinical, biochemical, and neuroimaging techniques to determine cellular stress indicators and related effects (Sharma and Padwad 2020). Cognitive function, memory performance, executive function, and total brain health are among the clinical outcomes studied in human research. Commonly studied cellular stress biomarkers include oxidative stress markers, inflammatory markers, neurotrophic factors, and signs of neuronal damage (Więdłocha et al. 2023).

Human studies have found that hormetic phytochemicals improve cognitive performance in the aging brain. Intervention studies using cognitive tests such as the mini-mental state examination (MMSE), the montreal cognitive assessment (MoCA), and particular memory tasks have shown that phytochemical therapies increase cognitive function (Gluhm et al. 2013). These enhancements include a wide range of cognitive domains, including attention, memory, processing speed, and executive function.

Human studies have found hormonal phytochemicals to modify oxidative stress indicators such as reactive oxygen species (ROS), lipid peroxidation products, and antioxidant enzyme activities (Sharifi-Rad et al. 2020). Following phytochemical treatments, these investigations found lower levels of oxidative stress and improved antioxidant defense systems. Chronic inflammation contributes significantly to the etiology of neurodegenerative disorders (Zhang et al. 2015). Human studies have shown that hormetic phytochemicals have antiinflammatory effects by measuring inflammatory indicators such as cytokines, chemokines, and C-reactive protein (CRP) (Sproston and Ashworth 2018). Phytochemical therapies have been demonstrated to lower systemic and neuroinflammatory indicators, suggesting that they have the capacity to control cellular stress associated with inflammation in the aging brain (Forni et al. 2019; Satyam and Bairy 2022). Neurotrophic factors are important in neuronal survival, synaptic plasticity, and neurogenesis. Hormetic phytochemicals have been studied in humans for their effects on neurotrophic factors such as brain-derived neurotrophic factor (BDNF) and nerve growth factor (NGF) (Miranda et al. 2019). Following phytochemical treatments, circulating levels of neurotrophic factors increased, indicating a possible function in enhancing neural resilience and lowering cellular stress (Bathina and Das 2015).

Magnetic resonance imaging (MRI) and positron emission tomography (PET) have offered useful insights into the impact of hormetic phytochemicals on brain structure and function (Shan et al. 2022). Human neuroimaging studies have found that phytochemical therapies increase brain volume, cortical thickness, and functional connectivity (Morita et al. 2016). These findings show that hormetic phytochemicals may have neuroprotective effects on cellular stress in the aging brain.

#### 4.4.1.2 Potential Mechanisms of Action

Several methods have been proposed to explain how hormetic phytochemicals affect cellular stress in brain aging. Hormetic phytochemicals' effects on cellular stress in brain aging are mediated by a variety of putative modes of action, including antioxidant defense, antiinflammatory pathways, protein homeostasis modulation, mitochondrial function augmentation, and synaptic plasticity control (Murugaiyah and Mattson 2015).

For example, polyphenols, flavonoids, and curcuminoids are renowned for their powerful antioxidant effects. They have the ability to scavenge free radicals, limit ROS formation, and boost the activity of natural antioxidant enzymes (Lobo et al. 2010). The activation of antioxidant defense mechanisms, notably the nuclear factor erythroid 2-related factor 2 (Nrf2) pathway, has been hypothesized as a crucial mechanism driving hormetic phytochemicals' antioxidative actions (Kang 2020). When Nrf2 is activated, antioxidant enzymes such as superoxide dismutase (SOD), catalase, and glutathione peroxidase are activated, which reduces oxidative stress and increases cellular resilience (Ngo and Duennwald 2022).

Inflammation is a hallmark of brain aging and neurodegenerative diseases. Hormetic phytochemicals possess antiinflammatory properties and can modulate key inflammatory signaling pathways, including nuclear factor-kappa B (NF-κB), mitogen-activated protein kinases (MAPKs), and cyclooxygenases (COX) (Corbi et al. 2016). By inhibiting the activation of these proinflammatory pathways, phytochemicals can attenuate the production of inflammatory mediators and reduce neuroinflammation, thereby protecting against cellular stress (Saleh et al. 2021).

The buildup of misfolded proteins, such as amyloid-beta and tau, is a hallmark of neurodegenerative disorders (Sweeney et al. 2017). Hormetic phytochemicals have been demonstrated to improve protein homeostasis by controlling the activities of chaperones, proteasomes, and autophagy (Murakami 2013). Phytochemical treatments increase the clearance of misfolded proteins and prevent their aggregation by activating heat shock proteins (HSPs) and the ubiquitin-proteasome system, lowering the cellular stress associated with protein misfolding (Penke et al. 2018).

Mitochondrial dysfunction is a critical contributor to cellular stress in brain aging. Hormetic phytochemicals, such as resveratrol and quercetin, have enhanced mitochondrial function by improving energy production, reducing ROS production, and promoting mitochondrial biogenesis (Deledda et al. 2022). These compounds activate key regulators of mitochondrial health, including sirtuins (e.g., SIRT1) and peroxisome proliferator-activated receptor gamma coactivator 1-alpha (PGC-1α). By preserving mitochondrial function, hormetic phytochemicals enhance cellular bioenergetics and protect against age-related oxidative damage (Wang et al. 2015).

Hormetic phytochemicals have been shown to increase neurotrophic factor secretion, including brain-derived neurotrophic factor (BDNF) and nerve growth factor (NGF) (Michael et al. 1997). These elements are essential for neuronal survival, synaptic plasticity, and neurogenesis. Phytochemical therapies increase neuronal resilience and synaptic connection by increasing neurotrophic support, reducing cellular stress, and sustaining cognitive function (Yoon et al. 2023).

These studies show that phytochemical therapies can enhance cognitive outcomes by modulating oxidative stress, inflammation, protein misfolding, mitochondrial function, and neuronal integrity. The fundamental processes include antioxidant defense, antiinflammatory pathways, protein homeostasis modulation, mitochondrial function augmentation, and synaptic plasticity regulation (Rahman et al. 2020). More study is required to identify the particular molecular targets and signaling pathways involved in the response to cellular stress and to confirm the translational potential of hormetic phytochemicals as therapeutic therapies for brain aging.

### 4.4.2 Comparison with Other Natural Products Used for Cellular Stress in Brain Aging

While there is evidence to support the effects of hormetic phytochemicals on cellular stress in brain aging, it is necessary to compare their efficacy to that of other natural products used to regulate cellular stress (Calabrese et al. 2010). Comparing the effects and efficacy of hormetic phytochemicals with the effects and efficacy of these natural items can give useful insights into their distinct features and possible benefits.

Vitamins, including vitamins E, C, and B complex, have long been known for their antioxidant qualities and potential advantages in lowering cellular stress. These vitamins work as free radical scavengers and aid in the reduction of oxidative stress in the brain (Traber and Stevens 2011). While vitamins are important for cellular health, hormetic phytochemicals provide extra advantages. Unlike vitamins, which are largely antioxidants, phytochemicals have pleiotropic effects such as anti-inflammatory capabilities, modulation of cellular signaling pathways, and gene expression control (Bevere et al. 2022). Furthermore, phytochemicals frequently have a larger variety of molecular targets than individual vitamins, allowing for more thorough cellular stress regulation (Mucha et al. 2021).

Omega-3 fatty acids, notably EPA and DHA, have been intensively researched for their possible neuroprotective benefits in brain aging. These fatty acids are necessary components of cell membranes and play an important role in membrane fluidity and integrity (Dyall 2015). Omega-3 fatty acids are antiinflammatory and linked to better cognitive performance. While omega-3 fatty acids are widely known for their neuroprotective properties, hormetic phytochemicals provide additional modes of action (Calder 2010). Phytochemicals have the ability to influence oxidative stress, inflammation, protein misfolding, and mitochondrial function, allowing for a multitargeted strategy to reduce cellular stress in the aging brain (Limanaqi et al. 2020).

Herbal extracts made from plants have been used in traditional medicine for ages to enhance health and well-being. Many plant extracts include bioactive chemicals such as polyphenols, flavonoids, alkaloids, and terpenoids, which have various pharmacological effects (Tungmunnithum et al. 2018). Ginkgo biloba, Bacopa monnieri, and Panax ginseng have all been researched for their possible neuroprotective effects. While these herbal extracts show promise regarding cellular stress reduction, hormetic phytochemicals provide a larger spectrum of bioactive components and modes of action (Landucci et al. 2019). Furthermore, as compared to plant extracts, hormetic phytochemicals frequently contain a greater concentration of certain bioactive substances, allowing for a more focused and effective effect on cellular stress in the aging brain (Qiao et al. 2023).

In comparison with individual vitamins, omega-3 fatty acids, herbal extracts, and hormetic phytochemicals offer a more complete approach to modulating oxidative stress, inflammation, protein homeostasis, mitochondrial function, and synaptic plasticity (Fila et al. 2021). More study is needed to determine the individual effects and combinations of various natural products and hormetic phytochemicals to optimize their usage in lowering cellular stress and enhancing brain function throughout aging.

## 4.5 Molecular Mechanisms of Hormetic Phytochemicals on Cellular Stress in Brain Aging

### 4.5.1 Overview of Molecular Mechanisms of Hormesis

Hormetic phytochemicals influence cellular stress in brain aging via a variety of molecular pathways. These processes include manipulating critical cellular signaling networks, regulating gene expression, and activating stress response pathways (Cornelius et al. 2013). The following section offers an outline of the molecular processes underpinning hormesis.

#### 4.5.1.1 Activation of Nrf2 Pathway

Hormetic phytochemicals have the ability to activate the nuclear factor erythroid 2-related factor 2 (Nrf2) pathway, which is a master regulator of the antioxidant response (Gupte et al. 2013). Normal circumstances cause Nrf2 to stay coupled to its cytoplasmic repressor, Kelch-like ECH-associated protein 1 (Keap1), resulting in proteasomal degradation (Tian et al. 2018). Hormetic phytochemicals, on the other hand, can disrupt the Nrf2-Keap1 complex, allowing Nrf2 to translocate into the nucleus and bind to antioxidant response elements (ARE) in target gene promoter regions (Yu and Xiao 2021). The Nrf2 pathway activates phase II detoxification enzymes, antioxidant enzymes, and other cytoprotective proteins, lowering oxidative stress and increasing cellular resilience (Zhang et al. 2013).

#### 4.5.1.2 Modulation of Inflammatory Pathways

Hormetic phytochemicals exhibit antiinflammatory effects by modulating key inflammatory signaling pathways, such as nuclear factor-kappa B (NF-κB) and mitogen-activated protein kinases (MAPKs). These phytochemicals can inhibit the activation of NF-κB, a transcription factor involved in the production of pro-inflammatory cytokines and mediators (Jayakumar et al. 2021). By suppressing NF-κB activity, hormetic phytochemicals reduce the expression of inflammatory genes and mitigate neuroinflammation. Furthermore, phytochemicals can influence MAPK pathways such as extracellular signal-regulated kinase (ERK), c-Jun N-terminal kinase (JNK), and p38 MAPK, all of which are implicated in the control of inflammatory responses (Saleh et al. 2021).

#### 4.5.1.3 Regulation of Protein Homeostasis

Protein homeostasis, also known as proteostasis, is critical for cellular function and avoiding the buildup of misfolded proteins. Hormetic phytochemicals have the ability to influence protein homeostasis by influencing the activity of molecular chaperones, proteasomes, and autophagy processes (Chen et al. 2023). These phytochemicals, for example, can stimulate the production of heat shock proteins (HSPs) such as Hsp70 and Hsp90, which play important roles in protein folding and breakdown (Zügel and Kaufmann 1999). They can also stimulate the ubiquitin-proteasome system, which is in charge of the selective destruction of misfolded and damaged proteins. Hormetic phytochemicals can also trigger autophagy, a mechanism through which cells destroy and recycle damaged biological components such as aggregated proteins (Li et al. 2022a, b; Kiruthiga et al. 2020).

#### 4.5.1.4 Enhancement of Mitochondrial Function

Mitochondrial dysfunction is a sign of aging in the brain and adds to cellular stress. Hormetic phytochemicals can improve mitochondrial function by encouraging mitochondrial biogenesis, increasing energy production, and decreasing oxidative stress (Chakrabarti et al. 2011). These phytochemicals activate key regulators of mitochondrial health, such as sirtuins (e.g., SIRT1) and peroxisome proliferator-activated receptor gamma coactivator 1-alpha (PGC-1$\alpha$) (Wang et al. 2015). When activated, these components result in increased mitochondrial biogenesis, better respiratory chain activity, and enhanced antioxidant defenses inside mitochondria (Bhatti et al. 2017).

#### 4.5.1.5 Modulation of Synaptic Plasticity and Neurotrophic Factors

Hormetic phytochemicals have the capacity to influence synaptic plasticity, a key mechanism underpinning learning and memory. These phytochemicals have been shown to increase the production of neurotrophic factors, including brain-derived neurotrophic factor (BDNF) and nerve growth factor (NGF) (Colucci-D'Amato et al. 2020). Neurotrophic factors are important in neuronal survival, synaptic connection, and neurogenesis. Hormeticphytochemicals increase synaptic plasticity, neuronal resilience, and cognitive performance through increasing neurotrophic support (Venkatesan et al. 2015).

### 4.5.2 Description of Specific Molecular Mechanisms of Hormetic Phytochemicals in Brain Aging

Hormetic phytochemicals reduce cellular stress in the brain during aging via distinct molecular pathways that target important cellular functions. These mechanisms include antioxidant activity, antiinflammatory effects, cellular signaling system modulation, gene expression control, and epigenetic changes (Limanaqi et al. 2020). Understanding these precise processes is critical for determining how hormetic phytochemicals affect cellular stress in brain aging.

#### 4.5.2.1 Antioxidant Activity

One of the key methods by which hormetic phytochemicals decrease cellular stress in brain aging is through antioxidant activity. These phytochemicals serve as free radical scavengers, neutralizing reactive oxygen species (ROS) and reducing oxidative damage (Poljsak 2011). Polyphenols, for example, have powerful antioxidant capabilities and can directly quench ROS or boost the activity of endogenous antioxidant enzymes such as superoxide dismutase (SOD), catalase, and glutathione peroxidase (GPx) (Kurutas 2015). Hormetic phytochemicals help preserve cellular homeostasis and protect against age-related neurodegenerative disorders by lowering oxidative stress (Calabrese et al. 2012).

#### 4.5.2.2 Antiinflammatory Effects

Chronic inflammation is linked to the advancement of brain aging and neurodegenerative diseases. Hormetic phytochemicals reduce inflammation by regulating several inflammatory signaling pathways (Limanaqi et al. 2020). For example, polyphenols, such as curcumin and resveratrol, inhibit the activation of nuclear factor-kappa B (NF-κB), a transcription factor involved in the expression of pro-inflammatory cytokines and mediators (Gonzales and Orlando 2008). These phytochemicals can suppress NF-κB activity by inhibiting its translocation into the nucleus or by preventing its DNA binding (Chauhan et al. 2022). Hormetic phytochemicals can also influence other critical inflammatory pathways, such as mitogen-activated protein kinases (MAPKs) and signal transducer and activator of transcription 3 (STAT3), reducing neuroinflammation in the brain throughout aging (Davinelli et al. 2016).

#### 4.5.2.3 Modulation of Cellular Signaling Pathways

Hormetic phytochemicals can influence cellular signaling pathways that are important for stress response and neuronal health (Calabrese et al. 2010). Resveratrol, for example, stimulates sirtuin proteins, notably SIRT1, which regulate cellular metabolism, DNA repair, and stress tolerance. Resveratrol activation of SIRT1 stimulates mitochondrial biogenesis, boosts antioxidant defenses, and improves cellular energy metabolism, all of which lead to neuroprotection (Lagouge et al. 2006). Hormetic phytochemicals can also target other signaling pathways involved in cellular survival, longevity, and stress response, such as the PI3K/Akt pathway, the mammalian target of rapamycin (mTOR) pathway, and the adenosine monophosphate-activated protein kinase (AMPK) pathway (Salminen and Kaarniranta 2012).

#### 4.5.2.4 Regulation of Gene Expression

Hormetic phytochemicals have the ability to alter gene expression in the brain throughout aging, resulting in the overexpression of neuroprotective genes and the downregulation of harmful genes (Calabrese et al. 2010). These phytochemicals can interact with certain transcription factors, coactivators, and epigenetic modifiers to affect gene expression in oxidative stress response, inflammation, apoptosis, synaptic plasticity, and neuronal survival (Sivandzade et al. 2019). Curcumin, for example, can activate the nuclear factor erythroid 2-related factor 2 (Nrf2) pathway, thereby increasing the production of antioxidant enzymes and phase II detoxifying enzymes (Kou et al. 2013). Resveratrol can influence gene expression and chromatin structure by activating the silent mating-type information regulation 2 homolog 1 (SIRT1)-dependent deacetylation of histones (Ghosh et al. 2013).

#### 4.5.2.5 Epigenetic Modifications

Hormetic phytochemicals can cause epigenetic changes in the brain, influencing gene expression patterns and cellular stress responses. These alterations include DNA methylation, histone acetylation, histone methylation, and microRNA expression (Baccarelli and Bollati 2009). For example, resveratrol has been demonstrated to impact the expression of genes involved in inflammation, apoptosis, and oxidative stress through modulating DNA methylation patterns (Fernandes et al. 2017). Curcumin can also decrease the activity of histone acetyltransferases (HATs) and histone deacetylases (HDACs), resulting in a change in histone acetylation status and associated gene expression alterations (Reuter et al. 2011). Hormetic phytochemical-induced epigenetic changes can have long-term consequences for cellular stress response and brain health.

### 4.5.3 Role of Hormetic Phytochemicals in Modulating Cellular Signaling Pathways

Hormetic phytochemicals work to reduce cellular stress in the brain by altering numerous cellular signaling pathways. These phytochemicals exert neuroprotective effects, decrease neuroinflammation, counteract oxidative stress, improve cellular stress resistance, and promote cellular homeostasis by targeting pathways such as PI3K/Akt/mTOR, NF-κB, Nrf2/ARE, MAPK, and AMPK (Murugaiyah and Mattson 2015).

#### 4.5.3.1 PI3K/Akt/mTOR Pathway

The phosphoinositide 3-kinase (PI3K)/Akt/mammalian target of rapamycin (mTOR) pathway is a key cellular signaling pathway involved in cell growth, survival, and metabolism (Martelli et al. 2010). Hormetic phytochemicals such as resveratrol and curcumin have been shown to modulate this pathway in the context of brain aging. These phytochemicals can activate Akt, leading to the inhibition of mTOR and subsequent activation of autophagy, a process involved in the clearance of damaged cellular components (Rahman et al. 2020). By modulating the PI3K/Akt/mTOR pathway, hormetic phytochemicals can promote cellular health, reduce oxidative stress, and enhance neuronal survival in the aging brain (Murugaiyah and Mattson 2015).

#### 4.5.3.2 NF-κB Signaling Pathway

The nuclear factor-kappa B (NF-κB) signaling pathway plays a crucial role in inflammation and immune responses. Activation of NF-κB leads to the transcriptional upregulation of pro-inflammatory genes, contributing to neuroinflammation and cellular stress in brain aging (Shih et al. 2015). Hormetic phytochemicals such as curcumin, resveratrol, and epigallocatechin gallate (EGCG) have been shown to inhibit NF-κB signaling by suppressing the nuclear translocation of NF-κB or inhibiting the activation of its upstream regulators, such as IκB kinase (IKK) (Corbi et al. 2016). By modulating NF-κB signaling, hormetic phytochemicals can mitigate neuroinflammation and reduce cellular stress in the aging brain (Calabrese et al. 2012).

#### 4.5.3.3 Nrf2/ARE Pathway

A critical cellular defense mechanism against oxidative stress is the nuclear factor erythroid 2-related factor 2 (Nrf2)/antioxidant response element (ARE) pathway (Kang 2020). Nrf2 is a transcription factor that controls the production of antioxidant enzymes and phase II detoxification enzymes (He et al. 2020). Hormetic phytochemicals, such as sulforaphane, curcumin, and resveratrol, can activate the Nrf2/ARE pathway by increasing Nrf2 nuclear translocation and, as a result, activating antioxidant and detoxifying genes (Houghton et al. 2016). Hormetic phytochemicals boost cellular antioxidant capacity, combat oxidative stress, and protect against age-related neurodegenerative disorders by increasing the Nrf2/ARE pathway (Hannan et al. 2020).

#### 4.5.3.4 MAPK Signaling Pathway

The MAPK pathway is a family of serine/threonine kinases involved in cellular stress response, inflammation, apoptosis, and neuronal survival. Hormetic phytochemicals, including resveratrol, curcumin, and EGCG, have been shown to alter MAPK signaling in the aging brain (Leri et al. 2020). MAPKs such as extracellular signal-regulated kinase (ERK), c-Jun N-terminal kinase (JNK), and p38 MAPK can be activated by these phytochemicals (Muniyappa and Das 2008). These MAPKs can be activated to control gene expression, boost cell survival, modify inflammation, and improve cellular stress tolerance. Hormetic phytochemicals exert neuroprotective properties and lower cellular stress in the brain during aging by regulating MAPK signaling.

#### 4.5.3.5 AMPK Signaling Pathway

The AMPK pathway, which is triggered by adenosine monophosphate, is a master regulator of cellular energy balance and stress response. AMPK activation promotes energy savings, improves mitochondrial function, and increases autophagy (Mihaylova and shaw 2011). Hormetic polyphenols like resveratrol and quercetin may activate AMPK, resulting in greater mitochondrial biogenesis, improved energy metabolism, and better cellular stress tolerance (García-Aguilar et al. 2021). Hormetic phytochemicals can boost cellular health and protect against age-related neurodegenerative disorders by regulating AMPK signaling (Murugaiyah and Mattson 2015).

## 4.6 Safety and Potential Adverse Effects of Hormetic Phytochemicals

Hormetic phytochemicals are plant-derived bioactive substances that have positive effects on cellular stress response pathways. These chemicals have received a lot of interest because of their possible involvement in supporting healthy brain aging (Liu et al. 2022). However, assessing their safety profile and potential side effects is critical before using them in brain health therapies.

### 4.6.1 Overview of Potential Adverse Effects

While hormetic phytochemicals are usually thought to be harmless, it is vital to be aware of the potential negative effects that may arise when these substances are eaten in large quantities or specific groups (Son et al. 2008). GGI abnormalities, allergic responses, hepatotoxicity, nephrotoxicity, neurotoxicity, and reproductive toxicity are all possible side effects (Sahebnasagh et al. 2022). However, it is important to remember that when hormetic phytochemicals are ingested within the approved dose range, deleterious effects are relatively uncommon. Depending on the phytochemical ingredient, dose, and individual characteristics, the particular side effects and their severity may vary (Jodynis-Liebert and Kujawska 2020).

#### 4.6.1.1 Gastrointestinal Disturbances

Certain hormetic phytochemicals can produce gastrointestinal problems such as nausea, vomiting, abdominal discomfort, and diarrhea. These symptoms are frequently moderate and temporary, but they might be more severe in those who have preexisting gastrointestinal disorders or are particularly sensitive to these substances (Bao et al. 2014). To reduce the risk of gastrointestinal side effects, start with low dosages and gradually raise the dosage (Martel et al. 2019).

#### 4.6.1.2 Allergic Reactions

Certain hormetic phytochemicals may cause allergic reactions in certain people. These responses can cause skin rashes, irritation, swelling, and, in severe cases, anaphylaxis. Individuals who have a history of allergies or a known sensitivity to certain plants or phytochemicals should exercise caution and seek the advice of a healthcare expert before taking these components (Gezer 2018).

#### 4.6.1.3 Hepatotoxicity and Nephrotoxicity

Excessive ingestion of some hormetic phytochemicals may result in hepatotoxicity or nephrotoxicity in rare situations. These negative consequences arise mostly when high quantities of certain substances are ingested over a lengthy period of time (Lambert et al. 2010). When utilizing hormetic phytochemicals, it is critical to check the liver and kidney function on a regular basis, especially in patients with preexisting liver or kidney disorders (Sharifi-Rad et al. 2020).

#### 4.6.1.4 Neurotoxicity

Although hormetic phytochemicals might provide neuroprotection, excessive or extended exposure to particular substances can cause neurotoxicity. Neurotoxicity symptoms include dizziness, headaches, disorientation, and even convulsions (Kumar and Khanum 2012). Individuals with neurological problems or using drugs that alter brain function should use hormetic phytochemicals with caution and seek medical advice before using them (Salehi et al. 2020).

#### 4.6.1.5 Reproductive Toxicity

Certain hormetic phytochemicals have been linked to reproductive harm, especially when taken in large dosages or at critical stages of pregnancy. Before introducing hormetic phytochemicals into their routine, pregnant women, women attempting to conceive, and people with reproductive issues should contact a healthcare expert (Khan et al. 2021).

### 4.6.2 Factors Affecting Safety and Dosage

Several variables can impact the safety and dose of hormetic phytochemicals in brain-aging therapies. Individual differences in reaction to these substances, such as age, gender, heredity, and underlying health issues, can influence their toxicity (Mao and Franke 2013). These considerations must be considered while establishing the proper dosage and considering the possible dangers of hormetic phytochemical use.

### 4.6.3 Drug Interactions and Contraindications

Hormetic phytochemicals may interact with some drugs, resulting in undesirable consequences or decreased efficacy. To guarantee the safe use of these drugs, it is critical to recognize potential drug interactions and contraindications, especially for those taking medicines for brain-related diseases (Smoliga et al. 2012). Some phytochemicals may disrupt the metabolism or efficacy of specific medications, while others may stimulate or inhibit the action of drug-metabolizing enzymes. Before introducing hormetic phytochemicals into brain health therapies, it is critical to consult with a healthcare expert to assess potential medication interactions and contraindications (Gómez-Garduño et al. 2022).

### 4.6.4 Detoxification of Hermetic Phytochemicals

When hormetic phytochemicals are overexposed or unfavorable effects are noticed, detoxification procedures can be used to remove or lower the amount of these substances in the body. Enhancing the body's natural elimination mechanisms, such as hepatic metabolism, renal excretion, and gastrointestinal elimination, is frequently involved in detoxification procedures (Calabrese and Kozumbo 2021).

Adequate hydration, fiber-rich food consumption, and supporting liver function through particular dietary measures and supplements can help detoxify. However, it is crucial to emphasize that detoxification treatments should be used under the supervision of healthcare specialists and suited to individual needs (Ioniță-Mîndrican et al. 2022).

Understanding possible side effects, variables influencing safety and dose, medication interactions, and detoxification procedures is critical for reaping the advantages of hormetic phytochemicals while minimizing the risk of side effects. Hormetic phytochemicals can be used effectively in brain aging therapies using a careful and individualized approach.

## 4.7 Conclusion and Future Perspectives

Throughout this chapter, we have investigated the response of cellular stress to hormetic phytochemicals in the context of brain aging. According to existing data, these plant-derived bioactive chemicals have the ability to activate cellular stress response pathways, resulting in neuroprotective effects and supporting healthy brain aging. These phytochemicals' hormetic dose-response relationship emphasizes the significance of optimum dosage to maximize their therapeutic benefits while minimizing potential harmful consequences.

Hormetic phytochemicals have been proven in studies to influence oxidative stress, inflammation, and cellular signaling pathways involved in neuronal survival, synaptic plasticity, and neurogenesis. These systems help to improve cognitive performance, increase neural resilience, and slow the neurodegenerative processes associated with brain aging. Hormetic phytochemicals have also been shown to have antioxidant, antiinflammatory, and anti-apoptotic activities, protecting against oxidative damage, neuroinflammation, and cell death. Despite the positive results, various routes for further study are needed to understand better the response of cellular stress to hormetic phytochemicals in brain aging. To begin, identifying the particular mechanisms of action and signaling pathways involved in these phytochemicals' neuroprotective properties would give useful insights into their prospective therapeutic uses. Furthermore, studying the effects of various combinations of hormetic phytochemicals and their interactions may reveal synergistic effects and improve therapeutic approaches.

Furthermore, well-designed clinical trials and longitudinal investigations are essential for determining hormetic phytochemical therapies' effectiveness, safety, and long-term consequences in brain aging. To establish the potential preventative and therapeutic uses of hormetic phytochemicals, these studies should include a wide population, including people with neurodegenerative illnesses. Hormetic phytochemicals have therapeutic potential as adjuvant therapy for neurodegenerative illnesses such as Alzheimer's disease, Parkinson's disease, and age-related cognitive decline. They may also have preventative benefits by encouraging healthy brain aging and lowering the risk of neurodegenerative diseases. More study is needed, however, to optimize dosage regimens, identify the most beneficial phytochemical combinations, and establish standardized formulations for clinical usage.

The rising volume of information on hormetic phytochemicals and their potential advantages in brain aging has important public health and policy implications. To begin, encouraging the intake of a wide variety of plant-based foods high in hormetic phytochemicals can improve brain function and lower the burden of neurodegenerative illnesses in aging populations. Public health campaigns should emphasize the significance of a balanced and diverse diet that includes various plant sources to guarantee an appropriate intake of these beneficial chemicals. Furthermore, incorporating scientific study findings into public health guidelines and policies can assist in evidence-based brain health promotion recommendations. This involves making information about the possible advantages and hazards of hormetic phytochemicals more available and educating healthcare practitioners and politicians about their role in brain aging therapies. Collaboration among academics, healthcare providers, politicians, and the food sector is critical for translating scientific information into practical strategies and policies to promote brain health and lower the burden of age-related cognitive decline.

Finally, the response of cellular stress to hormetic phytochemicals in brain aging opens new pathways for managing and preventing neurodegenerative disorders. The existing data support these drugs' neuroprotective benefits, but further study is needed to optimize their usage, find the most beneficial combinations, and evaluate long-term safety and efficacy. We can promote healthy brain aging and improve the quality of life for aging populations by accepting these scientific results and turning them into public health initiatives and policies. While preclinical studies offer promising results, the potential impact of hormetic phytochemicals on human brain health requires further investigation. This chapter examines the challenges of translating hormetic concepts into effective clinical interventions. It considers factors such as optimal dosages, personalized responses, and potential interactions with existing medications in the context of aging populations.

## References


Ademuyiwa OH, Fasogbon BM, Adebo OA (2021) The potential role of Piper guineense (black pepper) in managing geriatric brain aging: a review. Crit Rev Food Sci Nutr 63:2840–2850

Aguiar S, Borowski T (2013) Neuropharmacological review of the nootropic herb Bacopa monnieri. Rejuvenation Res 16(4):313–326

Ahmad A, Abuzinadah MF, Alkreathy HM, Banaganapalli B, Mujeeb M (2018) Ursolic acid rich Ocimum sanctum L leaf extract loaded nanostructured lipid carriers ameliorate adjuvant induced arthritis in rats by inhibition of COX-1, COX-2, TNF-α and IL-1: pharmacological and docking studies. PLoS One 13(3):e0193451

Almalki AFY, Arabdin M, Khan A (2021) The role of heat shock proteins in cellular homeostasis and cell survival. Cureus 13(9):e18316

Aqil F, Munagala R, Jeyabalan J, Vadhanam MV (2013) Bioavailability of phytochemicals and its enhancement by drug delivery systems. Cancer Lett 334(1):133–141

Ayala A, Muñoz MF, Argüelles S (2014) Lipid peroxidation: production, metabolism, and signaling mechanisms of malondialdehyde and 4-hydroxy-2-nonenal. Oxid Med Cell Longev 2014: 360438

Baccarelli A, Bollati V (2009) Epigenetics and environmental chemicals. Curr Opin Pediatr 21(2): 243

Balkrishna A, Thakur P, Varshney A (2020) Phytochemical profile, pharmacological attributes and medicinal properties of convolvulus prostratus – a cognitive enhancer herb for the management of neurodegenerative etiologies. Front Pharmacol 11:171

Bao Y, Wang W, Zhou Z, Sun C (2014) Benefits and risks of the hormetic effects of dietary isothiocyanates on cancer prevention. PLoS One 9(12):e114764

Bathina S, Das UN (2015) Brain-derived neurotrophic factor and its clinical implications. Arch Med Sci 6:1164–1178

Batiha GES, Beshbishy AM, Ikram M, Mulla ZS, El-Hack MEA, Taha AE, Algammal AM, Elewa YHA (2020) The pharmacological activity, biochemical properties, and pharmacokinetics of the major natural polyphenolic flavonoid: quercetin. Foods 9(3):374

Benameur T, Soleti R, Porro C (2021) The potential neuroprotective role of free and encapsulated quercetin mediated by mirna against neurological diseases. Nutrients 13(4):1318

Bevere M, Di Cola G, Santangelo C, Grazioli E, Marramiero L, Pignatelli P, Bondi D, Mrakic-Sposta S (2022) Redox-based disruption of cellular hormesis and promotion of degenerative pathways: perspectives on aging processes. J Gerontol A 77(11):2195–2206

Bhakta-Guha D, Efferth T (2015) Hormesis: decoding two sides of the same coin. Pharmaceuticals 8(4):865–883

Bhatti JS, Bhatti GK, Reddy PH (2017) Mitochondrial dysfunction and oxidative stress in metabolic disorders—a step towards mitochondria based therapeutic strategies. Biochim Biophys Acta 1863(5):1066–1077

Birben E, Sahiner UM, Sackesen C, Erzurum S, Kalayci O (2012) Oxidative stress and antioxidant defense. World Allergy Organ J 5(1):9–19

Bishop NA, Lu T, Yankner BA (2010) Neural mechanisms of ageing and cognitive decline. Nature 464(7288):529–535

Boyanapalli SSS, Paredes-Gonzalez X, Fuentes F, Zhang C, Guo Y, Pung D, Saw CLL, Kong ANT (2014) Nrf2 knockout attenuates the anti-inflammatory effects of phenethyl isothiocyanate and curcumin. Chem Res Toxicol 27(12):2036–2043

Calabrese EJ (2014) Hormesis: a fundamental concept in biology. Microb Cell 1(5):145

Calabrese EJ, Kozumbo WJ (2021) The hormetic dose-response mechanism: Nrf2 activation. Pharmacol Res 167:105526

Calabrese EJ, Mattson MP (2017) How does hormesis impact biology, toxicology, and medicine? NPJ Aging Mech Dis 3(1):13

Calabrese V, Cornelius C, Dinkova-Kostova AT, Calabrese EJ, Mattson MP (2010) Cellular stress responses, the hormesis paradigm, and vitagenes: novel targets for therapeutic intervention in neurodegenerative disorders. Antioxid Redox Signal 13(11):1763–1811

Calabrese V, Cornelius C, Dinkova-Kostova AT, Iavicoli I, Di Paola R, Koverech A, Cuzzocrea S, Rizzarelli E, Calabrese EJ (2012) Cellular stress responses, hormetic phytochemicals and vitagenes in aging and longevity. Biochim Biophys Acta 1822(5):753–783

Calabrese E, Iavicoli I, Calabrese V (2013) Hormesis: its impact on medicine and health. Hum Exp Toxicol 32(2):120–152

Calder PC (2010) Omega-3 fatty acids and inflammatory processes. Nutrients 2(3):355–374

Cascella R, Cecchi C (2021) Calcium dyshomeostasis in Alzheimer's disease pathogenesis. Int J Mol Sci 22(9):4914

Chakrabarti S, Munshi S, Banerjee K, Thakurta IG, Sinha M, Bagh MB (2011) Mitochondrial dysfunction during brain aging: role of oxidative stress and modulation by antioxidant supplementation. Aging Dis 2(3):242–256

Chauhan A, Islam AU, Prakash H, Singh S (2022) Phytochemicals targeting NF-κB signaling: potential anti-cancer interventions. J Pharm Anal 12(3):394–405

Chen X, Guo C, Kong J (2012) Oxidative stress in neurodegenerative diseases. Neural Regen Res 7(5):376
Chen L, Deng H, Cui H, Fang J, Zuo Z, Deng J, Li Y, Wang X, Zhao L (2018) Inflammatory responses and inflammation-associated diseases in organs. Oncotarget 9(6):7204–7218
Chen XQ, Shen T, Fang SJ, Sun XM, Li GY, Li YF (2023) Protein homeostasis in aging and cancer. Front Cell Dev Biol 11:1143532
Chi H, Chang HY, Sang TK (2018) Neuronal cell death mechanisms in major neurodegenerative diseases. Int J Mol Sci 19(10):3082
Chovatiya R, Medzhitov R (2014) Stress, inflammation, and defense of homeostasis. Mol Cell 54(2):281–288
Cianciulli A, Calvello R, Ruggiero M, Panaro MA (2022) Inflammaging and brain: curcumin and its beneficial potential as regulator of microglia activation. Molecules 27(2):341
Colucci-D'Amato L, Speranza L, Volpicelli F (2020) Neurotrophic factor BDNF, physiological functions and therapeutic potential in depression, neurodegeneration and brain cancer. Int J Mol Sci 21(20):7777
Concetta Scuto M, Mancuso C, Tomasello B, Laura Ontario M, Cavallaro A, Frasca F, Maiolino L, TrovatoSalinaro A, Calabrese EJ, Calabrese V (2019) Curcumin, hormesis and the nervous system. Nutrients 11(10):2417
Cooper AJL, Hanigan MH (2018) Metabolism of glutathione S-conjugates: multiple pathways. In: Comprehensive toxicology, vol 10, 3rd edn, pp 363–406. https://doi.org/10.1016/B978-0-12-801238-3.01973-5
Corbi G, Conti V, Davinelli S, Scapagnini G, Filippelli A, Ferrara N (2016) Dietary phytochemicals in neuroimmunoaging: a new therapeutic possibility for humans? Front Pharmacol 7:364
Cornelius C, Perrotta R, Graziano A, Calabrese EJ, Calabrese V (2013) Stress responses, vitagenes and hormesis as critical determinants in aging and longevity: mitochondria as a "chi". Immun Ageing 10(1):15
Dattilo S, Mancuso C, Koverech G, Di MP, Ontario ML, Petralia CC, Petralia A, Maiolino L, Serra A, Calabrese EJ, Calabrese V (2015) Heat shock proteins and hormesis in the diagnosis and treatment of neurodegenerative diseases. Immun Ageing 12(1):20
da Silva AB, Coelho PLC, das Neves Oliveira M, Oliveira JL, Amparo JAO, da Silva KC, Costa SL (2020) The flavonoid rutin and its aglycone quercetin modulate the microglia inflammatory profile improving antiglioma activity. Brain Behav Immun 85:170–185
Davinelli S, Maes M, Corbi G, Zarrelli A, Willcox DC, Scapagnini G (2016) Dietary phytochemicals and neuro-inflammaging: from mechanistic insights to translational challenges. Immun Ageing 13:1–17
De Vries JX, Raedsch R, Völker U, Walter-Sack I, Weber E (1988) Biliary excretion of phenprocoumon and metabolites. Eur J Clin Pharmacol 35(4):433–436
Dei Cas M, Ghidoni R (2019) Dietary curcumin: correlation between bioavailability and health potential. Nutrients 11(9):2147
Deledda A, Giordano E, Velluzzi F, Flore G, Franceschelli S, Speranza L, Ripari P (2022) Mitochondrial aging and senolytic natural products with protective potential. Int J Mol Sci 23(24):16219
Doyle KM, Kennedy D, Gorman AM, Gupta S, Healy SJM, Samali A (2011) Unfolded proteins and endoplasmic reticulum stress in neurodegenerative disorders. J Cell Mol Med 15(10): 2025–2039
Duan X, Li Y, Xu F, Ding H (2021) Study on the neuroprotective effects of Genistein on Alzheimer's disease. Brain Behav 11(5):e02100
Dyall SC (2015) Long-chain omega-3 fatty acids and the brain: a review of the independent and shared effects of EPA, DPA and DHA. Front Aging Neuroscience 7:52
Fernandes GFS, Silva GDB, Pavan AR, Chiba DE, Chin CM, Dos Santos JL (2017) Epigenetic regulatory mechanisms induced by resveratrol. Nutrients 9(11):1201
Figueira I, Garcia G, Pimpão RC, Terrasso AP, Costa I, Almeida AF, Tavares L, Pais TF, Pinto P, Ventura MR, Filipe A, McDougall GJ, Stewart D, Kim KS, Palmela I, Brites D, Brito MA,

Brito C, Santos CN (2017) Polyphenols journey through blood-brain barrier towards neuronal protection. Sci Rep 7(1):11456
Fila M, Chojnacki C, Chojnacki J, Blasiak J (2021) Nutrients to improve mitochondrial function to reduce brain energy deficit and oxidative stress in migraine. Nutrients 13(12):4433
Forni C, Facchiano F, Bartoli M, Pieretti S, Facchiano A, D'Arcangelo D, Norelli S, Valle G, Nisini R, Beninati S, Tabolacci C, Jadeja RN (2019) Beneficial role of phytochemicals on oxidative stress and age-related diseases. Biomed Res Int 2019:8748253
Franco R, Navarro G, Martínez-Pinilla E (2019) Hormetic and mitochondria-related mechanisms of antioxidant action of phytochemicals. Antioxidants 8(9):373
Fulda S, Gorman AM, Hori O, Samali A (2010) Cellular stress responses: cell survival and cell death. Int J Cell Biol 2010:214074
García-Aguilar A, Palomino O, Benito M, Guillén C (2021) Dietary polyphenols in metabolic and neurodegenerative diseases: molecular targets in autophagy and biological effects. Antioxidants 10(2):142
García-García FJ, Monistrol-Mula A, Cardellach F, Garrabou G (2020) Nutrition, bioenergetics, and metabolic syndrome. Nutrients 12(9):2785
Gessner DK, Ringseis R, Eder K (2017) Potential of plant polyphenols to combat oxidative stress and inflammatory processes in farm animals. J Anim Physiol Anim Nutr 101(4):605–628
Gezer C (2018) Stress response of dietary phytochemicals in a hormetic manner for health and longevity. In: Gene expression and regulation in mammalian cells-transcription toward the establishment of novel therapeutics. IntechOpen, London
Ghosh S, Liu B, Zhou Z (2013) Resveratrol activates SIRT1 in a Lamin A-dependent manner. Cell Cycle 12(6):872–876
Gluhm S, Goldstein J, Loc K, Colt A, Liew CV, Corey-Bloom J (2013) Cognitive performance on the mini-mental state examination and the montreal cognitive assessment across the healthy adult lifespan. Cogn Behav Neurol 26(1):1–5
Gómez-Garduño J, León-Rodríguez R, Alemón-Medina R, Pérez-Guillé BE, Soriano-Rosales RE, González-Ortiz A, Rivera-Espinosa L (2022) Phytochemicals that interfere with drug metabolism and transport, modifying plasma concentration in humans and animals. Dose-Response 20(3):15593258221120485
Gonzales AM, Orlando RA (2008) Curcumin and resveratrol inhibit nuclear factor-kappaB-mediated cytokine expression in adipocytes. Nutr Metab 5:1–13
Gorman AM (2008) Neuronal cell death in neurodegenerative diseases: recurring themes around protein handling. J Cell Mol Med 12(6):2263–2280
Grabowska W, Sikora E, Bielak-Zmijewska A (2017) Sirtuins, a promising target in slowing down the ageing process. Biogerontology 18(4):447–476
Grimm A, Eckert A (2017) Brain aging and neurodegeneration: from a mitochondrial point of view. J Neurochem 143(4):418–431
Gu C, Yang X, Huang L (2016) Cistanches Herba: a neuropharmacology review. Front Pharmacol 7:289
Guo C, Sun L, Chen X, Zhang D (2013) Oxidative stress, mitochondrial damage and neurodegenerative diseases. Neural Regen Res 8(21):2003
Guo H, Cao H, Cui X, Zheng W, Wang S, Yu J, Chen Z (2019) Silymarin's inhibition and treatment effects for Alzheimer's disease. Molecules 24(9):1748
Gupta A, Singh AK, Kumar R, Jamieson S, Pandey AK, Bishayee A (2021) Neuroprotective potential of ellagic acid: a critical review. Adv Nutr 12(4):1211–1238
Gupte AA, Lyon CJ, Hsueh WA (2013) Nuclear factor (erythroid-derived 2)-like-2 factor (Nrf2), a key regulator of the antioxidant response to protect against atherosclerosis and nonalcoholic steatohepatitis. Curr Diab Rep 13(3):362–371
Hajialyani M, Hosein Farzaei M, Echeverría J, Nabavi SM, Uriarte E, Sobarzo-Sánchez E (2019) Hesperidin as a neuroprotective agent: a review of animal and clinical evidence. Molecules 24(3):648

Hannan MA, Dash R, Sohag AAM, Haque MN, Moon IS (2020) Neuroprotection against oxidative stress: phytochemicals targeting trkb signaling and the nrf2-are antioxidant system. Front Mol Neurosci 13:116

Hasday JD, Singh IS (2000) Fever and the heat shock response: distinct, partially overlapping processes. Cell Stress Chaperones 5(5):471

Haskell-Ramsay C, Schmitt J, Actis-Goretta L (2018) The impact of epicatechin on human cognition: the role of cerebral blood flow. Nutrients 10(8):986

He F, Ru X, Wen T (2020) NRF2, a transcription factor for stress response and beyond. Int J Mol Sci 21(13):4777

Hewlings S, Kalman D (2017) Curcumin: a review of its effects on human health. Foods 6(10):92

Hordyjewska A, Ostapiuk A, Horecka A, Kurzepa J (2019) Betulin and betulinic acid: triterpenoids derivatives with a powerful biological potential. Phytochem Rev 18:929–951

Houghton CA, Fassett RG, Coombes JS (2016) Sulforaphane and other nutrigenomic Nrf2 activators: can the clinician's expectation be matched by the reality? Oxid Med Cell Longev 2016:7857186

Ighodaro OM, Akinloye OA (2018) First line defence antioxidants-superoxide dismutase (Sod), catalase (Cat) and glutathione peroxidase (Gpx): their fundamental role in the entire antioxidant defence grid. Alexandria J Med 54(4):287–293

Imran M, Ghorat F, Ul-Haq I, Ur-Rehman H, Aslam F, Heydari M, Rebezov M (2020) Lycopene as a natural antioxidant used to prevent human health disorders. Antioxidants 9(8):706

Ioniță-Mîndrican CB, Ziani K, Mititelu M, Oprea E, Neacșu SM, Moroșan E, Negrei C (2022) Therapeutic benefits and dietary restrictions of fiber intake: a state of the art review. Nutrients 14(13):2641

Jaiswal YS, Williams LL (2017) A glimpse of Ayurveda – the forgotten history and principles of Indian traditional medicine. J Tradit Complement Med 7(1):50–53

Jayakumar T, Lin KC, Chang CC, Hsia CW, Manubolu M, Huang WC, Sheu JR, Hsia CH (2021) Targeting mapk/nf-κb pathways in anti-inflammatory potential of rutaecarpine: impact on src/fak-mediated macrophage migration. Int J Mol Sci 23(1):92

Jęśko H, Wencel P, Strosznajder RP, Strosznajder JB (2017) Sirtuins and their roles in brain aging and neurodegenerative disorders. Neurochem Res 42(3):876–890

Jodynis-Liebert J, Kujawska M (2020) Biphasic dose-response induced by phytochemicals: experimental evidence. J Clin Med 9(3):718

Kandlur A, Satyamoorthy K, Gangadharan G (2020) Oxidative stress in cognitive and epigenetic aging: a retrospective glance. Front Mol Neurosci 13:41

Kang TC (2020) Nuclear factor-erythroid 2-related factor 2 (Nrf2) and mitochondrial dynamics/mitophagy in neurological diseases. Antioxidants 9(7):617

Khan MF, Alqahtani AS, Almarfadi OM, Ullah R, Nasr FA, Noman OM, Ahamad SR (2021) The reproductive toxicity associated with Dodonaea viscosa, a folk medicinal plant in Saudi Arabia. Evid Based Complement Alternat Med 2021:6689110

Kiruthiga C, Devi KP, Nabavi SM, Bishayee A (2020) Autophagy: a potential therapeutic target of polyphenols in hepatocellular carcinoma. Cancers 12(3):562

Kou X, Kirberger M, Yang Y, Chen N (2013) Natural products for cancer prevention associated with Nrf2–ARE pathway. Food Sci Human Wellness 2(1):22–28

Kumar G, Khanum F (2012) Neuroprotective potential of phytochemicals. Pharmacogn Rev 6(12):81

Kumar A, Kumar M, Jose A, Tomer V, Oz E, Proestos C, Oz F (2023) Major phytochemicals: recent advances in health benefits and extraction method. Molecules 28(2):887

Kurutas EB (2015) The importance of antioxidants which play the role in cellular response against oxidative/nitrosative stress: current state. Nutr J 15(1):1–22

Lagouge M, Argmann C, Gerhart-Hines Z, Meziane H, Lerin C, Daussin F, Auwerx J (2006) Resveratrol improves mitochondrial function and protects against metabolic disease by activating SIRT1 and PGC-1α. Cell 127(6):1109–1122

Lambert JD, Kennett MJ, Sang S, Reuhl KR, Ju J, Yang CS (2010) Hepatotoxicity of high oral dose (−)-epigallocatechin-3-gallate in mice. Food Chem Toxicol 48(1):409–416
Landucci E, Pellegrini-Giampietro DE, Bilia AR, Bergonzi MC (2019) Enhanced neuroprotective effects of panax ginseng g115® and ginkgo biloba gk501® combinations in vitro models of excitotoxicity. Int J Mol Sci 20(23):5872
Leak RK (2014) Heat shock proteins in neurodegenerative disorders and aging. J Cell Commun Signal 8(4):293–310
Lee J, Jo DG, Park D, Chung HY, Mattson MP (2014) Adaptive cellular stress pathways as therapeutic targets of dietary phytochemicals: focus on the nervous system. Pharmacol Rev 66(3):815–868
Lee J, Hong SW, Kwon H, Park SE, Rhee EJ, Park CY, Oh KW, Park SW, Lee WY (2019) Resveratrol, an activator of SIRT1, improves ER stress by increasing clusterin expression in HepG2 cells. Cell Stress Chaperones 24(4):825–833
Leri M, Scuto M, Ontario ML, Calabrese V, Calabrese EJ, Bucciantini M, Stefani M (2020) Healthy effects of plant polyphenols: molecular mechanisms. Int J Mol Sci 21(4):1250
Li LR, Sethi G, Zhang X, Liu CL, Huang Y, Liu Q, Tang FR (2022a) The neuroprotective effects of icariin on ageing, various neurological, neuropsychiatric disorders, and brain injury induced by radiation exposure. Aging (Albany NY) 14(3):1562
Li Y, Li S, Wu H (2022b) Ubiquitination-proteasome system (Ups) and autophagy two main protein degradation machineries in response to cell stress. Cells 11(5):851
Liguori I, Russo G, Curcio F, Bulli G, Aran L, DellaMD AP (2018) Oxidative stress, aging, and diseases. Clin Interv Aging 13:757–772
Limanaqi F, Biagioni F, Mastroiacovo F, Polzella M, Lazzeri G, Fornai F (2020) Merging the multi-target effects of phytochemicals in neurodegeneration: from oxidative stress to protein aggregation and inflammation. Antioxidants 9(10):1022
Lin JH, Walter P, Yen TSB (2008) Endoplasmic reticulum stress in disease pathogenesis. Annu Rev Pathol 3(1):399–425
Lindholm D, Wootz H, Korhonen L (2006) ER stress and neurodegenerative diseases. Cell Death Differ 13(3):385–392
Lindholm D, Korhonen L, Eriksson O, Kõks S (2017) Recent insights into the role of unfolded protein response in er stress in health and disease. Front Cell Dev Biol 5:48
Liu Y, Chen Z, Li A, Liu R, Yang H, Xia X (2022) The phytochemical potential for brain disease therapy and the possible nanodelivery solutions for brain access. Front Oncol 12:936054
Lobo V, Patil A, Phatak A, Chandra N (2010) Free radicals, antioxidants and functional foods: impact on human health. Pharmacogn Rev 4(8):118
Lovinger DM (2008) Communication networks in the brain: neurons, receptors, neurotransmitters, and alcohol. Alcohol Res Health 31(3):196–214
Lushchak VI, Duszenko M, Gospodaryov DV, Garaschuk O (2021) Oxidative stress and energy metabolism in the brain: midlife as a turning point. Antioxidants 10(11):1715
Ma Q (2013) Role of nrf2 in oxidative stress and toxicity. Annu Rev Pharmacol Toxicol 53(1): 401–426
Madabhushi R, Pan L, Tsai LH (2014) Dna damage and its links to neurodegeneration. Neuron 83(2):266–282
Maiti P, Dunbar G (2018) Use of curcumin, a natural polyphenol for targeting molecular pathways in treating age-related neurodegenerative diseases. Int J Mol Sci 19(6):1637
Mao L, Franke J (2013) Hormesis in aging and neurodegeneration—a prodigy awaiting dissection. Int J Mol Sci 14(7):13109–13128
Marrocco I, Altieri F, Peluso I (2017) Measurement and clinical significance of biomarkers of oxidative stress in humans. Oxid Med Cell Longev 2017:6501046
Martel J, Ojcius DM, Ko YF, Ke PY, Wu CY, Peng HH, Young JD (2019) Hormetic effects of phytochemicals on health and longevity. Trends Endocrinol Metab 30(6):335–346
Martelli AM, Evangelisti C, Chiarini F, Grimaldi C, Cappellini A, Ognibene MCJA (2010) The emerging role of the phosphatidylinositol 3-kinase/Akt/mammalian target of rapamycin

signaling network in normal myelopoiesis and leukemogenesis. Biochim Biophys Acta 1803(9): 991–1002
Martino G, Pluchino S, Bonfanti L, Schwartz M (2011) Brain regeneration in physiology and pathology: the immune signature driving therapeutic plasticity of neural stem cells. Physiol Rev 91(4):1281–1304
Mattson MP (2008) Hormesis defined. Ageing Res Rev 7(1):1–7
Medina Dos Santos N, Berilli Batista P, Batista ÂG, Maróstica Júnior MR (2019) Current evidence on cognitive improvement and neuroprotection promoted by anthocyanins. Curr Opin Food Sci 26:71–78
Mensch A, Zierz S (2020) Cellular stress in the pathogenesis of muscular disorders—from cause to consequence. Int J Mol Sci 21(16):5830
Michael GJ, Averill S, Nitkunan A, Rattray M, Bennett DLH, Yan Q, Priestley JV (1997) Nerve growth factor treatment increases brain-derived neurotrophic factor selectively in trka-expressing dorsal root ganglion cells and in their central terminations within the spinal cord. J Neurosci 17(21):8476–8490
Mihaylova MM, Shaw RJ (2011) The AMPK signalling pathway coordinates cell growth, autophagy and metabolism. Nat Cell Biol 13(9):1016–1023
Miranda M, Morici JF, Zanoni MB, Bekinschtein P (2019) Brain-derived neurotrophic factor: a key molecule for memory in the healthy and the pathological brain. Front Cell Neurosci 13:363
Misrani A, Tabassum S, Yang L (2021) Mitochondrial dysfunction and oxidative stress in Alzheimer's disease. Front Aging Neurosci 13:617588
Mitra S, Chakraborty AJ, Tareq AM, Emran TB, Nainu F, Khusro A, Idris AM, Khandaker MU, Osman H, Alhumaydhi FA, Simal GJ (2022) Impact of heavy metals on the environment and human health: novel therapeutic insights to counter the toxicity. J King Saud Univ Sci 34(3): 101865
Mittal M, Siddiqui MR, Tran K, Reddy SP, Malik AB (2014) Reactive oxygen species in inflammation and tissue injury. Antioxid Redox Signal 20(7):1126–1167
Mohd Sahardi NFN, Makpol S (2019) Ginger (Zingiber officinale Roscoe) in the prevention of ageing and degenerative diseases: review of current evidence. Evid Based Complement Alternat Med 2019:5054395
Mokra D, Joskova M, Mokry J (2022) Therapeutic effects of green tea polyphenol (–)-epigallocatechin-3-gallate (EGCG) in relation to molecular pathways controlling inflammation, oxidative stress, and apoptosis. Int J Mol Sci 24(1):340
Morita T, Asada M, Naito E (2016) Contribution of neuroimaging studies to understanding development of human cognitive brain functions. Front Hum Neurosci 10:464
Moura C, Lollo P, Morato P, Amaya FJ (2018) Dietary nutrients and bioactive substances modulate heat shock protein (Hsp) expression: a review. Nutrients 10(6):683
Mucha P, Skoczyńska A, Małecka M, Hikisz P, Budzisz E (2021) Overview of the antioxidant and anti-inflammatory activities of selected plant compounds and their metal ions complexes. Molecules 26(16):4886
Muniyappa H, Das KC (2008) Activation of c-Jun N-terminal kinase (JNK) by widely used specific p38 MAPK inhibitors SB202190 and SB203580: a MLK-3-MKK7-dependent mechanism. Cell Signal 20(4):675–683
Murakami A (2013) Modulation of protein quality control systems by food phytochemicals. J Clin Biochem Nutr 52(3):215–227
Murugaiyah V, Mattson MP (2015) Neurohormetic phytochemicals: an evolutionary–bioenergetic perspective. Neurochem Int 89:271–280
Navarrete-Yañez V, Garate-Carrillo A, Rodriguez A, Mendoza-Lorenzo P, Ceballos G, Calzada-Mendoza C, Hogan MC, Villarreal F, Ramirez-Sanchez I (2020) Effects of (−)-epicatechin on neuroinflammation and hyperphosphorylation of tau in the hippocampus of aged mice. Food Funct 11(12):10351–10361

Navarro-Cruz AR, Ramírez y Ayala R, Ochoa-Velasco C, Brambila E, Avila-Sosa R, Pérez-Fernández S, Aguilar-Alonso P (2017) Effect of chronic administration of resveratrol on cognitive performance during aging process in rats. Oxid Med Cell Longev 2017:8510761

Ngo V, Duennwald ML (2022) Nrf2 and oxidative stress: a general overview of mechanisms and implications in human disease. Antioxidants 11(12):2345

Nishihira J, Nishimura M, Kurimoto M, Kagami-Katsuyama H, Hattori H, Nakagawa T, Kobori M (2021) The effect of 24-week continuous intake of quercetin-rich onion on age-related cognitive decline in healthy elderly people: a randomized, double-blind, placebo-controlled, parallel-group comparative clinical trial. J Clin Biochem Nutr 69(2):203–215

Pandey KB, Rizvi SI (2009) Plant polyphenols as dietary antioxidants in human health and disease. Oxid Med Cell Longev 2(5):270–278

Payne A, Nahashon S, Taka E, Adinew GM, Soliman KF (2022) Epigallocatechin-3-Gallate (EGCG): new therapeutic perspectives for neuroprotection, aging, and neuroinflammation for the modern age. Biomolecules 12(3):371

Penke B, Bogár F, Crul T, Sántha M, Tóth M, Vígh L (2018) Heat shock proteins and autophagy pathways in neuroprotection: from molecular bases to pharmacological interventions. Int J Mol Sci 19(1):325

Peters R (2006) Ageing and the brain. Postgrad Med J 82(964):84–88

Petrovic S, Arsic A, Ristic MD, Cvetkovic Z, Vucic V (2020) Lipid peroxidation and antioxidant supplementation in neurodegenerative diseases: a review of human studies. Antioxidants 9(11): 1128

Picard M, McEwen BS (2018) Psychological stress and mitochondria: a conceptual framework. Psychosom Med 80(2):126–140

Picca A, Calvani R, Coelho-Junior HJ, Landi F, Bernabei R, Marzetti E (2020) Mitochondrial dysfunction, oxidative stress, and neuroinflammation: intertwined roads to neurodegeneration. Antioxidants 9(8):647

Pizzino G, Irrera N, Cucinotta M, Pallio G, Mannino F, Arcoraci V, Squadrito F, Altavilla D, Bitto A (2017) Oxidative stress: harms and benefits for human health. Oxid Med Cell Longev 2017: 8416763

Poljsak B (2011) Strategies for reducing or preventing the generation of oxidative stress. Oxid Med Cell Longev 2011:194586

Qiao J, Wang C, Chen Y, Yu S, Liu Y, Yu S, Jiang L, Jin C, Wang X, Zhang P, Zhao D, Wang J, Liu M (2023) Herbal/natural compounds resist hallmarks of brain aging: from molecular mechanisms to therapeutic strategies. Antioxidants 12(4):920

Rahman MA, Rahman MH, Biswas P, Hossain MS, Islam R, Hannan MA, Uddin MJ, Rhim H (2020) Potential therapeutic role of phytochemicals to mitigate mitochondrial dysfunctions in Alzheimer's disease. Antioxidants 10(1):23

Rao RV, Bredesen DE (2004) Misfolded proteins, endoplasmic reticulum stress and neurodegeneration. Curr Opin Cell Biol 16(6):653–662

Remondelli P, Renna M (2017) The endoplasmic reticulum unfolded protein response in neurodegenerative disorders and its potential therapeutic significance. Front Mol Neurosci 10:187

Reuter S, Gupta SC, Park B, Goel A, Aggarwal BB (2011) Epigenetic changes induced by curcumin and other natural compounds. Genes Nutr 6(2):93–108

Ristow M, Schmeisser K (2014) Mitohormesis: promoting health and lifespan by increased levels of reactive oxygen species (ROS). Dose Response 12(2):288

Sahebnasagh A, Eghbali S, Saghafi F, Sureda A, Avan R (2022) Neurohormetic phytochemicals in the pathogenesis of neurodegenerative diseases. Immun Ageing 19(1):36

Saleh HA, Yousef MH, Abdelnaser A (2021) The anti-inflammatory properties of phytochemicals and their effects on epigenetic mechanisms involved in tlr4/nf-κb-mediated inflammation. Front Immunol 12:606069

Salehi B, Mishra A, Nigam M, Sener B, Kilic M, Sharifi-Rad M, Fokou P, Martins N, Sharifi-Rad J (2018) Resveratrol: a double-edged sword in health benefits. Biomedicines 6(3):91

Salehi B, Venditti A, Sharifi-Rad M, Kręgiel D, Sharifi-Rad J, Durazzo A, Martins N (2019) The therapeutic potential of apigenin. Int J Mol Sci 20(6):1305

Salehi B, Machin L, Monzote L, Sharifi-Rad J, Ezzat SM, Salem MA, Merghany RM, El Mahdy NM, Kılıç CS, Sytar O, Sharifi-Rad M, Sharopov F, Martins N, Martorell M, Cho WC (2020) Therapeutic potential of quercetin: new insights and perspectives for human health. ACS Omega 5(20):11849–11872

Salim S (2017) Oxidative stress and the central nervous system. J Pharmacol Exp Ther 360(1): 201–205

Salminen A, Kaarniranta K (2012) AMP-activated protein kinase (AMPK) controls the aging process via an integrated signaling network. Ageing Res Rev 11(2):230–241

Sánchez-Martínez JD, Valdés A, Gallego R, Suárez-Montenegro ZJ, Alarcón M, Ibañez E, Alvarez-Rivera G, Cifuentes A (2022) Blood–brain barrier permeability study of potential neuroprotective compounds recovered from plants and agri-food by-products. Front Nutr 9: 924596

Santín-Márquez R, Alarcón-Aguilar A, López-Diazguerrero NE, Chondrogianni N, Königsberg M (2019) Sulforaphane—role in aging and neurodegeneration. GeroScience 41(5):655–670

Satyam SM, Bairy LK (2022) Neuronutraceuticals combating neuroinflammaging: molecular insights and translational challenges—a systematic review. Nutrients 14(15):3029

Sazonova MA, Sinyov VV, Ryzhkova AI, Sazonova MD, Kirichenko TV, Khotina VA, Khasanova ZB, Doroschuk NA, Karagodin VP, Orekhov AN, Sobenin IA (2021) Some molecular and cellular stress mechanisms associated with neurodegenerative diseases and atherosclerosis. Int J Mol Sci 22(2):699

Schieber M, Chandel NS (2014) Ros function in redox signaling and oxidative stress. Curr Biol 24(10):453–462

Schirrmacher V (2021) Less can be more: the hormesis theory of stress adaptation in the global biosphere and its implications. Biomedicines 9(3):293

Shahcheraghi SH, Salemi F, Peirovi N, Ayatollahi J, Alam W, Khan H, Saso L (2021) Nrf2 regulation by curcumin: molecular aspects for therapeutic prospects. Molecules 27(1):167

Shan Y, Wang Z, Song S, Xue Q, Ge Q, Yang H, Cui B, Zhang M, Zhou Y, Lu J (2022) Integrated positron emission tomography/magnetic resonance imaging for resting-state functional and metabolic imaging in human brain: what is correlated and what is impacted. Front Neurosci 16:824152

Sharifi-Rad M, Anil Kumar NV, Zucca P, Varoni EM, Dini L, Panzarini E, Rajkovic J, Tsouh Fokou PV, Azzini E, Peluso I, Prakash Mishra A, Nigam M, El Rayess Y, Beyrouthy ME, Polito L, Iriti M, Martins N, Martorell M, Docea AO, Sharifi-Rad J (2020) Lifestyle, oxidative stress, and antioxidants: back and forth in the pathophysiology of chronic diseases. Front Physiol 11:694

Sharma R, Padwad Y (2020) Nutraceuticals-based immunotherapeutic concepts and opportunities for the mitigation of cellular senescence and aging: a narrative review. Ageing Res Rev 63: 101141

Shih RH, Wang CY, Yang CM (2015) NF-kappaB signaling pathways in neurological inflammation: a mini review. Front Mol Neurosci 8:77

Si H, Liu D (2014) Dietary antiaging phytochemicals and mechanisms associated with prolonged survival. J Nutr Biochem 25(6):581–591

Silva dos Santos J, Goncalves Cirino JP, de Oliveira CP, Ortega MM (2021) The pharmacological action of kaempferol in central nervous system diseases: a review. Front Pharmacol 11:565700

Silvestro S, Sindona C, Bramanti P, Mazzon E (2021) A state of the art of antioxidant properties of curcuminoids in neurodegenerative diseases. Int J Mol Sci 22(6):3168

Singh BN, Shankar S, Srivastava RK (2011a) Green tea catechin, epigallocatechin-3-gallate (EGCG): mechanisms, perspectives and clinical applications. Biochem Pharmacol 82(12): 1807–1821

Singh N, Bhalla M, De JP, Gilca M (2011b) An overview on ashwagandha: a rasayana (Rejuvenator) of ayurveda. Afr J Tradit Complement Altern Med 8(5S):208

Singh A, Kukreti R, Saso L, Kukreti S (2019) Oxidative stress: a key modulator in neurodegenerative diseases. Molecules 24(8):1583

Sivandzade F, Prasad S, Bhalerao A, Cucullo L (2019) NRF2 and NF-κB interplay in cerebrovascular and neurodegenerative disorders: molecular mechanisms and possible therapeutic approaches. Redox Biol 21:101059

Smoliga JM, Vang O, Baur JA (2012) Challenges of translating basic research into therapeutics: resveratrol as an example. J Gerontol A Biomed Sci Med Sci 67(2):158–167

Socała K, Szopa A, Serefko A, Poleszak E, Wlaź P (2020) Neuroprotective effects of coffee bioactive compounds: a review. Int J Mol Sci 22(1):107

Sohn SI, Priya A, Balasubramaniam B, Muthuramalingam P, Sivasankar C, Selvaraj A, Valliammai A, Jothi R, Pandian S (2021) Biomedical applications and bioavailability of curcumin—an updated overview. Pharmaceutics 13(12):2102

Somero GN (2020) The cellular stress response and temperature: function, regulation, and evolution. J Exp Zool A Ecol Integr Physiol 333(6):379–397

Son TG, Camandola S, Mattson MP (2008) Hormetic dietary phytochemicals. Neuromolecular Med 10(4):236–246

Speciale A, Cimino F, Saija A, Canali R, Virgili F (2014) Bioavailability and molecular activities of anthocyanins as modulators of endothelial function. Genes Nutr 9(4):404

Springer M, Moco S (2019) Resveratrol and its human metabolites—effects on metabolic health and obesity. Nutrients 11(1):143

Sproston NR, Ashworth JJ (2018) Role of c-reactive protein at sites of inflammation and infection. Front Immunol 9:754

Stetler RA, Gan Y, Zhang W, Liou AK, Gao Y, Cao G, Chen J (2010) Heat shock proteins: cellular and molecular mechanisms in the central nervous system. Prog Neurobiol 92(2):184–211

Sweeney P, Park H, Baumann M, Dunlop J, Frydman J, Kopito R, McCampbell A, Leblanc G, Venkateswaran A, Nurmi A, Hodgson R (2017) Protein misfolding in neurodegenerative diseases: implications and strategies. Transl Neurodegener 6(1):6

Tedeschi P, Nigro M, Travagli A, Catani M, Cavazzini A, Merighi S, Gessi S (2022) Therapeutic potential of allicin and aged garlic extract in Alzheimer's disease. Int J Mol Sci 23(13):6950

Teleanu RI, Niculescu AG, Roza E, Vladâcenco O, Grumezescu AM, Teleanu DM (2022) Neurotransmitters—key factors in neurological and neurodegenerative disorders of the central nervous system. Int J Mol Sci 23(11):5954

Thornton T, Mills D, Bliss E (2023) Capsaicin: a potential treatment to improve cerebrovascular function and cognition in obesity and ageing. Nutrients 15(6):1537

Tian W, Rojo De La Vega M, Schmidlin CJ, Ooi A, Zhang DD (2018) Kelch-like ECH-associated protein 1 (Keap1) differentially regulates nuclear factor erythroid-2–related factors 1 and 2 (Nrf1 and nrf2). J Biol Chem 293(6):2029–2040

Traber MG, Stevens JF (2011) Vitamins C and E: beneficial effects from a mechanistic perspective. Free Radic Biol Med 51(5):1000–1013

Tungmunnithum D, Thongboonyou A, Pholboon A, Yangsabai A (2018) Flavonoids and other phenolic compounds from medicinal plants for pharmaceutical and medical aspects: an overview. Medicines 5(3):93

Turner PV, Brabb T, Pekow C, Vasbinder MA (2011) Administration of substances to laboratory animals: routes of administration and factors to consider. J Am Assoc Lab Anim Sci 50(5): 600–613

Ullah A, Munir S, Badshah SL, Khan N, Ghani L, Poulson BG, Emwas AH, Jaremko M (2020) Important flavonoids and their role as a therapeutic agent. Molecules 25(22):5243

Uttara B, Singh AV, Zamboni P, Mahajan R (2009) Oxidative stress and neurodegenerative diseases: a review of upstream and downstream antioxidant therapeutic options. Curr Neuropharmacol 7(1):65–74

Vaiserman A, Koliada A, Zayachkivska A, Lushchak O (2020) Nanodelivery of natural antioxidants: an anti-aging perspective. Front Bioeng Biotechnol 7:447

Vauzour D (2012) Dietary polyphenols as modulators of brain functions: biological actions and molecular mechanisms underpinning their beneficial effects. Oxid Med Cell Longev 2012: 914273

Venkatesan R, Ji E, Kim SY (2015) Phytochemicals that regulate neurodegenerative disease by targeting neurotrophins: a comprehensive review. BioMed Res Int 2015:814068

Vesely O, Baldovska S, Kolesarova A (2021) Enhancing bioavailability of nutraceutically used resveratrol and other stilbenoids. Nutrients 13(9):3095

Voth W, Jakob U (2017) Stress-activated chaperones: a first line of defense. Trends Biochem Sci 42(11):899–913

Wang SJ, Zhao XH, Chen W, Bo N, Wang XJ, Chi ZF, Wu W (2015) Sirtuin 1 activation enhances the PGC-1α/mitochondrial antioxidant system pathway in status epilepticus. Mol Med Rep 11(1):521–526

Webster JM, Darling AL, Uversky VN, Blair LJ (2019) Small heat shock proteins, big impact on protein aggregation in neurodegenerative disease. Front Pharmacol 10:1047

Wellen KE, Thompson CB (2010) Cellular metabolic stress: considering how cells respond to nutrient excess. Mol Cell 40(2):323–332

Więdłocha M, Zborowska N, Marcinowicz P, Dębowska W, Dębowska M, Zalewska A, Maciejczyk M, Waszkiewicz N, Szulc A (2023) Oxidative stress biomarkers among schizophrenia inpatients. Brain Sci 13(3):490

Xu Z, Feng W, Shen Q, Yu N, Yu K, Wang S, Guo Y (2017) Rhizoma coptidis and berberine as a natural drug to combat aging and aging-related diseases via anti-oxidation and AMPK activation. Aging Dis 8(6):760

Yang YY, Tsai TH (2019) Enterohepatic circulation and pharmacokinetics of genistin and genistein in rats. ACS Omega 4(19):18428–18433

Yenari MA (2002) Heat shock proteins and neuroprotection. Adv Exp Med Biol 13:281–299

Yoon S, Iqbal H, Kim SM, Jin M (2023) Phytochemicals that act on synaptic plasticity as potential prophylaxis against stress-induced depressive disorder. Biomol Ther 31(2):148–160

Young PNE, Estarellas M, Coomans E, Srikrishna M, Beaumont H, Maass A, Venkataraman AV, Lissaman R, Jiménez D, Betts MJ, McGlinchey E, Berron D, O'Connor A, Fox NC, Pereira JB, Jagust W, Carter SF, Paterson RW, Schöll M (2020) Imaging biomarkers in neurodegeneration: current and future practices. Alzheimers Res Ther 12(1):49

Yu C, Xiao JH (2021) The keap1-nrf2 system: a mediator between oxidative stress and aging. Oxid Med Cell Longev 2021:6635460

Zhang M, An C, Gao Y, Leak RK, Chen J, Zhang F (2013) Emerging roles of Nrf2 and phase II antioxidant enzymes in neuroprotection. Prog Neurobiol 100:30–47

Zhang YJ, Gan RY, Li S, Zhou Y, Li AN, Xu DP, Li HB (2015) Antioxidant phytochemicals for the prevention and treatment of chronic diseases. Molecules 20(12):21138–21156

Zhang L, ChengX XS, Bao J, Yu H (2018) Curcumin induces endoplasmic reticulum stress-associated apoptosis in human papillary thyroid carcinoma BCPAP cells via disruption of intracellular calcium homeostasis. Medicine 97(24):11095

Zhuang W, Yue L, Dang X, Chen F, Gong Y, Lin X, Luo Y (2019) Rosenroot (Rhodiola): potential applications in aging-related diseases. Aging Dis 10(1):134

Zorov DB, Juhaszova M, Sollott SJ (2014) Mitochondrial reactive oxygen species (Ros) and ros-induced ros release. Physiol Rev 94(3):909–950

Zügel U, Kaufmann SHE (1999) Role of heat shock proteins in protection from and pathogenesis of infectious diseases. Clin Microbiol Rev 12(1):19–39

# Synergistic Effects of Multifunctional Phytochemicals Modulating the Cellular Signal Pathways – as Neuroprotective Therapeutic Approach for Neurodegenerative Disorders

5

Rajalakshmi Manikkam, Indu Sabapathy, and Vijayalakshmi Periyasamy

**Abstract**

In the recent years, cognitive dysfunction and neuropathological diseases are significant public health concerns, and such neurodegenerative diseases (NDDs), including anxiety, depression dementia, Alzheimer's and Parkinson's diseases, seizures and ischaemic stroke disorders, can be severely functionally crippling as well as represent the major global cause of death. Furthermore, the lack of effective therapeutic alternatives creates challenges in minimizing the neuronal damages brought by the NDDs. There is a surge of interest in herbal medications, which include phytochemical components that may have long-term health promoting or therapeutic effects. Neural problems have traditionally been treated with herbal medication, and several of them have been demonstrated to have antioxidant and/or anti-inflammatory effects, now being explored to have neuroprotective potentials against central nervous system disorders. This chapter discusses the significance of phytochemicals for neuroprotective therapeutic potentials and other related disorders.

R. Manikkam (✉)
PG and Research Department of Zoology, Holy Cross College (Autonomous), Affiliated to Bharathidasan University, Tiruchirappalli, Tamil Nadu, India

PG and Research Department of Biotechnology and Bioinformatics, DBT-BIF Centre, Holy Cross College (Autonomous), Affiliated to Bharathidasan University, Tiruchirappalli, Tamil Nadu, India
e-mail: rajalakshmi@hcctrichy.ac.in

I. Sabapathy · V. Periyasamy
PG and Research Department of Biotechnology and Bioinformatics, DBT-BIF Centre, Holy Cross College (Autonomous), Affiliated to Bharathidasan University, Tiruchirappalli, Tamil Nadu, India



S. Pathak, A. Banerjee (eds.), *Neuroprotective Effects of Phytochemicals in Brain Ageing*, https://doi.org/10.1007/978-981-99-7269-2_5

**Keywords**

Neurodegenerative diseases · Alzheimer's and Parkinson's diseases · Seizures and ischemic stroke · Herbal medicines · Phytochemicals

## 5.1 Introduction

Neurodegeneration may result from the progressive abnormalities of the peripheral and central nervous systems (CNSs). Although several factors are known to directly contribute to the beginning of neurodegeneration, the key causative element is the production of free radicals by reactive nitrogen species (RNS), reactive oxygen species (ROS) and neuroinflammatory processes. Over 600 neurological illnesses have been documented globally (Di Meo et al. 2016). By the year 2040, neurodegenerative disorders are anticipated as a leading cause of death in the older age group (Mensah-Kane and Sumien 2023). The term "neuroprotection" implies the mechanisms that can protect the CNS from the neuronal damage caused by neuropsychiatric and neurodegenerative disorders, including Parkinson's disease (PD), Alzheimer's disease (AD), cerebrovascular impairment, seizures, anxiety and stress (Kumar 2006). Numerous types of synthetic and natural neuroprotective agents have been suggested as a cure for neurodegenerative disorders; however, the synthetic neuroprotective agents may cause adverse effects such as fatigue, tiredness, lethargy, anxiety, depression and imbalances. Herbal medicine, also known as phytotherapy, is the practice of using certain plant parts for therapeutic purposes, and in recent years, there has been a lot of attention paid to possible phytochemicals that can modify neuronal function and act as a preventative measure against neurodegeneration. This chapter will delve into the synergistic effects of multifunctional phytochemicals and their potential to modulate cellular signaling pathways involved in neurodegeneration.

## 5.2 Neurodegenerative Diseases (NDDs) Affecting the Cognitive Functions

Parkinson's disease (PD) is the most common type of human brain degenerative disorder that causes a major motor disability (Mhyre et al. 2012). The most ubiquitous cause of the recognizable symptoms of PD is the death of dopaminergic neurons in the substantia nigra and deficiency in cholinergic neurotransmission. A neurodegenerative condition called AD caused by mitochondrial dysfunction is characterized by impairment of memory and learning as well as cognitive disorders like depression, agitation and psychosis. AD is characterized by two distinct medical hallmarks: neurofibrillary tangles (NFTs) and hyperphosphorylation of tau proteins produce intracellular deposits of amyloid-(Aβ) peptide and senile plaques, which are extracellular deposits of Aβ peptide (Murphy and LeVine 3rd. 2010). Amyloid protein aggregation, mitochondrial dysfunction, proteolytic stress, oxidative stress, neuroinflammation, calcium load, glutamate toxicity and ageing are the molecular and cellular pathologies that are common to NDDs (Gan et al. 2018). The most prevalent neurological condition, epilepsy, affects about 50 million people globally. It is characterized by the sporadic occurrence of epileptic seizures or the recurrent involuntary contraction of striated muscle. A chain of events that results in neuronal damage and cell death in epilepsy may be sparked by abnormal voltage-sensitive ion channel function ($Na^+/Ca^{2+}$ ion channels), deprived Gamma-aminobutyric acid (GABA)-mediated inhibition and excessive glutamate-mediated neurotransmission (Cano et al. 2021). Degeneration of the spinal motor neurons of the cerebral cortex (upper) and those of the brainstem (lower) occurs in a fatal neurodegenerative condition, ALS (Marques et al. 2021). In reported cases, ALS occurred sporadically while a meager population of ALS is caused by superoxide dismutase gene mutations. ALS is characterized by mitochondrial dysfunction, ER stress, ROS-mediated oxidative stress, lipid peroxidation and changes in mitochondrial DNA and neuroinflammation (Chen et al. 2012). The genetic mutation in the huntingtin (HTT) gene causes HD and begins with neuronal degeneration in the striatum before causing damage to the cerebral cortex and thalamus. This causes the mutant huntingtin protein to have an excessively long repeat of polyglutathione. Personality changes, psychiatric disorders, and reduced motor and cognitive abilities are the key characteristics of HD (Sari 2011).

## 5.3 Role of Risk Factors in Neurodegenerative Diseases

There are specific pathological characteristics that distinguish the NDDs, such as Alzheimer's (AD), Parkinson's (PD), Huntington's (HD) and amyotrophic lateral sclerosis (ALS), including abnormal protein dynamics with impaired protein degradation and aggregation, oxidative stress and free radical formation, impaired mitochondrial biogenesis and mitochondrial dysfunction and exposure to metal toxicity and pesticides.

### 5.3.1 Environmental Contaminants: Heavy Metals, Neurotoxins, Pesticides

Metals play a crucial part in neurodegeneration. Accumulation of metal changes the homeostasis of the body and leads to an implied production of free radicals, for example, mitochondrial dysfunction is interconnected to metal-mediated oxidative stress and increased Fe levels in the brain are correlated with NDDs. Inside the cell, Fe is bounded by ferritin to an insoluble ferrihydrite core. The Fe is reduced to the ferrous state, thus releasing it from the ferritin core by the interaction of a neurotoxic

6-hydroxydopamine (6-OHDA). This reaction substantially accelerates lipid peroxidation and releases free radicals in the presence of ferritin. Thus, Fe accumulation plays a significant role in the oxidative damage of AD, whereas dietary fat and a deficient Fe metabolism cause lipid peroxidation in PD (Levi and Tiranti 2019; Xu et al. 2008). Heavy metals contribute to the production of ROS/RNS and oxidative stress either directly or indirectly. This results in an anomaly redox status of the cell, which has a detrimental effect on biomolecules and vital organs like the liver, kidney and CNS (Singh et al. 2017). Various reports have shown that oxidative stress is caused by exposure to high concentrations of heavy metals like lead and mercury as well as a lack of critical metals like selenium and zinc. Mercury (Hg) is neurotoxic, and long-term exposure has a negative impact on the brain (Tchounwou et al. 2003). Lead (Pb) exhibited strong binding affinity with amino acids and metal cofactors through the SH group, which interferes with decreased antioxidant enzyme activity (Singh et al. 2018). Arsenic (As) has been linked to cytotoxicity and genotoxicity (Sun et al. 2014). The experimental evidence of recent research shows that pesticide exposure may be linked to the pathogenesis of AD and PD.

## 5.4 Role of ROS Mediated Oxidative Stress and Inflammation in Neurodegenerative Diseases

The brain is highly prone to oxidative stress reactions, since it is the most physically active and vital organ in the body. The brain uses 20% more oxygen than other bodily regions since it has a greater oxygen need. Ageing, genetic and environmental factors greatly persuade the onset of NDDs (Bertram and Tanzi 2005). The NDDs attributed to oxidative stress-mediated calcium dysregulation, membrane depolarization and mitochondrial function impairment. Apoptosis/necrosis and malfunctioning of neuronal cells are particular characteristics of neurodegenerative illnesses, which have a negative impact on the nervous system. A sudden surge in the production of ROS is linked to mitochondrial dysfunction (Zorov et al. 2014; Guo et al. 2013). It has also been suggested that a rise in reactive oxygen species (ROS) and reactive nitrogen species (RNS) has a significant role in many neurodegenerative diseases (Singh et al. 2019). In general metabolism, the TCA cycle metabolic intermediates are used by mitochondria to produce ATP molecules during oxidative phosphorylation. Oxygen will occasionally absorb electrons throughout the electron transport chain, creating superoxide anion radicals ($O^{2-}$). Manganese superoxide dismutase works inside the mitochondria to change these superoxide radicals into $H_2O_2$. Glutathione peroxidase or peroxiredoxins work in the mitochondria to convert $H_2O_2$ to $H_2O$. The inhibition of electron transport chain proteins can result in an increase in ROS, which then causes a drop in the potential of the mitochondrial membrane, ATP loss, which leads to cell death (Zhao et al. 2019). Oxidative stress is the outcome of an imbalance between ROS generation and antioxidant enzyme activity (Liguori et al. 2018). Thus, oxidative stress has the potential to result in neuronal ageing, specifically neurodegenerative disorders. Restrained ROS generation has physiological roles, particularly in regulating intracellular signal transmission and managing cellular redox homeostasis (Schieber and Chandel 2014). Redox control of neuronal survival at the transcriptional level included changes in transcription factors like NFκB, activating protein-1, nuclear factor erythroid 2 (Nrf2), hypoxia-inducible factor (Tonelli et al. 2018), redox-sensitive apoptotic effectors, like caspases, anti-apoptotic proteins and cytochrome c. Redox-active metals (Cu & Fe) that aggressively contribute to ROS production are also abundant in the brain. The PUFA content of the brain cell membranes makes them more vulnerable to lipid peroxidation. The antioxidant glutathione, when present in the right amounts, plays a crucial part in the detoxification of ROS species in brain cells. NDDs like AD, PD and HD are brought on by an imbalance between GSH and ROS in the brain (Uttara et al. 2009; Johnson et al. 2012).

## 5.5 Neuroprotective Efficacies of Diverse Classes of Phytochemicals

There are several pharmacological substances on the market that work by changing the balance of specific brain-functioning molecules (neurotransmitters) to have neuroprotective effects. Some medications improve blood flow, $O_2$ consumption and glucose metabolism of CNS in chronically compromised human brain function, including stroke, poor brain blood flow, dementia and pseudo-dementia. A number of drugs that are made from medicinal plants have active phytochemical components that have been found to have memory-enhancing qualities (Kumar and Khanum 2012). Reduced cholinergic activity in the brain has been proposed as one of the causes of dementia. As a result, cholinergic medications (of plant origin) such as nicotine, muscarinic agonists (such as arecoline, pilocarpine, etc.) and cholinesterase inhibitors (such as huperzine) can help improve memory (Broadley and Kelly 2001). As "disease-modifying" therapy for AD, PD, neuroprotection aims to prevent the death of nerve cells, repair neural network and alleviate cerebral dysfunction. Numerous substances, including antioxidants, monoamine oxidase (MAO) inhibitors, coenzyme Q and anti-inflammatory drugs, have been reported to be neuroprotective, yet clinical studies are done to analyse the progression of the diseases (Naoi et al. 2022; Mallah et al. 2020). Numerous phytochemicals and herbal medicines may help cure AD, according to a variety of studies. These phytochemicals may have therapeutic benefits due to their ability to reduce the toxicity of Aβ while also serving as an antioxidant and an anti-inflammatory. Excessive $H_2O_2$ released by mitochondrial malfunction eventually causes irreparable cellular dysfunction and damage in the brain (Li et al. 2003). Clinical experiments have suggested that the flavonoids present in high concentrations in fruits, vegetables and resveratrol concentrated in red wine – improve cognitive performance; however, this has not yet been fully proven (Cicero et al. 2019). By directly scavenging ROS & RNS, upregulating ROS-neutralization and downregulating free radical activity, flavonoids operate as potent antioxidants. However, the antioxidant activity is not a direct determinant of the neuroprotective action, and novel molecular mechanisms are presently being postulated to support the survival of neurons.

Numerous plant sources contain a variety of phytochemicals that exhibit a variety of pharmacological activities, including neuroprotection (Rehman et al. 2019; Kumar and Khanum 2012), apoptotic induction, autophagy activation, antioxidant and anti-inflammatory potentials (Limanaqi et al. 2020). There is plenty of evidence to support flavonoids' ability to reduce oxidant-induced neuronal damage by suppressing caspase-3 activation. The epicatechin and 3'-O-methyl-epicatechin (flavanols) inhibit JNK, C-gun and pro-caspase-3 and preserve neurons from oxidative injury (Vauzour et al. 2008). Hesperetin and 5-Nitro-hespertin, two flavanones, have been found to prevent succumb OS-induced death of neurons by triggering crucial protein and lipid kinase signalling pathways (Tamilselvam et al. 2013; Evans et al. 2022). Targeting mitochondrial proteins can be used to create a potent treatment approach against AD. Altering mitochondrial mobility has a detrimental effect on mitochondrial function and is a key factor in the pathophysiology of AD. Numerous investigations have documented the antioxidant and mitochondrial targeting properties of carnitine, vitamins C& E and alpha-lipoic acid in AD (Pagano et al. 2014).

Quercetin therapy has shown distinctive functions of restoring ATP synthesis, mitochondrial membrane potential, and reducing the formation of ROS. Additionally, this therapy greatly increased AMPK expression, reduced the development of dispersed senile plaques and prohibited memory loss (Rahman et al. 2020). Since the 1990s, phenolic chemicals found in food have gained a lot of attention. Apple, grape and citrus fruit juice polyphenols have been found to have a higher neuroprotective effect than antioxidant vitamins in clinical investigation reports. Polyphenols including catechins exhibited anti-neurodegenerative properties that are partially mediated through the activation protein kinase C (PKC) and transcription factors thus promoting the activity of genes involved in cell survival. It has been proposed that catechins can protect neurons from Alzheimer's disease processes through their anti-inflammatory and cell-survival effects (Ide et al. 2018). However, the Mediterranean diets have demonstrated antioxidant and anti-inflammatory functions of flavonoids and polyphenols in human studies on type 2 diabetes & cancer prevention, CVD and stroke. In animal models of AD, a diet rich in high polyphenols content has been shown to reduce tau hyperphosphorylation and Aβ aggregation (Phan et al. 2019; Freyssin et al. 2018). A soy isoflavonoid called genistein shows therapeutic potential for treating a number of clinical disorders associated with ageing-related mitochondrial dysfunctions, such as neuroinflammation, oxidative stress and Aβ aggregate formation in AD. Genistein's therapeutic effects were related to its capacity to reduce mitochondrial dysfunctions brought on by Aβ aggregates (Rahman et al. 2020). Plant polyphenols increase mitochondrial biogenesis and reduce mitochondrial dysfunction in AD conditions (Teixeira et al. 2018). Resveratrol increases cAMP through the stimulation of the cAMP/CaMK/AMPA pathway while suppressing the activity of cAMP phosphodiesterases. Resveratrol has also been shown to increase NAD+ levels and stimulate mitochondrial dynamics, biogenesis and function through the activation of AMPK and PKC-ε (Zhou et al. 2021; Wan et al. 2016). Several flavones, including wogonin, quercetin and baicalein, increased the expression of Sirt1/AMPA/PGC-1 to boost the biogenesis of mitochondrial functions (Naoi et al. 2019). Oleuropein, found in extra virgin olive oil, reduces oxidative stress and controls mitochondrial biogenesis and mitogenesis in AD (Blanco-Benítez et al. 2022). Therefore, it has been predicted that pharmacological intervention using polyphenols will be a promising therapeutic strategy for neurodegenerative illnesses linked to mitochondrial dysfunction. Resveratrol, quercetin and rosmarinic acids reduced permeabilization of mitochondrial membrane and defended cells from apoptosis, Aβ aggregates, and ischemia (George et al. 2011). Curcumin in particular interacted directly with the voltage-dependent anion channel and phosphorylation of the ion channel by glycogen synthase kinase 3 and pore opening. Flavonoids like naringenin, kaempferol, quercetin, bailarein and epigallocatechin gallate) activated signal transductions leading to upregulation of Bcl-2 and Bcl-xL (anti-apoptotic proteins) and repression of Bax and Bak (pro-apoptotic proteins) and also altered the mitochondrial apoptosis pathway (Fig. 5.1; Table 5.1).

## 5.6 Conclusion and Future Perspectives

Plants are a boundless source of molecules that can be used to enhance human health in the form of foods, spices and herbs. Numerous phytochemicals were previously demonstrated to exert neuroprotective benefits. Although there is an increasing need for phytotherapeutic drugs, it is possible that several phytochemicals have additive, complementary and synergistic effects that activate/inhibit their activity. Additionally, a crucial quality for a potential neuroprotective drug is its capacity to permeate the BBB and reach the desired CNS locations. Natural phytochemicals have anti-amyloid, anti-inflammatory, antioxidant, anti-depressive and anti-convulsant activities in addition to regulating the neurotransmitter system. It has been claimed that choosing natural phytochemicals as anti-neuropsychotic and anti-neurodegenerative medications over synthetic medications for the treatment of NDDs is safer. Thus, exploring the molecular pathways and interacting receptors via high throughput screening, computational and wet-lab experimental techniques, and construction of clinical studies is vital in the process of drug development of therapeutically potential natural phytochemicals.

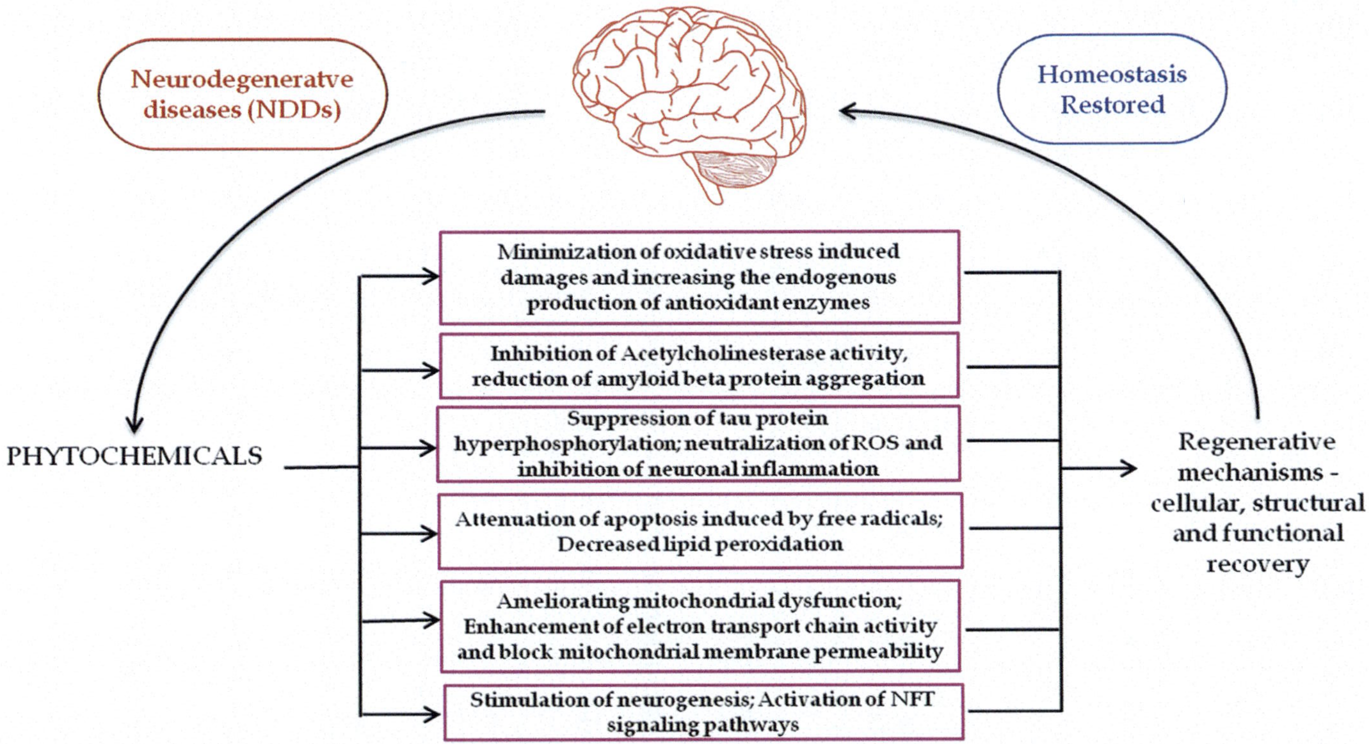


**Fig. 5.1** Phytochemicals modulating the cellular and molecular alterations exhibiting anti-neurodegenerative activity

**Table 5.1** Therapeutic mechanisms of medicinal plants and their phytochemicals for neuroprotective activity

| Medicinal plant | Biological activity | Phytochemicals reported | Probable mechanism of neuroprotective activity | References |
|---|---|---|---|---|
| *Acorus calamus* | Anti-stress and antioxidant activity | Monoterpenes-limonene, Xanthone glycosides, Triterpenoid saponins, flavones-galangin, polyphenolic compounds, steroids-β sitosterol | Restoring to normal and reducing the levels of stress markers; increasing the endogenous production of antioxidant enzymes | Esfandiari et al. (2018), Ukkirapandian et al. (2022) |
| | Anticonvulsant potential | | | |
| *Aegle marmelos* | Anti-oxidant, Alzheimer's disease treatment | Flavanoids- marmelosin, umbelliferone, rutin and Para-coumaric acid | Minimizing oxidative stress and inflammatory responses and significantly inhibiting acetylcholinesterase (AChE) activity, reduction of Aβ toxicity | Keowkase et al. (2020), Monika et al. (2023), Gautam and Ramanathan (2021), Raheja et al. (2019) |
| *Allium sativum* | Anti-oxidative action | Polyphenols, allicin, S-allyl cysteine, flavonoids and carotenoids, diallyldisulphide, diallyltrisulphide | Stimulating anti-oxidant enzyme levels and ameliorates dysfunction of mitochondria | Hazzaa et al. (2020), Saini et al. (2021) |
| | Alzheimer's disease treatment | | | |
| *Alpinia galangal* | Anti-oxidant and anti-amnesic potentials | Camphor, methyl cinnamate, 1,8-cineole, α-fenchyl acetate | Improves cognition by enhancing cholinergic transmission through AChE activity | Hanish Singh et al. (2011) |
| *Apium graveolens* | Antioxidant, anti-inflammatory and anti-stroke activities | Flavonoids quercetin3O-beta-Dglucuronide and isoharmentin3O-beta-Dglucuronide | Cerebroprotective effect via reduced infarction areas in the brain, increased monoamine oxidase A and B activity | Chonpathompikunlert et al. (2018), Jittiwat et al. (2021) |
| *Argyreia speciosa* | Antiamyloidogenic, anticholinesterase, anti-anxiety and anticonvulsant actvities | Scopoletin, festuclavine and ergometrine, alkaloids—ergoline, flavonoids, triterpenoids, saponin and steroids | Increases the endogenous levels of antioxidant enzymes, inhibits Aβ aggregation into insoluble fibrils | Kashyap et al. (2020), Jaiswal and Tailang (2018) |

(continued)

**Table 5.1** (continued)

| Medicinal plant | Biological activity | Phytochemicals reported | Probable mechanism of neuroprotective activity | References |
|---|---|---|---|---|
| *Artemisia absinthium*<br>*Artemisia nilagirica* | Anti-oxidant, anti-inflammatory and free radical neutralizing effects, anti-epileptic activity | Flavonoids- ferulic acid, luteolin, caffeic acid, quercetin, and apigenin, sesquiterpenoids- Artemisinin, essential oils, tannins, phenols and saponins | Focal ischemia and reperfusion attenuated by modulation of oxidative stress, mitochondrial death pathway and Nfr2 signaling pathway | Sailike et al. (2022), Bora and Sharma (2010), Bisht et al. (2021), Albaqami et al. (2022) |
| *Azadirachta indica* | Anti-oxidant, anti-apoptotic and anti-inflammatory activities | Flavonoids, phenols, imonoids, catechins, polyphenols, gallic acid, nimbins | Neuroprotective efficacy against AD & PD, ischemic-reperfusion injury through down-regulation of AChE activity and COX-2 protein expression levels | Xiang et al. (2018), Sandhir et al. (2021) |
| *Bacopa monnieri* | Anti-stress, anti-anxiety and memory and cognitive effects | Alkaloids- herpestine, hersaponin and monnierin, saponins- bacosides A and B | Repair of damaged neurons; improves nerve impulse transmission and restores synaptic activity | Abdul Manap et al. (2019), Fatima et al. (2022) |
| *Bellis perennis* | Anti-oxidant and neuroprotective effects | Flavanoids- apigenin-7-O-ghicopyranoside, saponins, glycosides | Decreased acetylcholinesterase activity, reduced lipid peroxidation | Marques et al. (2013) |
| *Berberis vulgaris* | Antioxidant, anti-inflammatory, anti-cholinergic | Alkaloids- Berberine, tannins, phenolic compounds, sterols, triterpenes | Decreasing neuronal inflammation, protects against apoptosis induced by $H_2O_2$ | Rad et al. (2022), Mokhber-Dezfuli et al. (2014) |
| *Calendula officinalis* | Anti-oxidant and alleviates metal-induced neurotoxicity | Triterpenes, flavonoids, phenolic acids, quinones and coumarins and carotenoids | Increased antioxidant defense enzyme levels with reduced LPO | Shivasharan et al. (2013), Wang et al. (2023) |
| *Canscora decussata* | Immunomodulatory, analgesic, anticonvulsant, anti-inflammatory efficacies, anti-Parkinson's activity | Phenolic compounds, xanthones, and triterpenoids | Improves endogenous production of antioxidant oxidant enzymes, reduces lipid peroxidation | Rehman et al. (2019) |

| | | | | |
|---|---|---|---|---|
| *Cardiospermum halicacabum* | Anti-convulsant and anti-inflammatory activities | Scopolamine, β-sitosterol, Apigenin, luteolin, apigenin-7-o-glucuronide, linoleic acid | Restores impaired cholinergic function and reduces cognitive dysfunction and improves memory, inhibits cyclooxygenase activity | Menichini et al. (2014), Kukkar et al. (2014) |
| *Carthamus tinctorius* | Anti-Alzheimer's disease drug, free radical scavenging activity, anti-inflammatory action | Flavanoids- kaempferol 3-O-rutinoside and anhydrosafflor yellow B, Carthamidin, isocarthamidin, luteolin | Anti-Aβ aggregation, suppressing tau protein hyperphosphorylation and ameliorating neuroinflammation | Lei et al. (2020), Liang and Lin (2022), Delshad et al. (2018) |
| *Cassia fistula* | Anti-oxidant activity | Flavonoids, anthraquinones, procyanidins, coumarins, alkaloids and glycosides | Inhibits anti-Aβ aggregation and inhibition of 5-lipoxygenase pathway and AchE activity | Thabit et al. (2018) |
| *Celastrus paniculatus* | Neuroprotective for cognitive dysfunctions and antioxidant, anti-anxiety, anti-epilepsy activities | Terpenoid, alkaloid, carbohydrates, saponin, and phenolic compounds. | Attenuates oxidative stress-induced damages in the brain tissue, restores neurobehavioral alterations | Bhagya et al. (2016), Malik et al. (2017) |
| *Centella asiatica* | Anti-psychotics | Triterpenoid saponins-asiaticoside, brahminoside, asiatic acid, brahmic acid, isobrahmic acid and betulic acid | Improves cognitive functions and memory; ameliorates dementia via reducing LPO and restoring GSH activity | Lokanathan et al. (2016), Teerapattarakan et al. (2018) |
| *Clitoria ternatea* | Neuroprotection against memory impairment | Quercetin, kaempferol, anthocyanins, and myricetin glycosides | Decreases lipid peroxidation levels and AChE activity | Jiji and Muralidharan (2021), Talpate et al. (2014), Shahnas and Akhila (2014), Damodaran et al. (2018) |
| *Coriandrum sativum* | Anti-oxidant, anti-anxiety and anticonvulsant effects | Linalool, Beta-myrcene | Improves endogenous production of antioxidant oxidant enzymes, reduces lipid peroxidation | Hosseini et al. (2021), Vekaria et al. (2012) |

(continued)

**Table 5.1** (continued)

| Medicinal plant | Biological activity | Phytochemicals reported | Probable mechanism of neuroprotective activity | References |
|---|---|---|---|---|
| *Crocus sativus* | Anti-neuropsychiatric and anti-neurodegenerative, antioxidant, anti-inflammatory effects | Carotenoids- crocin, glycosides picrocrocin, monoterpenoids and flavonoids | Reduces synaptic loss, declined phosphorylation of tau and inhibited Aβ aggregation | Bian et al. (2020), Soeda et al. (2016) |
| *Cucumis melo* | Anti-inflammatory and antioxidant activities | Phenolic compounds, alkaloids, flavonoids | Atteneuated oxidative stress and suppressed the inflammatory cytokines release | Ibrahim (2017), Park et al. (2009) |
| *Curcuma longa* | Anti-inflammatory, antioxidant, and anti-Aβ-aggregate activities | Curcumin | Stimulates neurogenesis and replaces the lost functional neurons by enhancing Trk/PI3K signalling and reduced tumour necrosis factor (TNF) and caspase activity levels | Cole et al. (2007), Nebrisi (2021), Sutalangka and Wattanathorn (2017) |
| *Cyperus rotundus* | Anti-dementia effect | Alkaloids, carbohydrates, glycosides, flavonoids, resins, saponins, steroids, tannins | Increases cholinergic neurotransmitters through inhibition of AChE, activity, improves memory | Hemanth Kumar et al. (2013) |
| *Dalbergia sissoo* | Neuroprotective for cognitive dysfunctions and antioxidant, anti-anxiety potentials | Biochanin A, isocaviumin, tectorigenin, tannins | Attenuates oxidative stress-induced damages in the brain tissue, restores neurobehavioral alterations | Raheja et al. (2021), Swaroop et al. (2014) |
| *Ginkgo biloba* | Anti-oxidant and free radical scavenging activity | Terpenoids- ginkgolides, flavonoids- quercetin, isorhamnetin), steroids-sitosterol and stigmasterol and shikimic acid, vanillic acid & benzoic acid | Neuroprotective effect against Aβ and NO-induced toxicity | Xie et al. (2022), Bastianetto et al. (2000), Chandrasekaran et al. (2001) |

| | | | | |
|---|---|---|---|---|
| *Glyzyrrhiza glabra* | Anti-oxidant, anti-inflammatory and anti-aging properties | Triterpenoid sapanonin-glycyrrhizin, glabrin A & B | Inhibition of HMGB1 proinflammatory cytokine-like activity | Paudel et al. (2020), Pastorino et al. (2018) |
| *Gymnema sylvestre* | Anti-oxidant and anti-inflammatory activity | Saponins, terpenoids, flavonoids, tannins, steroids, and alkaloids | Neuroprotective effect through regulation of oxidative stress mediators and LPO biomarkers | Fatani et al. (2015) |
| *Huperzia serrata* | Schizophrenia, cognitive dysfunction, anti-inflammatory activity | Alkaloid- Huperzine A, phenolic acid- ferulic acid and caffeic acid | AChE inhibitor, depreciates Aβ-mediated oxidative stress, apoptosis and mitochondrial dysfunction | Callizot et al. (2021) |
| *Hypericum perforatum* | Antidepressant agent, anti-inflammatory activity | Hypericin, pseudohypericin and related naphthodianthrones | Reduces ROS production through restoration of antioxidant defense mechanism, restoration of $Na^+$/$K^+$-ATPase activity | Khan et al. (2011), Karakas et al. (2017) |
| *Juglans regia* | Antioxidant and anti-cholinesterase activities, anti-Alzheimer's disease | Ecosa pentanoic acid, and docosa pentaenoic & hexaenoic acids | Inhibits AChE, butyrylcholinesterase (BChE) to manage neurodegeneration | Ilah et al. (2017), Orhan et al. (2011) |
| *Lawsonia inermis* | Antioxidant effect, anti-Alzheimer's activity | 3-O-β-acetyloleanolic acid, 3-O-(Z)-coumaroyl oleanolic acid, betulinic acid, and oleanolic acid | Modulates oxidative stress markers and AChE, butyrylcholinesterase (BChE), and β-secretase 1 (BACE-1) inhibition | Balaei-Kahnamoei et al. (2021), Mir et al. (2019) |
| *Leucojum aestivum* | Anti-Alzheimer's disease activity | Lycorine, haemanthamine and homolycorine type alkaloids | Inhibits AChE, BChE, and BACE-1 activities and prevents Aβ-mediated neurotoxicity | Georgieva et al. (2007) |
| *Lycium barbarum* | Antioxidant and anti-inflammatory effects | Olysaccharides and flavonoids | Inhibits cellular apoptosis, ameliorates mitochondrial dysfunction modulating mitochondrial apoptotic pathway | Guo et al. (2021), Hu et al. (2018) |

(continued)

**Table 5.1** (continued)

| Medicinal plant | Biological activity | Phytochemicals reported | Probable mechanism of neuroprotective activity | References |
|---|---|---|---|---|
| *Madhuca longifolia* | Anti-Alzheimer's disease activity | Steroid, triterpenoid, flavonoids, tannin, cardiac glycosides | Restores memory dysfunction | Khare et al. (2022) |
| *Mangifera indica* | Antioxidant and anti-inflammatory activities | Mangiferin, phenolic acids, benzophenones, flavonoids, ascorbic acid, carotenoids, and tocopherols | Inhibits AChE, BChE, and BACE-1 activities and prevents Aβ-mediated neurotoxicity | Al Omairi et al. (2018), Temviriyanukul et al. (2022), Lum et al. (2021) |
| *Medicago sativa* | Antioxidant activity | Polysaccharides, triterpenoid saponins | Alleviates the oxidative stress enzyme levels and surpasses H2O2-induced neuronal damages | Li et al. (2020), Liu et al. (2020) |
| *Mentha longifolia* | Anti-Alzheimer's disease activity | Menthol, menthone, isomenthone and piperitenone | Downregulates oxidative stress and neuroinflammatory markers, malondialdehyde (MDA), reduced GSH &NO and downregulated nuclear factor-κB (NF-κB) | Elshamy et al. (2021) |
| *Mirabilia jalaba* | Anticonvulsant and antioxidant activities | Tannins, phenolic compounds and flavanoids | Alleviates the oxidative stress enzyme levels and inhibits AChE activity | Babu et al. (2017) |
| *Momordica charantia* | Antioxidant and anticonvulsant activities | Polysaccharides, flavonoids, triterpenes, saponins, ascorbic acid and steroids | Increases cholinergic neurotransmitters through inhibition of AChE, activity, improves memory | Malik et al. (2011), Huang et al. (2018) |
| *Moringa oleifera* | Antioxidant and anti-ROS activity | Flavonoids, β-carotene, quercetin, rhamnose, kaempferol, glycosylates | Neuroprotection via modulation of NF-kB/Nrf2/HO-1 signaling pathway | Hashim et al. (2021), Mundkar et al. (2022) |

| | | | | |
|---|---|---|---|---|
| *Morus nigra* | Antidepressant and antioxidant effects | Phenol- syringic acid, flavonoids, flavonols, anthocyanins | Depresses oxidative and nitrative stress via nitro-oxidative system, improves cognitive deficits | Dalmagro et al. (2017), Turgut et al. (2016) |
| *Mucuna pruriens* | PD treatment, anti-acetylcholinesterase activity | Cardiac glycosides | Controlled stimulation of acetylcholinesterase activity, inhibits apoptotic pathways of dopaminergic neurons | Manyam et al. (2004), Kamkaen et al. (2022) |
| *Myrtus communis* | Antioxidant and anti-inflammatory activities, radical-scavenging effect, | Polyphenols, α-pinene, limonene, linalool, myrtenyl acetate, and α-terpinolene | Improves cognitive dysfunction through control of the elevated levels of hippocampal MMP-13, AChE activities and increases Na+,K+-ATPase levels | Cevikelli-Yakut et al. (2020), Tumen et al. (2012) |
| *Nardostachys jatamansi* | Anti-ischemic, antioxidant, anticonvulsant activities and amnesia therapy | Jatamansinol, polyphenols, flavonoids and sesquiterpenes | Improves cholinergic neurotransmission in brain cells | Joshi and Parle (2006) |
| *Nelumbo nucifera* | Antioxidant potential, Alzheimer's therapy | Flavonoids- naringenin, quercetin, isorhamnetin, kaempferol, luteolin, myricetin, cyanidin, and delphinidin | Increases cholinergic neurotransmitters through inhibition of AChE, BChE, and BACE-1 activities, improves cognitive functions | Kim et al. (2014), Temviriyanukul et al. (2020) |
| | | Phenolic acids- gallic acid, ferulic acid, and p-coumaric acid | | |
| *Nerium oleander* | Anti-stroke activity | Oleandrin, cardiac glycosides | Cerebroprotective effect via reduced infarction areas in the brain, increased monoamine oxidase A and B activity | Dunn et al. (2011) |

(continued)

**Table 5.1** (continued)

| Medicinal plant | Biological activity | Phytochemicals reported | Probable mechanism of neuroprotective activity | References |
|---|---|---|---|---|
| *Nigella sativa* | Anxiolytic, locomotor activity, anti-depressant activity | Thymoquinone, thymol, alkaloids like nigellidine, nigellimine and nigellicine | Stress reduction through inhibition of NO–cGMP pathway and modulation of serotonin levels, attenuates neuroinflammation by repressing pro-inflammatory cytokine expression | Islam et al. (2015), Farkhondeh et al. (2018) |
| *Ocimum basilicum* | Anticonvulsants, anti-stress and anti-inflammatory, anti-epileptic activities | β-Sitosterol,eugenol, luteolin, rosmarinic acid, apigenin | Neuroprotective effect by reducing MDA levels and increasing antioxidant enzymes level | Mansouri et al. (2022), Seyed et al. (2021) |
| *Oroxylum indicum* | Cognitive enhancement agents | Flavanoids- Oroxylin A, Baicalein and Chrysin | Attenuates Aβ-induced CNS toxicity via oxidative stress suppression and neutralization of ROS in Akt/CREB/Bcl-2 pathway improves mitochondrial function | Pondugula et al. (2021), Mairuae et al. (2019) |
| *Oxalis corniculata* | Parkinson's disease treatment, anti-stress and anti-anxiety activities | L-dopa, polyphenols and flavonoids, terpenoid compounds- β–sitosterol, caemphesterol | Increases motor coordination, exhibits anxiolytic action through activation of GABA chloride channel | Aruna et al. (2017) |
| *Panax ginseng* | Anti-stress activity | Ginsenosides | ROS scavenging activity, increasing the expressions Bax/Bcl-2 activity | Huang et al. (2019), Cho (2012), Li et al. (2021) |
| *Ricinus communis* | Anti-Alzheimer's disease activity | Undecylenic acid | Attenuates cerebral ischemia, neuroprotection via modulation of Aβ- and hyperphosphorylated tau-mediated pathological pathways | Ahmad et al. (2021), Lee et al. (2012) |

| | | | | |
|---|---|---|---|---|
| *Rosa* damascena | Anticonvulsant and anti-seizure activity | Quercetin, kaempferol, rutin, and myricetin | Exhibits cognitive improvement via destruction of ROS and reduced NO production, suppression of TNF-α and IL-1β | Esfandiary et al. (2015), Homayoun et al. (2015) |
| *Salvia officinalis* | Antioxidant, anti-dementia and anti-inflammatory activities | Phenolic acids, flavonoids and terpenoids, linalool, α-pinene and cineole, humulene, limonene, and thujone | Increases cholinergic neurotransmitters through inhibition of AChE, BChE, and BACE-1 activities, improves memory and cognitive functions | Ayoub et al. (2022), Faridzadeh et al. (2022), Lopresti (2017) |
| *Terminalia arjuna* | Anti-oxidant, anti-inflammatory activities | Triterpenoids and saponins | Ameliorates metal-induced toxicity and attenuates cerebral ischaemia by restoring damaged blood-brain barrier | Yaidikar and Thakur (2015), Kaliyappan et al. (2021) |
| *Terminalia chebula* | Anti-Alzheimer's disease activity | Ellagic acid | Neuroprotection attained by reduction of Aβ toxicity | Shen et al. (2017), Zhao et al. (2021) |
| *Vitex negundo* | Anti-oxidant, anti-inflammatory activities | Triterpenes, diterpenes, sesquiterpenes, lignan, flavonoids | Alleviates the oxidative stress enzyme levels | Siddiqui et al. (2018) |
| *Vitis vinifera* | Anti-oxidant, anti-inflammatory effects, anti-Alzheimer's disease activity | Stilbenoids, phenolic compounds, flavonoids, proanthocyanidins | Inhibits Aβ aggregation, suppressing tau protein hyperphosphorylation and ameliorating neuroinflammation | Rapaka et al. (2019), Lakshmi et al. (2014) |
| *Withania somnifera* | Anti-AD/PD/HD potentials, anti-oxidant, anti-inflammatory activities | Withaferin A, withanolide A& D | Oxidative stress (OS) management Vis modulation of OS response proteins: VGF, vimentin and peroxiredoxin-I | Wongtrakul et al. (2021), Dar and Ahmad (2020) |

## References


Abdul Manap AS, Vijayabalan S, Madhavan P, Chia YY, Arya A, Wong EH, Rizwan F, Bindal U, Koshy S (2019) Bacopa monnieri, a neuroprotective lead in Alzheimer disease: a review on its properties, mechanisms of action, and preclinical and clinical studies. Drug Target Insights 13: 1177392819866412. https://doi.org/10.1177/1177392819866412

Ahmad N, Mishra A, Ahsan F et al (2021) Evaluation of cerebroprotective effect of Ricinus communis leaves against ischemia reperfusion injury in rats. Clin Phytosci 7:85. https://doi.org/10.1186/s40816-021-00323-y

Al Omairi NE, Radwan OK, Alzahrani YA, Kassab RB (2018) Neuroprotective efficiency of Mangifera indica leaves extract on cadmium-induced cortical damage in rats. Metab Brain Dis 33(4):1121–1130. https://doi.org/10.1007/s11011-018-0222-6

Albaqami JJ, Benny TP, Hamdi H, Altemimi AB, Kuttithodi AM, Job JT, Sasidharan A, Narayanankutty A (2022) Phytochemical composition and in vitro antioxidant, anti-inflammatory, anticancer, and enzyme-inhibitory activities of artemisia nilagirica (C.B. Clarke) pamp. Molecules 27(20):7119. https://doi.org/10.3390/molecules27207119

Aruna K, Rajeswari PDR, Sankar SR (2017) The effect of Oxalis corniculata extract against the behavioural changes induced by 1-methyl-4-phenyl-1,2,3,6-tetrahydropyridine (MPTP) in mice. J Appl Pharm Sci 7(3):148–153

Ayoub IM, George MY, Menze ET, Mahmoud M, Botros M, Essam M, Ashmawy I, Shendi P, Hany A, Galal M, Ayman M, Labib RM (2022) Insights into the neuroprotective effects of Salvia officinalis L. and Salvia microphylla Kunth in the memory impairment rat model. Food Funct 13(4):2253–2268. https://doi.org/10.1039/d1fo02988f

Babu S, Swain S, Renuka K (2017) Neuroprotective activity of fractional flower extracts of mirabilis Jalapa against aluminium hydrochloride induced neurotoxicity in male Wister rats. Int J Pharm Pharm Sci 9:216. https://doi.org/10.22159/ijpps.2017v9i5.17584

Balaei-Kahnamoei M, Saeedi M, Rastegari A, Shams Ardekani MR, Akbarzadeh T, Khanavi M (2021) Phytochemical analysis and evaluation of biological activity of Lawsonia inermis seeds related to Alzheimer's disease. Evid Based Complement Alternat Med 2021:5965061. https://doi.org/10.1155/2021/5965061

Bastianetto S, Zheng WH, Quirion R (2000) The Ginkgo biloba extract (EGb 761) protects and rescues hippocampal cells against nitric oxide-induced toxicity: involvement of its flavonoid constituents and protein kinase C. J Neurochem 74(6):2268–2277. https://doi.org/10.1371/journal.pone.0002935.39

Bertram L, Tanzi RE (2005) The genetic epidemiology of neurodegenerative disease. J Clin Invest 115(6):1449–1457. https://doi.org/10.1172/JCI24761

Bhagya V, Christofer T, Shankaranarayana Rao BS (2016) Neuroprotective effect of Celastrus paniculatus on chronic stress-induced cognitive impairment. Indian J Pharmacol 48(6):687–693. https://doi.org/10.4103/0253-7613.194853

Bian Y, Zhao C, Lee SM-Y (2020) Neuroprotective potency of saffron against neuropsychiatric diseases, neurodegenerative diseases, and other brain disorders: from bench to bedside. Front Pharmacol 11:579052. https://doi.org/10.3389/fphar.2020.579052

Bisht D, Kumar D, Kumar D, Dua K, Chellappan DK (2021) Phytochemistry and pharmacological activity of the genus artemisia. Arch Pharm Res 44(5):439–474. https://doi.org/10.1007/s12272-021-01328-4

Blanco-Benítez M, Calderón-Fernández A, Canales-Cortés S, Alegre-Cortés E, Uribe-Carretero E, Paredes-Barquero M, Gimenez-Bejarano A, Duque González G, Gómez-Suaga P, Ortega-Vidal J, Salido S, Altarejos J, Martínez-Chacón G, Niso-Santano M, Fuentes JM, González-Polo RA, Yakhine-Diop SMS (2022) Biological effects of olive oil phenolic compounds on mitochondria. Mol Cell Oncol 9(1):2044263. https://doi.org/10.1080/23723556.2022.2044263

Bora KS, Sharma A (2010) Neuroprotective effect of Artemisia absinthium L. on focal ischemia and reperfusion-induced cerebral injury. J Ethnopharmacol 129(3):403–409. https://doi.org/10.1016/j.jep.2010.04.030

Broadley KJ, Kelly DR (2001) Muscarinic receptor agonists and antagonists. Molecules 6(3): 142–193. https://doi.org/10.3390/60300142

Callizot N, Campanari ML, Rouvière L, Jacquemot G, Henriques A, Garayev E, Poindron P (2021) Huperzia serrata extract 'NSP01' with neuroprotective effects-potential synergies of Huperzine a and polyphenols. Front Pharmacol 12:681532. https://doi.org/10.3389/fphar.2021.681532

Cano A, Fonseca E, Ettcheto M, Sánchez-López E, de Rojas I, Alonso-Lana S, Morató X, Souto EB, Toledo M, Boada M, Marquié M, Ruíz A (2021) Epilepsy in neurodegenerative diseases: related drugs and molecular pathways. Pharmaceuticals (Basel) 14(10):1057. https://doi.org/10.3390/ph14101057

Cevikelli-Yakut ZA, Ertas B, Sen A, Koyuncuoglu T, Yegen BC, Sener G (2020) Myrtus communis improves cognitive impairment in renovascular hypertensive rats. J Physiol Pharmacol 71(5). https://doi.org/10.26402/jpp.2020.5.07

Chandrasekaran K, Mehrabian Z, Spinnewyan B, Drieu K, Fiskum G (2001) Neuroprotective effects of bilobalide, a component of Ginkgo biloba extract (EGb 761), in gerbil global brain ischemia. Brain Res 922(2):282–292. https://doi.org/10.1016/S0006-8993(01)03188-2

Chen X, Guo C, Kong J (2012) Oxidative stress in neurodegenerative diseases. Neural Regen Res 7(5):376–385. https://doi.org/10.3969/j.issn.1673-5374.2012.05.009

Cho IH (2012) Effects of Panax ginseng in neurodegenerative diseases. J Ginseng Res 36(4): 342–353. https://doi.org/10.5142/jgr.2012.36.4.342

Chonpathompikunlert P, Boonruamkaew P, Sukketsiri W, Hutamekalin P, Sroyraya M (2018) The antioxidant and neurochemical activity of Apium graveolens L. and its ameliorative effect on MPTP-induced Parkinson-like symptoms in mice. BMC Complement Altern Med 18(1):103. https://doi.org/10.1186/s12906-018-2166-0

Cicero AFG, Ruscica M, Banach M (2019) Resveratrol and cognitive decline: a clinician perspective. Arch Med Sci 15(4):936–943. https://doi.org/10.5114/aoms.2019.85463

Cole GM, Teter B, Frautschy SA (2007) Neuroprotective effects of curcumin. Adv Exp Med Biol 595:197–212. https://doi.org/10.1007/978-0-387-46401-5_8

Dalmagro AP, Camargo A, Zeni ALB (2017) Morus nigra and its major phenolic, syringic acid, have antidepressant-like and neuroprotective effects in mice. Metab Brain Dis 32(6): 1963–1973. https://doi.org/10.1007/s11011-017-0089-y

Damodaran T, Tan BWL, Liao P, Ramanathan S, Lim GK, Hassan Z (2018) Clitoria ternatea L. root extract ameliorated the cognitive and hippocampal long-term potentiation deficits induced by chronic cerebral hypoperfusion in the rat. J Ethnopharmacol 224:381–390. https://doi.org/10.1016/j.jep.2018.06.020

Dar NJ, Ahmad M (2020) Neurodegenerative diseases and Withania somnifera (L.): an update. J Ethnopharmacol 256:112769. https://doi.org/10.1016/j.jep.2020.112769

Delshad E, Yousefi M, Sasannezhad P, Rakhshandeh H, Ayati Z (2018) Medical uses of Carthamus tinctorius L. (Safflower): a comprehensive review from Traditional Medicine to Modern Medicine. Electron Physician 10(4):6672–6681. https://doi.org/10.19082/6672

Di Meo S, Reed TT, Venditti P, Victor VM (2016) Role of ROS and RNS sources in physiological and pathological conditions. Oxidative Med Cell Longev 2016:1245049. https://doi.org/10.1155/2016/1245049

Dunn DE, He DN, Yang P, Johansen M, Newman RA, Lo DC (2011) In vitro and in vivo neuroprotective activity of the cardiac glycoside oleandrin from Nerium oleander in brain slice-based stroke models. J Neurochem 119(4):805–814. https://doi.org/10.1111/j.1471-4159.2011.07439.x

Elshamy S, Abdel Motaal A, Abdel-Halim M, Medhat D, Handoussa H (2021) Potential neuroprotective activity of Mentha longifolia L. in aluminum chloride-induced rat model of Alzheimer's disease. J Food Biochem 45(4):1770. https://doi.org/10.1111/jfbc.13644

Esfandiari E, Ghanadian M, Rashidi B, Mokhtarian A, Vatankhah AM (2018) The effects of Acorus calamus L. in preventing memory loss, anxiety, and oxidative stress on Lipopolysaccharide-induced neuroinflammation rat models. Int J Prev Med 9:85

Esfandiary E, Karimipour M, Mardani M, Ghanadian M, Alaei HA, Mohammadnejad D, Esmaeili A (2015) Neuroprotective effects of Rosa damascena extract on learning and memory in a rat model of amyloid-β-induced Alzheimer's disease. Adv Biomed Res 4:131. https://doi.org/10.4103/2277-9175.161512

Evans JA, Mendonca P, Soliman KFA (2022) Neuroprotective effects and therapeutic potential of the citrus flavonoid hesperetin in neurodegenerative diseases. Nutrients 14(11):2228. https://doi.org/10.3390/nu14112228

Faridzadeh A, Salimi Y, Ghasemirad H, Kargar M, Rashtchian A, Mahmoudvand G, Karimi MA, Zerangian N, Jahani N, Masoudi A, Sadeghian Dastjerdi B, Salavatizadeh M, Sadeghsalehi H, Deravi N (2022) Neuroprotective potential of aromatic herbs: rosemary, sage, and lavender. Front Neurosci 16:909833. https://doi.org/10.3389/fnins.2022.909833

Farkhondeh T, Samarghandian S, Shahri AMP, Samini F (2018) The neuroprotective effects of thymoquinone: a review. Dose Response 16(2):1559325818761455. https://doi.org/10.1177/1559325818761455

Fatani AJ, Al-Rejaie SS, Abuohashish HM, Al-Assaf A, Parmar MY, Ola MS, Ahmed MM (2015) Neuroprotective effects of Gymnema sylvestre on streptozotocin-induced diabetic neuropathy in rats. Exp Ther Med 9(5):1670–1678. https://doi.org/10.3892/etm.2015.2305

Fatima U, Roy S, Ahmad S, Al-Keridis LA, Alshammari N, Adnan M, Islam A, Hassan MI (2022) Investigating neuroprotective roles of Bacopa monnieri extracts: mechanistic insights and therapeutic implications. Biomed Pharmacother 153:113469. https://doi.org/10.1016/j.biopha.2022.113469

Freyssin A, Page G, Fauconneau B, Rioux BA (2018) Natural polyphenols effects on protein aggregates in Alzheimer's and Parkinson's prion-like diseases. Neural Regen Res 13(6):955–961. https://doi.org/10.4103/1673-5374.233432

Gan L, Cookson MR, Petrucelli L et al (2018) Converging pathways in neurodegeneration, from genetics to mechanisms. Nat Neurosci 21:1300–1309. https://doi.org/10.1038/s41593-018-0237-7

Gautam M, Ramanathan M (2021) Ameliorative potential of flavonoids of Aegle marmelos in vincristine-induced neuropathic pain and associated excitotoxicity. Nutr Neurosci 24(4):296–306. https://doi.org/10.1080/1028415X.2019.1627768

George J, Singh M, Srivastava AK, Bhui K, Roy P, Chaturvedi PK, Shukla Y (2011) Resveratrol and black tea polyphenol combination synergistically suppress mouse skin tumors growth by inhibition of activated MAPKs and p53. PLoS One 6(8):e23395. https://doi.org/10.1371/journal.pone.0023395. Retraction in: PLoS One. 2019;14(4):e0215980

Georgieva L, Berkov S, Kondakova V, Bastida J, Viladomat F, Atanassov A, Codina C (2007) Alkaloid variability in Leucojum aestivum from wild populations. Z Naturforsch C J Biosci 62(9-10):627–635. https://doi.org/10.1515/znc-2007-9-1002

Guo C, Sun L, Chen X, Zhang D (2013) Oxidative stress, mitochondrial damage and neurodegenerative diseases. Neural Regen Res 8(21):2003–2014. https://doi.org/10.3969/j.issn.1673-5374.2013.21.009

Guo L, Du QQ, Cheng PQ, Yang TT, Xing CQ, Luo XZ, Peng XC, Qian F, Huang JR, Tang FR (2021) Neuroprotective effects of Lycium barbarum berry on neurobehavioral changes and neuronal loss in the hippocampus of mice exposed to acute ionizing radiation. Dose Response 19(4):15593258211057768. https://doi.org/10.1177/15593258211057768

Hanish Singh JC, Alagarsamy V, Diwan PV, Sathesh Kumar S, Nisha JC, Narsimha RY (2011) Neuroprotective effect of Alpinia galanga (L.) fractions on Aβ(25-35) induced amnesia in mice. J Ethnopharmacol 138(1):85–91. https://doi.org/10.1016/j.jep.2011.08.048

Hashim J, Vichitphan S, Boonsiri P, Vichitphan K (2021) Neuroprotective assessment of moringa oleifera leaves extract against oxidative-stress-induced cytotoxicity in SHSY5Y neuroblastoma cells. Plants (Basel) 10(5):889. https://doi.org/10.3390/plants10050889

Hazzaa SM, Abdelaziz SAM, Abd Eldaim MA, Abdel-Daim MM, Elgarawany GE (2020) Neuroprotective potential of allium sativum against monosodium glutamate-induced excitotoxicity: impact on short-term memory, gliosis, and oxidative stress. Nutrients 12(4): 1028. https://doi.org/10.3390/nu12041028

Hemanth Kumar K, Tamatam A, Pal A, Khanum F (2013) Neuroprotective effects of Cyperus rotundus on SIN-1 induced nitric oxide generation and protein nitration: ameliorative effect against apoptosis mediated neuronal cell damage. Neurotoxicology 34:150–159. https://doi.org/10.1016/j.neuro.2012.11.002

Homayoun M, Seghatoleslam M, Pourzaki M, Shafieian R, Hosseini M, Ebrahimzadeh BA (2015) Anticonvulsant and neuroprotective effects of Rosa damascena hydro-alcoholic extract on rat hippocampus. Avicenna J Phytomed 5(3):260–270

Hosseini M, Boskabady MH, Khazdair MR (2021) Neuroprotective effects of Coriandrum sativum and its constituent, linalool: a review. Avicenna J Phytomed 11(5):436–450. https://doi.org/10.22038/AJP.2021.55681.2786

Hu X, Qu Y, Chu Q, Li W, He J (2018) Investigation of the neuroprotective effects of Lycium barbarum water extract in apoptotic cells and Alzheimer's disease mice. Mol Med Rep 17(3): 3599–3606. https://doi.org/10.3892/mmr.2017.8310

Huang HJ, Chen SL, Chang YT, Chyuan JH, Hsieh-Li HM (2018) Administration of Momordica charantia enhances the neuroprotection and reduces the side effects of LiCl in the treatment of Alzheimer's disease. Nutrients 10(12):1888. https://doi.org/10.3390/nu10121888

Huang X, Li N, Pu Y, Zhang T, Wang B (2019) Neuroprotective effects of ginseng phytochemicals: recent perspectives. Molecules 24(16):2939. https://doi.org/10.3390/molecules24162939

Ibrahim DS (2017) Neuroprotective effect of Cucumis melo Var. flexuosus leaf extract on the brains of rats with streptozotocin-induced diabetes. Metab Brain Dis 32(1):69–75. https://doi.org/10.1007/s11011-016-9886-y

Ide K, Matsuoka N, Yamada H, Furushima D, Kawakami K (2018) Effects of tea catechins on Alzheimer's disease: recent updates and perspectives. Molecules 23(9):2357. https://doi.org/10.3390/molecules23092357

Ilah A, uz Zaman A, Bano S, Ismail F, Subhan S (2017) Walnut (Juglans regia L) and its neuroprotective action mechanism on brain: a review. In: Subhan S, Bagchi M (eds) Phytopharmaceuticals for brain health. CRC Press, Boca Raton. https://doi.org/10.1201/9781315152998-15

Islam MH, Ahmad IZ, Salman MT (2015) Neuroprotective effects of Nigella sativa extracts during germination on central nervous system. Pharmacogn Mag 11(Suppl 1):S182–S189. https://doi.org/10.4103/0973-1296.157729

Jaiswal B, Tailang M (2018) Protective effect of ethanolic extract from the root of argyreia speciosa against global cerebral ischemic reperfusion injury in rats. J Drug Deliv Ther 8:8–15. https://doi.org/10.22270/jddt.v8i6.1990

Jiji KN, Muralidharan P (2021) Neuroprotective effects of Clitoria ternatea L against propionic acid-induced behavior and memory impairment in autistic rat model. Futur J Pharm Sci 7:163. https://doi.org/10.1186/s43094-021-00314-3

Jittiwat J, Chonpathompikunlert P, Sukketsiri W (2021) Neuroprotective effects of Apium graveolens against focal cerebral ischemia occur partly via antioxidant, anti-inflammatory, and anti-apoptotic pathways. J Sci Food Agric 101(6):2256–2263. https://doi.org/10.1002/jsfa.10846

Johnson WM, Wilson-Delfosse AL, Mieyal JJ (2012) Dysregulation of glutathione homeostasis in neurodegenerative diseases. Nutrients 4(10):1399–1440. https://doi.org/10.3390/nu4101399

Joshi H, Parle M (2006) Nardostachys jatamansi improves learning and memory in mice. J Med Food 9(1):113–118. https://doi.org/10.1089/jmf.2006.9.113

Kaliyappan K, Sathyamoorthy Y, Nambi P, Radhakrishnan R (2021) Aqueous extract of Terminalia arjuna bark attenuates blood brain barrier disruption in rat model of transient focal cerebral ischemia. Phytomed Plus 1(3):100092. https://doi.org/10.1016/j.phyplu.2021.100092

Kamkaen N, Chittasupho C, Vorarat S, Tadtong S, Phrompittayarat W, Okonogi S, Kwankhao P (2022) Mucuna pruriens seed aqueous extract improved neuroprotective and acetylcholinesterase inhibitory effects compared with synthetic L-dopa. Molecules 27(10):3131. https://doi.org/10.3390/molecules27103131

Karakas FP, Turker AU, Karakas A, Mshvildadze V, Pichette A, Legault J (2017) In vitro cytotoxic, antibacterial, anti-inflammatory and antioxidant activities and phenolic content in wild-grown flowers of common daisy—a medicinal plant. J Herbal Med 8:31–39. https://doi.org/10.1016/j.hermed.2016.11.003

Kashyap P, Ram H, Shukla SD, Kumar S (2020) Scopoletin: antiamyloidogenic, anticholinesterase, and neuroprotective potential of a natural compound present in Argyreia speciosa roots by in vitro and in Silico study. Neurosci Insights 15:2633105520937693. https://doi.org/10.1177/2633105520937693

Keowkase R, Kijmankongkul N, Sangtian W et al (2020) Protective effect and mechanism of fruit extract of aegle marmelos against amyloid-β toxicity in a transgenic caenorhabditis elegans. Nat Prod Commun 15(7):1934578X2093351. https://doi.org/10.1177/1934578X20933511

Khan A, Vaibhav K, Javed H, Khan MM, Tabassum R, Ahmed ME, Raza S, Ashafaq M, Khuwaja G, Gupta V, Deora AK, Pant A, Siddiqui M, Islam F (2011) Neuroprotective effect of Bellis perennis and Hypericum perforatum on PC 12 cells. Indian J Res Homeopathy 5:27–35. https://doi.org/10.53945/2320-7094.1725

Khare P, Kishore K, Sharma DK (2022) Investigation of memory enhancing activity of Madhuca longifolia leaf extract against colchicine: an experimental study and biochemical alterations in mice. Indian J Pharm Educ Res 56(2s):s303–s311. https://doi.org/10.5530/ijper.56.2s.101

Kim ES, Weon JB, Yun BR, Lee J, Eom MR, Oh KH, Ma CJ (2014) Cognitive enhancing and neuroprotective effect of the embryo of the nelumbo nucifera seed. Evid Based Complement Alternat Med 2014:869831. https://doi.org/10.1155/2014/869831

Kukkar MR, Saluja AK, Sachdeva PD, Kukkar RR (2014) In vivo investigation of the neuroprotective potential of Cardiospermum halicacabum Linn. Int J Pharm Pharm Sci 6:64–66

Kumar V (2006) Potential medicinal plants for CNS disorders: an overview. Phytother Res 20: 1023–1035

Kumar GP, Khanum F (2012) Neuroprotective potential of phytochemicals. Pharmacogn Rev 6(12):81–90. https://doi.org/10.4103/0973-7847.99898

Lakshmi BV, Sudhakar M, Anisha M (2014) Neuroprotective role of hydroalcoholic extract of Vitis vinifera against aluminium-induced oxidative stress in rat brain. Neurotoxicology 41:73–79. https://doi.org/10.1016/j.neuro.2014.01.003

Lee E, Eom JE, Kim HL, Kang DH, Jun KY, Jung DS, Kwon Y (2012) Neuroprotective effect of undecylenic acid extracted from Ricinus communis L. through inhibition of μ-calpain. Eur J Pharm Sci 46(1-2):17–25. https://doi.org/10.1016/j.ejps.2012.01.015

Lei H, Ren R, Sun Y, Zhang K, Zhao X, Ablat N, Pu X (2020) Neuroprotective effects of safflower flavonoid extract in 6-hydroxydopamine-induced model of Parkinson's disease may be related to its anti-inflammatory action. Molecules 25(21):5206. https://doi.org/10.3390/molecules25215206

Levi S, Tiranti V (2019) Neurodegeneration with brain iron accumulation disorders: valuable models aimed at understanding the pathogenesis of iron deposition. Pharmaceuticals 12(1):27. https://doi.org/10.3390/ph12010027

Li JM, Zhou H, Cai Q, Xiao GX (2003) Role of mitochondrial dysfunction in H2O2-induced apoptosis of intestinal epithelial cells. World J Gastroenterol 9(3):562–567. https://doi.org/10.3748/wjg.v9.i3.562

Li D, Liu D, Lv M, Gao P, Liu X (2020) Isolation of triterpenoid saponins from Medicago sativa L. with neuroprotective activities. Bioorgan Med Chem Lett 30:126956. https://doi.org/10.1016/j.bmcl.2020.126956

Li J, Huang Q, Chen J, Qi H, Liu J, Chen Z, Zhao D, Wang Z, Li X (2021) Neuroprotective potentials of Panax Ginseng against Alzheimer's disease: a review of preclinical and clinical evidences. Front Pharmacol 12:688490. https://doi.org/10.3389/fphar.2021.688490

Liang Y, Lin W (2022) Carthamus tinctorius L.: a natural neuroprotective source for anti-Alzheimer's disease drugs. J Ethnopharmacol 298:115656. https://doi.org/10.1016/j.jep.2022.115656

Liguori I, Russo G, Curcio F, Bulli G, Aran L, Della-Morte D, Gargiulo G, Testa G, Cacciatore F, Bonaduce D, Abete P (2018) Oxidative stress, aging, and diseases. Clin Interv Aging 13:757–772. https://doi.org/10.2147/CIA.S158513

Limanaqi F, Biagioni F, Mastroiacovo F, Polzella M, Lazzeri G, Fornai F (2020) Merging the multi-target effects of phytochemicals in neurodegeneration: from oxidative stress to protein aggregation and inflammation. Antioxidants (Basel) 9(10):1022. https://doi.org/10.3390/antiox9101022

Liu X, Xu S, Ding X, Yue D, Bian J, Zhang X, Zhang G, Gao P (2020) Structural characteristics of Medicago Sativa L. polysaccharides and se-modified polysaccharides as well as their antioxidant and neuroprotective activities. Int J Biol Macromol 147:1099–1106. https://doi.org/10.1016/j.ijbiomac.2019.10.078

Lokanathan Y, Omar N, Ahmad Puzi NN, Saim A, Hj IR (2016) Recent updates in neuroprotective and neuroregenerative potential of Centella asiatica. Malays J Med Sci 23(1):4–14

Lopresti AL (2017) Salvia (sage): a review of its potential cognitive-enhancing and protective effects. Drugs R D 17(1):53–64. https://doi.org/10.1007/s40268-016-0157-5

Lum PT, Sekar M, Gan SH, Pandy V, Bonam SR (2021) Protective effect of mangiferin on memory impairment: a systematic review. Saudi J Biol Sci 28(1):917–927. https://doi.org/10.1016/j.sjbs.2020.11.037

Mairuae N, Connor JR, Buranrat B, Lee SY (2019) Oroxylum indicum (L.) extract protects human neuroblastoma SH-SY5Y cells against β-amyloid-induced cell injury. Mol Med Rep 20(2):1933–1942. https://doi.org/10.3892/mmr.2019.10411

Malik ZA, Singh M, Sharma PL (2011) Neuroprotective effect of Momordica charantia in global cerebral ischemia and reperfusion induced neuronal damage in diabetic mice. J Ethnopharmacol 133(2):729–734. https://doi.org/10.1016/j.jep.2010.10.061

Malik J, Karan M, Dogra R (2017) Ameliorating effect of Celastrus paniculatus standardized extract and its fractions on 3-nitropropionic acid induced neuronal damage in rats: possible antioxidant mechanism. Pharm Biol 55(1):980–990. https://doi.org/10.1080/13880209.2017.1285945

Mallah K, Couch C, Borucki DM, Toutonji A, Alshareef M, Tomlinson S (2020) Anti-inflammatory and neuroprotective agents in clinical trials for CNS disease and injury: where do we go from here? Front Immunol 11:2021. https://doi.org/10.3389/fimmu.2020.02021

Mansouri S, Hosseini M, Alipour F, Beheshti F, Rakhshandeh H, Mohammadipour A, Anaeigoudari A, Jalili-Nik M, Khazdair MR, Jahani A (2022) Neuroprotective effects of the fractions of Ocimum basilicum in seizures induced by pentylenetetrazole in mice. Avicenna J Phytomed 12(6):614–626. https://doi.org/10.22038/AJP.2022.20470

Manyam BV, Dhanasekaran M, Hare TA (2004) Neuroprotective effects of the antiparkinson drug Mucuna pruriens. Phytother Res 18(9):706–712. https://doi.org/10.1002/ptr.1514

Marques T, Melo C, Carvalho R, Costa L, Araujo de Souza A, David J, David J, Freitas R (2013) Phytochemical profile and qualification of biological activity of an isolated fraction of bellis perennis. Biol Res 46:231–238. https://doi.org/10.4067/S0716-97602013000300002

Marques C, Burg T, Scekic-Zahirovic J, Fischer M, Rouaux C (2021) Upper and lower motor neuron degenerations are somatotopically related and temporally ordered in the Sod1 mouse model of amyotrophic lateral sclerosis. Brain Sci 11(3):369. https://doi.org/10.3390/brainsci11030369

Menichini F, Losi L, Bonesi M, Pugliese A, Loizzo MR, Tundis R (2014) Chemical profiling and in vitro biological effects of Cardiospermum halicacabum L. (Sapindaceae) aerial parts and seeds for applications in neurodegenerative disorders. J Enzyme Inhib Med Chem 29(5):677–685. https://doi.org/10.3109/14756366.2013.840614

Mensah-Kane P, Sumien N (2023) The potential of hyperbaric oxygen as a therapy for neurodegenerative diseases. Geroscience 45(2):747–756. https://doi.org/10.1007/s11357-022-00707-z

Mhyre TR, Boyd JT, Hamill RW, Maguire-Zeiss KA (2012) Parkinson's disease. Subcell Biochem 65:389–455. https://doi.org/10.1007/978-94-007-5416-4_16

Mir NT, Saleem U, Anwar F, Ahmad B, Ullah I, Hira S, Ismail T, Ali T, Ayaz M (2019) Lawsonia inermis markedly improves cognitive functions in animal models and modulate oxidative stress markers in the brain. Medicina (Kaunas) 55(5):192. https://doi.org/10.3390/medicina55050192

Mokhber-Dezfuli N, Saeidnia S, Gohari AR, Kurepaz-Mahmoodabadi M (2014) Phytochemistry and pharmacology of berberis species. Pharmacogn Rev 8(15):8–15. https://doi.org/10.4103/0973-7847.125517

Monika S, Thirumal M, Kumar PR (2023) Phytochemical and biological review of Aegle marmelos Linn. Future Sci OA 9(3):FSO849. https://doi.org/10.2144/fsoa-2022-0068

Mundkar M, Bijalwan A, Soni D, Kumar P (2022) Neuroprotective potential of Moringa oleifera mediated by NF-kB/Nrf2/HO-1 signaling pathway: a review. J Food Biochem 46(12):e14451. https://doi.org/10.1111/jfbc.14451

Murphy MP, LeVine H 3rd. (2010) Alzheimer's disease and the amyloid-beta peptide. J Alzheimers Dis 19(1):311–323. https://doi.org/10.3233/JAD-2010-1221

Naoi M, Wu Y, Shamoto-Nagai M, Maruyama W (2019) Mitochondria in neuroprotection by phytochemicals: bioactive polyphenols modulate mitochondrial apoptosis system, function and structure. Int J Mol Sci 20(10):2451. https://doi.org/10.3390/ijms20102451

Naoi M, Maruyama W, Shamoto-Nagai M (2022) Neuroprotective function of rasagiline and selegiline, inhibitors of type B monoamine oxidase, and role of monoamine oxidases in synucleinopathies. Int J Mol Sci 23(19):11059. https://doi.org/10.3390/ijms231911059

Nebrisi EE (2021) Neuroprotective activities of Curcumin in Parkinson's disease: a review of the literature. Int J Mol Sci 22(20):11248. https://doi.org/10.3390/ijms222011248

Orhan IE, Suntar IP, Akkol EK (2011) In vitro neuroprotective effects of the leaf and fruit extracts of Juglans regia L. (walnut) through enzymes linked to Alzheimer's disease and antioxidant activity. Int J Food Sci Nutr 62(8):781–786. https://doi.org/10.3109/09637486.2011.585964

Pagano G, Aiello Talamanca A, Castello G, Cordero MD, d'Ischia M, Gadaleta MN, Pallardó FV, Petrović S, Tiano L, Zatterale A (2014) Current experience in testing mitochondrial nutrients in disorders featuring oxidative stress and mitochondrial dysfunction: rational design of chemo-prevention trials. Int J Mol Sci 15(11):20169–20208. https://doi.org/10.3390/ijms151120169

Park S-S, Park N-O, Kang J-U, Shin S-C, Lee D-U (2009) Cognition enhancing effect of musk-melon (cucumis melo) extracts on scopolamine-induced memory impairment in mice. J Life Sci 19:688–691. https://doi.org/10.5352/JLS.2009.19.5.688

Pastorino G, Cornara L, Soares S, Rodrigues F, Oliveira MBPP (2018) Liquorice (Glycyrrhiza glabra): a phytochemical and pharmacological review. Phytother Res 32(12):2323–2339. https://doi.org/10.1002/ptr.6178

Paudel YN, Angelopoulou E, Semple B, Piperi C, Othman I, Shaikh MF (2020) Potential neuroprotective effect of the HMGB1 inhibitor glycyrrhizin in neurological disorders. ACS Chem Neurosci 11(4):485–500. https://doi.org/10.1021/acschemneuro.9b00640

Phan HTT, Samarat K, Takamura Y, Azo-Oussou AF, Nakazono Y, Vestergaard MC (2019) Polyphenols modulate Alzheimer's amyloid beta aggregation in a structure-dependent manner. Nutrients 11(4):756. https://doi.org/10.3390/nu11040756

Pondugula SR, Majrashi M, Almaghrabi M, Ramesh S, Abbott KL, Govindarajulu M, Gill K, Fahoury E, Narayanan N, Desai D, Ren J, Nadar R, McElroy T, Moore T, Majeed M, Kalyanam N, Dhanasekaran M (2021) Oroxylum Indicum ameliorates chemotherapy induced cognitive impairment. PLoS One 16(6):e0252522. https://doi.org/10.1371/journal.pone.0252522

Rad ES, Eidi A, Minai-Tehrani D, Bonakdar S, Shoeibi S (2022) Neuroprotective effect of root extracts of berberis vulgaris (barberry) on oxidative stress on SH-SY5Y cells. J Pharmacopuncture 25(3):216–223. https://doi.org/10.3831/KPI.2022.25.3

Raheja S, Girdhar A, Kamboj A, Lather V, Pandita D (2019) Aegle marmelos leaf extract ameliorates the cognitive impairment and oxidative stress induced by intracerebroventricular streptozotocin in male rats. Life Sci 221:196–203. https://doi.org/10.1016/j.lfs.2019.02.032

Raheja S, Girdhar A, Kamboj A, Lather V, Pandita D (2021) Protective effect of Dalbergia sissoo extract against amyloid-β (1-42)-induced memory impairment, oxidative stress, and neuroinflammation in rats. Turk J Pharm Sci 18(1):104–110. https://doi.org/10.4274/tjps.galenos.2020.04379

Rahman MA, Rahman MH, Biswas P, Hossain MS, Islam R, Hannan MA, Uddin MJ, Rhim H (2020) Potential therapeutic role of phytochemicals to mitigate mitochondrial dysfunctions in Alzheimer's disease. Antioxidants (Basel) 10(1):23. https://doi.org/10.3390/antiox10010023

Rapaka D, Bitra VR, Vishala TC, Akula A (2019) Vitis vinifera acts as anti-Alzheimer's agent by modulating biochemical parameters implicated in cognition and memory. J Ayurveda Integr Med 10(4):241–247. https://doi.org/10.1016/j.jaim.2017.06.013

Rehman MU, Wali AF, Ahmad A, Shakeel S, Rasool S, Ali R, Rashid SM, Madkhali H, Ganaie MA, Khan R (2019) Neuroprotective strategies for neurological disorders by natural products: an update. Curr Neuropharmacol 17(3):247–267. https://doi.org/10.2174/1570159X16666180911124605

Sailike B, Omarova Z, Jenis J, Adilbayev A, Akbay B, Askarova S, Jin W-L, Tokay T (2022) Neuroprotective and anti-epileptic potentials of genus Artemisia L. Front Pharmacol 13:1021501. https://doi.org/10.3389/fphar.2022.1021501

Saini N, Kadian M, Khera A, Aggarwal A, Kumar A (2021) Therapeutic potential of Allium Sativum against the Aβ(1-40)-induced oxidative stress and mitochondrial dysfunction in the Wistar rats. Am J Neurodegener Dis 10(2):13–27

Sandhir R, Khurana M, Singhal NK (2021) Potential benefits of phytochemicals from Azadirachta indica against neurological disorders. Neurochem Int 146:105023. https://doi.org/10.1016/j.neuint.2021.105023

Sari Y (2011) Huntington's disease: from mutant Huntingtin protein to neurotrophic factor therapy. Int J Biomed Sci 7(2):89–100

Schieber M, Chandel NS (2014) ROS function in redox signaling and oxidative stress. Curr Biol 24(10):R453–R462. https://doi.org/10.1016/j.cub.2014.03.034

Seyed MA, Ayesha S, Azmi N et al (2021) The neuroprotective attribution of Ocimum basilicum: a review on the prevention and management of neurodegenerative disorders. Futur J Pharm Sci 7:139. https://doi.org/10.1186/s43094-021-00295-3

Shahnas N, Akhila S (2014) Phytochemical, in vitro and in silico evaluation on clitoria ternatea for Alzheimer's disease. PharmaTutor 2(9):135–149

Shen YC, Juan CW, Lin CS, Chen CC, Chang CL (2017) Neuroprotective effect of terminalia chebula extracts and ellagic acid in PC12 cells. Afr J Tradit Complement Altern Med 14(4):22–30. https://doi.org/10.21010/ajtcam.v14i4.3

Shivasharan BD, Nagakannan P, Thippeswamy BS, Veerapur VP (2013) Protective effect of Calendula officinalis L. flowers against monosodium glutamate induced oxidative stress and excitotoxic brain damage in rats. Indian J Clin Biochem 28(3):292–298. https://doi.org/10.1007/s12291-012-0256-1

Siddiqui S, Saify ZS, Jamali KS, Tufail P, Kanwal A, Kamal A, Khan F (2018) Neuroprotective capabilities of Vitex negundo in primary hippocampal neurons. Pak J Pharm Sci 31(1 Suppl):341–344

Singh N, Gupta VK, Kumar A, Sharma B (2017) Synergistic effects of heavy metals and pesticides in living systems. Front Chem 5:1–9. https://doi.org/10.3389/fchem.2017.00070

Singh N, Kumar A, Gupta VK, Sharma B (2018) Biochemical and molecular bases of lead-induced toxicity in mammalian systems and possible mitigations. Chem Res Toxicol 31:1009–1021. https://doi.org/10.1021/acs.chemrestox.8b00193

Singh A, Kukreti R, Saso L, Kukreti S (2019) Oxidative stress: a key modulator in neurodegenerative diseases. Molecules 24(8):1583. https://doi.org/10.3390/molecules24081583

Soeda S, Aritake K, Urade Y, Sato H, Shoyama Y (2016) Neuroprotective activities of Saffron and Crocin. Adv Neurobiol 12:275–292. https://doi.org/10.1007/978-3-319-28383-8_14

Sun HJ, Rathinasabapathi B, Wu B, Luo J, Pu LP, Ma LQ (2014) Arsenic and selenium toxicity and their interactive effects in humans. Environ Int 69:148–158. https://doi.org/10.1016/j.envint.2014.04.019

Sutalangka C, Wattanathorn J (2017) Neuroprotective and cognitive-enhancing effects of the combined extract of Cyperus rotundus and Zingiber officinale. BMC Complement Altern Med 17(1):135. https://doi.org/10.1186/s12906-017-1632-4

Swaroop TV, Banerjee S, Handral M (2014) Neuroprotective evaluation of leaf extract of dalbergia sissoo in 3-nitropropionic acid induced neurotoxicity in rats. Int J Pharm Sci Drug Res 6(1): 41–47. https://www.ijpsdr.com/index.php/ijpsdr/article/view/294

Talpate KA, Bhosale UA, Zambare MR, Somani RS (2014) Neuroprotective and nootropic activity of Clitorea ternatea Linn.(Fabaceae) leaves on diabetes induced cognitive decline in experimental animals. J Pharm Bioallied Sci 6(1):48–55. https://doi.org/10.4103/0975-7406.124317

Tamilselvam K, Braidy N, Manivasagam T, Essa MM, Prasad NR, Karthikeyan S, Thenmozhi AJ, Selvaraju S, Guillemin GJ (2013) Neuroprotective effects of hesperidin, a plant flavanone, on rotenone-induced oxidative stress and apoptosis in a cellular model for Parkinson's disease. Oxid Med Cell Longev 2013:102741. https://doi.org/10.1155/2013/102741

Tchounwou PB, Ayensu WK, Ninashvili N, Sutton D (2003) Environmental exposure to mercury and its toxicopathologic implications for public health. Environ Toxicol 18:149–175. https://doi.org/10.1002/tox.10116

Teerapattarakan N, Benya-aphikul H, Tansawat R, Wanakhachornkrai O, Tantisira MH, Rodsiri R (2018) Neuroprotective effect of a standardized extract of Centella asiatica ECa233 in rotenone-induced parkinsonism rats. Phytomedicine 44:65–73. https://doi.org/10.1016/j.phymed.2018.04.028

Teixeira J, Deus CM, Borges F, Oliveira P (2018) Mitochondria: targeting mitochondrial reactive oxygen species with mitochondriotropic polyphenolic-based antioxidants. Int J Biochem Cell Biol 97:98–103. https://doi.org/10.1016/j.biocel.2018.02.007

Temviriyanukul P, Sritalahareuthai V, Promyos N, Thangsiri S, Pruesapan K, Srinuanchai W, Nuchuchua O, Siriwan D, On-Nom N, Suttisansanee U (2020) The effect of sacred lotus (nelumbo nucifera) and its mixtures on phenolic profiles, antioxidant activities, and inhibitions of the key enzymes relevant to Alzheimer's disease. Molecules 25(16):3713. https://doi.org/10.3390/molecules25163713

Temviriyanukul P, Kittibunchakul S, Trisonthi P, Kunkeaw T, Inthachat W, Siriwan D, Suttisansanee U (2022) Mangifera indica 'Namdokmai' prevents neuronal cells from amyloid peptide toxicity and inhibits BACE-1 activities in a drosophila model of Alzheimer's amyloidosis. Pharmaceuticals (Basel) 15(5):591. https://doi.org/10.3390/ph15050591

Thabit S, Handoussa H, Roxo M, El Sayed NS, Cestari de Azevedo B, Wink M (2018) Evaluation of antioxidant and neuroprotective activities of Cassia fistula (L.) using the Caenorhabditis elegans model. PeerJ 6:e5159. https://doi.org/10.7717/peerj.5159

Tonelli C, Chio IIC, Tuveson DA (2018) Transcriptional regulation by Nrf2. Antioxid Redox Signal 29(17):1727–1745. https://doi.org/10.1089/ars.2017.7342

Tumen I, Senol FS, Orhan IE (2012) Inhibitory potential of the leaves and berries of Myrtus communis L. (myrtle) against enzymes linked to neurodegenerative diseases and their antioxidant actions. Int J Food Sci Nutr 63(4):387–392. https://doi.org/10.3109/09637486.2011.629178

Turgut NH, Mert DG, Kara H, Egilmez HR, Arslanbas E, Tepe B, Gungor H, Yilmaz N, Tuncel NB (2016) Effect of black mulberry (Morus nigra) extract treatment on cognitive impairment and oxidative stress status of d-galactose-induced aging mice. Pharm Biol 54(6):1052–1064. https://doi.org/10.3109/13880209.2015.1101476

Ukkirapandian K, Kayalvizhi E, Udaykumar KP, Kandhi S, Muthulakshmi R (2022) The neuroprotective role of acorus calamus in developmental and histopathological changes in autism-induced wistar rats. Cureus 14(9):e29717. https://doi.org/10.7759/cureus.29717

Uttara B, Singh AV, Zamboni P, Mahajan RT (2009) Oxidative stress and neurodegenerative diseases: a review of upstream and downstream antioxidant therapeutic options. Curr Neuropharmacol 7(1):65–74. https://doi.org/10.2174/157015909787602823

Vauzour D, Vafeiadou K, Rodriguez-Mateos A, Rendeiro C, Spencer JP (2008) The neuroprotective potential of flavonoids: a multiplicity of effects. Genes Nutr 3(3-4):115–126. https://doi.org/10.1007/s12263-008-0091-4

Vekaria R, Patel M, Bhalodiya P, Patel V, Desai T, Tirgar P (2012) Evaluation of neuroprotective effect of Coriandrum sativum linn. Against ischemic-reperfusion insult in brain. Int J Phytopharmacol 3:186–193

Wan D, Zhou Y, Wang K, Hou Y, Hou R, Ye X (2016) Resveratrol provides neuroprotection by inhibiting phosphodiesterases and regulating the cAMP/AMPK/SIRT1 pathway after stroke in rats. Brain Res Bull 121:255–262. https://doi.org/10.1016/j.brainresbull.2016.02.011

Wang M, Ye H, Jiang P, Liu J, Wang B, Zhang S, Sik A, Li N, Liu K, Jin M (2023) The alleviative effect of Calendula officinalis L. extract against Parkinson's disease-like pathology in zebrafish via the involvement of autophagy activation. Front Neurosci 17:1153889. https://doi.org/10.3389/fnins.2023.1153889

Wongtrakul J, Thongtan T, Kumrapich B, Saisawang C, Ketterman AJ (2021) Neuroprotective effects of Withania somnifera in the SH-SY5Y Parkinson cell model. Heliyon 7(10):e08172. https://doi.org/10.1016/j.heliyon.2021.e08172

Xiang X, Wu L, Mao L, Liu Y (2018) Anti-oxidative and anti-apoptotic neuroprotective effects of Azadirachta indica in Parkinson-induced functional damage. Mol Med Rep 17:7959–7965. https://doi.org/10.3892/mmr.2018.8815

Xie L, Zhu Q, Lu J (2022) Can we use *Ginkgo biloba* extract to treat Alzheimer's disease? Lessons from preclinical and clinical studies. Cell 11(3):479. https://doi.org/10.3390/cells11030479

Xu J, Knutson MD, Carter CS, Leeuwenburgh C (2008) Iron accumulation with age, oxidative stress and functional decline. PLoS One 3(8):e2865. https://doi.org/10.1371/journal.pone.0002865

Yaidikar L, Thakur S (2015) Arjunolic acid, a pentacyclic triterpenoidal saponin of Terminalia Arjuna bark protects neurons from oxidative stress associated damage in focal cerebral ischemia and reperfusion. Pharmacol Rep 67(5):890–895. https://doi.org/10.1016/j.pharep.2015.02.003

Zhao RZ, Jiang S, Zhang L, Yu ZB (2019) Mitochondrial electron transport chain, ROS generation and uncoupling (review). Int J Mol Med 44(1):3–15. https://doi.org/10.3892/ijmm.2019.4188

Zhao L, Duan Z, Yu W, Wang M, Liu Y, Wang X, Li H (2021) Protective effect of Terminalia chebula Retz. extract against Aβ aggregation and Aβ-induced toxicity in Caenorhabditis elegans. J Ethnopharmacol 268:113640. https://doi.org/10.1016/j.jep.2020.113640

Zhou J, Yang Z, Shen R, Zhong W, Zheng H, Chen Z, Tang J, Zhu J (2021) Resveratrol improves mitochondrial biogenesis function and activates PGC-1α pathway in a preclinical model of early brain injury following subarachnoid hemorrhage. Front Mol Biosci 8:620683. https://doi.org/10.3389/fmolb.2021.620683

Zorov DB, Juhaszova M, Sollott SJ (2014) Mitochondrial reactive oxygen species (ROS) and ROS-induced ROS release. Physiol Rev 94(3):909–950. https://doi.org/10.1152/physrev.00026.2013

# Phytochemical Supplementation Regulates Immune Function and Modulates Cell Survival Signaling, Enhancing Neuroprotection and Cognitive Functions in Aging

6

Ramasamy Vasantharekha, Barathi Seetharaman, and Winkins Santosh

**Abstract**

Phytochemicals present in plant sources, including fruits and vegetables, both fresh and processed, have been known for their positive effects on human health. Phytochemical supplementation regulates immune functions through immunomodulation, controls cell survival signaling through its antioxidant properties, and acts as a selective modulator of signal transduction. Their roles may be stimulatory or suppressive, depending on the quantity of phytochemicals consumed. Phytochemicals in green tea can improve both cellular and humoral immune responses. Resveratrol exhibits anti-inflammatory and immunomodulatory properties. Carotenoids boost both specific and nonspecific immune activity. Flavonoids have antioxidant, antiallergic, anti-inflammatory, antithrombotic, antibacterial, antiviral, amyloidogenic, hepatoprotective, neurotrophic protective, anticarcinogenic, and immunomodulatory properties. Quercetin and curcumin, which are dietary polyphenols, significantly impact the modulation of numerous cell signaling networks and promote healthy aging. The critical signaling pathways that increase the prolonged existence of the organism in aging are AMPK, Sirtuin 1 (SIRT1), mTOR, Akt, IGF, SKA, and PKA. The phytochemicals modulated growth factors and proteins involved in inflammatory pathways and apoptosis, including NF-κB, Bad, caspase 9, Fast, and Bcl-2. Aging is a risk factor for most diseases, and phytochemicals act as immunomodulators by diminishing immunosenescence and enhancing

R. Vasantharekha · B. Seetharaman
Endocrine Disruption and Reproductive Toxicology Laboratory, Department of Biotechnology, School of BioEngineering, SRM Institute of Science & Technology, Kattankulathur, India

W. Santosh (✉)
PG & Research Department of Advanced Zoology & Biotechnology, Government Arts College for Men, Nandanam, Chennai, Tamilnadu, India



S. Pathak, A. Banerjee (eds.), *Neuroprotective Effects of Phytochemicals in Brain Ageing*, https://doi.org/10.1007/978-981-99-7269-2_6

antioxidant activity. They also help decrease oxidative stress on cells and modify cellular senescence, ultimately reducing the rate of diseases associated with aging and promoting healthy aging. This review will concentrate on the different sources of phytochemicals, their properties, and the defensive mechanisms of aging.

**Keywords**

Phytochemicals · Neuroprotection · Cognitive functions · Healthy ageing

## 6.1 Introduction

Aging is a leading risk factor for several potentially fatal diseases of the nervous, immune, endocrine, and metabolic dysregulation. The diseases that are more prone to aging are neurological diseases, cardiovascular disease (CVD), autoimmune diseases, endocrine disorders, cancer, and diabetes. An increase in age reduces proteostasis network activity due to accumulating stressors and damaged biomolecules. There is growing evidence that certain phytochemicals can retain redox homeostasis and have potential as therapeutic agents due to their bio-affinity, bio-durability, bio-transformation, bio-degradation, harmlessness, and lower side effects. Polyphenols, flavonoids, carotenoids, terpenoids, resveratrol, and catechins are the phytochemicals present as secondary metabolites in plants. Surprisingly, they exhibit antioxidant and prooxidant effects that influence cell growth and apoptosis. Many animal models, in vivo, epidemiological, and clinical trial studies have shown that herbs, fruits, vegetables, and foods are abundant sources of polyphenols, which show antioxidant and anti-inflammatory properties. Phytochemicals present in bacosides (*Bacopa monnieri*), rosemary acid (R.A.), Quercetin (Qu), and soy isoflavones (SIF) can consequently alter the antioxidant environment through the alterations of Nrf2, NQO1, and HO-1 signaling pathway in many neurodegenerative diseases like cancer, diabetes, spinal cord injury, and Parkinson's, and Alzheimer's disease (Singh et al. 2012; Sun et al. 2019; Ke et al. 2020; Ma et al. 2020). LPS mediates inflammation through pathways like NF-κB, ERK1/2, and Akt. The anti-inflammatory properties of many bioactive compounds can restore oxidative stress by silencing inflammatory mediators and enzymes like ROS, NO, COXs, iNOS, NF-κB, and AP-1. Most active compounds have shown some beneficial effects on several neurodegenerative diseases and were recommended to have various therapeutic effects through their oxidative stress-related activities. Dietary supplements or genetic interventions that alter oxidative stress-related pathways, such as antioxidant response element (ARE), Nrf2, etc., can slow down the aging process by reducing the influence of stressors and the level at which impaired biomolecule accumulation occurs.

## 6.2 Phytochemicals: Sources and Properties

The essential sources of phytochemicals are fruits and vegetables in the human diet, whether fresh or processed, highlighting their immense importance (Oz and Kafkas 2017). Approximately 200,000 bioactive compounds have been identified, with around 20,000 traced back to fruits, vegetables, chocolate, teas, grains, and legumes, making them excellent sources of these beneficial compounds. The primary sources of carotenoids are plants and vegetables, and these vegetable phytochemicals are either free compounds or bound to carbohydrate structures (Olsson et al. 2010; Patra 2012; Aguilera and Benítez 2021). A wide variety of phytochemicals, including acetophenones, phenolics, alkaloids, terpenoids, flavonoids, stilbenes, fiber, and condensed tannins, as well as less MW phenolic compounds like hydroxybenzoic and hydroxycinnamic acids, can be produced and stored by citrus plants (Okwu 2008; Issinger and Guerra 2021). Plants synthesize phytochemicals as a defense mechanism against environmental challenges, including fluctuations in water levels and microorganisms. These compounds also contribute to the distinctive color, flavor, aroma, and texture exhibited by plants. Further, recent studies have revealed that phytochemicals significantly affect human health (Tiwari et al. 2015) (Fig. 6.1).

Dietary phytochemicals found in plant-based foods possess medicinal and nutritional properties that aid in preventing different illnesses and promoting overall human health. As plant secondary metabolites, these phytochemicals are crucial in supporting the optimal functioning of the human body, particularly as antioxidants (Islam et al. 2020). By providing a protective shield against the harmful impact of free radicals, they serve as a barrier for cells. These radicals are unstable molecules that cause oxidative stress, potentially damaging cells and increasing the threat of chronic diseases like CVD, diabetes, and cancer. Some phytochemicals, such as carotenoids and polyphenols, possess potent antioxidant properties, effectively countering free radicals and alleviating oxidative stress. They help decrease oxidative damage to cells, diminishing the onset of aging (Shigenaga et al. 1994).

## 6.3 Phytochemicals Modulate Immune Function

The immune system deteriorates and changes with aging, which has detrimental effects, including immunosenescence, chronic inflammation, and immunological dysfunction (Chintapula et al. 2023). Substances known as immunomodulators are capable of regulating the immune system. There are three types: immunosuppressants, immunoadjuvants, and immunostimulants. Phytochemicals have become increasingly popular over time due to their diverse pharmacological effects, including immunomodulation (Behl et al. 2021) (Table 6.1).

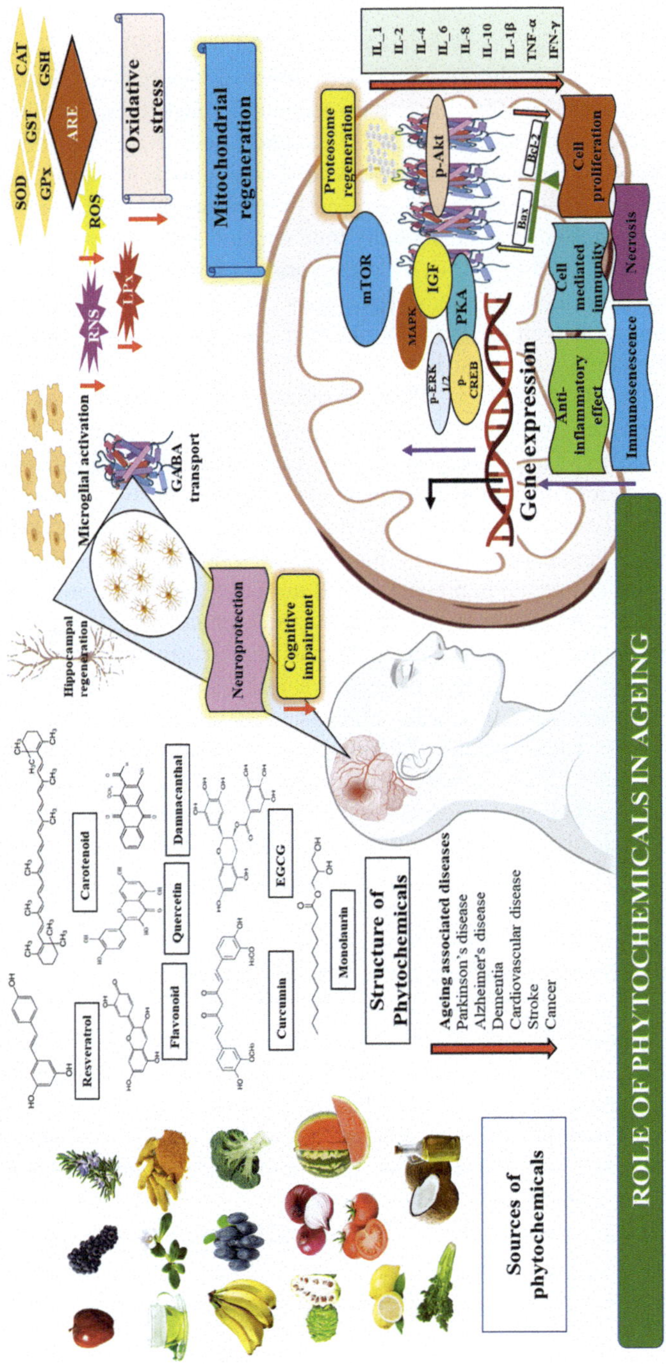


**Fig. 6.1** Source, structure and role of phytochemicals in ageing

**Table 6.1** Sources of phytochemicals and their functions (Howes et al. 2020)

| Phytochemicals | Source | Function |
|---|---|---|
| Reservatrol | Merlot, Black Grapes, Peanut Products, etc. | Modulates cell growth and anticarcinogenic properties (Martin et al. 2016).Immunomodulator/ anti-inflammatory agent (Gao et al. 2001) |
| Carotenoids | Banana fruit (Sidhu and Zafar 2018) oranges (Silalahi 2002) Watermelon (Kulczyński et al. 2017) | Anticarcinogenic, unique antioxidant properties |
| Flavanoids | Citrus fruits and flavones in celery | antioxidant, antithrombotic, antibacterial, anti-viral, antiallergic, anti-inflammatory, hepatoprotective, anticarcinogenic,and immunomodulatory properties (Cushnie and Lamb 2005) |
| Quercetin | Apples, grapes, tea, berries, brassica, capers, shallots, onions, tomatoes, many seeds, nuts, barks, flowers, and leaves | Treating inflammatory conditions, cardiovascular diseases, and respiratory infections (Li et al. 2016) |
| Curcumin | Curcuminoid powder (*Curcuma longa*) | Anti-cancer<br>Anti-arthritic, Analgesic |
| Terpenoids | M. koenigii Linn | Anxiolytic properties (Lu et al. 2006) |
| Damnacanthal | Noni (Pratap et al. 2016) | Hamsters fed a high-fat/cholesterol diet exhibit lower levels of TBARS, indicating a reduction in serum lipid peroxidation. (Lin et al. 2012) |

### 6.3.1 Green Tea

The contemporary utilization of phytochemicals to enhance immune function and combat diseases stems from ancient traditions that have been practiced for centuries. The significant role of green tea and its components in the treatment and prevention of different types of cancer through dietary interventions has been emphasized by recent developments. There is a belief that green tea has the potential to boost both humoral and cellular immunity, thereby reducing the likelihood of specific types of cancer. Additionally, it may offer advantages in the treatment of inflammatory disorders. Epigallocatechin gallate (EGCG) compounds present in tea leaves induce apoptosis, suppress cell growth, alter the cell cycle regulatory proteins expression, trigger caspases involved in cell death, and inhibit the triggering of NF-κB. Green tea also enhances and regulates IL-23-dependent DNA repair mechanisms and improves the performance of cytotoxic T cells in the tumor microenvironment. By modifying the signal transduction pathways responsible for cell growth, differentiation, inflammatory response, and metastasis, green tea can prevent the development of cancer (Butt and Sultan 2009). Polyphenols and EGCG are the main ingredients present in green tea have some anticancerous properties and have been studied for the past few decades. Studies on both intervention and epidemiology have provided evidence of the clinical importance and significant impact of green tea intake in preventing cancer. The chemopreventive effects of GTPs have been demonstrated in different types of cancers like prostrate (Singh et al. 2011), breast (Belguise et al. 2007), lung (Lu et al. 2006), and colorectal cancers (Larsen et al. 2009). GTPs have been found to have anticancer efficacy in a variety of malignant cell lines and non-human tumor models, including those of the oral cavity, stomach, skin, lung, colon, small intestine, esophagus, liver, pancreas, and mammary gland (Yang et al. 2002).

Green tea catechins (GTC) showed inhibitory effects on specific stages of carcinogenesis. Insights into the mechanisms by which green tea catechins regulate mammary carcinogenesis have been revealed through in vitro studies. Further, exploration through human trials that integrate these in-vitro findings could assess the relevance and applicability of these mechanisms in humans (Yiannakopoulou 2014). GTC has also been observed to suppress tumor development induced by DEN and NNK in the lung and forestomach of the mouse, as well as mouse lung carcinogenesis (Hirose et al. 1994).

### 6.3.2 Resveratrol

Resveratrol (RES) is a polyphenolic phytoalexin compound known as trans-3.4′,5-trihydroxystilbene, that is small and not classified as a flavonoid. It is present in wine (especially the skin of grapes), fruits, peanuts, and certain plants. It demonstrates effects on modulating cell growth and possesses anticarcinogenic properties (Martin et al. 2016). The metabolism of RES primarily takes place in the liver, resulting in the production of numerous detectable metabolites that is found in urine (Boocock et al. 2007). Considering its many biological activities, RES is recommended as a nutritional supplement for human use to treat immunological disorders (Walle 2011). RES inhibits the spleen cell proliferation produced by concanavalin A (ConA), IL-2, or alloantigens, acting as an immunomodulator and anti-inflammatory drug.

Moreover, it inhibits the production of IL-2 and IFN-γ by lymphocyte activation, as well as the TNF-α or IL-12 production by macrophage activation (Gao et al. 2001), it also interacts with various molecular targets and regulates innate and acquired immunity. At low doses, RES acts as an immunostimulant, whereas at high doses, it can act as an immunosuppressant. When tested on rats, RES showed the ability to decrease inflammation in the peritoneum membrane, enhance the body's defenses against cancerous cells, and reverse immunosenescence in older rats.

Furthermore, RES has been discovered in the activation of macrophages, T cells, and N.K. cells while also having the ability to regulate the suppressive functions of CD4+ CD25+ regulatory T-cells (Falchetti et al. 2001). The primary mechanisms of action through which RES exhibits anti-aging effects include the reduction of free radicals, COX inhibition, and the stimulation of various anti-inflammatory pathways and Sirt1 (Wiciński et al. 2017). Specifically, Sirt1 plays a crucial role in inhibiting signaling pathways like NF-κB/TLR4/STAT, decreasing cytokine secretion from quiescent immune cells. It also modulates the production of pro-inflammatory factors, including histamine, platelet-activating factor (PAF), and TNF-α, derived from macrophages and mast cells (Wiciński et al. 2018). The transcription factor STAT3 can be deacetylated by RES, interfering with nuclear factors essential for differentiating Th17 cells (Delmas et al. 2020). Recent research has established a correlation between the antitumor effects of resveratrol and its capacity to bolster antitumor immunity and counteract the immunosuppressive environment within tumors. RES acts explicitly by restoring the function of T-cells and targeting the immune barrier signaling pathway PD-1/PD-L1, thus exerting an immunomodulatory mechanism. Additionally, resveratrol can upregulate c-Myb through AKT- and mTORC2-mediated pathways, activating N.K. cells (Chhabra et al. 2021).

### 6.3.3 Carotenoids

Banana fruit contains carotenoids such as α-carotene, β-carotene, isolectin, and β-cryptoxanthin. These carotenoids have been found to worsen the risk of emerging eye and heart diseases. Additionally, they function as antioxidants and enhance the immune system (Sidhu and Zafar 2018). The carotenoids found in oranges improve both specific and nonspecific immune activity. According to studies, they also play a part in the secretion of immunomodulatory lipid molecules like leukotrienes and prostaglandins, neutralize reactive species produced by immune cells, eliminate immunosuppressive peroxides, and help preserve essential membrane receptors involved in immune functions (Silalahi 2002). Watermelon contains phytochemicals, including β-carotene, which promote platelet accumulation, and this process is vital for repairing and rebuilding blood vessels (Kulczyński et al. 2017). One of the primary phytochemicals with potential anticancer properties found in tomato products is lycopene, which is the main carotenoid present in tomatoes (Clinton et al. 1996). The carotenoids and polyphenols found in tomatoes possess unique antioxidant properties that help neutralize free radicals. These phytochemicals in tomatoes have been recognized for their ability to impede tumor growth by interfering with cancer initiation, promotion, or development (Martí et al. 2016).

### 6.3.4 Flavonoids

Flavonoids have demonstrated a wide range of effects in various in vitro and in vivo studies, such as antioxidant, antithrombotic, antibacterial, antiviral, antiallergic, anti-inflammatory, hepatoprotective, anticarcinogenic, and immunomodulatory activities (Cushnie and Lamb 2005). Flavonoids have been recognized for their ability to modulate various biological processes, including oxidation, detoxification enzymes, apoptosis, and the host immune system. Their likely influence on cellular activities, including safeguarding genomic integrity, indicates that flavonoids could be utilized as dietary compounds to prevent cancer and chronically worsening diseases (Tapas et al. 2008). In in vivo studies, it has been observed that flavonoids exhibit the ability to inhibit the degranulation of mast cells, basophils, and neutrophils. Additionally, by lowering lipid peroxidation and nitric oxide production, these flavonoids may be able to shield the rat brain against LPS-induced shock (Abd El-Gawad and Khalifa 2001). Studies have demonstrated that plant flavonoids can activate bacterial genes involved in N2 fixing, specifically in the instance of Rhizobium. This raises the possibility of a connection between certain flavonoids and mammalian gene expression.

Additionally, epidemiological studies have suggested that these phenolic chemicals may be necessary for promoting health and hindering long-lasting illnesses. Isoflavones are anticipated to bind to estrogen receptors due to their structural similarity. However, their effects are more nuanced and behave as partial estrogen agonists and antagonists. This concept has garnered interest from steroid biochemists and endocrinologists, who see the ability to use isoflavones to treat hormone-related disorders (Mendelson 1996).

The immune system is a well-defined defense mechanism present in all vertebrates that protects the host from various infectious agents, such as viruses, bacteria, and fungi, along with other harmful factors present in the environment. Flavonoids, dietary phytochemicals with antioxidant mechanisms, have been recognized for enhancing the immune response across all vertebrate taxa. In addition to boosting the immune system, nutritional polyphenols modulate detoxification enzymes, scavenge oxidative agents, and control gene expression in cells. In addition to increasing the immune system, dietary polyphenols modulate detoxification enzymes, scavenging ROS, and control cell expression (Catoni et al. 2008). Consuming purple sweet potato leaves enriched with flavonoids led to a notable growth in the proliferation and reaction of PBMC cells. Enhanced secretion of IL-2 and IL-4 immunoreactive cytokines, an increase in N.K. cells lytic activity, and salivary IgA secretion have also been studied (Chen et al. 2005).

Curcumin is one of turmeric's primary polyphenol-derived bioactive compounds (Curcuma longa). Several animal models and *in vivo* studies show that curcumin behaves as an effective antioxidant by scavenging ROS levels, hydroxyl radicals, SOD, CAT, MDA, NO, H2O2, and peroxynitrite and modulating the expression of HO-1, SOD, and GPx by increasing transcription factors (Nrf2) in normal cells. It mediates inflammatory and pro-inflammatory effects by suppressing NFkβ, PPAR-γ, Akt, and AP-1 are such transcription factors are involved in the induction of cytokines in the pro-inflammatory pathway like TNF-α, IL-1, IL-6, or chemokines (MIP, MCP-1) and also diminishing prostaglandin metabolism (COX-2) with critical roles in the pathogenesis of various disease complications. Curcumin might be used as an antidepressant and neuroprotectant by reducing inflammatory responses like pro-inflammatory and anti-inflammatory factors like NFKβ and AP-1 via the TLR-4 signaling pathway, PPAR-γ, and Akt.

## 6.4 Other Dietary Polyphenols

Polyphenols, secondary metabolites derived from natural sources, can commonly be present in vegetables, fruits, cereals, herbs, and beverages. The polyphenol content varies among fruits and spices, as berries and apples reportedly have approximately 200–300 mg of polyphenols per 100 g of fresh weight (Arts and Hollman 2005; Graf et al. 2005; Habauzit and Morand 2012; Tangney and Rasmussen 2013). Quercetin and curcumin are the dietary polyphenols play a significant role in the modulation of many inflammatory cell signaling networks (Jantan et al. 2021). Plantago species can be used to treat cancer, infection, and inflammation. Five of the phenolic compounds were extracted and examined for cytokine expression on human PBMCs in order to check the immunomodulatory effects of this species. Data showed that treatment with other phytochemicals such as aucubin, p-coumaric acid, chlorogenic acid, ferulic acid, and vanillic acid, significantly increased lymphocyte proliferative activity and IFN- production (Chiang et al. 2003).

Quercetin (Qu) is a very beneficial polyphenol that is known for its immunomodulatory activity and several inflammatory conditions in humans, such as prostatitis, cystitis, R.A., hypercholesterolemia, cancers, CVD, metabolic syndrome, and inflammation brought on by strenuous workouts and respiratory illness, have all been tested for with mixed results for quercetin's use as a preventative measure or treatment. (Li et al. 2016). Qu inhibits the production of ICAM-1 in IL-1-stimulated A549 lung epithelial cells through the MAPK pathway (Ying et al. 2009). Inflammatory cytokines such as TNF-$\alpha$, IL-1$\beta$, and IFN-$\gamma$ are the levels reduced by Qu in A549 alveolar epithelial cells exposed to benzo(a)pyrene, suggesting that quercetin is a viable option for preventing and treating inflammatory diseases (Günay et al. 2016). Qu has been shown to effectively reduce TNF-$\alpha$ levels in LPS activation on macrophages of the murine model (RAW 264.7) (Ghosh 1999).

Moreover, the impact of lemon polyphenol (LPP) on aging was investigated in senescence-accelerated prone mice 1 (SAMP1) and senescence resistance mice 1 (SAMR1), focusing on age-related scores, motor activity, cognitive function, and the gut microbiome. The findings revealed that consuming LPP increased the lifespan of the mice by roughly 3 weeks and mitigated the rise in age-related scores (such as periocular lesions) and motor atrophy. Additionally, long-term consumption of LPP potentially exhibited anti-aging effects on various characteristics and the intestinal environment (Shimizu et al. 2019).

## 6.5 Virgin Coconut Oil

One of the best sources of PUFA is virgin coconut oil (VCO), which has had several nutritional and therapeutic benefits for several years. Due to its high concentration of fatty acids, VCO has been utilized in Southeast Asian diets and medicine for several centuries and can be produced through the dry processing method. Coconut oils are of two types, based on the extraction process. They are (1) VCO and (2) refined, bleached, and deodorized copra oil (RCO). VCO is a native production method, i.e., directly from the fresh coconut endosperm, while RCO is obtained directly from the copra after processing using chemical treatment and heat. Both have equivalent fatty acid profiles. It contains bioactive compounds such as tocopherols, tocotrienols, and polyphenols, which have been responsible for antioxidant, anti-inflammatory, anti-pyretic, analgesic, hypolipidemic, and antihypertensive properties. One of the glycerol derivatives, monolaurin (a medium-chain fatty acid), has been shown to have immunomodulatory and antimicrobial properties, as well as the ability to regulate the inflammatory process. Polyphenols present in the VCO can attenuate cadmium-induced dyslipidemia and oxidative stress from the cardiovascular disease model rat. Studies conducted on rat models have demonstrated that antioxidants can reverse the decline in baroreflex sensitivity associated with hypertension. However, while refining removes some of these chemicals, VCO still has more beneficial substances, such as tocopherols, sterols, and polyphenols (Joshi et al. 2020). VCO can potentially modify several gene expressions associated with the inflammatory response and beneficial effects on the skin (Varma et al. 2019). Monocytes secrete diverse cytokines such as IL-1, IL-5, IL-6, IL-8, TNF-α, IFN-γ, and IL-10, which play crucial roles in both normal immune response and immunopathogenesis (Dooley et al. 1994; Ponzoni et al. 1992; Baggiolini et al. 1993). In a study, VCO demonstrated the ability to suppress the stimulation of pro-inflammatory cytokines induced by LPS in Human monocytic leukemia (THP-1) cells.

Furthermore, it was revealed that VCO effectively inhibited the expression of pro-inflammatory cytokines at both the protein and gene expression levels (Varma et al. 2019). TNF-α is an important cytokine that plays a crucial role in initiating inflammatory responses, and it has been identified as a vital therapeutic target for numerous inflammatory diseases (Neurath et al. 1997). Stimulating THP-1 cells with LPS to induce the release of TNF-α provides an effective model system for evaluating the potential anti-inflammatory effects of compounds. VCO significantly reduces the production of TNF-α in LPS-stimulated THP-1 cells. Moreover, VCO substantially reduces TNF-α levels in the cell culture medium, effectively inhibiting both secretion of TNF-α and expression of mRNA (Varma et al. 2019).

In a rat model, we administer VCO-elevated p-mTOR, SIRT1, p-LKB1, p-ERK, and p-CREB expression levels in the spleen. VCO's potential health benefits may stem from its ability to boost Th1 immunity by promoting the activation of survival signaling pathways, reducing the production of free radicals in the spleen, and its capacity to induce hypolipidemia.

## 6.6 Phytochemicals as Neuroprotectants

In specific brain pathologies, such as neurodegenerative illnesses, some polyphenols may have significant neuroprotective benefits, according to recent studies (Pocernich et al. 2011; Bhullar and Rupasinghe 2013; Shay et al. 2015). Neuroprotective mechanisms at the molecular level involve the regulation of the apoptotic cascade in mitochondria, which is delicately controlled by the balance between antiapoptotic and proapoptotic proteins like Bcl-2 and Bax. Phytochemicals can provide neuroprotection by promoting the introduction of antiapoptotic Bcl-2, thereby inhibiting apoptosis (Kinarivala et al. 2017). By triggering diverse intracellular signaling pathways that contribute to neuroprotection, bioactive polyphenols can produce neuromodulatory effects (Kinarivala et al. 2017). By upregulating the Bcl-2 expression, the PI3K/Akt pathway provides protection against apoptosis.

Conversely, the PKC-ERK1/2 pathway reduces the expression of antiapoptotic proteins like Bcl-xL, Bcl-2, and Bcl-w. Additionally, the Akt-ERK1/2 pathway inhibits the proapoptotic effects of Bad and Bim while activating caspases 9 and 3 (Nakaso et al. 2008; Vauzour et al. 2007). Polyphenols have neuroprotective effects by promoting the synthesis of neurotrophic factors like BDNF, NGF, and GDN. Upon binding to their respective receptors, these factors activate downstream neuroprotective pathways, thereby promoting neurotrophic activity (Moosavi et al. 2015). Polyphenols play a neuroprotective role by activating pathways regulating transcription, translation, proliferation, growth, and cell survival. These pathways include c, MAPK Akt-ERK1/2, PKC-ERK1/2, and MAPK. By binding to tyrosine receptor kinases (Trk), polyphenols initiate protein kinase cascades that ultimately activate CREB. Due to this, the expression of neurotrophic factors and antiapoptotic proteins like Bcl-2 and Bcl-xL is elevated (Huang and Reichardt 2003). Promoting receptor dimerization and autophosphorylation by this connection also initiates downstream signaling cascades that support the survival of neurons in spinal ganglion cells, neuronal cells in the hippocampus, and motor neurons (Liu et al. 2009). Polyphenols exhibit neuroprotective effects in the brain through their anti-inflammatory actions. They specifically suppress pro-inflammatory transcription factors like NF-B and prevent the production of cytokines from activated glial cells (Spencer et al. 2012). The potential of polyphenols to provide diverse and valuable effects on the nervous system makes them a promising resource for developing new drugs aimed at managing neurodegenerative diseases. Apart from the signaling pathways already discussed, the neurotrophic effects of polyphenols may be influenced by several other mechanisms. An increase in the expression of GAP-43 resulting from daidzein consumption has been demonstrated to promote axonal outgrowth in cultured hippocampal neurons.

Interestingly, pretreatment with antagonists that target PKC and estrogen receptors stopped the daidzein-induced activation of PKC and GAP-43. These results imply that daidzein-induced axonal expansion may be mediated via ER-mediated PKC phosphorylation of GAP-43 (Wang et al. 2008).

LY294002, a PI3-K inhibitor, and PD98059, a MAPK inhibitor, are examples of cell signaling cascade inhibitors having structural similarities to flavonoids. Regarding neuroinflammation, studies have demonstrated that PD98059 effectively inhibits the expression of iNOS and NO production in activated glial cells (Bhat et al. 1998). The protective effects against neuronal cell death induced by oxidized LDL are prominently demonstrated by epicatechin and its metabolites in the in vivo studies, 30-O-methyl-epicatechin. These compounds effectively inhibit the activation of c-jun, JNK, and caspase-3, thereby protecting against cell death (Schroeter et al. 2001). Hesperetin, a flavanone found in citrus fruits, activates the Akt signaling pathway at 0.1 and 0.3 nM concentrations. This activation of Akt results in the inhibition of proapoptotic proteins, including apoptosis signal-regulating kinase 1 (ASK1), Bad, caspase-9, and caspase-3 in cortical neurons ( Vauzour et al. 2007). Researchers have employed other neurotoxins to create experimental models to evaluate the neuroprotective potential of polyphenolic compounds. For instance, CAPE (caffeic acid phenethyl ester) has demonstrated its ability to protect PC12 cells against the dopaminergic neurotoxin MPP+ (1-methyl-4-phenylpyridinium) (dos Santos et al. 2014). The therapeutic potential of polyphenols has been demonstrated using in vivo models of neurodegenerative disorders brought on by various neurotoxins.

## 6.7 Phytochemicals Enhance Cognitive Functioning

Cognitive decline and age-related processes have both been linked to the incidence of oxidative stress. Phytochemicals that exhibit effects relevant to cognitive functions encompass various chemical structures, including alkaloids, multiple flavonoids, terpenoids, and other chemical classes. Considering the likelihood of different chemicals applying to distinct types of age-related mental illness, including food plants containing an array of neuroactive phytochemicals, could be advantageous for cognitive protection.

Flavonoids, captivating polyphenols derived from various plant species, have garnered interest due to their potential influence on the adult brain. A study was done to look into the impact of flavonoids on the limbic system's morphological, molecular, and electrophysiological changes, as well as on cognitive health, precisely behavior. It is worth noting that a single member of the vast nootropic family can have varying effects, such as improved cognitive function and increased neurogenesis, in both healthy and pathological conditions. (Bakoyiannis et al. 2019).

Several phytochemicals have cognitive and alertness/arousal properties that may help athletes perform better. The ecological roles played by various phytochemical groups can be connected to these consequences. The phenolic family and the polyphenol subgroup are currently the subject of intensive investigation. While these compounds are already recognized for contributing to overall health and cardiac function when consumed as part of a regular diet, the evidence supporting their short-term supplementation for additional cognitive and mood benefits is currently limited. This functional profile aligns with their natural role in providing general protection within plants.

Several phytochemicals possess properties that can enhance cognitive function and alertness/arousal, potentially contributing to improved sporting performance. The ecological roles of these phytochemical groups are closely associated with the observed effects. Specifically, phenolics and their polyphenol subgroup have become a focal point of research due to their significance in promoting good health and supporting cardiovascular function when incorporated into the regular diet (Kennedy 2019).

Triterpenes exhibit a wide range of ecological roles within plants, including significant antifeedant/anti-herbivore properties and the management of soil microorganisms. Many of these functions can be attributed to the specific chemical structure of triterpenes, which is also highly favored for hormonal signaling across different species. For instance, this structure is found in estrogen/testosterone-like human sex hormones and cortisol, invertebrate hormones, insect ecdysteroids, and plant brassinosteroids. Due to the structural resemblances, triterpenes present in the plant can disrupt ecdysteroid function and alter the insect life cycle that feeds on plants (Kennedy 2014). Additionally, there is potential for triterpenes to disrupt the endocrine function of invertebrates (Janer and Porte 2007) due to these structural resemblances.

Rosemary extracts have the properties like analgesic and antispasmodic. They are thought to be attributed to interactions with an opioid, GABA, serotonin (5-HT) receptors, or muscarinic (Martinez et al. 2009; Hosseinzadeh and Nourbakhsh 2003; Ventura-Martinez et al. 2011). On the other hand, the psychoactive properties of rosemary may be associated with its ability to bind to cholinergic receptors and inhibit cholinesterase activity (Perry et al. 1996; Orhan et al. 2008; Machado et al. 2012).

## 6.8 Phytochemicals as Antioxidants in Aging

Flavonoids possess a highly reactive O.H. group that allows them to donate a hydrogen atom, effectively diminishing potent free radicals that cause oxidation. The primary pathway targeted by these compounds concerning their anti-aging effects is the IIS pathway. For instance, Tambulin, a flavanol with hydroxy substitution, enhances stress tolerance and durability while mitigating the development of Parkinsonism in the *C. elegans* model (Banjarnahor and Artanti 2014). A natural polyphenol known as rosmarinic acid (R.A.) has demonstrated the ability to enhance the average lifespan of *C. elegans*. This is done by expressing daf-16, ins-18 activating the MAPK pathway through sek-1 and skn-1, and upregulating stress resistance and genes such as sod-3, ctl-1, and sod-5 involved antioxidants mechanisms (Lin et al. 2019). Curcumin has been found to extend the lifespan of worms by relying on the activities of unc-43, age-1, osr-1, sek-1, sir-2.1, skn-1, and mek-1, which are associated with the IGF, MAPK, and JNK signaling pathways (Liao et al. 2011). Additionally, the longevity-enhancing effects of polyphenols and chlorogenic acid occur through activating the IGF pathway, relying on the functions of hsf-1 daf-16, and skn-1 (Zheng et al. 2017). Further, these phytochemicals, EGCG and E.C., are the active compounds present in green tea, safeguard and promote the major gene expressed in the IGF pathway, effectively shielding *C. elegans* from oxidative stress and ultimately increasing their lifespan (Abbas and Wink 2009, 2010; Ayuda-Durán et al. 2019).

Moreover, the phytochemicals found in Noni, namely damnacanthal, myricetin, and ursolic acid, exhibit inhibitory solid effects on ERK. These compounds bind to the catalytic and phosphorylation sites of the ERK molecule, indicating their potential role in inhibiting ERK activity in splenocytes treated with NWS (Pratap et al. 2016). In mice, administering a polysaccharide-rich extract derived from Noni stimulated various immune molecules, including IFN-γ, IL-1β, IL-10, IL-12, TNF-α, p70, and NO. Concurrently, it suppressed the levels of IL-4, promoting antitumor effects (Hirazumi and Furusawa 1999). The effects of Noni fruit juices (NSL and NWS) on CAT and GPx activities and ERK 1/2 expression were found to vary based on age groups and dosage. Noni was observed to enhance the activities of SOD and GST, as well as the expression of Akt, while decreasing the expression of CREB, irrespective of the presence or absence of seedsNSL and NWS were found to increase antioxidant enzyme activities in splenocytes from young male Wistar rats during an initial investigation (Pratap et al. 2016). In general, the consumption of Noni juice can lead to a decrease in serum lipid peroxidation, as evidenced by reduced levels of TBARS in hamsters that are fed a diet high in fat and cholesterol. This result was ascribed to an increase in decreased GSH levels in the serum and an improvement in the total antioxidant capacity of the serum. (Lin et al. 2012).

Brahmi, a natural herb extensively utilized in Ayurvedic medicine, finds widespread application in treating anxiety, memory impairment, and cognitive deficits (Singh and Dhawan 1997; Singh et al. 2008). Brahmi's beneficial impact on the CNS may be due to its capacity to increase the activities of antioxidant enzymes like SOD, CAT, and GPx. (Bhattacharya et al. 2000; Shobana et al. 2012). Brahmi demonstrates the ability to selectively increase NO production and decrease the extent of lipid peroxidation depending on the age group. Additionally, Brahmi increases the expression of p-ERK1/2 and p-CREB, along with increasing p-Akt expression in the lymphocytes of early middle-aged and old rats. Brahmi can reverse the decline in CAT, GPx, and GST enzyme activities typically occurring with aging.

Moreover, it effectively suppresses the age-related increase in lipid peroxidation in splenocytes. This suggests that the enhanced antioxidant status induced by brahmi facilitates the clearance of free radicals, which may contribute to improved cell-mediated immune responses in aged rats. The observation of Brahmi increasing the phosphorylation of ERK 1/2, CREB, and Akt in lymphocytes from early middle-aged and old rats suggests that the ERK pathway might facilitate the translocation of CREB and Akt to the nucleus (Priyanka et al. 2013). The translocation may facilitate the binding of Brahmi to the antioxidant response element (ARE), thereby promoting its therapeutic effects on the immune system and antioxidant enzyme systems. (Priyanka et al. 2013). The elevated levels of antioxidant enzymes and cytokines induced by brahmi in the lymphocytes of aged rats may have occurred via the Nrf-2 pathway, potentially involving the modulation of Keap1 expression and potential interaction with PI3K, leading to enhanced p-Akt/Akt levels (Priyanka et al. 2013).

## 6.9 Conclusion

Studies on phytochemicals that delay aging are limited. Many phytochemicals present in these sources remain uninvestigated. It is anticipated that advances in our knowledge of the genetics of aging, combined with significant technological advancements in the extraction, characterization, and tracking of bioactive components in natural products, will enable us to identify native natural compounds with properties that extend lifespan and improve health, which will have significant positive control on human well-being. The anticancerous and antioxidant properties of natural phytochemicals are associated with various pathways, including oxidative stress, regulators of the cell cycle, proapoptotic and antiapoptotic factors, and signaling pathways that promote cell survival. Despite their potential benefits, the clinical application of natural phytochemicals is hindered by factors such as low solubility and bioavailability.

Further investigation is necessary to identify efficient transport mechanisms for these compounds in different types of cancer cells. Large-scale studies must be conducted to establish the pharmacokinetic interactions of these phytochemicals both alone and in combination with other drugs and identify their active site. These natural phytochemicals may treat various metabolic, endocrine, and neurological diseases and enhance immune surveillance and signaling.

## References


Abbas S, Wink M (2009) Epigallocatechin gallate from green tea (Camellia sinensis) increases lifespan and stress resistance in Caenorhabditis elegans. Planta Med 75(03):216–221

Abbas S, Wink M (2010) Epigallocatechin gallate inhibits beta amyloid oligomerization in Caenorhabditis elegans and affects the daf-2/insulin-like signaling pathway. Phytomedicine 17(11):902–909

Abd El-Gawad HM, Khalifa AE (2001) Quercetin, coenzyme Q10, and L- canavanine as protective agents against lipid peroxidation and nitric oxide generation in endotoxin-induced shock in rat brain. Pharmacol Res 43(3):257–263

Aguilera Y, Benítez V (2021) Phytochemicals: dietary sources, innovative extraction, and health benefits. Foods 11(1):72

Arts IC, Hollman PC (2005) Polyphenols and disease risk in epidemiologic studies. Am J Clin Nutr 81(1):317S–325S

Ayuda-Durán B, González-Manzano S, Miranda-Vizuete A, Dueñas M, Santos-Buelga C, González-Paramás AM (2019) Epicatechin modulates stress-resistance in C. elegans via insulin/IGF-1 signaling pathway. PLoS One 14(1):e0199483

Baggiolini M, Dewald B, Moser B (1993) Lnterleukin-8 and related chemotactic cytokines—CXC and CC chemokines. Adv Immunol 55:97–179

Bakoyiannis I, Daskalopoulou A, Pergialiotis V, Perrea D (2019) Phytochemicals and cognitive health: are flavonoids doing the trick? Biomed Pharmacother 109:1488–1497

Banjarnahor SD, Artanti N (2014) Antioxidant properties of flavonoids. Med J Indones 23(4): 239–244

Behl T, Kumar K, Brisc C, Rus M, Nistor-Cseppento DC, Bustea C et al (2021) Exploring the multifocal role of phytochemicals as immunomodulators. Biomed Pharmacother 133:110959

Belguise K, Guo S, Sonenshein GE (2007) Activation of FOXO3a by the green tea polyphenol epigallocatechin-3-gallate induces estrogen receptor α expression reversing invasive phenotype of breast cancer cells. Cancer Res 67(12):5763–5770

Bhat NR, Zhang P, Lee JC, Hogan EL (1998) Extracellular signal-regulated kinase and p38 subgroups of mitogen-activated protein kinases regulate inducible nitric oxide synthase and tumor necrosis factor-α gene expression in endotoxin-stimulated primary glial cultures. J Neurosci 18(5):1633–1641

Bhattacharya SK, Bhattacharya A, Kumar A, Ghosal S (2000) Antioxidant activity of Bacopa monniera in rat frontal cortex, striatum and hippocampus. Phytother Res 14:174–179

Bhullar KS, Rupasinghe HP (2013) Polyphenols: multipotent therapeutic agents in neurodegenerative diseases. Oxidative Med Cell Longev 2013:891748

Boocock DJ, Patel KR, Faust GE, Normolle DP, Marczylo TH, Crowell JA et al (2007) Quantitation of trans-resveratrol and detection of its metabolites in human plasma and urine by high performance liquid chromatography. J Chromatogr B 848(2):182–187

Butt MS, Sultan MT (2009) Green tea: nature's defense against malignancies. Crit Rev Food Sci Nutr 49(5):463–473

Catoni C, Peters A, Schaefer M (2008) Lifehistory trade-offs are influenced by the diversity, availability and interactions of dietary antioxidants. Anim Behav 76(4):1107–1119

Chen CM, Li SC, Lin YL, Hsu CY, Shieh MJ, Liu JF (2005) Consumption of purple sweet potato leaves modulates human immune response: T-lymphocyte functions, lytic activity of natural killer cell and antibody production. World J Gastroenterol 11(37):5777–5781

Chhabra G, Singh CK, Amiri D, Akula N, Ahmad N (2021) Recent advancements on immunomodulatory mechanisms of resveratrol in tumor microenvironment. Molecules 26(5):1343

Chiang LC, Ng LT, Chiang W, Chang MY, Lin CC (2003) Immunomodulatory activities of flavonoids, monoterpenoids, triterpenoids, iridoid glycosides and phenolic compounds of Plantago species. Planta Med 69(07):600–604

Chintapula U, Chikate T, Sahoo D, Kieu A, Guerrero Rodriguez ID, Nguyen KT, Trott D (2023) Immunomodulation in age-related disorders and nanotechnology interventions. Wiley Interdiscip Rev Nanomed Nanobiotechnol 15(1):e1840

Clinton SK, Emenhiser C, Schwartz SJ, Bostwick DG, Williams AW, Moore BJ, Erdman JW Jr (1996) *cis-trans* Lycopene isomers, carotenoids, and retinol in the human prostate. Cancer Epidemiol Biomark Prev 5:823–833

Cushnie TPT, Lamb AJ (2005) Antimicrobial activity of flavonoids. Int J Antimicrob Agents 26(5): 343–356

Delmas D, Limagne E, Ghiringhelli F, Aires V (2020) Immune Th17 lymphocytes play a critical role in the multiple beneficial properties of resveratrol. Food Chem Toxicol 137:111091

Dooley DP, Cox RA, Hestilow KL, Dolan MJ, Magee DM (1994) Cytokine induction in human coccidioidomycosis. Infect Immun 62(9):3980–3983

Falchetti R, Fuggetta MP, Lanzilli G, Tricarico M, Ravagnan G (2001) Effects of resveratrol on human immune cell function. Life Sci 70(1):81–96

Gao X, Xu YX, Janakiraman N, Chapman RA, Gautam SC (2001) Immunomodulatory activity of resveratrol: suppression of lymphocyte proliferation, development of cell-mediated cytotoxicity, and cytokine production. Biochem Pharmacol 62(9):1299–1308

Ghosh B (1999) Quercetin inhibits LPS-induced nitric oxide and tumor necrosis factor-α production in murine macrophages. Int J Immunopharmacol 21(7):435–443

Graf BA, Milbury PE, Blumberg JB (2005) Flavonols, flavones, flavanones, and human health: epidemiological evidence. J Med Food 8(3):281–290

Günay E, Celik S, Sarinc-Ulasli S, Özyürek A, Hazman Ö, Günay S et al (2016) Comparison of the anti-inflammatory effects of proanthocyanidin, quercetin, and damnacanthal on benzo (a) pyrene exposed A549 alveolar cell line. Inflammation 39:744–751

Habauzit V, Morand C (2012) Evidence for a protective effect of polyphenols-containing foods on cardiovascular health: an update for clinicians. Ther Adv Chronic Dis 3(2):87–106

Hirazumi A, Furusawa E (1999) An immunomodulatory polysaccharide-rich substance from the fruit juice of Morinda citrifolia (Noni) with antitumour activity. Phytother Res 13(5):380–387

Hirose M, Hoshiya T, Akagi K, Futakuchi M, Ito N (1994) Inhibition of mammary gland carcinogenesis by green tea catechins and other naturally occurring antioxidants in female Sprague-Dawley rats pretreated with 7, 12-dimethylbenz [a] anthracene. Cancer Lett 83(1–2): 149–156

Hosseinzadeh H, Nourbakhsh M (2003) Effect of *Rosmarinus officinalis* L. aerial parts extract on morphine withdrawal syndrome in mice. Phytother Res 17(8):938–941

Howes MR, Perry NSL, Vásquez-Londoño C, Perry EK (2020) Role of phytochemicals as nutraceuticals for cognitive functions affected in ageing. Br J Pharmacol 177(6):1294–1315. https://doi.org/10.1111/bph.14898. Epub 2020 Feb 3. PMID: 31650528; PMCID: PMC7056459

Huang EJ, Reichardt LF (2003) Trk receptors: roles in neuronal signal transduction. Annu Rev Biochem 72(1):609–642

Islam SU, Ahmed MB, Ahsan H, Islam M, Shehzad A, Sonn JK, Lee YS (2020) An update on the role of dietary phytochemicals in human skin cancer: new insights into molecular mechanisms. Antioxidants 9(10):916

Issinger OG, Guerra B (2021) Phytochemicals in cancer and their effect on the PI3K/AKT-mediated cellular signalling. Biomed Pharmacother 139:111650

Janer G, Porte C (2007) Sex steroids and potential mechanisms of non-genomic endocrine disruption in invertebrates. Ecotoxicology 16(1):145–160

Jantan I, Haque MA, Arshad L, Harikrishnan H, Septama AW, Mohamed-Hussein ZA (2021) Dietary polyphenols suppress chronic inflammation by modulation of multiple inflammation-associated cell signaling pathways. J Nutr Biochem 93:108634

Joshi S, Kaushik V, Gode V, Mhaskar S (2020) Coconut oil and immunity: what do we really know about it so far. J Assoc Physicians India 68(7):67–72

Ke B, Zhang T, An T, Lu R (2020) Soy isoflavones ameliorate the cognitive dysfunction of Goto-Kakizaki rats by activating the Nrf2-HO-1 signalling pathway. Aging (Albany NY) 12(21): 21344–21354. https://doi.org/10.18632/aging.103877. Epub 2020 Nov 7. PMID: 33180745; PMCID: PMC7695387

Kennedy DO (2014) Plants and the human brain. Oxford University Press, New York

Kennedy DO (2019) Phytochemicals for improving aspects of cognitive function and psychological state potentially relevant to sports performance. Sports Med 49:39–58

Kinarivala N, Patel R, Boustany RM, Al-Ahmad A, Trippier PC (2017) Discovery of aromatic carbamates that confer neuroprotective activity by enhancing autophagy and inducing the antiapoptotic protein B-cell lymphoma 2 (Bcl-2). J Med Chem 60(23):9739–9756

Kulczyński B, Gramza-Michałowska A, Kobus-Cisowska J, Kmiecik D (2017) The role of carotenoids in the prevention and treatment of cardiovascular disease–current state of knowledge. J Funct Foods 38:45–65

Larsen CA, Bisson WH, Dashwood RH (2009) Tea catechins inhibit hepatocyte growth factor receptor (MET kinase) activity in human colon cancer cells: kinetic and molecular docking studies. J Med Chem 52(21):6543–6545

Li Y, Yao J, Han C, Yang J, Chaudhry MT, Wang S et al (2016) Quercetin, inflammation and immunity. Nutrients 8(3):167

Liao VHC, Yu CW, Chu YJ, Li WH, Hsieh YC, Wang TT (2011) Curcumin-mediated lifespan extension in Caenorhabditis elegans. Mech Ageing Dev 132(10):480–487

Lin C, Xiao J, Xi Y, Zhang X, Zhong Q, Zheng H et al (2019) Rosmarinic acid improved antioxidant properties and healthspan via the IIS and MAPK pathways in Caenorhabditis elegans. Biofactors 45(5):774–787

Lin YL, Chou CH, Yang DJ, Chen JW, Tzang BS, Chen YC (2012) Hypolipidemic and antioxidative effects of noni (Morinda citrifolia L.) juice on high-fat/cholesterol-dietary hamsters. Plant Foods Hum Nutr 67:294–302

Liu R, Gao M, Qiang GF, Zhang TT, Lan X, Ying J, Du GH (2009) The anti-amnesic effects of luteolin against amyloid β25–35 peptide-induced toxicity in mice involve the protection of neurovascular unit. Neuroscience 162(4):1232–1243

Lu G, Liao J, Yang G, Reuhl KR, Hao X, Yang CS (2006) Inhibition of adenoma progression to adenocarcinoma in a 4-(methylnitrosamino)-1-(3-pyridyl)-1-butanone–induced lung tumorigenesis model in a/J mice by tea polyphenols and caffeine. Cancer Res 66(23): 11494–11501

Ma Z, Lu Y, Yang F, Li S, He X, Gao Y, Zhang G, Ren E, Wang Y, Kang X (2020) Rosmarinic acid exerts a neuroprotective effect on spinal cord injury by suppressing oxidative stress and inflammation via modulating the Nrf2/HO-1 and TLR4/NF-κB pathways. Toxicol Appl Pharmacol 397:115014. https://doi.org/10.1016/j.taap.2020.115014. Epub ahead of print

Machado DG, Cunha MP, Neis VB, Balen GO, Colla AR, Grando J et al (2012) *Rosmarinus officinalis* L. hydroalcoholic extract, similar to fluoxetine, reverses depressive-like behavior without altering learning deficit in olfactory bulbectomized mice. J Ethnopharmacol 143(1): 158–169

Martí R, Roselló S, Cebolla-cornejo J (2016) Tomato as a source of carotenoids and polyphenols targeted to cancer prevention. Cancers 8(6):58

Martin LM, Johnson PJ, Amorim JR, Honaker AR, Donaldson RS, DeClue AE (2016) Investigation of the potential immunomodulatory effects of resveratrol on equine whole blood: An in vitro investigation. Res Vet Sci 106:97–99

Martinez AL, Gonzalez-Trujano ME, Pellicer F, Lopez-Munoz FJ, Navarrete A (2009) Antinociceptive effect and GC/MS analysis of *Rosmarinus officinalis* L. essential oil from its aerial parts. Planta Med 75(5):508–511

Mendelson CR (1996) Mechanism of steroid hormone action. In: Griffin JE, Ojeda SR (eds) Textbook of endocrine physiology, 3rd edn. Oxford University Press, New York, NY, pp 29–65

Moosavi F, Hosseini R, Saso L, Firuzi O (2015) Modulation of neurotrophic signaling pathways by polyphenols. Drug Des Devel Ther 10:23–42

Nakaso K, Ito S, Nakashima K (2008) Caffeine activates the PI3K/Akt pathway and prevents apoptotic cell death in a Parkinson's disease model of SH-SY5Y cells. Neurosci Lett 432(2): 146–150

Neurath MF, Fuss I, Pasparakis M, Alexopoulou L, Haralambous S, Meyer zum Büschenfelde KH et al (1997) Predominant pathogenic role of tumor necrosis factor in experimental colitis in mice. Eur J Immunol 27(7):1743–1750

Okwu DE (2008) Citrus fruits: a rich source of phytochemicals and their roles in human health. Int J Chem Sci 6(2):451–471

Olsson ME, Gustavsson KE, Vagen IM (2010) Quercetin and isorhamnetin in sweet and red cultivars of onion (Allium cepa L.) at harvest, after field curing, heat treatment, and storage. J Agric Food Chem 58(4):2323–2330

Orhan I, Aslan S, Kartal M, Sener B, Baser HC, K. (2008) Inhibitory effect of Turkish *Rosmarinus officinalis* L. on acetylcholinesterase and butyrylcholinesterase enzymes. Food Chem 108(2): 663–668

Oz AT, Kafkas E (2017) Phytochemicals in fruits and vegetables. *Waisundara V. Superfood and functional food. London: IntechOpen*, 175–184

Patra AK (ed) (2012) Dietary phytochemicals and microbes. Springer Science & Business Media

Perry N, Court G, Bidet N, Court J, Perry E (1996) European herbs with cholinergic activities: potential in dementia therapy. Int J Geriatr Psychiatry 11(12):1063–1069

Pocernich CB, Lange LB, Sultana MR, Butterfield AD (2011) Nutritional approaches to modulate oxidative stress in Alzheimer's disease. Curr Alzheimer Res 8(5):452–469

Ponzoni M, Casalaro A, Lanciotti M, Montaldo PG, Cornaglia-Ferraris P (1992) The combination of γ-interferon and tumor necrosis factor causes a rapid and extensive differentiation of human neuroblastoma cells. Cancer Res 52(4):931–939

Pratap UP, Anand K, Yasmine F, Hima L, Priyanka HP, Thyagarajan S (2016) Phytochemicals in Morinda citrifolia fruit selectively modulate age-associated immunity and antioxidant enzyme activities through ERK pathway in splenic lymphocytes of male F344 rats. J Recept Signal Transduct 36(2):139–151

Priyanka HP, Singh RV, Mishra M, ThyagaRajan S (2013) Diverse age-related effects of Bacopa monnieri and donepezil in vitro on cytokine production, antioxidant enzyme activities, and intracellular targets in splenocytes of F344 male rats. Int Immunopharmacol 15(2):260–274

dos Santos NAG, Martins NM, de Barros Silva R, Ferreira RS, Sisti FM, dos Santos AC (2014) Caffeic acid phenethyl ester (CAPE) protects PC12 cells from MPP+ toxicity by inducing the expression of neuron-typical proteins. Neurotoxicology 45:131–138

Schroeter H, Spencer JP, Rice-Evans C, Williams RJ (2001) Flavonoids protect neurons from oxidized low-density-lipoprotein-induced apoptosis involving c-Jun N-terminal kinase (JNK), c-Jun and caspase-3. Biochem J 358(3):547–557

Shay J, Elbaz HA, Lee I, Zielske SP, Malek MH, Hüttemann M (2015) Molecular mechanisms and therapeutic effects of (−)-epicatechin and other polyphenols in cancer, inflammation, diabetes, and neurodegeneration. Oxidative Med Cell Longev 2015:181260

Shigenaga MK, Hagen TM, Ames BN (1994) Oxidative damage and mitochondrial decay in aging. Proc Natl Acad Sci 91(23):10771–10778

Shimizu C, Wakita Y, Inoue T, Hiramitsu M, Okada M, Mitani Y et al (2019) Effects of lifelong intake of lemon polyphenols on aging and intestinal microbiome in the senescence-accelerated mouse prone 1 (SAMP1). Sci Rep 9(1):3671

Shobana C, Ramesh Kumar R, Sumathi T (2012) Alcoholic extract of bacopa monniera Linn. Protects against 6-hydroxydopamine-induced changes in behavioral and biochemical aspects: a pilot study. Cell Mol Neurobiol 32(7):1099–1112

Sidhu JS, Zafar TA (2018) Bioactive compounds in banana fruits and their health benefits. Food Qual Saf 2(4):183–188

Silalahi J (2002) Anticancer and health protective properties of citrus fruit components. Asia Pac J Clin Nutr 11(1):79–84

Singh BN, Shankar S, Srivastava RK (2011) Green tea catechin, epigallocatechin-3-gallate (EGCG): mechanisms, perspectives and clinical applications. Biochem Pharmacol 82(12): 1807–1821

Singh HK, Dhawan BN (1997) Neuropsychopharmacological effects of the ayurvedic nootropic bacopa monniera Linn (Brahmi). Indian J Pharmacol 29:S359–65

Singh M, Murthy V, Ramassamy C (2012) Standardized extracts of bacopa monniera protect against MPP+- and paraquat-induced toxicity by modulating mitochondrial activities, proteasomal functions, and redox pathways. Toxicol Sci 125(1):219–232. https://doi.org/10.1093/toxsci/kfr255

Singh RH, Narsimhamurthy K, Singh G (2008) Neuronutrient impact of ayurvedic Rasayana therapy in brain aging. Biogerontology 9:369–374

Spencer JP, Vafeiadou K, Williams RJ, Vauzour D (2012) Neuroinflammation: modulation by flavonoids and mechanisms of action. Mol Asp Med 33(1):83–97

Sun GY, Li R, Yang B, Fritsche KL, Beversdorf DQ, Lubahn DB, Geng X, Lee JC, Greenlief CM (2019) Quercetin potentiates docosahexaenoic acid to suppress lipopolysaccharide-induced oxidative/inflammatory responses, Alter lipid peroxidation products, and enhance the adaptive stress pathways in BV-2 microglial cells. Int J Mol Sci 20(4):932. https://doi.org/10.3390/ijms20040932. PMID: 30795510; PMCID: PMC6413212

Tangney CC, Rasmussen HE (2013) Polyphenols, inflammation, and cardiovascular disease. Curr Atheroscler Rep 15:1–10

Tapas AR, Sakarkar DM, Kakde RB (2008) Flavonoids as nutraceuticals: a review. Trop J Pharm Res 7(3):1089–1099

Tiwari BK, Brunton N, Brennan CS (2015) Handbook of plant food phytochemicals. Wiley-Blackwell

Varma SR, Sivaprakasam TO, Arumugam I, Dilip N, Raghuraman M, Pavan KB et al (2019) In vitro anti-inflammatory and skin protective properties of virgin coconut oil. J Tradit Complement Med 9(1):5–14

Vauzour D, Vafeiadou K, Rice-Evans C, Williams RJ, Spencer JP (2007) Activation of pro-survival Akt and ERK1/2 signalling pathways underlie the antiapoptotic effects of flavanones in cortical neurons. J Neurochem 103(4):1355–1367

Ventura-Martinez R, Rivero-Osorno O, Gomez C, Gonzalez-Trujano ME (2011) Spasmolytic activity of *Rosmarinus officinalis* L. involves calcium channels in the Guinea pig ileum. J Ethnopharmacol 137(3):1528–1532

Walle T (2011) Bioavailability of resveratrol. Ann N Y Acad Sci 1215(1):9–15

Wang P, Jeng CJ, Chien CL, Wang SM (2008) Signaling mechanisms of daidzein-induced axonal outgrowth in hippocampal neurons. Biochem Biophys Res Commun 366(2):393–400

Wiciński M, Malinowski B, Węclewicz MM, Grześk E, Grześk G (2017) Anti-atherogenic properties of resveratrol: 4-week resveratrol administration associated with serum concentrations of SIRT1, adiponectin, S100A8/A9 and VSMCs contractility in a rat model. Exp Ther Med 13(5):2071–2078

Wiciński M, Socha M, Walczak M, Wódkiewicz E, Malinowski B, Rewerski S et al (2018) Beneficial effects of resveratrol administration—focus on potential biochemical mechanisms in cardiovascular conditions. Nutrients 10(11):1813

Yang CS, Maliakal P, Meng X (2002) Inhibition of carcinogenesis by tea. Annu Rev Pharmacol Toxicol 42(1):25–54

Yiannakopoulou EC (2014) Effect of green tea catechins on breast carcinogenesis. Eur J Cancer Prev 23(2):84–89

Ying B, Yang T, Song X, Hu X, Fan H, Lu X et al (2009) Quercetin inhibits IL-1 beta-induced ICAM-1 expression in pulmonary epithelial cell line A549 through the MAPK pathways. Mol Biol Rep 36:1825–1832

Zheng SQ, Huang XB, Xing TK, Ding AJ, Wu GS, Luo HR (2017) Chlorogenic acid extends the lifespan of Caenorhabditis elegans via insulin/IGF-1 signaling pathway. J Gerontol Series A 72(4):464–472

# Impact of Aging, Lifestyle, and Phytochemicals on the Commencement of Neurological Disorders

7

Hadeer M. Hamdalla, Mohamed Y. Zaky, and Manal Abdul-Hamid

**Abstract**

Neurodegenerative diseases (NDs) are examples of neurological disorders. Common NDs include Alzheimer's disease (AD), Parkinson's disease (PD), Huntington's disease (HD), and multiple sclerosis (MS). NDs are caused by the progressive loss of function or structure of nerve cells in the central nervous system. NDs typically start slowly and subtly in mid-to-late life, and both hereditary and environmental variables might affect how quickly they proceed. Brain damage, age, ethnicity, a positive family history, an unhealthy lifestyle (e.g., high-fat and high-sugar diets, and addiction to alcohol and nicotine), related exaggerated neuroinflammation, oxidative stress, and endogenous variables are the main risk factors for commencement and progression of neurological disorders. NDs place a significant burden on families and healthcare systems as a leading cause of morbidity and cognitive decline in the elderly. The high levels of daily monitored care and support required, as well as the disability they induce, raise the cost and societal effect of NDs. However, it is now evident that consuming a diet rich in phytochemicals can affect neuroinflammation and may help to create innovative therapeutics that can stop the gradual malfunctioning of neuronal populations that underlie neurodegeneration.

H. M. Hamdalla · M. Abdul-Hamid (✉)
Cell Biology, Histology and Genetics Division, Department of Zoology, Faculty of Science, Beni-Suef University, Beni-Suef, Egypt
e-mail: manal.mohamed3@science.bsu.edu.eg

M. Y. Zaky
Molecular Physiology Division, Department of Zoology, Faculty of Science, Beni-Suef University, Beni-Suef, Egypt



S. Pathak, A. Banerjee (eds.), *Neuroprotective Effects of Phytochemicals in Brain Ageing*, https://doi.org/10.1007/978-981-99-7269-2_7

**Keywords**

Neurodegenerative diseases · Aging · Phytochemicals · Lifestyle

## Abbreviations

| | |
|---|---|
| (PGC)1α | Peroxisome proliferator-activated receptor gamma coactivator |
| AD | Alzheimer's disease |
| AMPK | 5'AMP-activated protein kinase |
| AUD | Alcohol use disorder |
| BER | Base excision repair |
| Cur | Curcumin |
| DSBR | DNA double-strand break repair |
| EGCG | Epigallocatechin-3-gallate |
| GSK-3 | Cur can also inhibit glycogen synthase kinase-3 |
| HD | Huntington's disease |
| HPA | Hypothalamic-pituitary-adrenal |
| IGF1 | Insulin-like growth factor 1 |
| IL-1β | Interleukin-one beta |
| MS | Multiple sclerosis |
| ND | Neurodegenerative diseases |
| NDs | Neurodegenerative diseases |
| NER | Nucleotide excision repair |
| Nrf1 | Nuclear respiratory factor 1 |
| Nrf2 | Nuclear erythroid 2-related factor 2 |
| PARP1 | Poly (ADP-ribose) polymerase |
| PD | Parkinson's disease |
| PSD95 | Synaptophysin and postsynaptic density protein 95 |
| RES | Resveratrol |
| RNS | Reactive nitrogen species |
| ROS | Endogenous reactive oxygen species |
| SASP | Senescence-associated secretory phenotype |
| TFAM | Mitochondrial transcription factor A |
| TNF-α | Tumor necrosis factor alpha |

## 7.1 Introduction

Health outcomes across the lifespan are assumed to be influenced by the delicate interactions between the biological, behavioral, and psychological components, as well as the social and physical environment. Numerous disorders, including neurological disorders like AD, PD, and others, have been studied as prospective targets for the prevention of lifestyle factors, such as food, exercise, and social interaction

(Popa-Wagner et al. 2020). Other risk and protective factors, like depression, metabolic syndrome, and cardiovascular disease, may also interact with these variables. On the other hand, numerous earlier studies discovered that regularly taking phytochemicals improved physical and mental performance, increased neuronal cell survival, and boosted the antioxidant system (Zhang et al. 2015). It has been shown that frequent, early use of phytochemicals and their derivatives can slow the spread of the illness. Additionally, phytochemicals and their derivatives can be used as both preventive and therapeutic agents by individuals suffering from depression, AD, PD, and other NDs. Currently, more precise and specific analyses of phytochemical metabolites quickly depict the distribution, metabolism, and bioactivity of phytochemicals in the brain (Naoi et al. 2019). Thus, this chapter aims to present the potential of phytochemicals, aging, and lifestyle in the commencement of NDs such as AD and PD. In addition, the effects of phytochemicals on NDs are discussed in this chapter.

## 7.2 Neurodegenerative Disorders

Neurodegenerative disorders encompass a wide range of debilitating diseases characterized by the progressive loss of neurons and their function in the central or peripheral nervous systems. These conditions, which include Alzheimer's disease, Parkinson's disease, Huntington's disease, and amyotrophic lateral sclerosis (ALS), often begin subtly but worsen over time. Neurodegenerative disorders cause a devastating range of symptoms, including memory loss, cognitive decline, movement disorders, difficulty speaking or swallowing, and behavioral changes. As these disorders progress, they rob individuals of their independence, quality of life, and ultimately, their lives. While the exact causes of many neurodegenerative disorders remain complex and not fully understood, factors like aging, genetics, oxidative stress, inflammation, and protein misfolding likely play key roles in their development (Daviglus et al. 2011).

### 7.2.1 Alzheimer's Disease

Alzheimer's disease (AD) is a progressive neurodegenerative disorder that gradually erodes memory, thinking skills, and the ability to perform everyday tasks. While its exact causes remain unknown, hallmarks of the disease include the buildup of amyloid-beta plaques and tangled tau proteins in the brain. These changes lead to the death of brain cells, causing brain shrinkage and the characteristic decline in cognitive function. Initially, Alzheimer's presents with mild memory loss, but as the disease progresses, it can severely impact communication, judgment, and self-care, ultimately leading to an inability to carry out basic daily activities. Their loss of critical thinking abilities, antagonistic behavior, and linguistic deficiencies cause disruption in their lives.

Although the main causes of AD are still being widely contested, some potential candidates include mitochondrial malfunction, cholinergic dysfunction, β amyloid (βA) protein plaques, tau aggregation, neuroinflammation, and DNA damage (Daviglus et al. 2011).

### 7.2.2 Parkinson's Disease

Parkinson's Disease (PD) is a chronic, progressive neurodegenerative disorder (ND) with both motor and nonmotor symptoms. Its hallmark symptoms include tremors, stiffness (rigidity), slowed movement (bradykinesia), and impaired balance. As the disease progresses, non-motor symptoms such as cognitive decline, sleep disturbances, depression, and anxiety can become increasingly prevalent. The root cause of Parkinson's is the degeneration of neurons in a brain region called the substantia nigra, which produces the neurotransmitter dopamine. This loss of dopamine disrupts communication within the brain, causing the characteristic movement-related symptoms and impacting a wide range of bodily functions. Through its progressive degenerative effects on muscular function and mobility, patients, families, and careers are all significantly impacted clinically by the condition. The motor symptoms of PD are believed to be produced by the death of striatal dopaminergic neurons, despite the fact that the presence of nonmotor symptoms also demonstrates neuronal loss in nondopaminergic sites (DeMaagd and Philip 2015).

### 7.2.3 Aging and the Onset of Neurodegenerative Disorders

Aging is a natural, irreversible time-dependent progressive process that causes recognizable and characterized transformations in physical structure and functionality in each organism. The most noticeable effects of aging include an ongoing decline in physiological function, cognitive impairments, deficits in learning and memory, a lessened capacity to adjust to environmental stimuli, increased susceptibility to diseases, and a rise in mortality (Farooqui 2012).

A substantial social and health burden is being imposed by the aging of the global population. For instance, the aging process itself is by far the most prominent risk factor for neurodegeneration among numerous others. Numerous factors contribute to the wide-ranging adverse consequences of aging on the brain and cognition. Therefore, in order to design effective therapeutics, it is imperative to consider the basic mechanisms of aging and their role in the initiation and progression of NDs (Hou et al. 2019).

López-Otín et al. (2013) identified nine so-called "hallmarks of aging" and categorized them into primary, antagonistic, and integrative hallmarks. AD and PD, the two most prevalent neurodegenerative disorders, are mostly seen in senior people, and the probability of developing these disorders rises with age (Liu 2022).

While antagonistic factors are defensive or hostile reactions to the initial disturbance. These reactions initially lessen the damage, but over time they may start to cause harm. The primary and antagonistic hallmarks cause cumulative harm that leads to integrative hallmarks, which are ultimately to blame for the functional deterioration that comes with aging (Hou et al. 2019).

## 7.3 Hallmarks Related to Age and Neurodegenerative Disorders

### 7.3.1 Primary Hallmarks

#### 7.3.1.1 DNA Damage and Genomic Instability

As a result of being exposed to possible environmental hazards like radiation, chemical mutagens, carcinogens, endogenous reactive oxygen species (ROS), or reactive nitrogen species (RNS), each human cell produces about 105 DNA damage every day (Lindahl 1993). The etiology of age-related neurodegenerative processes may be influenced by the fact that DNA damage increases with aging (Chow and Herrup 2015). Numerous types of DNA damage including bulky adducts, abasic sites, single-strand breaks, double-strand breaks, insertions, and deletions have been linked to NDs (Jeppesen et al. 2011).

Genomic instability is brought on by exponential DNA damage, which also sets off a signaling pathway that pervades the whole cell. An abundant protein called poly (ADP-ribose) polymerase (PARP1) promotes the PARylation process, which causes PAR polymers to build up at DNA damage sites and cleaves a significant amount of NADC (Bai 2015). To maintain DNA repair, mitophagy, and mitochondrial health, NADC is a crucial component. Constant DNA damage requires high energy levels to hasten the repair process. DNA damage can hasten cellular senescence and inflammation, resulting in age-related neurodegeneration. Lovell et al. (1999) have suggested that the progression of AD is accelerated by elevated DNA damage and impaired DNA repair. In AD patients, base excision repair protein expression may decline as the disease worsens since it is tissue-, cell-, and cell-age-specific (Weissman et al. 2007).

### 7.3.2 Telomere Attrition

Cellular senescence is driven by telomere shortening, which is also likely a factor in organismal aging. Defects in telomere maintenance accelerate aging in mice and humans (López-Otín et al. 2013) and, in patients with AD, are associated with cognitive impairment, amyloid pathology, and hyperphosphorylation of tau via oxidative stress and inflammation (Rossiello et al. 2022). Telomere-related pathophysiology in ND illnesses has been mentioned by Rossiello et al. (2022). In a mouse model of PD, telomerase activators reduced levels of pathogenic α-synuclein protein and alleviated motor symptoms, indicating a causative role for telomere disruption in the disease.

### 7.3.3 Epigenetic Alterations

Human disease processes associated with aging are influenced by epigenetic pathways. Epigenetic systems control a variety of phenotypic and biological functions, such as learning, memory, and behavior (López-Otín et al. 2013), and both epigenetic and environmental variables are thought to affect predisposition to sporadic AD. The epigenetic alterations include DNA and histone acetylation, PARylation, and methylation, all of which are crucial in AD (Hwang et al. 2017). The long duration of many forms of juvenile PD suggests that aging-related mechanisms that are observed in normal aging and other common hallmarks of ND combined to determine the speed of PD progression at different stages.

### 7.3.4 Proteostasis

The mechanisms that maintain proteome homeostasis and control protein turnover are referred to as proteostasis (Ross et al. 2021). Proteostasis mechanisms prevent protein misfolding and aggregation in animals. Proteostasis is regulated by a number of mechanisms in eukaryotic cells, such as autophagy, ubiquitination, and the lysosomal system. Autophagy is a degradation mechanism that accelerates the lysosomal process and reduces the levels of inflammatory cytokines brought on by misfolded proteins or damaged organelles (Tanaka and Matsuda 2014). Different NDs including AD and PD frequently aggregate misfolded proteins.

## 7.4 Antagonistic Hallmarks

### 7.4.1 Mitochondrial Dysfunction

High energy levels are needed by neural cells to carry out neuronal activities, and mitochondrial alterations have a significant impact. Individual mitochondria fission and fusion as well as mitophagy's elimination of damaged mitochondria serve to maintain a healthy mitochondrial pool. The generation of ROS is accelerated in damaged mitochondria and has been linked to both the common process of aging and the bulk of recognized neurodegenerative diseases (Johri and Beal 2012). Aside from ATP generation, mitochondria also play important roles in a number of intercellular pathways, such as lipid biosynthesis, calcium signaling, and cell death, all of which are crucial steps in the emergence of ND (Keogh and Chinnery 2015). Age-related impairments in brain mitochondrial function are thought to be a key cause of the earliest signs of aging. Age-related changes in brain mitochondrial function are regarded to be one of the main causes of most well-known NDs as well as the aging process.

Mitophagy is a quality control mechanism that keeps the mitochondria functioning efficiently and specifically destroys damaged mitochondria. However, age-related autophagy dysfunctions result in the buildup of aberrant mitochondria, which elevates oxidative stress (Tran and Reddy 2020). Mitophagy plays a crucial role in mitigating age-related ND diseases like AD (Fang et al. 2019). $A\beta_{1-42}$, a protein related to AD, and p-tau were found to be the main causes of mitophagy failure in a recent study. Mitophagy failure also affects ATP synthesis, which can lead to excessive mitochondrial fission and 5'AMP-activated protein kinase (AMPK) activation, which can both enhance tau phosphorylation (Fang et al. 2019). These findings indicate that mitophagy is essential for lowering inflammation, a critical factor in the emergence of NDs.

### 7.4.2 Cellular Senescence

A senescence-associated secretory phenotype (SASP), which is related to cellular senescence, is a stress-induced state of sustained cell cycle arrest that develops with aging (Hayflick 1965). SASP is a beneficial cell survival mechanism that helps remove damaged cells in high-stress situations (Kültz 2005). Although cellular senescence is normally a tumor-suppressing process, it may act as a tumor starter in elderly individuals (Loaiza and Demaria 2016).

NADC: NADH ratio reduction, growth arrest, and a reduction in IL-1β mediated SASP induction are all characteristics of cells going through mitochondrial-associated senescence (Wiley et al. 2016). Through the high mobility group A (HMGA)-nicotinamide phosphoribosyltransferase (NAMPT)-NADC signaling pathway, which controls proinflammatory SASP, NADC metabolism is a key SASP regulator (Nacarelli et al. 2019). High levels of DNA damage cause cells to become senescent and proliferate less actively. Due to an increased need for energy and an increase in ROS levels in neuronal cells, aging is linked to a decreased ability of DNA to repair DNA damage and an increase in DNA damage (Maynard et al. 2015). High levels of DNA damage cause cells to senescence, which is the cessation of cell division. Senescence-like neurons are neural cells that consistently exhibit cellular senescence. The CNS contains a variety of cell types, such as neurons, oligodendrocytes, microglia, and astrocytes. A senescent phenotype can develop in any of these cell types. Senescent cells in the CNS are related to tissue deterioration, diminished neurogenesis, and neuroinflammation (Baker and Petersen 2018). Therefore, senescent cell accumulation is now widely acknowledged as a fundamental phase in age-dependent tissue dysfunction and is linked to the pathogenesis of numerous degenerative disorders, including AD (Wissler Gerdes et al. 2020).

### 7.4.3 Deflated Metabolism and Disturbed Nutritional Sensing

In various species, including mice, calorie restriction lengthens lifespan and downregulates nutritional signaling pathways. It may also have neuroprotective effects in human tissues (Fontana et al. 2016). Major nutrient-sensing molecules and pathways include sirtuins (López-Otín et al. 2013), AMP-activated protein kinase (AMPK), mechanistic target of rapamycin (mTOR), insulin, insulin-like growth factor 1 (IGF1), and insulin. The potential as therapeutic targets for neurodegenerative illnesses includes mTOR, AMPK, and sirtuins. Patients with neurological diseases typically have a metabolic malfunction, which has been linked to oxidative stress, mitochondrial dysfunction, and decreased NAD+ abundance (Babbar and Sheikh 2013).

## 7.5 Integrative Hallmarks

### 7.5.1 Stem Cell Exhaustion

For optimum health in later life, functional stem cells are required. However, over the course of an organism's life, stem cell activity and proliferative potential drop (Ahmed et al. 2017). Age-related high levels of DNA damage, poor DNA repair capability, proteostasis issues, epigenetic deregulation, mitochondrial malfunction, telomerase inactivation, and/or cell senescence can all contribute to this functional loss (Oh et al. 2014). In AD patients with CNS stem cell exhaustion, there is a correlation between impaired cognitive function and decreased neurogenesis in the dentate gyrus (Vivar 2015). Increased expression of the transcription factors forkhead transcription factor FOXO4 or the heat shock protein HSP70, as well as rapamycin therapy, boosts stem cell function (Oh et al. 2014). Heterochronic parabiosis, which exposes aged animals to young blood, enhances stem cell function in the spinal cord, and brain, delays aging, and revives cognitive functions (Villeda et al. 2014). In order to treat age-related ND illnesses, it may be beneficial to target aged stem cell regeneration.

### 7.5.2 Immune System

Leptin, ghrelin, insulin, adiponectin, and IGF1 levels fluctuate with aging and have an impact on neuronal injury and degeneration. One of the most important changes that occurs as we age is dysregulation of the immune response, which causes a persistent inflammatory state. Among the dysregulated proinflammatory mediators, cytokines and chemokines play a crucial role in the development of chronic inflammation and the immunosenescence process (Chung et al. 2019). Unresolved and unmanaged chronic inflammation that has multiple low-grade, systemic, and aging-related responses is what is known as age-associated chronic inflammation. Because the immune system is crucial for forming the brain during development and because the immune system and the neurological system both alter with age, the loss of immune response control in the brain is probably a contributing factor to neurodegeneration (Amor and Woodroofe 2014). Immune dysfunction and neuroinflammation are features of AD pathology (Kinney et al. 2018). These data collectively imply that aging characteristics are all intimately associated with AD and other NDs (Fig. 7.1).

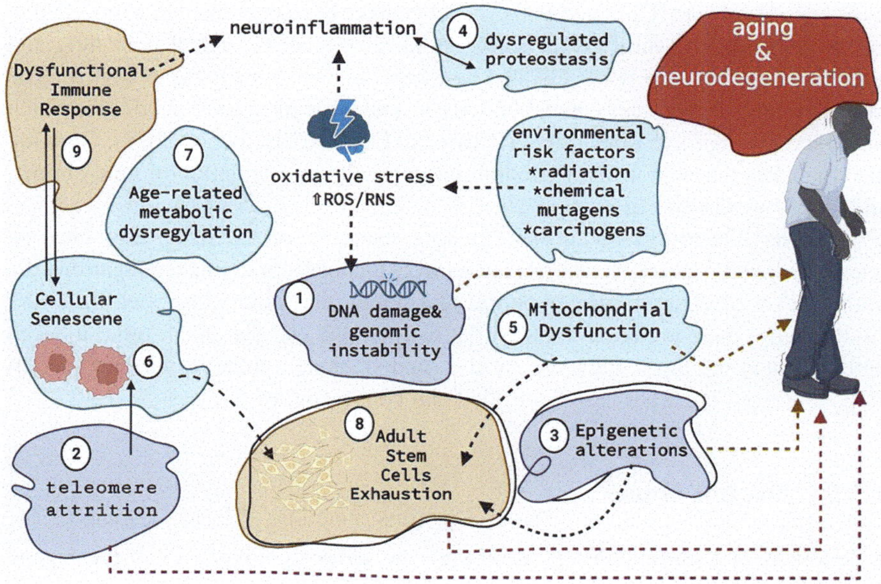


**Fig. 7.1** Aging and neurodegeneration pathways

### 7.5.3 Lifestyle Choices and Neurodegenerative Disorders

A growing body of research indicates that a sedentary lifestyle along with an unhealthy diet, and exposure to toxins can have significant effects on lifespan, health, and the emergence of neurodegenerative diseases, even though the molecular mechanisms underlying these effects are largely unknown (Popa-Wagner et al. 2020). The risk of NDs was much reduced in people who led healthy, low-risk lifestyles (no smoking, daily exercise, moderate alcohol use, and a moderate weight by the middle of their 40s). Therefore, daily risk factors including stress, a lack of exercise, poor diet, obesity, high plasma cholesterol levels, smoking, or drinking may have a deleterious impact on health, contributing in part to their relatively high incidence of NDs. Modifiable risk variables for AD and PD have been proposed to include diet/nutritional supplement intake, physical activity level, type 2 diabetes, alcohol consumption level, mood disorders, hypertension, and hypercholesterolemia (Daviglus et al. 2011). Accordingly, the current chapter focuses on the link between lifestyle including physical inactivity, diet habits, smoking and alcohol abuse, and the commencement of neurological diseases.

### 7.5.4 Smoking

Research studies show that smoking may hasten the aging of the brain (Ho et al. 2017). Metals contained in cigarette smoke, such as iron, copper, zinc, metal, manganese, and lead (Ghio and Hilborn 2017), have been documented to build up in tissues and fluids (Ashraf 2012; Saffari et al. 2014). Heavy metal poisoning and cognitive decline could be accelerated by metal buildup in the nervous system. The most frequent sign of heavy metal toxicity is a neurodegeneration condition, which is marked by cognitive impairment (Patwa and Flora 2020; Li et al. 2021). Growing data indicates that various NDs are influenced by the dysregulation of iron, copper, and zinc homeostasis (Li et al. 2021).

However, the results on former smoking status as an AD risk factor must be interpreted with caution, as several studies were inconsistent or unclear regarding the normal level of cigarette use or period of smoking cessation (Peters et al. 2008). Additionally, the correlation between the risk for AD and the use of other tobacco products such the pipes, cigars, cigarillos, and smokeless tobacco was not specifically addressed in earlier case-controlled and cohort studies.

### 7.5.5 Alcohol Abuse

Alcoholism and chronic heavy drinking can seriously impair the neurological system's ability to operate (Pervin and Stephen 2021). Along with altered emotions and personalities, these side effects can also impair perception, learning, and memory. Techniques in neuropathology and imaging have shown that alcoholics' physical brain abnormalities, such as nerve cell degeneration and brain shrinkage, exist. The hippocampus, frontal lobe, and corpus callosum are among the brain regions with white matter (WM) atrophy, axonal loss, and demyelination, which are typical signs of chronic excessive alcohol intake that results in cognitive (Kamal et al. 2020). The biggest modifiable risk factor for the onset of dementia is alcohol use disorder (AUD) (Schwarzinger et al. 2018). Moreover, AD and PD are two neurodegenerative illnesses that have been linked to chronic excessive alcohol consumption disorder (Kamal et al. 2020). Contrarily, recent research studies have focused on the neuroprotective effects of low to moderate alcohol usage (Zhang et al. 2020). According to Zhang et al. (2020), low to moderate drinking was associated with better overall cognitive performance among middle-aged or older men and women in the United States as well as better and more consistent cognition domain outcomes for word recall, mental status, and vocabulary. Low to moderate alcohol consumption was also associated with slower rates of cognitive decline in those categories.

### 7.5.6 Stress

Due to its strong connection to the modern lifestyle, stress is one of the most significant risk factors. Since chronic stress and visinin-like protein-1 (VILIP-1) levels in cerebrospinal fluid (CSF) were found to be correlated, it has been linked to generalized neurodegenerative processes in the brain. The hypothalamic–pituitary–adrenal (HPA) axis' malfunction, which results in cell changes may be the cause of the connection between stress and neurodegeneration (Peña-Bautista et al. 2020).

Chronic stress is seen as a key risk factor for the emergence of AD. It is postulated that stress in particular may quicken the course of AD via an insulin-dependent mechanism. Studies in AD mouse models revealed that stress may induce an increase in neuronal damage and senile plaque deposition. Additionally, early life stress and a rise in Aβ accumulation in the hippocampus appear to be associated with long-term survival (Peña-Bautista et al. 2020). Stress may speed the progression of PD (Peña-Bautista et al. 2020). For example, under chronic stress conditions, PD male mice showed a rise in α-synuclein buildup along with motor impairments, although the molecular processes behind gender differences have not yet been clarified.

### 7.5.7 Physical Inactivity

A sedentary lifestyle is considered one of the risk factors for a wide variety of diseases in the twenty-first century. Stroke, AD, and PD risk may increase with lifestyle without sufficient exercise training (Mattson 2012). Physical inactivity is a critical risk factor for cognitive decline in aging and for AD (Norton et al. 2014). In order to explain the underlying mechanism, Gaspar et al. (2011) proposed two hypotheses: (a) PE reduces chronic oxidative stress while promoting mitochondrial biogenesis and upregulating autophagy in PD and (b) exercise promotes the production of dopamine and trophic factors like glial-derived neurotrophic factor (GDNF), insulin-like growth factor-1 (IGF-1), brain-derived neurotrophic factor (BDNF), and fibroblast growth factor 2. Conversely, physical activity is a crucial element throughout human life. Regular exercise plays a crucial function in keeping a healthy lifestyle. In addition to supporting the circulatory, immunological, and metabolic systems, it also modifies the brain's internal environment (Speisman et al. 2013). Acute, high-intensity exercise and routine, moderate aerobic exercise have both been shown to boost neurotransmission and raise levels of circulating neurotrophic factors, improving mood and cognitive abilities (Jeon and Ha 2015; Basso and Suzuki 2017).

Data emerged in recent years suggest the preventive effects of both short-term and long-term physical activity on NDs (Vecchio et al. 2018; Fuller and Whitham 2020).

### 7.5.8 Dietary Habits

For many years, poor dietary habits (an unhealthy diet rich in saturated fat and cholesterol) have been believed as the main risk factor for hypertension, metabolic syndrome, atherosclerosis, and cerebrovascular disorders which are all caused by systemic, chronic inflammation, also known as metabolic inflammation (Hu and Willett 2002; Olefsky and Glass 2010).

Multiple pathophysiological pathways are shared by a variety of neurological illnesses, including increased oxidative stress, neuroinflammation, and disturbed

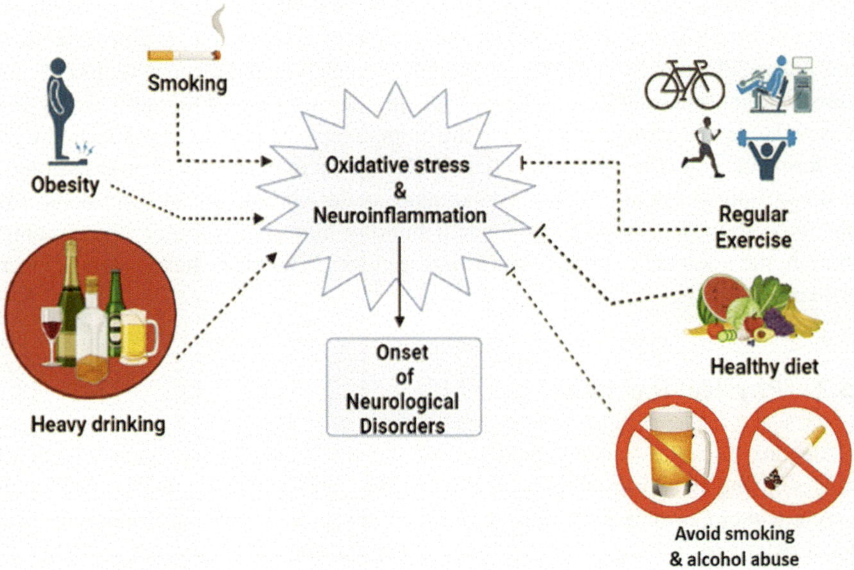


**Fig. 7.2** Lifestyle choices and neurological disorders

metabolism. A spectrum of healthy living choices should include adequate nutritional choices. For instance, diets based on plants, low in fat and saturated fat, high in fiber and antioxidants, as well as other lifestyle changes, may lessen the severity and disability of common neurological illnesses (Fig. 7.2). According to epidemiological research, eating well preserves cognitive and motor function, increases lifespan, and guards against depressive illnesses. Age-dependent neurodegeneration has been linked to interventions using substances derived from food. However, there is still some uncertainty regarding the positive effects of phytochemicals in clinical research (Naoi et al. 2019).

### 7.5.9 Phytochemicals and Neurodegenerative Disorders

In order to delay cognitive and functional deficits that come with aging and ensure a high quality of life, there has been an increase in interest in a number of phytochemicals (herbal medicine). Due to their high phytochemical content, fruits and vegetables are now urgently needed on a daily basis. Studies (Zhang et al. 2015; Karatoprak et al. 2022) proposed that phytochemicals possess many properties: (a) antioxidant, (b) antiallergic and antiinflammatory, (c) antiviral, (d) antiproliferative, and (e) anticarcinogenic.

It has been suggested that phytochemicals derived from plants, could be used to treat a variety of neurological diseases (Zhang et al. 2015). Contrary to vitamins and minerals, phytochemicals are not required for the preservation of cell viability, but they are crucial in preventing inflammation and oxidative stress linked to normal aging as well as acute and chronic age-related brain disorders (Kim et al. 2010).

The condition of NDs can be improved by using phytochemicals as nutraceuticals; however, the exact underlying mechanism is still unknown. According to one study, phytochemicals have the ability to scavenge free radicals and activate Ras-MEK-ERK and Keap-Nrf2-antioxidant response element pathways to increase the expression of antioxidant enzymes and fight oxidative damage, therefore improving the condition (Zhang et al. 2015). Although the precise cellular targets of phytochemicals are unknown, it has been suggested that these substances may work by activating prosurvival MAPK pathways, including PI3K/Akt and PKC (Naoi et al. 2019). These pathways modify the phosphorylation state of target molecules and modulate gene expression, which has an impact on cellular function, synaptic plasticity, and memory formation. An important trigger for MAPK cascades that activate prosurvival ERK, apoptosis-inducing Jun N-terminal kinase, and p38 MAPKs is oxidative stress. Through transcriptional overexpression of antiapoptotic Bcl-2 and Bcl-xL, and suppression of Fas-mediated apoptosis, ERK activation prevents the development of the death-inducing signaling complex and increases cell survival.

Studies on the biological actions and underlying health benefits of phytochemicals from various food sources, like fruits and vegetables, with regard to brain function are only now starting to receive more attention (Nan et al. 2021; Wang et al. 2022). This chapter discusses some of the most commonly used naturally available phytochemicals that can be used to treat neurological dysfunctions. Epigallocatechin-3-gallate (EGCG), curcumin (Cur), resveratrol (RES), baicalein, and genistein are typical phytochemicals that we encounter on a regular basis.

### 7.5.10 Epigallocatechin-3-Gallate

EGCG, the most prevalent polyphenol in tea leaves, has captured the attention of everyone in recent years due to its ability to slow down neuronal aging. Tea consumption is negatively linked with the prevalence of NDs. In numerous studies involving AD models, EGCG demonstrated effective outcomes. A research study has hypothesized that EGCG administration reduced β-and γ-secretases by inhibiting ERK and NF-κB, hence delaying the death of neuronal cells (Nan et al. 2021). According to another study, the administration of EGCG may have reduced amyloid plaques by cleaving APP via β-and γ-secretase. In a recent study, EGCG treatment diminished Tau hyperphosphorylation, BACE1 expression, and Aβ1–42 expression which enhanced the antioxidant system and reduced AchE activity, and ultimately led to an enhancement in the learning and memory function of AD rats (Nan et al. 2021). EGCG has also attracted great interest as a promising medication candidate for the treatment of PD (Wang et al. 2022).

### 7.5.11 Curcumin

Cur displays unusual photophysical and photochemical capabilities because of the presence of two phenols joined by a linear β-diketone linker. Cur is a crucial tool in the diagnosis of AD since it specifically binds to βA plaques and emits a highly fluorescent signal. By limiting how much it self-assembles and by causing fibrillar Aβ-40 to disaggregate, cur lowers A levels, deposits, and aggregation. Additionally, it inhibits fibrillar A production and weakens fibrillar Aβ40 and Aβ42 in vitro, providing protection against the toxicity of Aβ (Adami and Bottai 2022).

In a PD model, Cur has the ability to bind α-Syn, which makes up Lewis bodies, and prevent it from building up in dopaminergic (DA) neurons. Cur also rotational behavior diminished DA neuron mortality in the substantia nigra, and lowered the activation of astrocyte, possibly via the Wnt/β-Catenin signaling pathway in another rat PD model (Adami and Bottai 2022).

Although in vitro and in vivo investigations produced encouraging results, clinical tests revealed that Cur had limited absorption; as a result, low doses of the drug had no effect on patients (Karatoprak et al. 2022).

### 7.5.12 Resveratrol

RES is a naturally occurring nonflavonoid polyphenolic chemical found in rhubarb, mulberries, grapes, peanuts, and several other plants. Its consumption has been linked to a number of positive health effects, including anticarcinogenic, antiinflammatory, and brain-protective effects. The neuroprotective properties of RES in neurological conditions like AD and PD disease are connected to the defense of neurons against oxidative damage and toxicity as well as the prevention of apoptotic neuronal death.

Evidence suggested that RES inhibits or at least delays the aggregation of Aβ. RES treatment stopped the development of amyloid fibrils by binding to the N-terminus of A monomers. RES changed the shape of the $A\beta_{1-42}$ oligomers, prevented oligomer aggregation, hindered the development of $A\beta_{1-42}$ fibrils, and destabilized Aβ aggregates (Andrade et al. 2018). One of the primary contributors to the formation of ROS is mitochondrial complex I. Studies on the pathology and biochemistry of PD in animal models revealed disease-specific declines in protein levels or mitochondrial complex I activity in the substantia nigra pars compacta. Free radical production can be increased by changes in oxidative homeostasis caused by abnormalities in mitochondrial energy metabolism (Andrade et al. 2018). Furthermore, RES-encapsulated NPs have just recently been investigated in a few methods for the treatment of PD.

### 7.5.13 Genistein

Genistein is an isoflavone that occurs naturally and is primarily found in soybeans, while it is also present in green peas, lentils, and peanuts. Genistein can reduce the symptoms of postmenopausal conditions and prevent breast and prostate cancer. Researchers have also discovered that genistein can shield neurons from damage from cardiovascular disease, has a neuroprotective effect, and can cross the blood–brain barrier (Li et al. 2022).

Genistein may be an effective treatment option for AD because of its remarkable multipotent capacity to reduce inflammation and modulate oxidative stress, mitochondrial dysfunction, and aberrant protein aggregations (Li et al. 2022).

Due to its antiinflammatory and antioxidant qualities, genistein has been proposed to have therapeutic or preventative significance for Parkinson's disease (PD). Rotenone mimics PD in vitro in SH-SY5Y cells overexpressing mutant α-synuclein by significantly elevating mitochondrial oxidative stress, apoptosis, and cell death in comparison with the control. It is interesting to notice that rotenone's effects were lessened by preincubating with genistein (20 M) (Wu et al. 2018).

Genistein activates the Nrf2/HO-1 signaling pathway through the ER route, producing a neuroprotective effect. As indicated, genistein has been shown to have a comparable impact in the AD cell models (Li et al. 2022).

### 7.5.14 Baicalein

The root of Scutellaria baicalensis Georgi is a well-known and compatible ingredient in the herbal decoction used to treat illnesses of the central nervous system in traditional Chinese medicine. Modern pharmacokinetic studies have verified that the main bioactive flavone component of S. baicalensis Georgi's root is baicalein (5,6,7-trihydroxyflavone). Studies have revealed that baicalein has a number of important pharmacological properties, including those that lower oxidative stress, have antiinflammatory effects, prevent amyloid protein aggregation, reduce excitotoxicity, stimulate neurogenesis and differentiation, and have antiapoptotic effects. These characteristics suggest that baicalein has therapeutic potential for treating PD and AD.

By inducing the production of NTFs, members of Bcl2 protein family, and antioxidants, future prospective studies on phytochemicals and their derivatives will be used to develop new neuroprotective medications that have a better neuroprotective potency and a more stable bioavailability. More information about the molecular mechanisms underpinning phytochemicals' antiamyloidogenic properties is provided, and phytochemical derivatives are used to stop the development of harmful Aβ and α-Syn oligomers in the brains of AD and PD patients. Phytochemicals are used to convert mature protein aggregates into soluble, harmless forms in place of an earlier planned antibody treatment (Naoi et al. 2019).

## 7.6 Conclusion and Future Perspectives

When cells fail to adjust to age-related increases in oxidative, metabolic, and ionic stress, an excessive buildup of harmed proteins, DNA, and membranes could result in the onset of NDs such as AD and PD. Moreover, lifestyle factors like high-fat diets, high-sugar diets, and alcohol and cigarette addiction can induce oxidative stress and neuroinflammation, which have a deleterious impact on the onset and severity of NDs. However, fortunately, there are a number of advantageous dietary components, such as the antioxidants epigallocatechin-3-gallate, curcumin, resveratrol, baicalein, and genistein, as well as healthy diet plan, regular exercise, quitting smoking, and heavy drinking may prevent aging and associated NDs by increasing autophagy or neurogenesis in the adult brain. Further research is needed to determine the optimal dosages, combinations, and long-term safety profiles of these phytochemicals for maximizing their neuroprotective effects. Understanding how individual genetic and environmental factors influence the efficacy of these compounds will pave the way for personalized neuroprotective strategies.

## References

Adami R, Bottai D (2022) Curcumin and neurological diseases. Nutr Neurosci 25(3):441–461. https://doi.org/10.1080/1028415X.2020.1760531

Ahmed AS, Sheng MH, Wasnik S, Baylink DJ, Lau KW (2017) Effect of aging on stem cells. World J Exp Med 7(1):1–10. https://doi.org/10.5493/wjem.v7.i1.1

Amor S, Woodroofe MN (2014) Innate and adaptive immune responses in neurodegeneration and repair. Immunology 141(3):287–291. https://doi.org/10.1111/imm.12134

Andrade S, Ramalho MJ, Pereira MDC, Loureiro JA (2018) Resveratrol brain delivery for neurological disorders prevention and treatment. Front Pharmacol 9:1261. https://doi.org/10.3389/fphar.2018.01261

Ashraf MW (2012) Levels of heavy metals in popular cigarette brands and exposure to these metals via smoking. ScientificWorldJournal 2012:729430. https://doi.org/10.1100/2012/729430

Babbar M, Sheikh MS (2013) Metabolic stress and disorders related to alterations in mitochondrial fission or fusion. Mol Cell Pharmacol 5(3):109–133

Bai P (2015) Biology of poly(ADP-ribose) polymerases: the factotums of cell maintenance. Mol Cell 58(6):947–958. https://doi.org/10.1016/j.molcel.2015.01.034

Baker DJ, Petersen RC (2018) Cellular senescence in brain aging and neurodegenerative diseases: evidence and perspectives. J Clin Invest 128(4):1208–1216. https://doi.org/10.1172/jci95145

Basso JC, Suzuki WA (2017) The effects of acute exercise on mood, cognition, neurophysiology, and neurochemical pathways: a review. Brain Plast 2(2):127–152. https://doi.org/10.3233/bpl-160040

Chow HM, Herrup K (2015) Genomic integrity and the ageing brain. Nat Rev Neurosci 16(11):672–684. https://doi.org/10.1038/nrn4020

Chung HY, Kim DH, Lee EK, Chung KW, Chung S, Lee BYBP (2019) Redefining chronic inflammation in aging and age-related diseases: proposal of the senoinflammation concept. Aging Dis 10(2):367–382. https://doi.org/10.14336/ad.2018.0324

Daviglus ML, Plassman BL, Pirzada A, Bell CC, Bowen PE, Burke JR, Williams JW Jr (2011) Risk factors and preventive interventions for Alzheimer disease: state of the science. Arch Neurol 68(9):1185–1190. https://doi.org/10.1001/archneurol.2011.100

Demaagd G, Philip A (2015) Parkinson's disease and its management: part 1: disease entity, risk factors, pathophysiology, clinical presentation, and diagnosis. P T 40(8):504–532

Fang EF, Hou Y, Palikaras K, Adriaanse BA, Kerr JS, Yang B, Caponio D (2019) Mitophagy inhibits amyloid-β and tau pathology and reverses cognitive deficits in models of Alzheimer's disease. Nat Neurosci 22(3):401–412. https://doi.org/10.1038/s41593-018-0332-9

Farooqui AA (2012) Beneficial effects of flavonoids on neurological disorders. In: Farooqui AA (ed) Phytochemicals, signal transduction, and neurological disorders. Springer, New York, pp 83–115

Fontana L, Cummings NE, Arriola Apelo SI, Neuman JC, Kasza I, Schmidt BA, Lamming DW (2016) Decreased consumption of branched-chain amino acids improves metabolic health. Cell Rep 16(2):520–530. https://doi.org/10.1016/j.celrep.2016.05.092

Fuller OK, Whitham M (2020) The protective effect of exercise in neurodegenerative diseases: the potential role of extracellular vesicles. Cells 9(10):2182. https://doi.org/10.3390/cells9102182

Gaspar MJM, Amaral TF, Oliveira BM, Borges N (2011) Protective effect of physical activity on dissatisfaction with body image in children—a cross-sectional study. Psychol Sport Exerc 12(5):563–569

Ghio AJ, Hilborn ED (2017) Indices of iron homeostasis correlate with airway obstruction in an NHANES III cohort. Int J Chron Obstruct Pulmon Dis 12:2075–2084. https://doi.org/10.2147/copd.s138457

Hayflick L (1965) The limited in vitro lifetime of human diploid cell strains. Exp Cell Res 37:614–636. https://doi.org/10.1016/0014-4827(65)90211-9

Ho TT, Warr MR, Adelman ER, Lansinger OM, Flach J, Verovskaya EV, Figueroa ME, Passegué E (2017) Autophagy maintains the metabolism and function of young and old stem cells. Nature 543(7644):205–210. https://doi.org/10.1038/nature21388

Hou Y, Dan X, Babbar M, Wei Y, Hasselbalch SG (2019) Ageing as a risk factor for neurodegenerative disease. Nat Rev Neurol 15(10):565–581. https://doi.org/10.1038/s41582-019-0244-7

Hu FB, Willett WC (2002) Optimal diets for prevention of coronary heart disease. JAMA 288(20): 2569–2578. https://doi.org/10.1001/jama.288.20.2569

Hwang JY, Aromolaran KA, Zukin RS (2017) The emerging field of epigenetics in neurodegeneration and neuroprotection. Nat Rev Neurosci 18(6):347–361. https://doi.org/10.1038/nrn.2017.46

Jeon YK, Ha CH (2015) Expression of brain-derived neurotrophic factor, IGF-1 and cortisol elicited by regular aerobic exercise in adolescents. J Phys Ther Sci 27(3):737–741. https://doi.org/10.1589/jpts.27.737

Jeppesen DK, Bohr VA, Stevnsner T (2011) DNA repair deficiency in neurodegeneration. Prog Neurobiol 94(2):166–200. https://doi.org/10.1016/j.pneurobio.2011.04.013

Johri A, Beal MF (2012) Mitochondrial dysfunction in neurodegenerative diseases. J Pharmacol Exp Ther 342(3):619–630. https://doi.org/10.1124/jpet.112.192138

Kamal H, Tan GC, Ibrahim SF, Shaikh MF, Mohamed IN, Mohamed RMP, Kumar J (2020) Alcohol use disorder, neurodegeneration, Alzheimer's and Parkinson's disease: interplay between oxidative stress, neuroimmune response and excitotoxicity. Front Cell Neurosci 14: 282. https://doi.org/10.3389/fncel.2020.00282

Karatoprak G, Göger F, Çelik İ, Budak Ü, Akkol EK, Aschner M (2022) Phytochemical profile, antioxidant, antiproliferative, and enzyme inhibition-docking analyses of Salvia ekimiana Celep & Doğan. S Afr J Bot 146:36–47. https://doi.org/10.1016/j.sajb.2021.09.033

Keogh MJ, Chinnery PF (2015) Mitochondrial DNA mutations in neurodegeneration. Biochim Biophys Acta 1847(11):1401–1411. https://doi.org/10.1016/j.bbabio.2015.05.015

Kim J, Lee HJ, Lee KW (2010) Naturally occurring phytochemicals for the prevention of Alzheimer's disease. J Neurochem 112(6):1415–1430. https://doi.org/10.1111/j.1471-4159.2009.06562.x

Kinney JW, Bemiller SM, Murtishaw AS, Leisgang AM, Salazar AM, Lamb BT (2018) Inflammation as a central mechanism in Alzheimer's disease. Alzheimers Dement (N Y) 4:575–590. https://doi.org/10.1016/j.trci.2018.06.014

Kültz D (2005) Molecular and evolutionary basis of the cellular stress response. Annu Rev Physiol 67:225–257. https://doi.org/10.1146/annurev.physiol.67.040403.103635

Li H, Mu Q, Kang Y, Yang X, Shan L, Wang M, Wang F (2021) Association of cigarette smoking with male cognitive impairment and metal ions in cerebrospinal fluid. Front Psych 12:738358. https://doi.org/10.3389/fpsyt.2021.738358
Li R, Robinson M, Ding X, Geetha T, Al-Nakkash L (2022) Genistein: a focus on several neurodegenerative diseases. J Food Biochem 46(7):e14155. https://doi.org/10.1111/jfbc.14155
Lindahl T (1993) Instability and decay of the primary structure of DNA. Nature 362(6422): 709–715. https://doi.org/10.1038/362709a0
Liu R-M (2022) Aging, cellular senescence, and Alzheimer's disease. Int J Mol Sci 23(4):1989
Loaiza N, Demaria M (2016) Cellular senescence and tumor promotion: is aging the key? Biochim Biophys Acta 1865(2):155–167. https://doi.org/10.1016/j.bbcan.2016.01.007
López-Otín C, Blasco MA, Partridge L, Serrano M, Kroemer G (2013) The hallmarks of aging. Cell 153(6):1194–1217. https://doi.org/10.1016/j.cell.2013.05.039
Lovell MA, Gabbita SP, Markesbery WR (1999) Increased DNA oxidation and decreased levels of repair products in Alzheimer's disease ventricular CSF. J Neurochem 72(2):771–776. https://doi.org/10.1046/j.1471-4159.1999.0720771.x
Mattson MP (2012) Energy intake and exercise as determinants of brain health and vulnerability to injury and disease. Cell Metab 16(6):706–722. https://doi.org/10.1016/j.cmet.2012.08.012
Maynard S, Fang EF, Scheibye-Knudsen M, Croteau DL, Bohr VA (2015) DNA damage, DNA repair, aging, and neurodegeneration. Cold Spring Harb Perspect Med 5(10):a025130. https://doi.org/10.1101/cshperspect.a025130
Nacarelli T, Lau L, Fukumoto T, Zundell J, Fatkhutdinov N, Wu S, Noma KI (2019) NAD(+) metabolism governs the proinflammatory senescence-associated secretome. Nat Cell Biol 21(3): 397–407. https://doi.org/10.1038/s41556-019-0287-4
Nan S, Wang P, Zhang Y, Fan J (2021) Epigallocatechin-3-gallate provides protection against Alzheimer's disease-induced learning and memory impairments in rats. Drug Des Devel Ther 15:2013–2024. https://doi.org/10.2147/dddt.s289473
Naoi M, Shamoto-Nagai M, Maruyama W (2019) Neuroprotection of multifunctional phytochemicals as novel therapeutic strategy for neurodegenerative disorders: antiapoptotic and antiamyloidogenic activities by modulation of cellular signal pathways. Future Neurol 14(1):FNL9. https://doi.org/10.2217/fnl-2018-0028
Norton S, Matthews FE, Barnes DE, Yaffe K, Brayne C (2014) Potential for primary prevention of Alzheimer's disease: an analysis of population-based data. Lancet Neurol 13(8):788–794. https://doi.org/10.1016/s1474-4422(14)70136-x
Oh J, Lee YD, Wagers AJ (2014) Stem cell aging: mechanisms, regulators and therapeutic opportunities. Nat Med 20(8):870–880. https://doi.org/10.1038/nm.3651
Olefsky JM, Glass CK (2010) Macrophages, inflammation, and insulin resistance. Annu Rev Physiol 72:219–246. https://doi.org/10.1146/annurev-physiol-021909-135846
Patwa J, Flora SJS (2020) Heavy metal-induced cerebral small vessel disease: insights into molecular mechanisms and possible reversal strategies. Int J Mol Sci 21(11):3862. https://doi.org/10.3390/ijms21113862
Peña-Bautista C, Casas-Fernández E, Vento M, Baquero M, Cháfer-Pericás C (2020) Stress and neurodegeneration. Clin Chim Acta 503:163–168. https://doi.org/10.1016/j.cca.2020.01.019
Pervin Z, Stephen JM (2021) Effect of alcohol on the central nervous system to develop neurological disorder: pathophysiological and lifestyle modulation can be potential therapeutic options for alcohol-induced neurotoxication. AIMS Neurosci 8(3):390–413. https://doi.org/10.3934/Neuroscience.2021021
Peters R, Poulter R, Warner J, Beckett N, Burch L, Bulpitt C (2008) Smoking, dementia and cognitive decline in the elderly, a systematic review. BMC Geriatr 8:36. https://doi.org/10.1186/1471-2318-8-36
Popa-Wagner A, Dumitrascu DI, Capitanescu B, Petcu EB, Surugiu R, Fang WH, Dumbrava DA (2020) Dietary habits, lifestyle factors and neurodegenerative diseases. Neural Regen Res 15(3): 394–400. https://doi.org/10.4103/1673-5374.266045

Ross AB, Langer JD, Jovanovic M (2021) Proteome turnover in the spotlight: approaches, applications, and perspectives. Mol Cell Proteomics 20:100016. https://doi.org/10.1074/mcp.R120.002190

Rossiello F, Jurk D, Passos JF, D'adda Di Fagagna F (2022) Telomere dysfunction in ageing and age-related diseases. Nat Cell Biol 24(2):135–147. https://doi.org/10.1038/s41556-022-00842-x

Saffari A, Daher N, Ruprecht A, De Marco C, Pozzi P, Boffi R, Sioutas C (2014) Particulate metals and organic compounds from electronic and tobacco-containing cigarettes: comparison of emission rates and secondhand exposure. Environ Sci Process Impacts 16(10):2259–2267. https://doi.org/10.1039/c4em00415a

Schwarzinger M, Pollock BG, Hasan OSM, Dufouil C, Rehm J (2018) Contribution of alcohol use disorders to the burden of dementia in France 2008–13: a nationwide retrospective cohort study. Lancet Public Health 3(3):e124–e132. https://doi.org/10.1016/s2468-2667(18)30022-7

Speisman RB, Kumar A, Rani A, Foster TC, Ormerod BK (2013) Daily exercise improves memory, stimulates hippocampal neurogenesis and modulates immune and neuroimmune cytokines in aging rats. Brain Behav Immun 28:25–43. https://doi.org/10.1016/j.bbi.2012.09.013

Tanaka K, Matsuda N (2014) Proteostasis and neurodegeneration: the roles of proteasomal degradation and autophagy. Biochim Biophys Acta 1843(1):197–204. https://doi.org/10.1016/j.bbamcr.2013.03.012

Tran M, Reddy PH (2020) Defective autophagy and mitophagy in aging and Alzheimer's disease. Front Neurosci 14:612757. https://doi.org/10.3389/fnins.2020.612757

Vecchio LM, Meng Y, Xhima K, Lipsman N, Hamani C, Aubert I (2018) The neuroprotective effects of exercise: maintaining a healthy brain throughout aging. Brain Plast 4(1):17–52. https://doi.org/10.3233/bpl-180069

Villeda SA, Plambeck KE, Middeldorp J, Castellano JM, Mosher KI, Luo J (2014) Young blood reverses age-related impairments in cognitive function and synaptic plasticity in mice. Nat Med 20(6):659–663. https://doi.org/10.1038/nm.3569

Vivar C (2015) Adult hippocampal neurogenesis, aging and neurodegenerative diseases: possible strategies to prevent cognitive impairment. Curr Top Med Chem 15(21):2175–2192. https://doi.org/10.2174/1568026615666150610141524

Wang Y, Wu S, Li Q, Lang W, Li W, Jiang X, Wang H (2022) Epigallocatechin-3-gallate: a phytochemical as a promising drug candidate for the treatment of Parkinson's disease. Front Pharmacol 13:977521. https://doi.org/10.3389/fphar.2022.977521

Weissman L, Jo DG, Sørensen MM, De Souza-Pinto NC, Markesbery WR, Mattson MP, Bohr VA (2007) Defective DNA base excision repair in brain from individuals with Alzheimer's disease and amnestic mild cognitive impairment. Nucleic Acids Res 35(16):5545–5555. https://doi.org/10.1093/nar/gkm605

Wiley CD, Velarde MC, Lecot P, Liu S, Sarnoski EA, Freund A, Campisi J (2016) Mitochondrial dysfunction induces senescence with a distinct secretory phenotype. Cell Metab 23(2):303–314. https://doi.org/10.1016/j.cmet.2015.11.011

Wissler Gerdes EO, Zhu Y, Tchkonia T, Kirkland JL (2020) Discovery, development, and future application of senolytics: theories and predictions. FEBS J 287(12):2418–2427. https://doi.org/10.1111/febs.15264

Wu HC, Hu QL, Zhang SJ, Wang YM, Jin ZK, Lv LF, Cheng OM (2018) Neuroprotective effects of genistein on SH-SY5Y cells overexpressing A53T mutant α-synuclein. Neural Regen Res 13:1375. https://doi.org/10.4103/1673-5374.235250

Zhang YJ, Gan RY, Li S, Zhou Y, Li AN, Xu D-P, Li HB (2015) Antioxidant phytochemicals for the prevention and treatment of chronic diseases. Molecules 20(12):21138–21156

Zhang R, Shen L, Miles T, Shen Y, Cordero J, Qi Y, Li C (2020) Association of low to moderate alcohol drinking with cognitive functions from middle to older age among US adults. JAMA Netw Open 3(6):e207922. https://doi.org/10.1001/jamanetworkopen.2020.7922

# Ethnopharmacology: A Boon for Brain Aging

8

Mohamed Y. Zaky, Nahed S. Lamloum, and Osama M. Ahmed

**Abstract**

Aging is a universal process that is beyond the scope of human control. It cannot be stopped because it is a natural occurrence. Before the commencement of aging, there is the possibility of taking preventive actions for many disorders related to aging. Aging is a persistent, unstoppable, and degenerative process. Lifestyle has a significant impact on how we age, and a changing lifestyle is a primary factor in premature age-related issues. The term "ethno-medical knowledge" refers to the understanding of traditional medicine, which is frequently used by many indigenous populations and ethnic groups. A variety of anthropological and ethnopharmacological investigations could be used to help record the disorganized traditional medical system. Phytochemicals derived from medicinal plants help to preserve the brain's chemical balance by suppressing the activation of receptors for key inhibitory neurotransmitters. In traditional medicine, many plants have been recorded to alleviate age-related diseases. The chapter discusses the value of ethnopharmacological research in mitigating and preventing brain aging.

**Keywords**

Nootropics · Phytochemicals · Neuroprotection · Ethnopharmacology · Phenols

M. Y. Zaky (✉) · O. M. Ahmed
Molecular Physiology Division, Zoology Department, Faculty of Science, Beni-Suef University, Beni-Suef, Egypt
e-mail: mohamedzaki448@science.bsu.edu.eg

N. S. Lamloum
Department of Biochemistry, Faculty of Science, Beni-Suef University, Beni-Suef, Egypt





S. Pathak, A. Banerjee (eds.), *Neuroprotective Effects of Phytochemicals in Brain Ageing*, https://doi.org/10.1007/978-981-99-7269-2_8

## 8.1 Introduction

According to the World Health Organization (WHO), the elderly population will soon outnumber children, with more people reaching extreme old age than ever before. The fundamental reason for this is the significant increase in life expectancy and declining fertility rates, which may result in a continuation, if not acceleration, of population aging (Chen et al. 2019). Aging is the result of the accumulation of a broad range of cellular and molecular long-term damage. Aging leads to a continuous reduction in mental and physical capacity as well as an increased risk of diseases and, eventually, death (Tewari et al. 2018). Yet, these alterations are neither linear nor continuous, and their connection to a person's age in years is tenuous. At 70 years old, some individuals may enjoy excellent health and independence, while others may require significant care due to poor health (Chan 2019).

Neurodegeneration is a process that occurs as a result of both neuropathological disorders and brain aging. Brain pathology, in the form of cerebrovascular and neurodegenerative illnesses, is recognized as a primary cause of death worldwide, with an incidence of around 2/1000 and an 8% global death rate (Martínez-Hernández et al. 2023). The concept of "neuroprotection" refers to the techniques and defenses employed to shield the central nervous system (CNS) from damage to neurons brought on by both acute (such as a stroke or trauma) and chronic neurodegenerative diseases (such as Alzheimer's and Parkinson's diseases) (Li et al. 2023).

Recently, herbal therapies have gained popularity as conventional drugs like antibiotics, which once had almost universal success against serious diseases, are losing their effectiveness (Ugoeze 2022). About 2000 plant species are regarded as having therapeutic significance in Ayurveda (ancient Indian medicine), whereas the Chinese Pharmacopoeia contains about 5700 traditional medicines, the majority of which are derived from plants. Reverse pharmacology is used in Ayurveda-based drug development, where drug candidates are first found based on widespread population use and then verified in clinical studies. According to experts, this method can shorten the typical 12-year drug discovery process to 5 years or less at a fraction of the expense (Das et al. 2016). Modern psychopharmacology has a brief history, and its current notions are more "pharmaco-centric" than those of the majority of other modern medical specialties (Yildirim 2021). The first clinically beneficial psychoactive medications were not accidentally discovered until the 1950s. At this point, careful clinical observations were crucial. The field of psychopharmacology was eventually created as a result of continued, intensive research into the mechanisms of action and structure-activity correlations of some of these so-called first-generation psychotherapeutics (Brouillot et al. 2020). Thereby, the majority of currently available medications that are effective for treating mental health issues are the results of efforts to fully define, comprehend, and enhance the therapeutic potentials of medications whose clinical efficacy was already known or by extrapolating existing knowledge on the modes of action of known agents with clearly observable effects on mood, behavior, emotion, etc. (Husain et al. 2007). This chapter will focus on the neuroprotective functions of phytochemicals and neuroprotective herbs in aging and other associated illnesses, focusing on their mode of action and therapeutic potential.

## 8.2 Phytochemicals in Neuroprotection

Using phytoconstituents to treat or prevent neurodegenerative diseases has generated a lot of interest in the scientific community and the general public. Numerous phytochemicals have been demonstrated to have neuroprotective effects in neurological illness models in animals and cell cultures (Fig. 8.1). For instance, piceatannol can shield cultured neurons from A-induced death, while chalcone can shield neurons from ischemic brain damage. Studies in animal models of neurodegenerative illnesses and studies on human populations' epidemiology have shown that phytochemicals in fruits and vegetables actually shield the nervous system against illness (Gouvinhas et al. 2022). The majority of studies on the health advantages of phytochemicals have emphasized how many of the active molecules have antioxidant activities. Various phytochemicals' neuroprotective properties are linked to decreased oxidative stress. Reducing oxidative stress and protecting cultured hippocampal neurons from nitric oxide-mediated cell death, for instance, were achieved by resveratrol, quercetin, and catechins (Orlova et al. 2022). Numerous natural ingredients, including a-tocopherol, ginsenosides, resveratrol, lycopene, and *Ginkgo biloba*, have been demonstrated to possess neuroprotective characteristics (Jahangir et al. 2022).

### 8.2.1 Alkaloids

A large class of naturally occurring substances called alkaloids typically comprise oxygen, nitrogen, carbon, and hydrogen. Alkaloids can have an impact on the CNS, particularly the brain and spinal cord's nerve cells, which regulate a variety of direct bodily processes and behaviors. They may also have an impact on the autonomic nervous system, which controls respiration, heartbeat, circulation, and internal organ function. The indole carbon-nitrogen ring is found in indole alkaloids, which are also present in the psychedelic substances lysergic acid diethylamide and the psilocybin and ergine alkaloids from fungi, as well as the psychedelic substance lysergic acid diethylamide (LSD). These alkaloids can conflict with or complement serotonin's activity in the brain (Otimenyin 2022). The principal mechanism of action of ergot alkaloids was formerly believed to be their notable effects on blood flow. The spinal cord and CNS are impacted by tropane alkaloids found in Datura, including atropine, hyoscyamine, and scopolamine (Dey et al. 2020). A highly effective analgesic and narcotic, an isoquinoline alkaloid like morphine, separated from Papaver somniferum (opium poppy), is a drug. Gamma-aminobutyric acid (GABA) levels rise in the brain synapses as a result of morphine's strong binding to the μ-opioid receptor (MOR) in the CNS (Kumar and Khanum 2012). Vinpocetine is a highly effective vasodilator, an alkaloid derived from Vinca minor. The consequences of

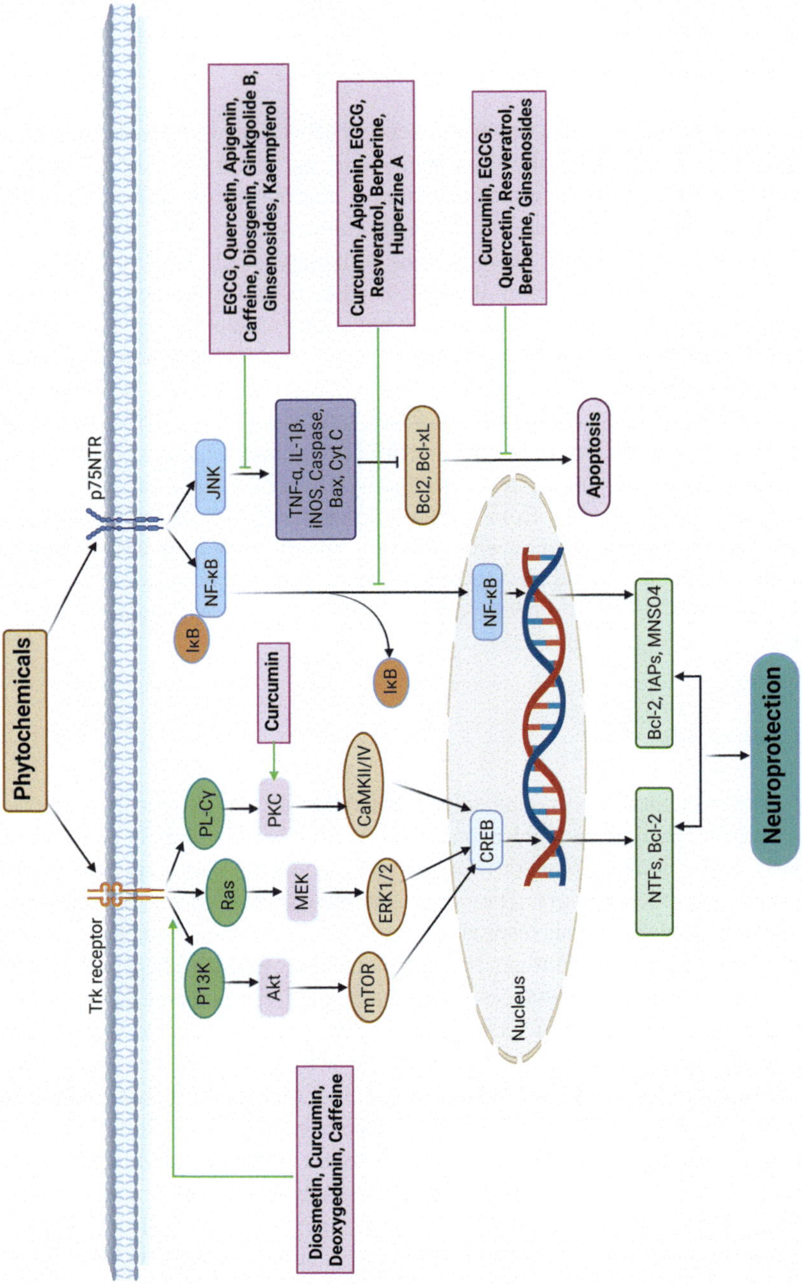


**Fig. 8.1** Neuroprotection mechanisms of phytochemicals

hypoxia and ischemia may be mitigated by vinpocetine's specific augmentation of cerebral blood flow and metabolism, involving increased absorption of glucose (Nutrition and Republic 2022). Galantamine is a tertiary alkaloid that can activate nicotinic receptors, which improve cognition and memory even more, and it is a member of the class of medications known as cholinesterase inhibitors (Reddy et al. 2019).

### 8.2.2 Flavonoids

The ability of flavonoids to regulate neuronal activity and stop age-related neurodegeneration has recently attracted a lot of attention. Studies on the effects of flavonoid-rich plant or food extracts on numerous mammalian species, including humans, have shown that they can enhance existing neuronal function, preserve susceptible neurons, or promote neuronal regeneration to improve memory and learning. Individual flavonoids, like the citrus flavonoid tangeretin, have been shown to preserve nigrostriatal integrity and functionality after 6-hydroxydopamine lesioning, implying that they may act as possible neuroprotective agents against the pathology that underlies Parkinson's disease (Kumar and Khanum 2012). In pertinent *in vitro* and *in situ* models, it has been demonstrated that dietary anthocyanins, such as cyanidin-3-rutinoside and pelargonidin-3-glucoside, as well as flavanones like hesperetin and naringenin, as well as their in vivo metabolites, can pass the blood-brain barrier (BBB) (Kumar and Khanum 2012).

### 8.2.3 Phenols

During the 1990s, there was a lot of interest in dietary phenolic compounds because of mounting proof of their positive effects on human health. Apple, grape, and citrus fruit juice polyphenols are more effective at protecting the nervous system than antioxidant vitamins, according to proof from in vivo, in vitro, and clinical trials (Niaz and Khan 2020). Green tea's phenolic components (catechins and epicatechins) can shield neurons from a wide range of oxidative and metabolic damage, including protecting dopaminergic neurons from harm brought on by 6-hydroxydopamine in a Parkinson's disease rat model (Arif et al. 2022). Stimulation of protein kinase C (PKC) and transcription factors that promote the production of cell-survival genes have a role in catechins' neuroprotective effects, which are polyphenols (Malar et al. 2020). It has been proposed that catechins can protect neurons from Alzheimer's disease (AD) processes and reduce the development of the disease. Many signaling pathways can be activated by catechins and their metabolites to boost cell survival and produce anti-inflammatory effects. Changes in the expression of pro- and anti-apoptotic proteins, as well as increased antioxidant protection, are examples of these processes (Afzal et al. 2022).

### 8.2.4 Fatty Acids

It has been demonstrated that both in animals and people, using monounsaturated fatty acids and polyunsaturated fatty acids (PUFAs) can reduce cognitive decline. The majority of research has been on the n23 and n26 PUFAs present in nuts, like walnuts, which also include the monounsaturated fatty acid oleic acid (8:1) and the n26 and n23 PUFAs linoleic acid (LA) and a-linolenic acid (ALA) (Mariamenatu and Abdu 2021). For neurons to connect with other cells, they need to have normal anchor receptors, ion channels, and the ability to liberate and reabsorb neurotransmitters that have not been metabolized. These requirements are all dependent on the structure of the cell. The neuronal membrane's fatty acid content affects these characteristics (Ameer 2016). While essential fatty acids have been proven to promote membrane fluidity and PUFA content, it has been shown that, as people age, the amount of fatty acids in neuronal membranes decreases. Along with having an impact on the biophysical characteristics of membranes, PUFAs may be directly involved in signaling cascades that improve synaptic plasticity, neuroprotection, and neuronal function when they are present in the form of phospholipids in neuronal membranes (Britten-Jones et al. 2023).

### 8.2.5 Terpenoids and Saponins

Since more than a century ago, many essential oils derived from plants, including wormwood, have been recognized to have convulsant effects. Valepotriates and sesquiterpenes (Valerian officinalis) are the two pharmacologically stimulating components found in the rhizome of valerian. The active substances have been shown to have sedating effects on mice. Nevertheless, it has been observed that valerian crude extract has GABA (B) receptor binding characteristics and decreases the rat synaptosomal absorption of GABA (Orhan 2021). Huperzine A, which was isolated from a Chinese medicinal plant, has a variety of neuroprotective properties. Huperzine A increased spatial working memory and helped with learning and memory problems (Cao et al. 2021). Triterpenoid, brahminoside, and monoterpenes, including β-pinene and γ-terpinene, are chemical compounds in Centella asiatica L. that are helpful in reviving and bolstering the neurological system (Kumar and Khanum 2012).

### 8.2.6 Resveratrol

Resveratrol is an antioxidant phytophenol found in high concentrations in red grapes. Recent research has demonstrated that resveratrol penetrates the CNS quickly after peripheral treatment and can shield brain and spinal cord neurons from ischemia damage. In a stroke model, resveratrol administration reduced ischemia damage to the brain and further shielded spinal cord neurons from ischemia damage (Ahmed et al. 2017). *Resveratrol* can rescue cultured neurons from oxidative damage caused by nitric oxide. Similarly, in midbrain slice cultures, resveratrol protected dopaminergic neurons from metabolic and oxidative stressors, a paradigm relevant to Parkinson's disease (Sahebnasagh et al. 2022). *Resveratrol* prevented neuronal cells from being destroyed by amyloid β-peptide in AD models and increased amyloid β-peptide elimination from cultured cells (Marambaud et al. 2005).

### 8.2.7 Allium and Allicin

Allium and allicin, two organosulfur compounds that are prevalent in garlic and onions, have been discovered to have neuroprotective properties. Apart from antioxidant activity, allyl-containing sulfides may promote stress-response pathways, causing the overexpression of neuroprotective proteins, including mitochondrial uncoupling proteins (Sunkaria et al. 2017). Furthermore, *allicin* stimulates transient receptor potential (TRP) ion channels in neurons' plasma membranes. Several additional phytochemicals, such as isothiocyanates, garlic alliums, and cannabinoids, stimulate TRP channels in neurons, leading to adaptive cellular stress responses (Kumar and Khanum 2012).

### 8.2.8 Curcumin

Curcumin has been shown to have several positive effects on the neurological system. Curcumin therapy shielded neurons from ischemia-induced cell death and improved behavioral impairments in an animal model of stroke (Pluta et al. 2022). Moreover, increasing data from cell culture and animal models suggest that dietary curcumin may be a promising candidate for either the avoidance or amelioration of severe, disabling age-related neurodegenerative diseases like Alzheimer's, Parkinson's, and stroke. Furthermore, curcumin has been proven in a depression animal model to boost the expression of brain-derived neurotrophic factor (BDNF) and counteract the chronic stress-induced reduction of hippocampal neurogenesis (Subhan et al. 2017).

### 8.2.9 Sulforaphane

Sulforaphane is an isothiocyanate found in large quantities in broccoli, Brussels sprouts, and other cruciferous vegetables. In animal models of neurodegenerative diseases, both chronic and acute, *sulforaphane* has been proven to offer neuroprotective advantages. In a mouse model of stroke, *sulforaphane* therapy decreased the degree of brain injury and cerebral edema while preserving the retinal pigment cells (Kim 2021). *Sulforaphane* has been demonstrated to guard dopaminergic neurons against mitochondrial toxins as well as oxidative stress in cultured neurons (Klomparens and Ding 2019).

### 8.2.10 Hypericin and Pseudohypericin

Hypericin and pseudohypericin, two naphthodianthrones, are the main constituents of *Hypericum perforatum*. There is compelling proof that *hypericin and pseudohypericin* aid in antidepressant activity. Certain antidepressants work by inhibiting monoamine oxidase, which is one mechanism by which they boost levels of neurotransmitters like serotonin, norepinephrine, or dopamine (Garg et al. 2023). This molecule appears to inhibit serotonin, dopamine, and norepinephrine synaptic reuptake. By preventing neurotransmitter reuptake, synaptic levels are increased. This is yet another way synthetic antidepressants may function (Huber et al. 2022).

## 8.3 The Role of Phytochemicals in Receptor Binding Activity

A number of phytochemicals have an impact on neuronal excitability by activating or inactivating specific ion channels or receptors. The brain contains about 50 neurotransmitters that belong to several chemical families. The first neurotransmitter to be identified, acetylcholine (Ach), has an important role in the brain and is currently known to be crucial for memory and learning (Akyuz et al. 2021).

A number of phytochemicals have an impact on neuronal excitability by activating or inactivating specific ion channels or receptors. The brain contains about 50 neurotransmitters that belong to several chemical families. The first neurotransmitter to be identified, acetylcholine (Ach), has an important role in the brain and is currently known to be crucial for memory and learning (Akyuz et al. 2021).

Capsaicin, the phytochemical responsible for the potent and unpleasant physiological impacts of hot peppers, is a well-known example. Capsaicin induces certain $Ca^{2+}$ channels known as vanilloid receptors. The principal inhibitory neurotransmitter in the brain, GABA, is now recognized to promote the action of ionotropic receptors in a broad range of natural compounds. Many plant-derived flavonoids, terpenes, and related compounds alter how ionotropic GABA receptors work (Curtis et al. 1970). These GABA modulators can be detected in a variety of foods and drinks, including tea, red wine, whisky, and herbal supplements like *ginseng* and *Ginkgo biloba*. Examples of these foods and drinks include grapefruit, onions, and numerous fruits and vegetables. The agonist muscimol (Johnston 2005) and antagonists bicuculline (Curtis et al. 1970) and picrotoxin (from *Anamirta cocculus*) (Jarboe et al. 1968) were some of the initial materials utilized to study ionotropic GABA receptors. Certain chemicals can pass the BBB and affect how the brain functions. Low levels of glutamate receptor activation can promote synaptic plasticity and protect neurons from dysfunction and degeneration, in part through increasing the production of neurotrophic factors. Moreover, a variety of phytochemicals open transient receptor potential (TRP) calcium channels in the membrane of nerve cells, which has a well-known impact on the neurological system. Tetrahydrocannabinol (THC), a potent component of garlic, capsaicin, and allicin, all open TRP channels (Johnston 2005).

## 8.4 Nootropics

The term "nootropics," which is derived from Greek and means "acting on the mind," is utilized by proponents of smart medications to refer to prescription medications and nutritional components that have a favorable impact on brain function (Modrzyński 2021). There are many pharmacological substances on the market that have been used for their ability to provide neuroprotection. Medications designed to enhance neurofunction typically function by varying the ratio of specific chemicals (neurotransmitters) in the brain (Farooqui 2019). Many medications made from medicinal plants have been shown to have bioactive phytochemical components that have been proven to improve. Decreased cholinergic activity in the brain has been proposed as one of the causes of dementia. Consequently, cholinergic medicines (of plant origin) can be used to enhance memory. These include muscarinic agonists, nicotinic agonists, and cholinesterase inhibitors (Farooqui 2019).

## 8.5 Neuroprotective Herbs

For primary healthcare for humans and animals, almost 80% of people in underdeveloped nations depend on phytomedicine (Khan and Ahmad 2019). Many plants have been employed in traditional medical practices to treat cognitive disorders, including problems with memory and neurodegenerative illnesses such as AD. Many plants have been employed in traditional medical practices to treat cognitive disorders, including memory-related disorders and neurodegenerative diseases like AD (Palimariciuc et al. 2023). Complex mixes of organic substances, including fatty acids, sterols, alkaloids, flavonoids, glycosides, saponins, tannins, terpenes, and others, can be found in herbal products. Supporters of herbal medicines assert that a plant's medicinal effectiveness originates from the synergistic impacts of its numerous constituents, as opposed to the individual compounds of conventional medications, which are isolated by pharmacologists. As a result, it is thought that traditional medicines are efficient and have few to no side effects. Between 25 and 50 percent of today's medications were first sourced from plants (Singh et al. 2023). In Asian nations, more than 120 traditional medicines are utilized to treat problems of the CNS (Abdolmaleki et al. 2020).

### 8.5.1 *Bacopa Monniera*

The Ayurvedic medical system has been using "*Brahmi*" for centuries. It was traditionally utilized as a brain tonic to improve learning and memory and to relieve anxiety and epilepsy in patients (Sendri and Bhandari 2023). The alkaloids brahmine and herpestine, saponins d-mannitol, hersaponin, and monnierin are just a few of the active components found in *Bacopa*. Beta-sitosterol, stigmasterol, betulic acid, various bacosides, and bacopasaponins are only a few of the other active ingredients that have since been discovered. Bacosides A and B are the ingredients in *bacopa* that have cognitive benefits (Sukumaran et al. 2019). To help injured neurons heal, the bacosides stimulate kinase activity, neuronal production, synaptic activity recovery, and ultimately nerve impulse transmission (Singh 2019).

### 8.5.2 *Ginkgo biloba L.*

*Ginkgo biloba* (*Ginkgoaceae*) is native to East Asia and is also known as the kew tree, yinhsing, ginkyo, and maiden hair tree. It has been utilized in conventional Chinese medicine to treat memory loss caused by blood circulation problems (Chandrasekaran et al. 2003)· By increasing oxygen delivery and assisting the body in removing free radicals, the herb acts to improve memory. Included among the phytoconstituents are the terpenoids bilobolide, ginkgolides, flavonoids, steroids, and organic acids. Leaf extract has a distinctive polyvalent pharmacological activity due to the presence of 24% flavonoids and 6% terpenic lactones. Quercetin, Keampferol, and Isorhamnetin make up the majority of the flavonoid fraction, while the terpenic derivatives include the ginkgolides A, B, C, J, and M and the bilobalide, a sesquiterpenic trilactone (Chandrasekaran et al. 2003).

### 8.5.3 *Centella asiatica*

*Centella asiatica* (CA) L. Urban, also known as *Hydrocotyle asiatica* L., is a psychotropic herb that belongs to the *Apiaceae* (*Umbelliferae*) family and has been used for thousands of years as a *medhya rasayana* in the Ayurvedic medical system (Mitra et al. 2007). It has been demonstrated to decrease the indicators of oxidative stress. Triterpenoid saponins like asiaticoside, centelloside, brahminoside, oxyasiaticoside, brahmoside, asiaticoside, brahminoside, thankunoside, isothankunoside, and related sapogenins make up the majority of this plant's bioactive components. Moreover, it contains triterpenoid acids, like betulic acid, isobrahmic acid, asiatic acid, madecassic acid, brahmic acid, and betulic acid (Sarma et al. 2016).

### 8.5.4 *Panax ginseng*

*Panax ginseng*, also known as Asian ginseng, is one of the most popular species and is also utilized in Korean traditional medicine and Kampo. Ginseng is a term for a collection of various species within the *Panax* genus and the family *Araliaceae* growing in northeastern Asia. It is a type of herb for raising energy that is most frequently used in Asian medicine. It has been used as a tonic and rejuvenating agent for thousands of years, in addition to being utilized as an anti-aging herb (Wu et al. 2020). An adaptogenic herb, *ginseng* root is known for its ginsenosides (triterpenic saponin complexes), which can boost the body's tolerance to stress, trauma, anxiety, and exhaustion by altering immunological function. Several pathways in *ginseng* may protect against neurodegeneration (Szczuka et al. 2019).

### 8.5.5 *Convolvulus Pluricaulis*

The herb *Convolvulus pluricaulis* (CP), commonly referred to as Shankhpushpi, has been utilized for hundreds of years in India to treat nerve problems like stress, anxiety, and insomnia. It fosters a sense of serenity and harmony and lessens tension, worries, and mental exhaustion (Firdaus and Singh 2021). In Ayurvedic medicine, it is frequently used to treat neurological diseases. According to herbalists, Shankhpushpi soothes nerves by controlling the body's synthesis of the stress chemicals cortisol and adrenaline. Also, it is thought that Shankhpushpi is an anti-aging treatment known as Rasayana in Ayurvedic medicine (Sonkar et al. 2022).

### 8.5.6 *Emblica officinalis*

The herb *Emblica officinalis*, also known as Amla, is widely distributed in India, Sri Lanka, Pakistan, Malaysia, China, and Bangladesh. The plant's fruits are utilized in Ayurveda as a powerful rasayana (Shrivastava et al. 2022). The rasayana are used to extend life and improve health by boosting the immune system, slowing down aging, and reviving the body in weakened states. In Ayurveda, the fruits of *E. officinalis* are highly regarded for their therapeutic usefulness, and Amla is known as a maharasayana (Zende and Amarprakash 2020). Chyavanprash has demonstrated strong adaptogenic, anti-stress immunopotentiating, and memory-enhancing properties, according to experimental research (Baliga et al. 2015). An herbal supplement called Immuplus, which contains *Emblica officinalis* as one of its constituents, has been demonstrated to exhibit immunomodulatory effects through humoral and cell-mediated immunity (Joshi et al. 2023).

### 8.5.7 *Withania somnifera*

The Solanaceae family member *Withania somnifera*, frequently known as Ashwagandha in Sanskrit, is grown in soils unsuited for other crops and require less upkeep. In Ayurveda and Unani practices in India, the plant is well known for its variety of therapeutic purposes (Alam et al. 2016). In Ayurveda, *Withania somnifera* is highly regarded for its rejuvenative and tonic characteristics, which are identical to those of the Asian herb ginseng. It has traditionally been used by Ayurveda to treat libido issues, exhaustion, getting better from long sickness, mental health issues, and as a rasayana (rejuvenator) (Mahomoodally et al. 2013). In healthy individuals of normal age, *Withania somnifera* root extracts have been shown in clinical investigations to considerably reduce the aging process (Kuppurajan et al. 1989). It has been shown that the cholinergic signal transduction cascade activities in the cortex and basal forebrain are greatly influenced by WS extract. The increase in cortical muscarinic acetylcholine receptor capacity generated by *Withania somnifera* may help explain how *Withania somnifera* extracts affect cognition and memory in humans and animals (Mandal and Reddy 2017). In an animal model of AD, the *Withania somnifera* glycowithanolides were investigated, and it was discovered that they could reverse the cognitive deficits and the abnormal central cholinergic markers brought on by the neurodegeneration carried out by the neurotoxins (Pandey et al. 2018). Many studies supported *Withania somnifera* role in neuroprotection and tardive dyskinesia (Subasini et al. 2022).

### 8.5.8 *Ocimum sanctum*

The majority of Hindus worship holi basil, or *Occimum sanctum,* as it is known in India as "Tulsi," every morning. It is considered one of India's holiest plants and is used in Ayurveda for its curative and life-giving qualities. There are other Tulsi cultivars that are sometimes referred to as "Holy," but *Ocimum sanctum* is the genuine kind (Priyadarshini 2013). A range of disorders can be treated using Tulsi. For its impact on the central nervous system, an ethanol extract of *Ocimum sanctum* leaves was tested. In "open-field" investigations, it decreased fighting time and ambulation caused by apomorphine, shortened the period of lost reflex in mice caused by pentobarbital, and lengthened the timeframe of electroshock and pentylenetetrazole-induced convulsions in mice (Mandal and Reddy 2015). *Ocimum sanctum* extract also has the potential to be a neuroprotective drug (preserving autobiographical memory), and in the rat model of neurodegeneration, In the CA1, CA3, and DG hippocampal areas, neuropeptide Y expression was stimulated, which led to neuronal growth (Mataram et al. 2021).

## 8.6 Conclusion

The ethnopharmacological applications, phytochemical components, and neuroprotective properties of various medicinal plants are discussed in this chapter, which highlights the significance of these substances in providing for the dietary and medical demands of elderly people. The main phytochemicals found are phenolic compounds, which have been shown to be potentially active substances and provide the foundation for further studies into the creation of potential novel medicines. Recent reports suggest that several phytochemicals have neuroprotective effects in a number of animal models of neurological illness. A number of these herbal medicines' benefits on brain function and neuroprotection are supported by these researchers' phytochemical findings. The current investigation has come to the conclusion that more clinical and phytochemical research is required. The identification of these promising phytochemicals lays the groundwork for translational research aimed at developing accessible and affordable neuroprotective interventions for the aging population. Furthermore, this work underscores the value of traditional medicine as a rich source of potential therapeutic compounds. Collaborative efforts between ethnopharmacologists, phytochemists, and clinicians are essential for harnessing the full potential of these natural compounds to combat age-related brain health challenges.

## References


Abdolmaleki A, Akram M, Saeed MM et al (2020) Herbal medicine as neuroprotective potential agent in human and animal models: a historical overview. J Pharm Care 8:75–82

Afzal O, Dalhat MH, Altamimi ASA et al (2022) Green tea catechins attenuate neurodegenerative diseases and cognitive deficits. Molecules 27:7604

Ahmed T, Javed S, Javed S et al (2017) Resveratrol and Alzheimer's disease: mechanistic insights. Mol Neurobiol 54:2622–2635

Akyuz E, Polat AK, Eroglu E et al (2021) Revisiting the role of neurotransmitters in epilepsy: an updated review. Life Sci 265:118826

Alam MK, Hoq MO, Uddin MS (2016) Therapeutic use of Withania somnifera. Asian J Med Biol Res 2:148–155

Ameer K (2016) Avocado as a major dietary source of antioxidants and its preventive role in neurodegenerative diseases. Adv Neurobiol 12:337–354

Arif MU, Khan MKI, Riaz S et al (2022) Role of fruits in aging and age-related disorders. Exp Gerontol 162:111763

Baliga MS, Meera S, Shivashankara AR et al (2015) The health benefits of Indian traditional ayurvedic Rasayana (anti-aging) drugs: a review. In: Watson RR (ed) Foods and dietary supplements in the prevention and treatment of disease in older adults. Academic Press, London, pp 151–161

Britten-Jones AC, Craig JP, Downie LE (2023) Omega-3 polyunsaturated fatty acids and corneal nerve health: current evidence and future directions. Ocul Surf 27:1–12

Brouillot P, Broussolle P, Greffe J et al (2020) The birth of psychopharmacotherapy: explorations in a new world—1952–1968. In: Healy D (ed) The Psychopharmacologists 3. CRC Press, Boca Raton, pp 1–53. https://doi.org/10.4324/9781003058892-1

Cao D, Sun P, Bhowmick S et al (2021) Secondary metabolites of endophytic fungi isolated from Huperzia serrata. Fitoterapia 155:104970

Chan EYY (2019) Disaster public health and older people. Routledge, London

Chandrasekaran K, Mehrabian Z, Spinnewyn B et al (2003) Neuroprotective effects of bilobalide, a component of Ginkgo biloba extract (EGb 761) in global brain ischemia and in excitotoxicity-induced neuronal death. Pharmacopsychiatry 36(Suppl 1):S89–S94. https://doi.org/10.1055/s-2003-40447

Chen R, Xu P, Song P et al (2019) China has faster pace than Japan in population aging in next 25 years. Biosci Trends 13:287–291

Curtis DR, Duggan AW, Felix D, Johnston GAR (1970) GABA, bicuculline and central inhibition. Nature 226:1222–1224

Das PK, Badore NS, Patel P, Deshmukh N (2016) Ethnomedicinal wound healing plant in Khargone District of Madhya Pradesh: a survey over Nimari communities. Pharm Biol Eval 3:388–399

Dey P, Kundu A, Kumar A et al (2020) Analysis of alkaloids (indole alkaloids, isoquinoline alkaloids, tropane alkaloids). Recent Adv Nat Prod Anal 2020:505–567. https://doi.org/10.1016/B978-0-12-816455-6.00015-9

Farooqui AA (2019) Molecular mechanisms of dementia: biomarkers, neurochemistry, and therapy. Academic Press, London

Firdaus Z, Singh TD (2021) An insight in pathophysiological mechanism of Alzheimer's disease and its management using plant natural products. Mini Rev Med Chem 21:35–57

Garg P, Alambayan J, Garg V (2023) Herbal approaches in the management of mental depression. CNS Neurol Disord Targets 22:98–124

Gouvinhas I, Garcia J, Granato D, Barros A (2022) Seed phytochemical profiling of three olive cultivars, antioxidant capacity, enzymatic inhibition, and effects on human neuroblastoma cells (SH-SY5Y). Molecules 27:5057. https://doi.org/10.3390/molecules27165057

Huber N, Korhonen S, Hoffmann D et al (2022) Deficient neurotransmitter systems and synaptic function in frontotemporal lobar degeneration—insights into disease mechanisms and current therapeutic approaches. Mol Psychiatry 27:1300–1309

Husain GM, Mishra D, Singh PN et al (2007) Ethnopharmacological review of native traditional medicinal plants for brain disorders. Pharmacogn Rev 1:19–29

Jahangir MA, Jain P, Verma R, Taleuzzaman M, Ahsan MJ, Chettupalli AK, Muheem A, Mirza MA (2022) Transdermal nutraceuticals delivery system for CNS disease. CNS Neurol Disord Drug Targets 21(10):977–993

Jarboe CH, Porter LA, Buckler RT (1968) Structural aspects of picrotoxinin action. J Med Chem 11:729–731

Johnston GAR (2005) GABAA receptor channel pharmacology. Curr Pharm Des 11:1867–1885

Joshi Y, Rani H, Kaur G et al (2023) Herbal and Ayurvedic plants as remedial approach for viral diseases with focus on COVID-19: a narrative review. Curr Tradit Med 9:66–86

Khan MSA, Ahmad I (2019) Herbal medicine: current trends and future prospects. In: Khan MSA, Ahmad I, Chattopadhyay D (eds) New look to phytomedicine. Elsevier, London, pp 3–13

Kim J (2021) Pre-clinical neuroprotective evidences and plausible mechanisms of Sulforaphane in Alzheimer's disease. Int J Mol Sci 22:2929

Klomparens EA, Ding Y (2019) The neuroprotective mechanisms and effects of sulforaphane. Brain Circ 5:74

Kumar GP, Khanum F (2012) Neuroprotective potential of phytochemicals. Pharmacogn Rev 6:81–90. https://doi.org/10.4103/0973-7847.99898

Kuppurajan K, Rajgopalan SS, Sitaram R et al (1989) Effects of Ashwagandha on the process of ageing on human volunteers. J Res Ayur Sid 1:247–258

Li X, Zhu Y, Wang Y et al (2023) Neural stem/progenitor cell-derived extracellular vesicles: a novel therapy for neurological diseases and beyond. Med Comm 4:e214

Mahomoodally MF, Bhugun V, Chutterdharry G (2013) Complementary and alternative medicines use against neurodegenerative diseases. Adv Pharmacol Pharm 1:103–123

Malar DS, Prasanth MI, Brimson JM et al (2020) Neuroprotective properties of green tea (Camellia sinensis) in Parkinson's disease: a review. Molecules 25:3926

Mandal A, Reddy PJM (2015) A review on phytochemical and pharmacological uses of Ocimum sanctum Linn. World J Pharm Res 4:1535–1549

Mandal A, Reddy JM (2017) A review on potential therapeutic uses of Withania somnifera. World J Pharm Res 6:846–860

Marambaud P, Zhao H, Davies P (2005) Resveratrol promotes clearance of Alzheimer's disease amyloid-β peptides. J Biol Chem 280:37377–37382

Mariamenatu AH, Abdu EM (2021) Overconsumption of omega-6 polyunsaturated fatty acids (PUFAs) versus deficiency of omega-3 PUFAs in modern-day diets: the disturbing factor for their "balanced antagonistic metabolic functions" in the human body. J Lipids 2021:1–15

Martínez-Hernández MI, Acosta-Saavedra LC, Hernández-Kelly LC et al (2023) Microglial activation in metal neurotoxicity: impact in neurodegenerative diseases. Biomed Res Int 2023: 7389508

Mataram MBA, Hening P, Harjanti FN et al (2021) The neuroprotective effect of ethanolic extract Ocimum sanctum Linn. in the regulation of neuronal density in hippocampus areas as a central autobiography memory on the rat model of Alzheimer's disease. J Chem Neuroanat 111:101885

Mitra R, Mitchell AB, Agricola S (2007) Medicinal plants of India. Asia Pacific Biotech News 11: 707–725

Modrzyński RJ (2021) Nootropic drugs method supporting memory and concentration at work and studies. Rozpr Społeczne 15:117–125

Niaz K, Khan F (2020) Analysis of polyphenolics. In: Silva AS, Nabavi SF, Saeedi M, Nabavi SM (eds) Recent advances in natural products analysis. Elsevier, London, pp 39–197. https://doi.org/10.1016/B978-0-12-816455-6.00003-2

Nutrition P, Republic C (2022) Nootropics as cognitive enhancers: types, dosage and side effects of smart drugs. Nutrients 14(16):3367

Orhan IE (2021) A review focused on molecular mechanisms of anxiolytic effect of Valerina officinalis L. in connection with its phytochemistry through in vitro/in vivo studies. Curr Pharm Des 27:3084–3090

Orlova A, Kysil E, Tsvetkova E et al (2022) Phytochemical characterization of water avens (geum rivale L) extracts: structure assignment and biological activity of the major phenolic constituents. Plants 11:2859. https://doi.org/10.3390/plants11212859

Otimenyin S (2022) Herbal biomolecules acting on central nervous system. In: Mandal SC, Nayak AK, Dhara AK (eds) Herbal biomolecules in healthcare applications. Academic Press, London, pp 475–523. https://doi.org/10.1016/B978-0-323-85852-6.00030-5

Palimariciuc M, Balmus I-M, Gireadă B et al (2023) The quest for neurodegenerative disease treatment—focusing on Alzheimer's disease personalised diets. Curr Issues Mol Biol 45:1519–1535

Pandey A, Bani S, Dutt P et al (2018) Multifunctional neuroprotective effect of Withanone, a compound from Withania somnifera roots in alleviating cognitive dysfunction. Cytokine 102: 211–221

Pluta R, Furmaga-Jabłońska W, Januszewski S, Czuczwar SJ (2022) Post-ischemic brain neurodegeneration in the form of Alzheimer's disease proteinopathy: possible therapeutic role of curcumin. Nutrients 14:248

Priyadarshini R (2013) Effectiveness of Tulsi leaves powder upon blood glucose level in diabetic patients

Reddy BN, Naganathan L, Sunitha ANV et al (2019) A review on nutraceuticals. J Pharmacogn Phytochem 8:735–744

Sahebnasagh A, Eghbali S, Saghafi F et al (2022) Neurohormetic phytochemicals in the pathogenesis of neurodegenerative diseases. Immun Ageing 19:1–16

Sarma K, Borkakoty B, Parida P, Bora SS (2016) Identification of natural lead molecules of Centella asiatica and Azadirachta indica targeting cholera toxin through structure based drug design. Indo Am J Pharm Res 6(08):6418–6429

Sendri N, Bhandari P (2023) Bacopa monnieri. In: Amalraj A, Matharu A, Karthik Varma AC, Kuttappan S (eds) Herbs, spices and their roles in nutraceuticals and functional foods. Elsevier, London, pp 111–131

Shrivastava S, Kaur J, Mehraj M et al (2022) *Emblica officinalis* (Amla): a comprehensive review of the miracle berry. Pharma Innov J 11(6):6–16

Singh VP (2019) A review on pharmacodynamics of Ashtamangal ghrita and its uses in mental and physical growth in children. J Pharmacogn Phytochem 8:3809–3812

Singh AK, Kumar P, Rajput VD et al (2023) Phytochemicals, antioxidant, anti-inflammatory studies, and identification of bioactive compounds using GC–MS of ethanolic novel polyherbal extract. Appl Biochem Biotechnol 195:4447–4468

Sonkar S, Singh AK, Husen A (2022) Convolvulus pluricaulis (Shankhpushpi) and Erythroxylum coca (coca plant). In: Husen A (ed) Herbs, shrubs, and trees of potential medicinal benefits. CRC Press, Boca Raton, pp 83–94

Subasini S, Ananthi P, Hemkumar K, Pius A (2022) Withania somnifera in Ayurveda. Chem Biol Act Ther Appl Med Plants Ayurveda, pp 89–107

Subhan S, Bagchi M, Ilah A (2017) Current perspectives on the beneficial role of curcuma longa and Centella asiatica Linn. in neurological disorders. In: Subhan S, Bagchi M (eds) Phytopharmaceuticals for brain health. CRC Press, Boca Raton, pp 141–152

Sukumaran NP, Amalraj A, Gopi S (2019) Neuropharmacological and cognitive effects of Bacopa monnieri (L.) Wettst–a review on its mechanistic aspects. Complement Ther Med 44:68–82

Sunkaria A, Yadav A, Sharma SK, Sandhir R (2017) Molecular mechanisms involved in the neuroprotective action of phytochemicals. In: Farooqui T, Farooqui AA (eds) Neuroprotective effects of phytochemicals in neurological disorders. Wiley, Hoboken, NJ, pp 515–559

Szczuka D, Nowak A, Zakłos-Szyda M et al (2019) American ginseng (Panax quinquefolium L.) as a source of bioactive phytochemicals with pro-health properties. Nutrients 11:1041

Tewari D, Sah AN, Patni P, Bawari S (2018) Ageing and ethnopharmacology. In: Ageing and healthcare in India issues & challenges. Indu Book Services, New Delhi, pp 99–108

Ugoeze KC (2022) Phytopharmaceuticals for treating sexually transmitted diseases. In: Singh I, Sindhu RK, Shirkhedkar A, Panichayupakaranant P (eds) Herbal drugs for the management of infectious diseases. Wiley, Hoboken, NJ, pp 179–261. https://doi.org/10.1002/9781119818779.CH8

Wu C, Lee S-L, Taylor C et al (2020) Scientific and regulatory approach to botanical drug development: a US FDA perspective. J Nat Prod 83:552–562

Yildirim A (2021) Medicinal and aromatic plants: economics production agricultural ultilization and other aspects. Iksad Publishing House, Ankara

Zende AK, Amarprakash D (2020) Role of Ayurveda in prevention of Covid-19 pandemic-an integrative overview. J Ayurveda Integr Med Sci 5:350–364

# Potential Effects of Antiaging Compounds: Implications on Brain Aging

9

G. P. Abinaya, B. K. Iyshwarya, and Ramakrishnan Veerabathiran


**Abstract**

A phytochemical is a natural substance extracted from plants that aids in the defense against multiple infections and prevents many life-threatening diseases. It can reduce inflammation and oxidative stress and is linked to pharmacological inhibition. The phytochemicals in herbal medicines play a crucial role in regulating the brain's chemical balance by influencing the activity of sensors for major neurotransmissions. In traditional medicine, various plants have been recognized to cure psychological impairments. Research has examined these drugs' potential benefits in treating neurodegenerative diseases in recent years. Alzheimer's disease and Parkinson's disease are diseases characterized by inflammatory responses, cognitive decline, and dysfunctional mitochondria. Brain aging is considered one of the major causes of these diseases due to neuronal loss, a multifactorial phenomenon. Amyloid beta peptide levels increase in individuals who have Alzheimer's disease as a result of the phosphorylation of tau protein. Parkinson's disease, a neurodegenerative disorder characterized by chronic inflammation, loss of dopaminergic neurons, is a severe form of neurodegeneration. Using phytochemicals can also alleviate the effects of mitochondrial dysfunction and oxidative stress resulting from these neurological disorders. These afflictions can be diagnosed using the following diagnostic tools: DNA methylation, delta age, and magnetic resonance imaging. In this chapter, we explore several phytochemicals that have been proven to have protective effects in various model systems of brain disorders. The identification of phytochemicals with neuroprotective properties opens up exciting avenues for



G. P. Abinaya · B. K. Iyshwarya · R. Veerabathiran (✉)
Human Cytogenetics and Genomics Laboratory, Faculty of Allied Health Sciences, Chettinad Hospital and Research Institute, Chettinad Academy of Research and Education, Kelambakkam, Tamilnadu, India







S. Pathak, A. Banerjee (eds.), *Neuroprotective Effects of Phytochemicals in Brain Ageing*, https://doi.org/10.1007/978-981-99-7269-2_9

developing novel therapeutic strategies for brain disorders. Further research into the mechanisms of action of these phytochemicals, their optimal delivery methods, and their potential synergistic effects when used in combination could lead to the development of more targeted and effective treatments for devastating neurological conditions.

**Keywords**

Neurodegenerative diseases · Phytochemicals · Alzheimer's disease · Parkinson's disease · DNA methylation

## Abbreviation

| | |
|---|---|
| AD | Alzheimer's disease |
| APP | Amyloid precursor protein |
| ATP | Adenosine triphosphate |
| BBB | Blood-brain barrier |
| CNS | Central nervous system |
| CVD | Cardiovascular diseases |
| DBS | Deep brain stimulation |
| DNA | Deoxyribonucleic acid |
| EEG | Electroencephalogram |
| EGCG | Epigallocatechin |
| MCI | Mild cognitive impairment |
| MRI | Magnetic resonance imaging |
| MYR | Myricetin |
| NLRP3 | Nucleotide-binding oligomerization domain-like receptor protein 3 |
| OSA | Obstructive sleep apnea |
| PD | Parkinson's disease |
| PET | Positron emission tomography |
| RONS | Reactive oxygen and nitrogen species |
| rRNA | Ribosomal ribonucleic acid |
| SNP | Single-nucleotide polymorphism |
| SNpc | Substantia nigra pars compacta |
| TNF-α | Tumor necrosis factor-α |

## 9.1 Introduction

Phytochemicals are chemical compounds extracted from plants that resist bacterial, viral, and fungal infections. They are used mainly for the prevention of cancer. Capsaicin, catechins, lycopene, and cucurbitacin B are some dietary phytochemicals used for chemoprevention (Ranjan et al. 2019). Phytochemicals such as taxol analogs and vinca alkaloids are novel drugs used for cancer therapy. They often modulate the pathways involved in the growth and advancement of cancer by inactivating carcinogens, inhibiting progression, and promoting the arrest of the cell cycle and apoptosis (Choudhari et al. 2020). Recently, many types of research mainly focus on age-related diseases, which has paved the way for the pharmacological inhibition of aging pathways to increase lifespan and antiaging. The phytochemicals, including resveratrol, curcumin, and quercetin, have proved to be one of the most effective therapies for antiaging when studied using animal models (Corrêa et al. 2018). The following phytochemicals are reviewed in this chapter and include flavonoids, curcumin, resveratrol, quercetin, epigallocatechin gallate, and myricetin.

Cytokines' inflammatory processes characterize neurodegenerative diseases such as tumor necrosis factor-alpha (TNF-α). TNF-α increases the production of Aβ peptides, which activates the macrophages and makes them lose their ability to phagocytosis, resulting in chronic inflammation. The elevated TNF-α is accompanied by neuronal loss and damage, higher Aβ production, and loss of synaptic function. Since plant derivatives are anti-inflammatory, they act as potential therapeutics for diseases, namely AD and Parkinson's disease. The major difficulty faced by the researchers is when the drugs have to cross the blood-brain barrier, and phytochemicals are proven advantageous with good bioavailability and affinity with the target receptors in the brain (Zahedipour et al. 2022).

Brain aging is a multifactorial phenomenon that occurs due to changes in the structure and functions of the neurons in the circuit as well as loss of dendritic arbors, and synaptic contact causes high neuronal loss. Aging aids in the advancement of cognitive decline. It is a complex process that gradually decreases physiological and behavioral functions. The characteristics assigned in aging are synaptic dystrophy, abnormalities in the cytoskeleton, and regression in pyramidal neurons. Though the characteristics have been identified, the causative molecular mechanism is still elusive (Zia et al. 2021). Reactive oxygen species or free radicals are by-products of mitochondrial respiration that are required for basic physiological activities but turn toxic if high levels gradually impair all cognitive and mitochondrial functions. It induces mitophagy, for instance, parkin in the case of Parkinson's disease.

Aging is marked by the reduction in the α mitochondrial F1 ATP synthase subunit, which undergoes oxidative phosphorylation to synthesize ATP. The marked deficits in the α subunit led to lesser production of ATP, synthesis of free radicals, and oxidation of DNA and lipids. Mitochondrial dysfunction is highly involved in the pathogenesis of AD. Age-related impairment of mitochondrial function is probably one of the first events causing the late onset of AD. It affects the expression of amyloid precursor protein (APP) and produces beta-amyloid oligomers that cumulate into plaque. The mitochondrial dysfunction is further worsened by the impaired glucose metabolism since the beta-amyloid proteins inhibit the glucose and glutamate transporters by decreasing the transmembrane potential, thus reducing the functional capacity of the mitochondria (Ionescu-Tucker and Cotman 2021).

Astrocytes are highly heterogeneous glial cells in the central nervous system (CNS) that are morphologically different from one another, even if present in the same region. When the protein expression profile is analyzed, there are higher chances of age-related diseases and cognitive decline due to heterogeneity. However, the underlying molecular mechanisms are still elusive. One of the dominant risk factors of neurological disorders is neuroinflammation, caused by the activation of glial cells, as these cells release cytokines, free radicals, and antioxidants (Matias et al. 2019).

## 9.2 Neurodegenerative Diseases

The two neurodegenerative diseases discussed in this chapter are Alzheimer's disease and Parkinson's disease. They are the most common and highly prevalent neurological diseases when compared with other ailments such as epilepsy and dementia. The preponderance of Alzheimer's disease is about 60–80% with more than 40 genes, and by 2050, it may rise 3 times globally (Scheltens et al. 2021). Parkinson's disease is considered a fast-growing condition since its characteristics resemble a pandemic with 3–5% of genetic factors (Bloem et al. 2021).

### 9.2.1 Alzheimer's Disease

Alzheimer's disease (AD) is a progressive neurological disease that causes atrophy and death of brain cells, leading to dementia. About 80% of dementia cases are due to AD. It affects other cognitive domains, such as thinking, learning, and behavior. Statistics show that the level of AD is increasing by 89% in the USA (Weller and Budson 2018). The initial symptoms of AD are characterized by memory loss and confusion. People often forget the conversations or misplace things. AD is caused mainly due to the hyperphosphorylation of tau protein. Tau proteins are the proteins that stabilize the internal cells of the brain. If the tau pathology is stronger in the frontoparietal regions of the brain, it acts as a hub for all cognitive functions, leading to the manifestation of the disease (Frontzkowski et al. 2022).

Some neuropathological features visible with normal aging are also observed in Alzheimer's disease and dementia. With the normal progression of age, many cognitive functions such as memory, processing emotions, and working abilities become relatively slower, sometimes leading to mild cognitive impairment (MCI), causing functional dependence. Natural aging is a decline of these cognitive abilities, whereas, in the case of AD, the same abilities have deteriorated with additional diminishment. At present, synaptic markers are widely used to determine the brain's synaptic density, representing the functional proteins in the pre-synaptic or post-synaptic terminals. In AD and dementia, the loss of proteins was observed more than the general loss of synapses (Denver and McClean 2018).

### 9.2.2 Parkinson's Disease

Parkinson's disease (PD) is a neurodegenerative disorder with difficulty in movements. Some of its common symptoms are bradykinesia, tremor, postural imbalance, and rigidity. The pathological features of PD include the reduction of dopaminergic neurons in the substantia nigra pars compacta (SNpc) and the accumulation of misfolded alpha-synuclein in Lewy bodies. The etiologic mechanism of PD is not known, and the treatment is based on the stages of the disease. Replacement of dopamine is the current method, and deep brain stimulation (DBS) is recommended for a later stage. The prevalence is estimated to be 0.3% in the general population. The incidence rate is 1% and 3% for people over 60 and 80 years of age, respectively. Age, environmental factors such as pesticides, exposure to toxic substances like manganese, and first-degree relatives in case of family history are some of the known hazards.

The loss of dopaminergic neurons and the presence of Lewy bodies are the hallmarks of PD. The major components of Lewy bodies are alpha-synuclein and ubiquitin. Upon misfolding of alpha-synuclein, it forms amyloid aggregates rich in β-sheet and forms an inclusion. It has been found that almost 60% of the dopaminergic nerves are lost at the time of diagnosis (Balestrino and Schapira 2020). Brain aging is characterized by mitochondrial aging with dysfunctioning of mitochondria, modified autophagy, and decreased oxidative stress. The changes in the mitochondrial activities and deletion in the mitochondrial DNA are observed in the substantia nigra, and it is analyzed in both aged and affected individuals (Rango and Bresolin 2018). Therefore, the phytochemicals, which are natural extracts of plants, are used as antiaging compounds in the above-mentioned neurodegenerative disorders with higher efficiency.

## 9.3 Phytochemicals as Antiaging Compounds in Brain Aging

### 9.3.1 Flavonoids

Flavonoids are an essential part of green plants distributed all over them. It is categorized into six different groups: flavanols, flavonols, isoflavones, anthocyanidins, flavanones, and flavones. From the groups mentioned above, flavones are the most abundantly found phytochemicals. In addition, it is also divided based on the extent to which it oxidizes or reduces the aglycon ring. Since flavonoids are ubiquitous, they are also used in nutraceuticals, pharmaceuticals, cosmetics, etc., and positively affect human health. Researchers have to focus on the absorption and dietary concentrations of flavonoids in the future. Flavonoids are known to be anti-mutagenic, anti-inflammatory, anti-allergic, and vasodilators.

The antioxidant activity of flavonoids is beneficial in preventing the production of free radicals and cellular oxidative stress related to many diseases, for instance, cardiovascular diseases (CVDs), diabetes, cancer, and neurological disorders. Flavonoids can cross the blood-brain barrier, protect the CNS from inflammation

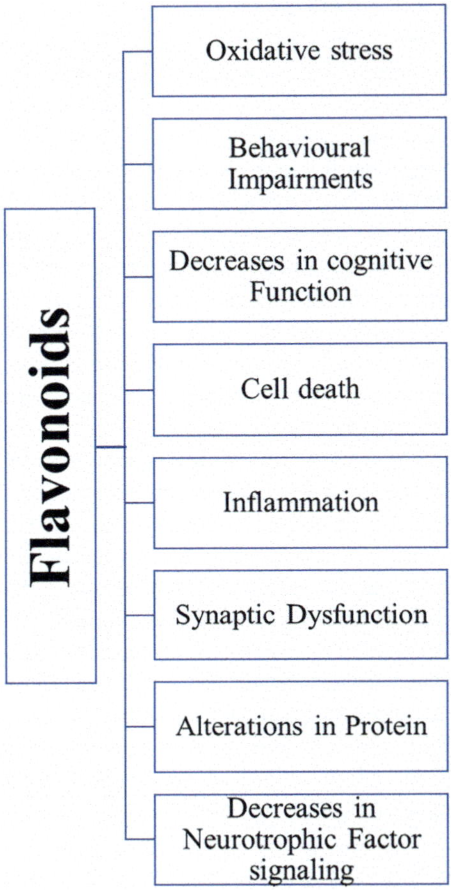


**Fig. 9.1** Neurodegenerative diseases and brain aging are linked to multiple pathways altered by flavonoids. Reduced oxidative stress, activation of neurotrophic factor signaling pathways, reduced protein processing, synaptic dysfunction, and inhibition of inflammation can all increase brain cell function and survival

and accumulation of transition metals, and decrease the preponderance of dementia and brain aging. It has also been identified as an antiaging and antidiabetic therapeutic (Minocha et al. 2022). A study has been conducted with *Alpinia galanga* (L.) to examine its antiaging properties. It was observed that the extract has higher antiaging potential because it has reduced the senescence of cells in normal fibroblasts normally induced by stressors such as doxorubicin (Tungmunnithum et al. 2020). The pathological features of AD include the amyloid and intracellular neurofibrillary tangles consisting of an abnormally phosphorylated tau protein. The amyloid peptide deposits led to synaptic and cognitive impairment and neuronal death. Experiments were conducted with animal models, providing us with various insights about inflammation, neurotoxicity, neurodegeneration, and lesions in the brain. The main target of action for the therapy to cure Alzheimer's disease is the reduction of free radicals. In the case of Parkinson's disease, flavonoids act by reducing the cellular oxidative potential and neuronal death mechanism (Fig. 9.1) (de Andrade Teles et al. 2018).

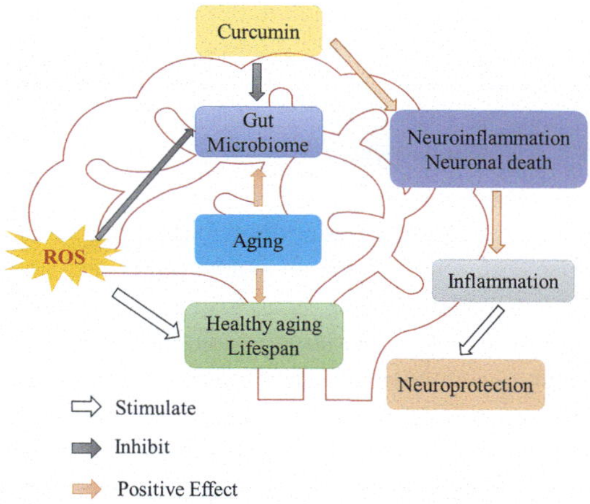


**Fig. 9.2** Curcumin is a powerful neuroprotectant and antiaging compound. The combination of curcumin supplementation along with regular physical exercise may have the potential to reduce the negative effects of oxidative stress during aging as well as age-related diseases

### 9.3.2 Curcumin

Curcumin is an Indian spice extracted from the rhizome of Curcumin longa. It is known as turmeric, a medicinal herb used to cure sprains, swelling, and abdominal pain. It contains two phenolic rings linked by a linear β-diketone linker that promotes keto-enol tautomerism. Commercially, curcumin has three forms: curcumin, desmethoxycurcumin, and bisdemethoxycurcumin. Among the three curcuminoids, curcumin is the major component that gives turmeric the yellow color and is used as an additive and food colorant. It is well known for its multitarget effects, which prevent chronic diseases such as hypertension, hyperlipidemia, and cerebrovascular diseases because a disease, for instance, diabetes, activates the receptors for glycosylation of end products on the surface of glial cells and neurons, which in turn activates the Aβ influx causing brain damage and impairs cognitive function. Consumption of solid lipid curcumin increases cognitive function, decreases stress, and minimizes stress's harmful impact on mood, enhancing the quality of living in the elderly population, thus decreasing the rate of brain aging (Fig. 9.2) (Chen et al. 2018).

Curcumin is a golden nutraceutical because it aids in the delayed onset of age-related issues as it is a good scavenger of reactive oxygen and nitrogen species (RONS), thereby preventing inflammation and oxidative stress (Benameur et al. 2021). Hence, this is used for the treatment of Alzheimer's disease. It has been demonstrated that curcumin has the potential of anti-inflammation, antioxidants, scavenging of reactive oxygen species, and iron chelation. Therefore, it is a promising therapy to cure Parkinson's disease (PD) (Nebrisi 2021). Accordingly, curcumin may be a valuable nutraceutical spice that may prevent aging and age-related diseases. Currently, there is no evidence indicating that any nutraceutical influences lifespan and aging, and a complete randomized clinical trial in humans will also be necessary to confirm whether curcumin is effective for preventing diseases involving inflammation, including cardiovascular disease, Alzheimer's disease, cardiometabolic syndrome, and aging.

### 9.3.3 Resveratrol

Resveratrol is a secondary metabolite, phytoalexin, from the polyphenol plant. It is most commonly found in grapevine, wine, and other sources, including spinach, peanuts, pomegranate, bananas, etc. The phenylpropanoid pathway synthesizes resveratrol by generating 4-coumaroyl-CoA, activating stilbene synthase and chalcone synthase, and producing a polyketide compound under biotic and abiotic conditions. The polyketide compound folds to form an aromatic ring that generates an unstable metabolite called stilbene-2-carboxylic acid. The final step involves a series of steps involving decarboxylation, dehydration, and enolization of the above metabolite, yielding a resveratrol molecule. Researchers found it difficult to make the resveratrol cross the blood-brain barrier (BBB) orally, which resulted in reduced stability, elevated photosensitivity, and increased metabolism. However, it was later detected in the plasma and the cerebrospinal fluid affecting the CNS.

Overexpression of SIRT-1 helps in neuroprotection by inhibiting NF-κB signaling; since resveratrol is an activator of SIRT-1, it ultimately cures neurodegenerative diseases like Alzheimer's disease and Parkinson's disease. SIRT-1 is a member of the sirtuin family that uses nicotinamide to catalyze the deacetylation of substrates. It has also been proven effective in regulating glial functions, uptake of glutamate, and decreased apoptosis (Gomes et al. 2018). Moreover, SIRT-1 is an antiaging factor that on activation downregulates mTOR pathways by inhibiting the production of reactive oxygen species, promoting mitochondrial biogenesis, and preventing aging (Zhou et al. 2021).

### 9.3.4 Quercetin

Quercetin, named after quercetum, is a polar inhibitor of auxin transport and a member of flavonols. The role of quercetin includes the regulation of signaling pathways, expression of genes, antioxidants, and elevation of mitochondrial function by suppressing the atrophic factors. It is majorly found in barks and rinds, and the most common glycosidic form is quercetin-3-rutinoside or rutin. The oral bioavailability (approx. 2%) is lower than the radiolabeled quercetin since it is estimated to be 44.8% and found to be effective. Subsequently, quercetin undergoes three phases of metabolism: modification, conjugation, and elimination. It can cross the blood-brain barrier and control the division of U251 cells, promoting apoptosis. However, the brain's efflux transporters like P-glycoprotein use quercetin as a substrate at lower concentrations. Despite that, quercetin, when conjugated with α-tocopherol, protects microvascular endothelial cells from β-amyloid-induced toxicity, eventually suppressing brain aging (Zaplatic et al. 2019).

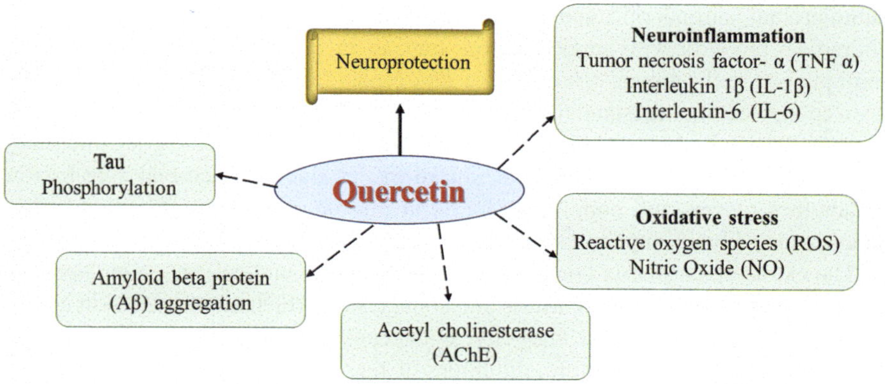


**Fig. 9.3** Neuroprotective effects of quercetin for Alzheimers's disease. It is possible that quercetin protects neurons from Alzheimer's disease by inhibiting oxidative stress and neuroinflammation downstream, which directly protects neurons, inhibits AChE enzymes, and results in an increase in acetylcholine levels while reducing Tau phosphorylation and Aβ aggregation

An experiment has been performed to study the antiaging characteristics of quercetin using human erythrocytes. The aging of erythrocytes decreases the cellular ability to decrease oxidative stress. The effects of oxidative stress include distorted calcium homeostasis and caspase-3 activation, thus worsening the aging process. However, natural aging initiates a non-enzymatic glycation reaction whose by-products induce modifications in the structure of proteins and change the biochemical functions. Quercetin is an anti-glycation compound that inhibits the formation of end-product glycation in erythrocytes (Remigante et al. 2022).

Quercetin is found to protect neuronal cells by reducing oxidative stress and neuroinflammation. The properties against Alzheimer's disease include inhibiting the accumulation of Aβ and phosphorylation of tau proteins (Fig. 9.3) (Khan et al. 2019). It was studied that the combination of quercetin and piperine raised the neuroprotective effect and contributed to higher impacts on anti-inflammation and neuroprotective effects against rats with iron and rotenone supplement-induced deficits (Sharma et al. 2020).

### 9.3.5 Epigallocatechin Gallate

Epigallocatechin gallate (EGCG) is an essential component of the polyphenolic part of green tea with numerous therapeutic uses. Approximately 30% of green tea consists of polyphenols, of which 60–80% are catechins, particularly epigallocatechin, with highly promoting health benefits. It regulates blood pressure and decreases cholesterol and body fat. EGCG is an odorless, white, or cream-colored powder soluble in water, ethanol, acetone, and tetrahydrofuran. It has three aromatic rings linked together by a pyran ring. Many experiments have been performed that study the pharmacological properties such as anti-oxidative, antimicrobial, anti-allergic, and antidiabetic. Oxidizing phenolic groups achieve the antiradical property with single electron transfer in B- and D-rings by producing semiquinone and quinone (Bartosikova and Necas 2018). Recently, studies have been conducted to understand the impact of green tea on neurological disorders. The level of EGCG in the systemic circulation is very low, and it undergoes microbial degeneration in the small and large intestines, producing microbial ring-fission metabolites as free and conjugated forms in plasma and urine. This aids in the suppression of brain impairment (Pervin et al. 2019).

The extracellular matrix constituents, such as elastin and collagen, are degraded by their respective enzymes, elastase and collagenase, which accelerates skin aging. Since elastin maintains skin elasticity, collagen keeps up both the strength and elasticity of the skin. Therefore, inhibitors of these two enzymes can be evolved into antiaging compounds. Experiments on the Marula stem ethanol extract found six compounds, including quinic acid, catechin, epigallocatechin gallate, and epicatechin gallate, potential antiaging compounds (Shoko et al. 2018). It has been studied that natural polyphenols such as flavonoids have the potential to regulate the neuropathological changes and cognitive decline linked to AD via several mechanisms, for instance, regulation of oxidation and inflammation as well as catabolism and oligomerization. The evidence suggests that the consumption of tea decreases the risk of AD. The administration of EGCG significantly decreases the deposition of Aβ and oxidative stress due to the property of free radical scavenging, anti-inflammation, and iron chelation activity (Lange et al. 2022).

Neuroinflammation is one of the essential mechanisms in advancing neurodegenerative diseases such as Parkinson's disease. It is caused by the selective damage of neurons due to the activation of microglial cells. EGCG downregulates the expression of nitric oxide synthase and TNF- α, which protects against inflammation-mediated dopaminergic neuron injury. To reduce neurotoxicity and microglial inflammation, EGCG weakens the expression of inflammatory cytokines. The rate-limiting enzyme for dopamine production is tyrosine hydroxylase, which can be regulated, and the catechins inhibit the proinflammatory enzymes in green tea. Furthermore, the upregulation of microRNA-190 reduces the level of tyrosine hydroxylase in substantial pars nigra compacta and inhibits the activation of microbial cells (Luo et al. 2021).

### 9.3.6 Myricetin

Myricetin (MYR) belongs to the polyphenolic group of flavonoid compounds found in various plants. It is isolated from Myrica rubra consisting of yellow crystals. MYR is soluble in solvents like methanol, acetonitrile, and ethanol. It is known to have a major role biologically. Moreover, it has many therapeutic effects on diseases such as thrombosis, diabetes, Alzheimer's disease, and atherosclerosis. It also regulates the expression of MAPK and AchE and suppression of cytokines to enhance any cardiac abnormality and aids as an adjuvant in case of neural injury and cancer, as well as a potential drug for viral infections such as COVID-19. It is a highly recommended chemo-preventive agent because it inhibits proteins and activates signaling pathways that promote cell division and apoptosis. MYR has numerous pharmacological effects that prevent inflammations, for instance, pneumonia, neuritis, and myocarditis. It is an activator of immune receptors (immunomodulator), but its impact on the nervous system is still not profoundly examined. It is a fruitful treatment for respiratory diseases, particularly COVID-19, because it ameliorates cardiac abnormalities and arrhythmia, reduces oxidative stress, and is antiviral (Song et al. 2021).

Aβ is the product of hydrolysis of the APP, which has two pathways: continuous hydrolysis of β-secretase (BACE-1) and gamma-secretase to synthesize Aβ. BACE-1 is the rate-limiting enzyme and a potential target for curing Alzheimer's disease. The formation and hydrolysis of Aβ are in dynamic equilibrium; if this is broken, the deposition of Aβ takes place in the brain, leading to AD. It is observed that MYR is an inhibitor of BACE-1, which breaks down the APP and disrupts the synthesis of Aβ. In addition, MYR elevates the level of alpha-secretase that breaks down APP into fragments, thus reducing the making of Aβ. MYR reduces the levels of inflammatory factors and inhibits the activation of microglia and nucleotide-binding oligomerization domain-like receptor protein 3 (NLRP3). The negative effect on NLRP3 reduces the brain's neuroinflammatory damage, thereby decreasing the level of Aβ.

Since MYR is a flavonoid, the pyrogallol in the B ring favors the role of an antioxidant. It obstructs any damage in the DNA and lipids by regulating the signaling pathways such as MAPK and P13K caused due to H2O2- induced cell injury. There is a rise in antiapoptotic factors such as Bcl-2 by inhibiting apoptosis caused by oxidative stress. Oxidative stress promotes the phosphorylation of tau proteins by inhibiting PP2A, which destroys mitochondria. Thus, MYR protects neuronal cells through its antioxidant nature (Liu et al. 2020).

## 9.4 Biomarkers for Brain Aging

Many biomarkers are used to diagnose brain aging and neurological disorders, such as epigenetic clocks, proteomic clocks, and neuroimaging. Another biomarker is telomere length, which was once a popular biomarker but has not been used recently because of its unpredictability of mortality or other impacts. For people with mental illness, the transcriptomic brain age is increased, which is detected by transcriptomics. Metabolism of serum and urine detects cardiovascular diseases, metabolic syndromes, and cognitive decline. The proportion of cell type changes with age that can be predicted using CNS and immune cell phenotypes. The sleep electroencephalogram (EEG) detects mortality. The clinical assessments of cognition include frailty indices and the ApoE4 allele (Higgins-Chen et al. 2021). Table 9.1 shows the possible biomarkers and their effects on phytochemicals in brain aging.

**Table 9.1** Functions of biomarkers and its effects

| S. no | Biomarkers | Source | Functions | Effect |
|---|---|---|---|---|
| 1. | DNA methylation | DNA molecule | Degrades proteins and inhibits protein synthesis associated with long-term memory | Parkinson's disease<br>Huntington's disease |
| 2. | Magnetic Resonance Imaging-based biomarkers | | Indicator of normal biological processes, pathogenic processes, or a response to a therapeutic intervention. | Dementia<br>Alzheimer's disease |
| 3. | Obstructive sleep apnea | Change in sleeping pattern | Helps in the diagnosis of Alzheimer's disease by disruption of sleep and hypocapnia | Alzheimer's disease<br>Epilepsy |
| 4. | Delta age | Bone mineral density | Captures both biological and chronological age. It is mostly associated with bone mineral density and cognitive function | Cognitive decline<br>Diabetes |
| 5. | Diet | Glucose and ketones | Plays a major role in influencing brain age. Ketones reduce inflammation and produce free radicals | Alzheimer's disease<br>Vascular dementia |

### 9.4.1 DNA Methylation

The chronological age is the accurate measure to estimate the age of an organ or a person where some regions of the genome undergo modifications in the age-related DNA methylation, indicating the existence of other methylating regions associated with aging. The studies have proved that DNA methylation is one of the most promising biomarkers correlated with chronological age. The method of DNA methylation uses mathematical algorithms and sets of CpG sites to calculate the biological age of the cells. The level of methylation consists of locus-specific hypermethylation and global hypomethylation. Since the CpG sites are different in all DNA sources, it is used to estimate both chronological and biological ages. The genes constantly undergo epigenetic changes in response to any environmental stress. DNA methylation occurs throughout the lifespan, sometimes leading to pathological changes with some negative impact. Therefore, novel targets are being identified to reverse the age-related phenotypes and increase lifespan.

Some factors are used to assess the age-related consequences in our bodies. DNA methylation is involved in the degradation of proteins, and its link with the senescent cells is beginning to unfold. The ribosomal RNAs (rRNAs) are used in translation, which is decreased due to physiological age. Studies have proved that methylation in rDNA inhibits transcription and stops the expression of 18S, 5.8S, and 28S rRNA. It has also been discovered that the methylation of DNA impairs the synthesis of proteins associated with long-term memory (Jiang and Guo 2020).

### 9.4.2 Magnetic Resonance Imaging-based biomarkers

The most widely used neuroimaging technique is magnetic resonance imaging (MRI), which broadly has two diagnostic criteria, namely, Alzheimer's disease and dementia with Lewy bodies. As per the guidelines of the National Institute on Aging-Alzheimer's Association (NIA-AA) for AD, disproportionate atrophy on MRI in the medial and basal lobes is used as a biomarker of neuronal degeneration. MRI can be a supportive biomarker in the case of dementia, where the guidelines state that the absence of minimal atrophy assists in diagnosis but lacks diagnostic specificity. If the patients have atrophy in the medial temporal lobe, it indicates dementia with Lewy bodies (Shimizu et al. 2018).

### 9.4.3 Obstructive Sleep Apnea

Obstructive sleep apnea (OSA) is a sleep disorder linked to changes in the pattern of breathing characterized by partial closure of the upper respiratory pathway by desaturation of blood oxygen. It can lead to disrupted sleep, re-oxygenation, and hypocapnia. Studies have shown that any interruption in sleep can be related to Alzheimer's disease (AD) and the development of dementia. The Analysis of the Alzheimer Disease Neuroimaging Initiative (ADNI) cohort found that the early onset of cognitive impairment is linked to OSA. They have suggested continuous pathway airway pressure (CPAP). It has been discovered that the changes in the level of beta-amyloid and tau protein present in the cerebrospinal fluid and blood serum are associated with OSA by positron emission tomography (PET). When groups were arranged according to the ApoE allele, the severity of OSA correlated positively with beta-amyloid phosphorylated tau, and total tau in ApoE3 carriers and ApoE2 carriers was negatively matched. Therefore, it was concluded that OSA raises the levels of AD biomarkers using PET, but the reversibility of the long-term effects of CPAP is elusive (Mullins et al. 2020).

### 9.4.4 Delta Age

Recently, the most commonly used biomarker is the delta age, which captures the differences between the biological age of the brain and the chronological age. The delta age is known to have been associated with lower bone mineral density (BMD), high blood pressure, diabetes, and cognitive decline. The BMD and cognitive function decrease with aging, and the cognitive decline is highly related to brain volume. MRI, genomics, blood-based biomarkers, and metabolomics measure the basis of delta age. The delta age is highly predicted in the lower part of the thalamus and lower volumes in the fornix. The genome-wide association tests showed that single-nucleotide polymorphism (SNP) is associated with gene regions such as KLF3-AS1 and STX1, and when the cognitive test results were combined, it was discovered that there were SNPs in MED8, COLEC10, and PLIN4 genes (Kim et al. 2023).

### 9.4.5 Diet

Diet has better control over all sorts of changes in our body. Likewise, it can also influence brain aging in its pre-symptomatic stage when the prevention is most effective. The data shows that the average age for destabilization of age is the late 40s, which is linked to cognitive decline and higher insulin resistance. Studies show that this biomarker is modulated by consuming a balanced diet; for instance, the stability of the brain is increased and decreased by glucose and ketones, respectively. It is found that ketones increase Gibbs's free energy by 27% compared to glucose. Experiments were conducted on the acute effects of modulating food controlled for the role of potential mechanisms associated with large-scale epidemiological studies comparing diets.

For instance, insulin resistance facilitates vascular dementia because higher glucose levels increase inflammation and block nitric oxide, thereby increasing blood viscosity. In the case of Alzheimer's disease, the insulin-degrading enzyme plays a major role in removing excess insulin and amyloid protein from the brain. Similarly, ketones are known to reduce inflammation and produce free radicals. This synergistic effect in regaining the cognitive function followed by a single dose of D-β-hydroxybutyrate monoester (D- βHb). As a result, dietary interventions can significantly impact neurobiological effects (Mujica-Parodi et al. 2020). The aging process raises the risk of a number of diseases, such as diabetes (Tamura et al. 2016), cancer (The ICGC/TCGA Pan-Cancer Analysis of Whole Genomes Consortium 2020), cardiovascular disease (Mukherjee et al. 2009), neurological disorders (Wu et al. 2021; Brown et al. 2018), osteoarthritis, joint pain, and frailty (McCulloch et al. 2017)and obstructive lung disease (Cordoba-Lanus et al. 2017), which place a burden on the healthcare system. Knowledge of how to prevent biological aging is therefore crucial both on an individual and community level (Chang et al. 2019; Needham et al. 2015). Researchers have found that biological age models may be able to predict morbidity and age-related mortality is more accurate than chronological age (Levine 2013; Ashapkin et al. 2019; Cole et al. 2018).

## 9.5 Conclusion

The phytochemicals are used to treat neurological illnesses such as Alzheimer's disease and Parkinson's disease from their capacity to lower inflammation and oxidative stress, both of which play essential roles in cancer treatment. It has been shown that brain aging is one of the critical elements that might contribute to these disorders. Aging is a complex biological process that manifests at the cellular and molecular levels, resulting in brain disorders, cardiovascular diseases, and other conditions that hereditary and environmental factors may impact. Understanding brain aging is still in its early stages, and more research is being conducted to produce medications and effective therapies. Plant extracts have recently been used to restore cognitive abilities. Flavonoids, curcumin, epigallocatechin gallate, quercetin, resveratrol, and myricetin are among the phytochemicals discussed in this chapter. Flavonoids are an effective treatment for Alzheimer's and Parkinson's disease, and plants such as A. galanga may be employed in future esthetic applications. Curcumin has greater specificity and sensitivity, although early illness identification remains challenging, and curcumin formulations are therapeutic.

Though quercetin is a promising phytochemical, it has a limited ability to penetrate the blood-brain barrier, limiting its efficacy. Therefore, research is done to increase its bioavailability and brain penetrability. Resveratrol is a practical approach for boosting health owing to its neuroprotective properties and anti-inflammatory nature, yet it is essential to enhance the tissue-targeting concentration for its usage in many biological applications. Epigallocatechin gallate (EGCG) inhibits brain dysfunction, and in the future, it would be vital to know the concentration of ECGC and its metabolites to brain function. The properties of myricetin comprise immune modulation and an antitumor nature that protects against these disorders. Moreover, it is a potential chemical for the treatment of COVID-19, but its safety and impacts are currently under investigation. Hence, neurodegenerative illnesses are treated by these substances, and they are identified using methods like DNA methylation, epigenetic clocks, delta ages, and proteomic clocks. Since aging is multidimensional and not a single process, multiple biomarkers and many more are being researched that may predict age-related mortality and morbidity. Future research must seek clinical validation of health assumptions based on in vivo and in vitro study findings. Human trials of various beneficial drugs and their compounds should have been done, incorporating vulnerability and toxicity evaluations to identify any harmful effects.

**Acknowledgments** The authors are grateful to the Chettinad Academy of Research Education for its continued support and encouragement.

## References

de Andrade Teles RB, Diniz TC, Costa Pinto TC, de Oliveira Junior RG, Gama e Silva M, de Lavor ÉM, Fernandes AW, de Oliveira AP, de Almeida Ribeiro FP, da Silva AA, Cavalcante TC (2018) Flavonoids as therapeutic agents in Alzheimer's and Parkinson's diseases: a systematic review of preclinical evidences. Oxidative Med Cell Longev 2018:7043213

Ashapkin VV, Kutueva LI, Vanyushin BF (2019) Epigenetic clock: just a convenient marker or an active driver of aging? Reviews on biomarker studies in aging and anti-aging research: 175–206

Balestrino R, Schapira AH (2020) Parkinson disease. Eur J Neurol 27(1):27–42

Bartosikova L, Necas J (2018) Epigallocatechin gallate: a review. Vet Med 63(10):443–467

Benameur T, Soleti R, Panaro MA, La Torre ME, Monda V, Messina G, Porro C (2021) Curcumin as prospective antiaging natural compound: focus on brain. Molecules 26(16):4794

Bloem BR, Okun MS, Klein C (2021) Parkinson's disease. Lancet 397(10291):2284–2303

Brown PJ, Wall MM, Chen C, Levine ME, Yaffe K, Roose SP, Rutherford BR (2018) Biological age, not chronological age, is associated with late-life depression. J Gerontol Series A 73(10): 1370–1376

Chang AY, Skirbekk VF, Tyrovolas S, Kassebaum NJ, Dieleman JL (2019) Measuring population ageing: an analysis of the Global Burden of Disease Study 2017. Lancet Public Health 4(3): e159-67

Chen M, Du ZY, Zheng X, Li DL, Zhou RP, Zhang K (2018) Use of curcumin in diagnosis, prevention, and treatment of Alzheimer's disease. Neural Regen Res 13(4):742

Choudhari AS, Mandave PC, Deshpande M, Ranjekar P, Prakash O (2020) Phytochemicals in cancer treatment: from preclinical studies to clinical practice. Front Pharmacol 10:1614

Cole JH, Ritchie SJ, Bastin ME, Hernández V, Muñoz Maniega S, Royle N, Corley J, Pattie A, Harris SE, Zhang Q, Wray NR (2018) Brain age predicts mortality. Mol Psychiatry 23(5): 1385–1392

Cordoba-Lanus E, Cazorla-Rivero S, Espinoza-Jimenez A, de-Torres JP, Pajares MJ, Aguirre-Jaime A, Celli B, Casanova C (2017) Telomere shortening and accelerated aging in COPD: findings from the BODE cohort. Respir Res 18(1):1–8

Corrêa RC, Peralta RM, Haminiuk CW, Maciel GM, Bracht A, Ferreira IC (2018) New phytochemicals as potential human antiaging compounds: reality, promise, and challenges. Crit Rev Food Sci Nutr 58(6):942–957

Denver P, McClean PL (2018) Distinguishing normal brain aging from the development of Alzheimer's disease: inflammation, insulin signaling and cognition. Neural Regen Res 13(10): 1719

Frontzkowski L, Ewers M, Brendel M, Biel D, Ossenkoppele R, Hager P, Steward A, Dewenter A, Römer S, Rubinski A, Buerger K (2022) Earlier Alzheimer's disease onset is associated with tau pathology in brain hub regions and facilitated tau spreading. Nat Commun 13(1):1–4

Gomes BA, Silva JP, Romeiro CF, Dos Santos SM, Rodrigues CA, Gonçalves PR, Sakai JT, Mendes PF, Varela EL, Monteiro MC (2018) Neuroprotective mechanisms of resveratrol in Alzheimer's disease: role of SIRT1. Oxidative Med Cell Longev 2018:8152373

Higgins-Chen AT, Thrush KL, Levine ME (2021) Aging biomarkers and the brain. Semin Cell Dev Biol 116:180–193. Academic Press

Ionescu-Tucker A, Cotman CW (2021) Emerging roles of oxidative stress in brain aging and Alzheimer's disease. Neurobiol Aging 107:86–95

Jiang S, Guo Y (2020) Epigenetic clock: DNA methylation in aging. Stem Cells Int 2020:1047896

Khan H, Ullah H, Aschner M, Cheang WS, Akkol EK (2019) Neuroprotective effects of quercetin in Alzheimer's disease. Biomol Ther 10(1):59

Kim J, Lee J, Nam K, Lee S (2023) Investigation of genetic variants and causal biomarkers associated with brain aging. Sci Rep 13(1):1526

Lange KW, Lange KM, Nakamura Y (2022) Green tea, epigallocatechin gallate and the prevention of Alzheimer's disease: clinical evidence. Food Sci Human Wellness 11(4):765–770

Levine ME (2013) Modeling the rate of senescence: can estimated biological age predict mortality more accurately than chronological age? J Gerontol Series A: Biomed Sci Med Sci 68(6): 667–674

Liu M, Guo H, Li Z, Zhang C, Zhang X, Cui Q, Tian J (2020) Molecular level insight into the benefit of myricetin and dihydromyricetin uptake in patients with Alzheimer's diseases. Front Aging Neurosci 12:601603

Luo M, Gan RY, Li BY, Mao QQ, Shang A, Xu XY, Li HY, Li HB (2021) Effects and mechanisms of tea on Parkinson's disease, Alzheimer's disease and depression. Food Rev Intl:1–29

Matias I, Morgado J, Gomes FC (2019) Astrocyte heterogeneity: impact to brain aging and disease. Front Aging Neurosci 11:59

McCulloch K, Litherland GJ, Rai TS (2017) Cellular senescence in osteoarthritis pathology. Aging Cell 16(2):210–218

Minocha T, Birla H, Obaid AA, Rai V, Sushma P, Shivamallu C, Moustafa M, Al-Shehri M, Al-Emam A, Tikhonova MA, Yadav SK (2022) Flavonoids as promising neuroprotectants and their therapeutic potential against Alzheimer's disease. Oxidative Med Cell Longev 2022: 6038996

Mujica-Parodi LR, Amgalan A, Sultan SF, Antal B, Sun X, Skiena S, Lithen A, Adra N, Ratai EM, Weistuch C, Govindarajan ST (2020) Diet modulates brain network stability, a biomarker for brain aging, in young adults. Proc Natl Acad Sci 117(11):6170–6177

Mukherjee M, Brouilette S, Stevens S, Shetty KR, Samani NJ (2009) Association of shorter telomeres with coronary artery disease in Indian subjects. Heart 95(8):669–673
Mullins AE, Kam K, Parekh A, Bubu OM, Osorio RS, Varga AW (2020) Obstructive sleep apnea and its treatment in aging: effects on Alzheimer's disease biomarkers, cognition, brain structure and neurophysiology. Neurobiol Dis 145:105054
Nebrisi EE (2021) Neuroprotective activities of curcumin in Parkinson's disease: a review of the literature. Int J Mol Sci 22(20):11248
Needham BL, Rehkopf D, Adler N, Gregorich S, Lin J, Blackburn EH, Epel ES (2015) Leukocyte telomere length and mortality in the National Health and Nutrition Examination Survey, 1999–2002. Epidemiology 26(4):528
Pervin M, Unno K, Takagaki A, Isemura M, Nakamura Y (2019) Function of green tea catechins in the brain: epigallocatechin gallate and its metabolites. Int J Mol Sci 20(15):3630
Rango M, Bresolin N (2018) Brain mitochondria, aging, and Parkinson's disease. Genes 9(5):250
Ranjan A, Ramachandran S, Gupta N, Kaushik I, Wright S, Srivastava S, Das H, Srivastava S, Prasad S, Srivastava SK (2019) Role of phytochemicals in cancer prevention. Int J Mol Sci 20(20):4981
Remigante A, Spinelli S, Basile N, Caruso D, Falliti G, Dossena S, Marino A, Morabito R (2022) Oxidation stress as a mechanism of aging in human erythrocytes: protective effect of quercetin. Int J Mol Sci 23(14):7781
Scheltens P, De Strooper B, Kivipelto M, Holstege H, Chételat G, Teunissen CE, Cummings J, van der Flier WM (2021) Alzheimer's disease. Lancet 397(10284):1577–1590
Sharma S, Raj K, Singh S (2020) Neuroprotective effect of quercetin in combination with piperine against rotenone-and iron supplement–induced Parkinson's disease in experimental rats. Neurotox Res 37:198–209
Shimizu S, Hirose D, Hatanaka H, Takenoshita N, Kaneko Y, Ogawa Y, Sakurai H, Hanyu H (2018) Role of neuroimaging as a biomarker for neurodegenerative diseases. Front Neurol 9:265
Shoko T, Maharaj VJ, Naidoo D, Tselanyane M, Nthambeleni R, Khorombi E, Apostolides Z (2018) Antiaging potential of extracts from Sclerocarya birrea (A. Rich.) Hochst and its chemical profiling by UPLC-Q-TOF-MS. BMC Complement Altern Med 18:1–4
Song X, Tan L, Wang M, Ren C, Guo C, Yang B, Ren Y, Cao Z, Li Y, Pei J (2021) Myricetin: a review of the most recent research. Biomed Pharmacother 134:111017
Tamura Y, Takubo K, Aida J, Araki A, Ito H (2016) Telomere attrition and diabetes mellitus. Geriatr Gerontol Int 16:66–74
The ICGC/TCGA Pan-Cancer Analysis of Whole Genomes Consortium (2020) Pan-cancer analysis of whole genomes. Nature 578(7793):82–93
Tungmunnithum D, Tanaka N, Uehara A, Iwashina T (2020) Flavonoids profile, taxonomic data, history of cosmetic uses, anti-oxidant and anti-aging potential of Alpinia galanga (L.) Willd. Cosmetics 7(4):89
Weller J, Budson A (2018) Current understanding of Alzheimer's disease diagnosis and treatment. F1000Res:7
Wu JW, Yaqub A, Ma Y, Koudstaal W, Hofman A, Ikram MA, Ghanbari M, Goudsmit J (2021) Biological age in healthy elderly predicts aging-related diseases including dementia. Sci Rep 11(1):15929
Zahedipour F, Hosseini SA, Henney NC, Barreto GE, Sahebkar A (2022) Phytochemicals as inhibitors of tumor necrosis factor alpha and neuroinflammatory responses in neurodegenerative diseases. Neural Regen Res 17(8):1675
Zaplatic E, Bule M, Shah SZ, Uddin MS, Niaz K (2019) Molecular mechanisms underlying protective role of quercetin in attenuating Alzheimer's disease. Life Sci 224:109–119
Zhou DD, Luo M, Huang SY, Saimaiti A, Shang A, Gan RY, Li HB (2021) Effects and mechanisms of resveratrol on aging and age-related diseases. Oxidative Med Cell Longev 2021:9932218
Zia A, Pourbagher-Shahri AM, Farkhondeh T, Samarghandian S (2021) Molecular and cellular pathways contributing to brain aging. Behav Brain Funct 17(1):1–30

# Phytochemicals and Natural Extracts, Secondary Metabolites of Plants and Improvement of Brain Function

10

Sridevi Gopathy, Srividya Seshadri, P. Amudha, R. Vidya, M. Jayalakshmi, Langeswaran Kulanthaivel, Muthuvel Raju, and Gowtham Kumar Subbaraj

**Abstract**

Memory, one of the most enhanced functions of human being, proclaims an activity-dependent synaptic plasticity at various synaptic levels in the brain. Among the most important functions of the brain, memory and language

---

Srividya Seshadri, P. Amudha, R. Vidya, M. Jayalakshmi and Langeswaran Kulanthaivel contributed equally with all other contributors.

---

S. Gopathy
Department of Physiology, SRM Dental College, Chennai, Tamilnadu, India
e-mail: sridevig1@srmist.edu.in

S. Seshadri
Department of Biochemistry, Sathyabama Dental College and Hospital, Chennai, Tamil Nadu, India

P. Amudha · R. Vidya
Department of Biochemistry, School of Life Sciences, Vels Institute of Science, Technology and Advanced Studies (VISTAS), Chennai, Tamil Nadu, India

M. Jayalakshmi
Department of Biochemistry, Mahalashmi Women's College of Arts and Science, Chennai, Tamilnadu, India

L. Kulanthaivel
Department of Biomedical Sciences, Science Campus, Alagappa University, Karaikudi, Tamil Nadu, India

M. Raju (✉)
Department of Biochemistry, Karuna Medical College and Hospital, Palakkad, Kerala, India

G. K. Subbaraj (✉)
Faculty of Allied Health Sciences, Chettinad Hospital and Research Institute, Chettinad Academy of Research and Education, Kelambakkam, Tamilnadu, India
e-mail: gowtham.genetics@care.edu.in





S. Pathak, A. Banerjee (eds.), *Neuroprotective Effects of Phytochemicals in Brain Ageing*, https://doi.org/10.1007/978-981-99-7269-2_10

processing and thought processes and intelligence play an important role in cognition. To improve cognitive functions in terms of general intelligence, enhance learning abilities, good thinking, and react faster, and boost memory, many groups of compounds in various combinations called nootropics are therapeutically used. Antipsychotics, acetylcholinesterase inhibitors, NMDAR drugs, and selective estrogen receptor modulators are commonly used allopathic drugs that improve brain cognition. Siddha and Ayurvedic medicinal products like Ashwagandha, *Centella asiatica*, *Terminalia chebula*, *Bacopa monnieri*, and Shankhpushpi were administered after scientific studies to boost memory and enhance the intellectual capacity in human beings. They contain polyunsaturated fatty acid (PUFA), polyphenols, phenols, flavonoids, tannins, terpenoids, alkaloids, steroids, glycosides, anthocyanins, etc. that help to modulate brain function at the molecular level.

**Keywords**

Acetylcholinesterase (Ach) inhibitors · Polyunsaturated fatty acids (PUFA) · Bioactive molecules · Nanoparticle-based delivery systems

## 10.1 Introduction

The brain being the most specially functioning organ in the human body works by complex mechanisms to act as an interpreter of peripheral senses, initiator of body movements by muscles, seat of super intelligence, and an emotional and behavioral controller. With its refined functions that discriminate a human being from other living things in the universe, the brain acts as a crown jewel to add beauty to the physiology of the human body. David J. Linden, in his book *'The accidental mind,'* considered the brain as an overrated organ with much of an engineering miracle in it. He explained that the listlessness of neurons behaving as computational devices with the active involvement of numerous neurons and synapses produces the brain's computing power. Though this complex function results from a complex circuitry that could be encoded in a genome, the functioning of a brain is increasingly dependent on experience-dependent plasticity to establish the memory pattern. The underlying cellular processes that develop synaptic plasticity could be co-opted by various learning systems followed by the brain. Memory, one of the most enhanced functions of human being, proclaims an activity-dependent synaptic plasticity at various synaptic levels in the brain so that encoding of information and storage of memory is mediated. Memory, an ability to retain and recall information is one of the important processes essential for the survival of man in the present world. These compounded memories also shape the identity of a person, guide our thoughts and decisions, and impact the emotional reactions of an individual.

The physiological connection between changes at synapse and the development of memory and its retention was experimentally proved by recordings from in vitro and in vivo electrophysiological tracings to elicit those activity-dependent long-lasting efficiency changes at the synaptic level. The changes promote long-term potentiation (LTP) and long-term depression (LTD) changes in the pre- and post-synaptic junctions of mammalian neurons. Also, substantial evidence from modern imaging technologies reveals morphological changes in the synaptic spines along with proxy from electrophysiological functions at the synapse. Reports suggested that molecular modifications at post-translational, translational, and transcriptional levels play a very important role in long-term memory formation.

Memory consolidation is a phenomenon by which temporary information with labile memory can be transformed into very stable, long-lasting memory in the brain. A common feature of long-term memory formation in the brain memory system initially exists in a brittle state and is capable of getting disrupted due to various interferences that happen at the molecular to functional behavioral level. On the progression of time, the memory develops to become stronger and irrepressible to any disruption. Thus, this process of strengthening and stabilization of information is called memory consolidation. After further consolidation, this information becomes fixated and stored in memory. Also reports reveal that such stored information that is resistant to interferences, even at the molecular level, can be retrieved after 15 years with good reactivation. Explicit memories related to events that have happened to a person are considered episodic and general facts, and information considered semantic is regulated by brain regions, namely the hippocampus, neocortex, and amygdala. Implicit memories related to motor functional memories are controlled by basal ganglia and cerebellum. Working memory which is short term is influenced by the prefrontal cortex.

Disorders of memory seem to be a common outcome of neurological damage and diseases. Alzheimer's disease is a progressive neurocognitive disorder that is initiated by a mild loss of memory to develop a stage of lack of conversation and response to the external environment. Researchers believe that genetic mechanisms may play a role in the development of Alzheimer's disease. This disease affects different parts of the brain involved in memory and language processing thought processes and intelligence. This neurodegenerative disease is caused by accumulation of extracellular masses of amyloid β (Aβ) plaques and intracellular neurofibrillary tangles developed from per-phosphorylated τ-protein in brain regions like cortical and limbic system of the human brain, thereby causing memory loss and progressive neurocognitive dysfunction. Studies postulated that Neurofibrillary tangles, neuronal plaques, and overloaded amyloid were reported in hippocampus sections at the level of the lateral geniculate body in many patients diagnosed with Alzheimer's disease.

Amnesia is a disorder related to long-term memory, where episodic memory is more evidently impaired compared to semantic or procedural memory, and the damage in such cases is confined to the frontal brain, which is responsible for the acquisition of new knowledge information. The normal physiological process of aging also results in the impairment of long-term memory, in particular for nonconcrete encoding tasks, and gradually progresses to degenerative dementias, where broader ranges of memory functions would be affected (Nagy et al. 1996). In

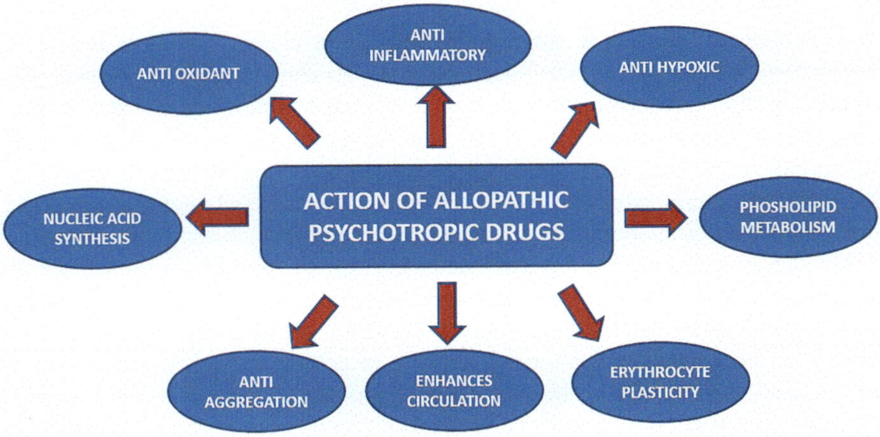


**Fig. 10.1** Mechanism of action of allopathic psychotropic drugs

modern medicine, various psychoactive drugs are being administered to treat various neurological disorders. A group of anticholinergic drugs interrupt with memory encoding process of new information into long-term memory, similar to Alzheimer's disease in normal elderly people. Another class of benzodiazepine drugs also selectively affect long-term storage, resulting in amnesic disorders. The catecholamines treated for Parkinson's disease and Korsakoff's syndrome also interfere with human memory functions.

To improve general intelligence, enhance learning abilities, good thinking, and react faster, and boost memory, many groups of compounds in various combinations called nootropics are available.

These substances are natural products that are ingested with food supplements. In patients with cognitive deficits, nootropics are typically well tolerated; adverse effects are rare and they improve cognitive functions like memory and learning, especially in patients suffering from memory loss. The mechanism by which nootropics is that it improves glucose and oxygen supply to brain, their anti-hypoxic effects, protection against neurotoxicity, enhances nucleic acid synthesis in neurons, and stimulates phospholipid metabolism in neurohormonal membranes. It is also reported that few kinds of nootropic medications are involved in the eradication of oxygen free radicals, with greater anti-aggregation effect, and erythrocyte plasticity, and improvement in blood flow to the brain (Fig. 10.1).

## 10.2 Potential Cognitive Enhancement with Drugs

Cognitive function involves both neurocognition and social cognition along with the acquisition of learned information and development of knowledge. These two cognitive functions predict the functional product of an individual with patients suffering from schizophrenia and any bipolar disorder. Reports confirmed evidences where patients with schizophrenia are greatly impaired in both neurocognition and social cognition. The mechanism underlying cognitive impairment varies among patients with schizophrenia, bipolar disorder, Alzheimer's disease, dementia, and Parkinson's disease. In schizophrenia, there is a significant loss of connectivity of schizophrenia with the striatum, thalamus, and parietal cortex along with alterations in the neurotransmitters namely glutamate, γ-aminobutyric acid (GABA), and dopamine. In Alzheimer's disease, Synaptic dysfunction with impaired glutamate system and damage to synaptic plasticity by β-amyloid protein resulted in cognitive deterioration. In Parkinson's disease, cognition is affected by the impaired metabolism of α-synuclein along with amyloid protein and cholinergic dysfunction. Different classes of compounds were used to treat those neurocognitive disorders based on the underlying causes.

Acetylcholinesterase (Ach) inhibitors: A group of Ach namely inhibitors, galantamine, donepezil, and rivastigmine, as approved by the US Food and Drug Administration (FDA), is administered for Alzheimer's disease. These drugs enhanced the concentrations of acetylcholine, inhibited the aggregation of β-amyloid plaques in the synapses, and improved the communication between neurons and cognition. Clinical studies also reported that these acetylcholinesterase inhibitors also enhanced cognition in patients with Parkinson's disease. Antipsychotics: These groups of medications act as major therapeutic remedies for schizophrenia or psychosis symptoms. Schizophrenia score assessed by Brief Assessment of Cognition improved the attention and processing speed item in showed greater improvement in the paliperidone palmitate long-acting injection administration compared to risperidone long-acting. Patients receiving final doses of lurasidone daily exhibited significant improvement in overall cognitive performances. NMDA receptors (NMDAR) happen to be the primary target for cognitive enhancement drugs because they play an important role in neural plasticity that activates long-term potentiation and long-term depression. Several classes of NMDA-enhancing drugs have been studied to evaluate their effect on cognitive impairment in dementia. One such drug is Sodium benzoate, which acts as an inhibitor of D-amino acids oxidase and can increase d-serine levels and improve activation of NMDAR. Sodium benzoate acts as an adjunctive to the administration of antipsychotics in chronic schizophrenia patients. Another drug, sildenafil, acts as a phosphodiesterase-5 inhibitor, increases the levels of cyclic guanosine monophosphate wrought by NMDARs inside the cell, and increases memory consolidation and long-term potentiation. LCarnosine, an effective antioxidant and a good antiglycation product protects the cultured neurons against hypoxia, hypoglycemia, and neurotoxicity related to NMDA.

Patients with schizophrenia show significant learning impairment with abnormal neural activity. In such cases, a selective estrogen receptor modulator (SERM) called raloxifene protects neural activity and learning with memory and increases cognition in healthy older men with schizophrenia. It is based on the physiological theory that circulating levels of estrogen regulate emotional responses, enhance working memory, and increase the activity of the frontal cortex. Estrogen therapy is administered to improve memory and fast recovery from basic psychotic and generalized anxiety symptoms in older females and women belonging to childbearing age (Fig. 10.2).

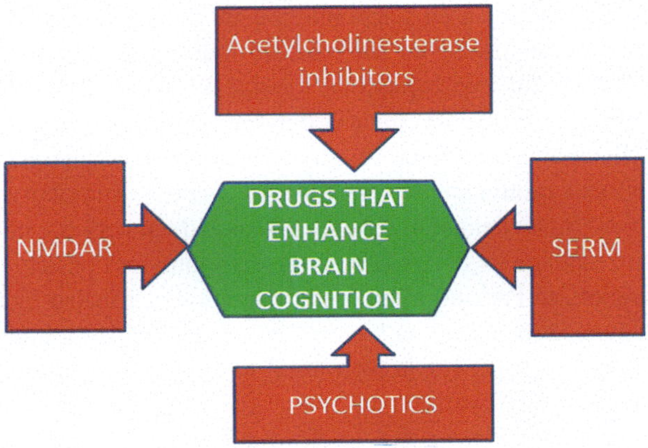


**Fig. 10.2** Drugs that enhance brain cognition

## 10.3 Herbal Drugs: Mechanisms and Interactions

Ayurveda and Siddha medicine are a personalized system of traditional medicine that is native to the Indian subcontinent. Ayurveda is based on a holistic therapeutic remedy that provides an optimistic balance between the body, mind, and soul. Various popular plants that take their origin from ancient man form a major contribution to the system of Ayurveda and many medicinal formulations are developed from plants to act as cognitive enhancers, reduce brain aging, and improve memory. Such formulations are made from Ashwagandha, Brahmi Turmeric, *Shankhpushpi*, *Convolvulus pluricaulis, Gotu kola*, *Evolvulus alsinoides*, *Centella asiatica,* Vacā, Kuṣṭha and Purāṇa Ghṛita. Certain popular Ayurvedic formulations including Brāhmī Ghṛita, Brento Forte, and Divya Medha Kwath were treated for loss of memory and learning deficits.

Ashwagandha (*Withania somnifera* (L.)) is a popular stress reliever known to boost memory in human beings. A double-blind, placebo-controlled clinical study reported that Ashwagandha root extract in a sustained-release capsule with a dose of 300 mg significantly increased cognitive functions, enhanced recall memory assessed by neuropsychological test reduced stress, improved sleep quality, and overall health in stressed subjects. Ashwagandha root works at the molecular level to treat Alzheimer's disease, by activating the NF-$\kappa$B and effectively hindering the production *of* $\beta$-amyloid and decreased apoptotic cell death. It thereby improves and restores synaptic function. It possesses antioxidant effects and migration of Nrf2 into the nucleus for scavenging the free radicals.

*Bacopa monnieri* is an Ayurvedic plant medicine used from ancient origin and commonly called Brahmi. *Bacopa* acts to improve general intelligence, learning capacity, and memory and offers protection of physiological functions of neurons in Alzheimer's disease, improves thinking skills, reduces anxiety, and treats attention deficit hyperactivity disorder in children. A randomized, placebo control study conducted between 40 and 65 years for memory and anxiety tests after regular

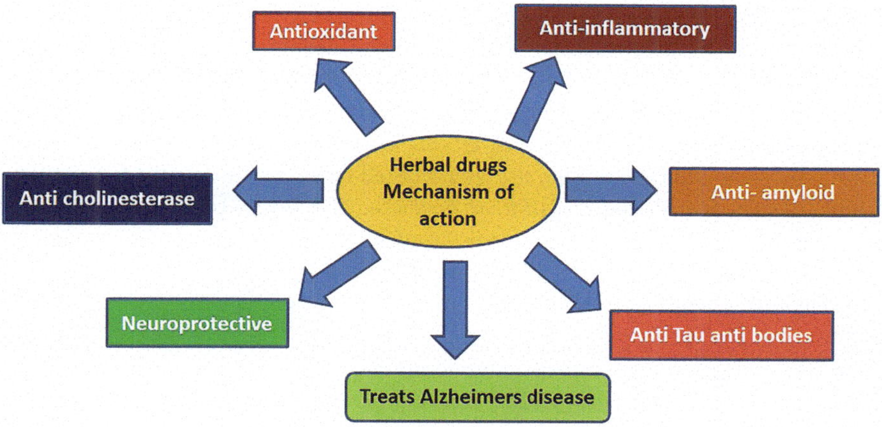


**Fig. 10.3** Mechanism of action in herbal drugs

administration of Brahmi for 6 weeks months revealed that the rate of forgetting new information was decreased. Another report revealed that *Brahmi* derived from the extract of *Bacopa monniera,* acts as an antioxidant, quenches the free radicals, and protects neurons in the hippocampus, prefrontal cortex, and striatum against DNA damage and cytotoxicity in Alzheimer's disease by decreasing lipid peroxidation, and increased glutathione peroxidase and chelation of iron. It also protects the cholinergic neurons and effectively reduces β-amyloid deposition and stress-induced alterations in the hippocampus. This effect is attributed to the nitric oxide (NO)-mediated cerebral vasodilation and thereby Brahmi helps to improve the total memory scores, a good significant increase in logical memory, and paired associate learning in human subjects (Fig. 10.3).

## 10.4 Phytochemicals: A Natural Approach to Cognitive Health

Phytochemicals are naturally occurring compounds found in plants, responsible for their vibrant colors, flavors, and aromas. Beyond our senses, a growing body of research suggests that these bioactive compounds may hold significant potential for protecting and improving brain health. Some phytochemicals have shown promise in combating the damaging effects of oxidative stress and inflammation, both of which are linked to cognitive decline. Others appear to interact with pathways involved in memory, learning, and the growth of new brain cells.

*Centella asiatica* is a small herb belonging to the family Apiceae, commonly called jalbrahmi. The leaves of the plant have been traditionally used as a memory enhancer and antiaging agent in Ayurvedic medicine. Owing to phyto-constituents like asiatic acid, asiaticosides, and madasiatic acid, madecassoside, it is well known to possess antioxidative stress, anti-inflammatory, antiulcer, antipsoriatic, hepatoprotective, antiapoptotic, antidiabetic, immunostimulant activities (Mehla et al. 2020). Reports revealed that *C. asiatica aqueous extract* significantly improved cognition and also reversed cognitive impairment developed by streptozotocin in rats in a dose-dependent manner from 100 to 300 mg/kg. Reports revealed that 200 mg/kg administration for 15 days of *C. asiatica* activated learning and memory changes with a significant increase in arborization of dendrites in the hippocampus during the postpartum period in rats, which lasted for at least 6 months postpartum. Studies also revealed that *C. asiatica* decreased the levels of β amyloid in the hippocampus of PSAPP transgenic mice and improved dopamine, noradrenaline, and 5 HT in rat brains.

*Terminalia chebula* Retz. Belonging to the family Combretaceae, it is yet another traditional herbal plant possessing numerous pharmacological benefits due to the presence of bioactive compounds like polyphenols, flavonoids, terpenes, alkaloids, anthocyanins, and glycosides. Under Unani, Ayurveda, and homeopathic forms of medicine, in spite of its antimicrobial, antidiabetic, anti-hyperlipidemic, anti-ulcer, and wound healing effects, *T.chebula* therapeutically benefitted geriatric diseases and enhanced learning, memory, and overall brain functions. Kumar et al. (2021) revealed that hydroalcoholic fruit extract of *Terminalia chebula* at administered at 1000 mg/kg showed 83.33% neuroprotection against MES-induced seizures and 66.66% neuroprotection in PTZ-induced seizures. With the preservation of neurological functions in seizure-induced oxidative stress and effective treatment of cognitive impairment as evaluated by elevated plus maze and passive avoidance test in rats. The enhancement of learning and memory after treatment with *Terminalia chebula* fruit extracts in ethanol-induced cognitive impairment versus diazepam-induced amnesia was analyzed. *T chebula* was able to significantly restrain oxidative stress and offer protection against oxidative damage and thus preferred as a good remedy for neurodegenerative diseases in rats.

Shankhpushpi, vernacularly called Shankhini, is a commonly practiced Ayurvedic plant medicine prepared to boost memory and enhance the intellectual capacity of Human beings. This plant, grown popularly in most houses, possesses significant neuroprotection and enhances neurological functions. Reports reveal that the plant increases the concentration and learning capabilities of human subjects and acts as a good remedy for mental fatigue, insomnia, stress-induced anxiety, irritability, and depression. Confounding evidence states that Shankhpushpi is known to combat anxiety and stress-related changes and increase the positivity and upliftment of mood and also reinforces the constructive effects of yoga and meditation. Since the olden days, ancient people believed that this plant works to improve the brain's functioning capacity, eradicate mental fatigue, increase the concentration capacity, and increase the efficiency of computer-related work and fast memorization as well.

Phytochemicals obtained from medicinal plants and herbs offer good protection for the physiological functions of the brain by regulating a good chemical balance in the brain tissues and prompting the neuronal synapses and receptors with functional neurotransmitters. Phytochemicals are bioactive compounds produced by plants. They are available as natural resources of medicinal compounds that play a very important role in the pharmacological benefits of various plants, herbs, and their products. Polyphenols, phenols, flavonoids, tannins, terpenoids, alkaloids, steroids, glycosides, anthocyanins, etc., are plant compounds found in many natural foods. These Phytochemicals offer protection against oxidative stress and inflammation, prevention of cancer, effective antioxidants, and antiaging effects.

Polyunsaturated fatty acids (PUFA) are extensively present in the cell membranes of the brain. As the brain requires higher amounts of PUFA to develop greater and more complex phospholipids and sphingophospholipids for their myelination, the brain tissues are specially premeditated to save such fats via their special membrane portals located in the blood-brain barrier. Two popularly known PUFAs that enhance brain health are omega-3 fatty acids and omega-6 fats. PUFAs, along with their mediators, perform various physiological functions like increased neurotransmission, improved cell survival rates, and enhanced neuroinflammatory processes, regulation of mood and cognition. Thus, PUFAs regulate signaling pathways to treat neuro deficits in Alzheimer's disease and depressive disorders. The monoaminergic systems like dopaminergic, noradrenergic, and serotonergic develop greatly in the presence of long-chain acting PUFA's. Reports also suggest that synaptic plasticity, learning, and memory are enhanced in the presence of PUFA, and a combination of arachidonic acid and docosahexaenoic acid also activates neurogenesis in hippocampal regions of the brain.

Polyphenols are compounds naturally derived from vegetables, fruits, cereals, etc. Polyphenols are reported to regulate brain homeostasis by their role in neuronal adaptation to stress responses and prevention of advancement of neurodegenerative diseases. Research on the benefits of polyphenols at the molecular level studies indicates that it is engaged in oxidative and inflammatory signaling pathways of stress, protective mechanisms, and neuro-hormetic effects with the encoding of antioxidant enzymes, neuro-growth factors, phase-2 enzymes and cyto-defensive proteins like sirtuin-FoxO pathway, Nrf-2/ARE pathway and the NF-κB pathways (Vauzour 2012). Reports reveal that polyphenols function as antioxidants that reduce the aging of neurons, execute autophagic processes, improve cognition, and scavenge stress-producing free radicals in the human body. For example, resveratrol, a well-known polyphenolic agent present in Mediterranean diets, favors good health of the intestine and brain and is known to preserve brain functions during neuroinflammation (Hornedo-Ortega et al. 2018).

Terpenes are compounds that contain units of volatile, organic, 5-carbon compound isoprene. It plays a dual role as an attractant and deterrent and appears less predominant in diets. Reports reveal that intake of single doses of monoterpenes obtained from edible herbs like *Mentha piperita*, *Salvia officinalis* and diterpenes from *Ginkgo biloba plant* extracts and triterpenes obtained from plants like *Panax ginseng*, *Bacopa monnieri*, etc., have been known to enhance cognition and learning, memory and alertness in the human brain. Monoterpenes act as inhibitors of acetyl cholinesterase and offer allosteric binding to gamma-aminobutyric acid (GABA) and nicotinic, as well as muscarinic receptors and thereby offering anxiolytic and sedative properties. Diterpenes act by greatly increasing the cerebral blow flow and improving cognitive function. Single administration and long-term intake of diterepenes attenuate attention task performances in the human brain. Triterpenes, commonly referred to as adaptogens in the stress axis, are reported to interact with

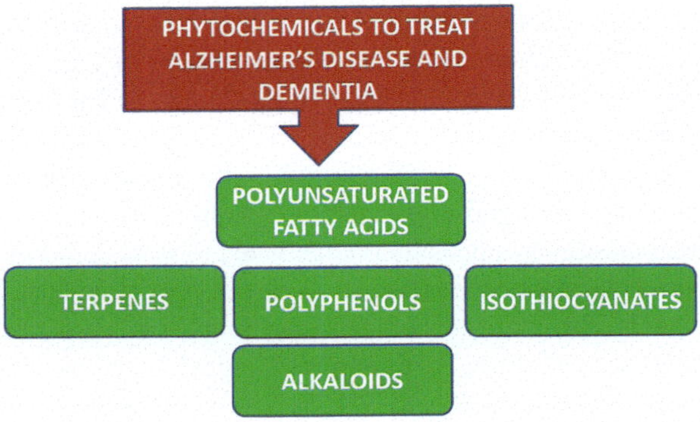


**Fig. 10.4** Phytochemical compounds reported to treat dementia and Alzheimer's disease

many mammalians' steroid hormone estrogen and glucocorticoid receptors, and their mechanism of action is multifarious. These compounds are known to enhance the physiological functioning of the central nervous system (CNS) by increasing Nitric Oxide (NO) synthesis and enhancing neurotransmission.

Isothiocyanates, commonly predominant in the family of the Brassicacae, contain sulfur-related phytochemicals. Consumption of Brassicaceae vegetables has been reported to reduce the incidence of cancers and many chronic neurodegenerative diseases. It is also suggested that the administration of Isothiocyanates helps to prevent and treat cognition related diseases owing to the antioxidant properties and anti-amyloidogenic nature. Recent research studies have also focused on the efficacy of Isothiocyanates as an effective phytochemical in the treatment of Dementia. Sulforaphane, a derivative of Isothiocyanates, was able to significantly decrease the translocation of NFκB and, affects the production of pro-inflammatory cytokines and reactive oxygen species, and inhibits neuronal apoptotic cell death in experimental models of neurodegeneration diseases. Another derivative of isothiocyanates, namely moringin (MG) or 4-(α-L-rhamnosyloxy) benzyl isothiocyanate, is present in fair amounts in certain vegetables like Moringa oleifera from tropical and equatorial areas of the world. Reports suggest that moringin possesses beneficial effects like antioxidant and anti-inflammatory activities and offers protection against chronic neurodegenerative disorders (Fig. 10.4).

Alkaloids are a group of organic compounds that contain nitrogen derived from various flowering plants like Papaveraceae, Solanaceae, Amaryllidaceae, and Ranunculaceae. Generally, Alkaloids are used therapeutically as an analgesic, anti-arrhythmic, anti-asthmatic antihypertensive, anticancer, antibacterial, antipyretics, and antihyperglycemic. Previous literature revealed that alkaloids have been used in the treatment of dementia for many years since 1960. The Food and Drug Administration (FDA) approval investigated and approved two alkaloid-based drugs, namely Galantamine and Rivastigmine, for Alzheimer's Disease, and that many intrinsic anticholinesterase compounds were present in alkaloids that make it a potential target for treating dementia-like diseases. Nicotinic acetylcholine receptors are actively involved in the process of learning. Alkaloids are known to regulate nAChR activity and improve memory function. A study investigated the acute effects of three known nicotinic alkaloids, nicotine, cotinine, and anatabine, by suppressing scopolamine-induced memory deficit in rats. They used two classic memory paradigms, namely Y-maze and novel object recognition in rodents. The study revealed that all three compounds were able to suppress scopolamine-induced spatial memory deficit in the Y-maze spontaneous alternation paradigm. However, only nicotine was able to suppress the short-term object memory deficit and regulate short-term spatial memory. This potential is attributed to the nicotinic AChR activity. Thus, nAChR-activating alkaloids could own distinct pro-cognitive properties, thereby improving the memory patterns in rodents.

Edible consumed food and natural extracts products, their metabolites, and their pharmacological values specific to brain function. Secondary metabolites found in nature have been shown to be the most productive source of new medication candidates. Recent use of them, however, has shown a reduction in interest in drug research and development. Opportunities for the discovery of mostly new low--molecular-weight lead compounds persist thanks to the structural variety offered by natural products compared to conventional combinatorial chemistry. There are certainly many more appealing natural lead compounds waiting to be identified, as only around 10% of the world's biodiversity has been examined for putative biological activity. Many bioactive natural items that have been mentioned throughout history as being used as medicines have yet to be discovered despite their widespread usage in traditional medicines, cures, potions, and oils. Through hundreds of years of trial and error, via palatability testing or untimely deaths, and seeking readily available foods for the treatment of ailments, man has learned the vast majority of natural product applications from medicinal plants.

## 10.5 Impact of Different Drug Formulations on Brain Function

The therapeutic properties of plants have been known for thousands of years, and their ethnopharmacological qualities were initially mined for potential treatments. Natural components were investigated for use in the development of antiretroviral medications in response to the 1980s AIDS crisis by the National Cancer Institute (NCI) and others. The discovery of calanolides, a new class of chemicals, is the most significant outcome of these studies. Sarawak Medichem Pharmaceuticals licensed and tested calanolide A in Phase II clinical studies, but no further drug development plans have been announced. Arteether and Artemotil are authorized antimalarial medicines produced from artemisinin, which was first used to treat fevers and chills in traditional Chinese medicine.

In Europe, other artemisinin compounds are in varying phases of clinical development as antimalarial medicines. Researchers are evaluating a synthetic trioxolane based on the pharmacophores in conjunction with piperaquine (a synthetic bisquinoline antimalarial medication). The indole alkaloids grandisines A and grandisines B were extracted from the leaves of the Elaeocarpus grandis tree, a native of the Australian rainforest. Grandisine A has a one-of-a-kind tetracyclic framework, while Grandisine B has an uncommon isoquinuclidinone and indolizidine ring structure. Galantamine hydrobromide, derived from the Galanthus nivalis plant, is used to treat neurological problems, including Alzheimer's disease. Recent studies have shown that marine actinomycetes provide abundant resources for novel natural compounds. At least 30,000 species of algae (macroalgae and seaweed) exist across the world. These algae are vital to the environment because they produce oxygen, offer survival for fish and people, act as a source of medication and fertilizer, and are a treasure trove of architecturally unique natural products. In the 1970s and 1980s, sea algae were the primary source for the family of chemicals known as terpenoids. Terpenoid-type structures have been studied chemically, leading to the isolation of several classes, such as brominated, nitrogen, and oxygen heterocycles, phenazine derivatives, sterols, amino acids, amines, and guanidine derivatives. Green, brown, and red algae have all been extensively tested for their antibacterial and antifungal capabilities with regard to biological activity.

Sedentary and without a central neurological, digestive, or circulatory system, sponges (phylum Porifera) rely on a steady stream of water passing through their bodies to sustain life. Sponge species universally feed on 'current' or 'filter' food and have no real defenses against predators. They date back around 500 million years, making them among the oldest multicellular creatures. The C-nucleosides spongouridine and spongothymidine, isolated and identified by Bergmann in the early 1950s from the Caribbean sponge Cryptotheca crypta, are largely considered the first major discovery of biologically active compounds from marine sources. It required 15 years following the discovery of the antiviral activity of these compounds by the synthesis of structural analogs when either cytosine arabinoside (Ara-C) or adenosine arabinoside (Ara-A) were developed as clinical anticancer medicines or antiviral agents, respectively. It was previously believed that the existence of a deoxyribose or ribose sugar group was necessary for a nucleoside to have its biological function. Through these analyses, the powerful antileukemic agent Ara-C was discovered.

Since inflammation is at the root of so many different types of human illness-from cancer to Alzheimer's to arthritis-there is a huge need for effective new anti-inflammatory medications. The wide variety of secondary metabolites produced by bacteria represents a promising new source of anti-inflammatory agents. Most marine creatures rely on their innate immune systems for defense; therefore, it seems sense that many marine bacteria produce anti-inflammatory chemicals. Therefore, marine bacteria may have evolved to be superior to their terrestrial counterparts at avoiding detection by the innate immune system, and the production of anti-inflammatory natural compounds may play a significant role in this process. Secondary metabolites of unique structural diversity and biological activity are produced by both cyanobacteria and heterotrophic marine bacteria, both of which play important roles globally in marine ecosystems.

Pro-inflammatory mediators can be activated by a wide range of stimuli, including those with antibacterial, immunological, and toxicological capabilities. The interleukin (IL) family of cytokines, prostaglandins, leukotrienes, nitric oxide, and TNF-alpha are all examples of mediators. Numerous inflammatory-related disorders have been linked to an overproduction of pro-inflammatory mediators. An appealing technique for treating these diseases would be to prevent the body from producing too much of these mediators. In recent years, researchers have isolated many marine natural compounds with anti-inflammatory properties. Sponge, mollusk, sea comb, algae, and bacterial cells have all been used to extract these chemicals. Natural marine chemicals having anti-inflammatory properties are produced by microalgae, cyanobacteria, and other marine microorganisms.

The filamentous cyanobacterium Lyngbya majuscule produced a lipopeptide called microcolin A, which had anti-inflammatory and immunosuppressive effects. They suppressed the proliferation of mouse splenocytes activated by mixed lymphocytes or anti-immunoglobulin M when used in conjunction with phorbol 12-myristate 13-acetate (PMA) and ionomycin. Stimulated splenocytes with concanavalin A, phytohemagglutinin, and lipopolysaccharide (LPS) had reduced IL-2 and IL-2 receptor expression. Concanavalin A downregulated IL-2 and IL-2 receptor expression in stimulated splenocytes. Human promyelocytic leukemia (HL-60) and Chinese hamster ovary (CHO) cells have shown that microcolin A prevents cell adhesion through the lymphocyte function-associated molecule-1 (LFA-1) and the intercellular adhesion molecule-1 (ICAM-1).

In a concentration-dependent manner, scytonemin, a yellow-green pigment isolated from the extracellular sheath of cyanobacteria, including Lyngbyamajuscula, reduced ear edema in mice. PKC activity promotes inflammation in a wide range of diseases, and Neovascularization is a hallmark of inflammatory diseases, including psoriasis and rheumatoid arthritis. It was also inhibited by scytonemin in human umbilical vein endothelial cells. Malyngamide F acetate, which is derived from the cyanobacterium Lyngbyamajuscula, suppressed LPS-induced NO production in RAW 264.9 cells. In the following experiments, researchers monitored the mRNA and protein expression levels of certain cytokines and found that IL-6 was downregulated while TNF-mRNA and protein expression levels were upregulated. Our group is now exploring this exciting observation to identify the specific pathway(s) that malyngamide F acetate inhibits. Lyngbyamajuscula is suspected to generate malyngamide S, which was first isolated from the sea hare Bursatellaleachii. Malyngamide S inhibited superoxide production in human peripheral blood neutrophils treated with inflammatory agents N-formyl-methionylleucyl-phenylalanine (fMLP) and PMA.

Some cultures have relied on the filamentous cyanobacterium spirulina for generations as a staple diet for humans and animals. In Arthrospira maxima, the protein-bound pigment phycocyanin makes up more than 20% of the dry weight. Phycocyanin is an open-chain tetrapyrrole. Using a mouse model, it was demonstrated that the biliprotein effectively reduced many forms of edema and scavenged radical oxygen species at a second-order rate constant comparable to that of other non-steroidal anti-inflammatory medicines like indomethacin and ibuprofen. The generation of prostaglandin E2, the activity of phospholipase A2, and the release of histamine may all be affected, contributing to its anti-inflammatory action. Mice with zymosan-induced arthritis showed a substantial decrease in glucuronidase levels after treatment with crude extracts of this cyanobacterium.

The number of marine-derived compounds with promising anticancer activity that have progressed to human clinical trials has increased dramatically since the early 1990s. The Caribbean marine tunicate Ecteinascidia turbinate was used to extract the chemical known as trabectedin (Yondelis). As a result of increased communication and cooperation between academic institutions, major pharmaceutical firms, and the National Cancer Institute, several natural product materials have been subjected to cutting-edge tests, leading to multiple therapeutic candidates being tested in human clinical.

In 2003, researchers initially identified salinosporamide A from Actinomyces sp. Salinisporatropica, an obligate marine bacterium. The chemical was shown to be a proteasome inhibitor after further investigation into its mode of action. As a novel and promising target for anticancer drugs, the proteasome is now solely treatable with bortezomib. Clinical studies using salinosporamide A to treat multiple myeloma have advanced to the phase I stage. Pettit et al. and co-workers discovered dolastatin 10 from the sea hare Dolabellaauricularia in the 1970s. Microtubule construction and tubulin-dependent GTP hydrolysis were both blocked by this compound, which occurred naturally in the cyanobacteria the seahorse consumed. The NCI initially tested dolastatin 10 in Phase I clinical studies with positive results. Around 40% of patients suffered from peripheral neuropathy, and patients with hormone-refractory metastatic malignancy or recurrent platinum-sensitive ovarian carcinoma had poor outcomes. Numerous dolastatin 10 analogs have been synthesized, with TZT-1027 being particularly non-toxic and antitumor active.

From the Lyngbyamajuscula/Schizothrix sp. assemblage of marine cyanobacteria, a new disulfide dimer called somocystinamide A was discovered. At nM concentrations, it is only marginally damaging the mouse neuro-2a neuroblastoma cells but is exceedingly lethal to caspase-8-containing leukemia cell lines. Studies have shown that it interferes with caspase 8, which has effects on angiogenesis and apoptosis. Nanoparticles have been created to further evaluate the molecule. Eight of the 10 amino acids in depsipeptidecoibamide A have N-methyl groups attached to them, forming a macrocycle. Leptolyngbyasp specimens from Panama yielded the chemical in question. Coibamide A's low GI50 (in the nM range) and unique profile in the NCI (US) 60-cell line assay indicate a novel method of action. The chemical synthesis of this promising lead molecule is underway to pave the way for further in vivo assessments and mechanisms of action research.

Several plant species have been discovered to have their own NPs synthesized from plant extracts. Fourier-transform infrared spectroscopy studies have been performed to better understand the biomolecules that may be involved in the ion-reduction and capping processes. The FT-IR analysis of the synthesized Ag NPs from the *Acacia mearnsii* bark extract showed that compounds with hydroxyl, alkyne, carboxyl, and amide groups were used. These chemicals were found in the monoterpenoids, sesquiterpenes, and phytols. In another investigation of green-synthesized Au NPs utilizing Suaedamonoica leaf extract, it was shown that biomolecules with carboxyl, amine, and hydroxyl functional groups were involved in the reduction of Au ions. Flavonoids, terpenoids, chlorogenic acid, and other isolated compounds have all been well employed in the green production of NPs.

The development of effective plant-based pharmaceuticals for boosting human brain function is hampered by a lack of standardization and the inability to adequately identify essential active components and understand synergy among them. Insect models can be used to study the neural and behavioral effects of individual phytochemicals and complex combinations in a way that is easy, affordable, time-efficient, and morally acceptable. Many terpene and phenolic extracts have been shown to have potential beneficial effects on central nervous system function additionally. The addition of challenges such as herbivore attack, salinity, UV radiation, bacteria, or fungus may be used to upregulate and standardize the quantities of desirable active components in carefully regulated environments. Most investigations of brain function have been conducted in a theoretical in vitro or in vivo context, using parent molecules or chemical amounts that are very implausible to be present in the human brain.

Enhancing the bioavailability of medications by enhancing their solubility, stability, and targeted delivery is largely dependent on drug delivery systems. According to a study by the World Health Organization, traditional medicine meets or helps to meet the basic health needs of about 80% of the people in poor countries. At the moment, scientists are studying the bioactive compounds, chemical makeup, and medicinal potential of different plant species to come up with new active ingredients that have fewer side effects than current molecules. Plants have been known for a long time to have a lot of natural chemicals with medical uses, and they still have a lot to offer when it comes to finding new, more effective drugs. Few drug delivery systems are commonly employed to increase bioavailability.

Lipid-based drug delivery systems, such as liposomes, micelles, and lipid nanoparticles, are effective in improving the bioavailability of poorly soluble drugs. These systems can solubilize lipophilic drugs, protect them from degradation, and facilitate their absorption in the gastrointestinal tract. Lipid-based formulations also offer the advantage of targeted delivery to specific tissues or cells. Polymeric nanoparticles, composed of biocompatible and biodegradable polymers, are widely used for drug delivery. These nanoparticles can encapsulate drugs and protect them from degradation, prolonging their circulation time in the body. Surface modification with ligands or antibodies enables targeted delivery to specific tissues or cells. Polymeric nanoparticles may be engineered to release medications in a regulated way, hence enhancing bioavailability and efficacy for therapy.

Nanocrystals are submicron-sized crystalline particles that enhance drug solubility and dissolution rate. By reducing drug particle size to the nanoscale, nanocrystals exhibit improved bioavailability due to increased surface area for absorption. Nanocrystals can be formulated into suspensions or solid dosage forms for oral or parenteral administration. Microspheres are spherical particles ranging from a few micrometers to a few millimeters in size. They can be made from biodegradable polymers or proteins and can encapsulate drugs. Microspheres provide sustained release of drugs, prolonging their therapeutic effect and reducing the frequency of dosing. They can be administered via various routes, including oral, injectable, or implantable routes.

Prodrugs are inactive or less active forms of drugs that undergo enzymatic or chemical conversion in the body to release the active drug. Prodrug approaches can improve the bioavailability of drugs by enhancing their solubility, stability, and permeability. They can also target specific tissues or cells and enable controlled drug release. Implantable devices, such as drug-eluting stents, implants, or patches, offer sustained and controlled drug release directly at the target site. These systems can provide localized therapy, reduce systemic side effects, and enhance the bioavailability of drugs by maintaining therapeutic drug levels for an extended period. The selection of a specific drug delivery system depends on factors such as drug properties, desired route of administration, target site, and therapeutic objectives. Each system has its advantages and limitations, and the optimal choice varies based on the specific drug and application. Nanotechnology-based drug delivery systems have shown promising potential in increasing the bioavailability of natural compounds and phytochemicals from plant extracts for brain functions. These approaches utilize nanoparticles and other nanoscale carriers to enhance the delivery of therapeutic compounds across the blood-brain barrier (BBB) and improve their effectiveness in targeting specific regions of the brain.

Nanoparticles such as liposomes, polymeric nanoparticles, and solid lipid nanoparticles can be used to encapsulate natural compounds and phytochemicals. These nanoparticles protect the compounds from degradation, improve their solubility, and facilitate their transport across the BBB. The surface of the nanoparticles can be modified with targeting ligands to enhance their specificity towards brain cells or regions. Nano-emulsions are colloidal dispersions of oil and water stabilized by surfactants. They can be used to encapsulate lipophilic natural compounds and phytochemicals, improving their solubility and stability. Nano-emulsions have shown potential in enhancing the delivery of bioactive compounds to the brain due to their small droplet size and ability to bypass the BBB through various transport mechanisms.

Dendrimers are highly branched synthetic polymers that can be functionalized with natural compounds and phytochemicals. They offer a high degree of control over their size, surface charge, and drug loading capacity. Dendrimers have been investigated for targeted delivery of therapeutic compounds to the brain, as their size allows them to cross the BBB, and their surface can be modified for specific interactions with brain cells. Nano-gels are three-dimensional networks of crosslinked polymers that can encapsulate both hydrophilic and hydrophobic natural compounds and phytochemicals. They have a high water content, which provides biocompatibility and helps in the controlled release of encapsulated compounds. Nano-gels can be designed to respond to external stimuli, enabling triggered release of drugs in specific brain regions.

Nanosuspensions are stable colloidal dispersions of nanosized drug particles in a liquid vehicle. They may increase the bioavailability of poorly soluble natural compounds and phytochemicals by improving their solubility and dissolving rate. Nanosuspensions can be administered via different routes, including intravenous injection or intranasal delivery, to target the brain effectively. These nanotechnology-based approaches offer several advantages for enhancing the bioavailability and targeted delivery of natural compounds and phytochemicals to the brain. However, it is important that further research is required to optimize these systems, assess their safety, and evaluate their efficacy in various brain functions and disease conditions. The brain is an important part of the human body, with neurons and neuroglia responsible for transmitting and receiving signals. Microglia and astrocytes are required to act quickly in response to neuronal damage or stress. Neuronal impulses play a crucial role in determining neuroglial activation, which can be compromised by chronically activated neuroglia. Neurodegeneration is a part of neuropathological diseases and the natural aging process of the brain, and it accounts for around 2% of all deaths and 8% of the global death rate.

Severe functional impairment may arise from neuropsychiatric and neurodegenerative conditions such as Parkinson's disease, schizophrenia, Alzheimer's, dementia, depression, cerebrovascular impairment, and seizure disorders. Neuroprotection is the process used to protect neurons in the central nervous system (CNS) against damage caused by both acute and chronic stressors. Stroke and dementia are important sources of individual and family disappointment due to a lack of effective therapy choices, leading to the search for novel drugs and insights into causing neuronal death. Herbal medications with phytochemical elements that may have long-term health-promoting or therapeutic effects have seen increased interest in the past decade. Several herbal plants may be treated for relieving certain health problems for short or long periods of duration, and their therapeutic effects are independent of the plant's nutritional benefit to humans. Vegetables and fruits contain phytochemicals that can help prevent common illnesses such as cardiovascular disease, cancer, and neurological disorders. A diet rich in vegetables and fruits may help protect against disorders related to neural dysfunction.

The public's perception of naturally produced chemicals and extracts as being more secure and acceptable than manufactured chemical goods has risen as a result of the 'Green' movement, leading to a rise in sales of herbal treatments. About 80% of people in low-income regions use phytomedicine as their first line of defense against illness. Herbal remedies contain fatty acids, glycosides, alkaloids, sterols, flavonoids, saponins, terpenes, and tannins. Herbal medicine advocates believe that plant therapeutic value stems obtained by the interplay between its many parts, leading many to believe that herbal remedies are safer and more effective than conventional pharmaceuticals. About a quarter to half of all modern medications are derived from plants.

A few years ago, it was a growing scientific approach for identifying and characterizing novel medicinal plants for treating neurodegenerative illnesses and brain damage caused by stroke. More than 120 different types of Asian traditional medicine are now being employed to treat CNS diseases. Several types of medical plants used in the medical system have been found in India to exhibit promising neuropsychopharmacological activity. *Allium sativum*, *Nicotiana tabacum*, *Bacopa monnierae*, *Ginkgo biloba*, *Centella asiatica*, Celastruspaniculatus, Withaniasomnifera, *Salvia officinalis*, *Ricinus communis*, *Angelica sinensis*, *Huperiza serrata* plants are utilized in traditional Chinese medicine for treatment.

Scientists and the public have paid a lot of attention to the fact that several plants have been shown to protect nerve cells in animal and cell culture models of neurological diseases. Chalcone (safflor yellow B) has been found to protect neurons from ischemic brain damage. The protective effects of phytochemicals in fruits and vegetables against neurodegenerative diseases have been demonstrated in both epidemiological studies and experimental models of these diseases in animals. Hundreds of studies have shown that phytochemicals, such as lycopene, alpha-tocopherol, resveratrol, ginsenosides, and ginkgo biloba, have neuroprotective properties. These antioxidants have been shown to lower oxidative stress and prevent nitric oxide-mediated cell death in cultured hippocampus neurons. Following a few neuroprotective herbs, combined with their main bioactive constituent and how they function.

## 10.6 Flavonoids and Their Role in Brain Function

Flavonoids and other phenolic chemicals have long been studied for their potential to treat and prevent illness in humans. The most important idea is that phytochemicals from medicinal plants have health-promoting properties, such as antioxidant, antibacterial, anticancer, cardioprotective, immune-boosting, anti-inflammatory, and UV-blocking properties. Catechins, a polyphenol, have been shown to have neuroprotective effects by activating protein kinase C (PKC) and also transcription factors. They have been proposed as a means to slow the progression of Alzheimer's disease and provide neuronal protection. Catechins and their metabolites have been demonstrated to activate signaling pathways, which can affect the production of anti-apoptotic and pro-apoptotic proteins, as well as upregulating antioxidant defense, to exert anti-inflammatory and cell-survival activities.

Flavonoids have been shown to regulate nerve cells and forestall neurodegeneration with age. Studies in animals have shown that citrus flavanone tangeretin can protect against the neurodegenerative pathology of Parkinson's disease by preserving nigro-striatal integrity and function after 6-hydroxydopamine-induced lesioning. These extracts can either protect neurons from damage, boost current neuronal function, or stimulate neuronal regeneration. The blood-brain barrier regulates the entrance of xenobiotics into the brain, such as flavonoids. Several dietary anthocyanins, such as pelargonidin-3-glucoside, and cyanidin-3-rutinoside and flavanones, such as hesperidin and naringenin, their in vivo metabolites, have been shown to cross the BBB in relevant in vitro and in situ models. Anthocyanins may be able to cross the monolayer tomentosa used in vitro to simulate the BBB.

The alkaloid family is a large class of organic compounds that includes many different types of compounds. Many bodily processes and behaviors are regulated by nerve cells of the central nervous system (CNS), which may be impacted by alkaloids. The autonomic nervous system, which controls functions including heart rate, blood pressure, and respiration, may also be impacted. The carbon-nitrogen ring is present in the indole alkaloids, psilocybin, ergine, LSD, and other compounds that change consciousness. Ergot alkaloids have been found to have other effects on the body, such as atropine, hyoscyamine, and scopolamine. Opium poppy is a source of the isoquinoline alkaloid morphine, which is a powerful analgesic and addictive medication. Vinpocetine is an extracted form of Vinca minor which acts as a powerful vasodilator. Galantamine is a tertiary alkaloid of the phenanthrene chemical family that can activate nicotinic receptors, improving cognition and memory. All of these alkaloids have been linked to serotonin's neurotransmitter role in the

Tannins are a diverse class of polyphenols produced by plants in response to biotic and abiotic stress. Their molecular weights (500–20,000 Da) and structural variety (phenolic rings and hydroxyl groups) demonstrate their antioxidant and protein-binding characteristics. Fraga-Corral Tannins are soluble in aqueous solutions due to their molecular makeup and high hydroxyl radical content, and they also have a strong propensity to form stable bonds with proteins and carbohydrates. However, their association with polysaccharides reduces the likelihood of interaction and connection with proteins. Tannins are compounds found in many plant species, including barks, seeds, fruit peels, and brown algae. They have been classified in a variety of ways based on molecular weight, characteristics, and origin, but the functional unit classification has gained the most acceptance. Phlorotannins (PT) have been recorded from brown macroalgae, while hydrolyzable tannins (HT), proanthocyanidins (CT), besides complex tannins (CoT), are all present in terrestrial plants. The majority of tannins are found in the earth's soil, with CT being the most frequent kind. Tannins appear to have exactly the same effects, such as antioxidants, antimicrobial activity, or predator deterrence, despite variations in concentration and class in different plant fractions.

Essential oils from plants, such as wormwood, have been traditionally used to treat seizures for more than a century. The rhizome of the plant Valeriana officinalis includes the active ingredients valepotriates and sesquiterpenes (valerenic acid and acetoxyvaleric acid), both of which have therapeutic use. Mice were shown to experience sedation after exposure to the active compounds. Yet research has revealed that a crude valerian extract may bind to GABA (B) receptor and decrease GABA intake in rat synaptosomes. The Chinese plant Huperia serrata is the source of the sesquiterpene alkaloid huperzine A., which has neuroprotective effects. Huperzine-A usage improved spatial working memory and helped to improve learning and memory. The brahminosides (Sakina and Dandiya 1990) and monoterpenes present in *Centella asiatica* L. triterpenoids that have been proven to stimulate and reinforce the neurological system are a-Pinene and α-Terpinene.

The most important therapeutic component of Paeonia lactiflora Pall is paeoniflorin (PF), a bicyclic monoterpene glycoside. Because of its neuroprotective properties, PF may be used to treat brain ischemia, epilepsy, and Parkinson's disease (PD). In vitro cell experiments by Zheng et al. demonstrate that PF protects neurons via a decreasing mechanism involving Bcl-2, Bax, and caspase-3. PF treatment reversed MPTP's effects on dopaminergic neurons in mouse models of Parkinson's disease, boosting the dopamine catabolism and restoring protein levels of striatal and melanotic dopaminergic transporters (DAT) and tyrosine hydroxylase (TH) serrate. The modification of survival signaling pathways for neurons has the potential to be effective in the treatment of central nervous system diseases. Many pathways enhance cell survival in the nervous system and beyond, including ERK1/2 and PI3K. These include cell proliferation, differentiation, survival, and modulation of responses to various growth factors, all of which have lately been connected to the ERK pathway, which is a subfamily of MAPKs. The upstream activator kinase mitogen-activated protein kinase (MEK) phosphorylates threonine and tyrosine residues on ERK1/2 to trigger the kinases. After activation, ERK1/2 relocates to different subcellular sites and phosphorylates a range of downstream substrates, including transcription regulators and protein cytoskeletal.

Several studies have indicated that PI3K and its downstream effector Akt contribute to the neurotrophic effects of polyphenols, which include neuronal survival, enhanced neurite outgrowth, and other advantages. Curcumin (Cur)-mediated neuroprotection against cognitive impairment in rats, according to Hoppe et al., happened through the PI3K/Akt pathway. Scutellarin, which promotes the neurotrophins through pCREB and pAkt signaling but not through MAPKs, has been reported to protect neurons exposed to hypoxia. A phenolic acid derivative, methyl 3,4-dihydroxybenzoate, improved neuronal survival and neurite formation in cultured primary cortical neurons through the PI3K/Akt signaling pathway, but a PI3K-specific inhibitor reversed these benefits. According to one research, oroxylin A activated the Akt pathway in primary cortical neurons, resulting in increased BDNF production and neuronal differentiation. Puerarin activated the ERK1/2 and PI3K/Akt pathways, protecting dopaminergic cells and increasing the impact of NGF on neuritogenesis in PC12 cells. Finally, the natural flavonoid astilbin has been demonstrated to reduce depressive-like behavior in mice models of depression by activating the MAPK/ERK and PI3K/Akt signaling pathways, which are BDNF's downstream signaling pathways.

The Nrf2 pathway is a major transcriptional regulator that binds to ARE in the regulatory domain of many genes, enabling the synthesis of various antioxidants and enzymes for detoxification to be activated. Evidence indicates that Nrf2 is important in maintaining intracellular redox homeostasis and the antioxidant response system since Nrf2-knockout animals have a severe lack of coordinated gene expression regulation and are more susceptible to oxidative damage. In central nervous system (CNS) diseases, this route and the detoxification enzymes it regulates have received special attention. In conclusion, previous research has shown that phytochemicals and natural extracts can have a variety of beneficial effects on human health (Paramita et al. 2022; Lal et al. 2022; Esmeeta et al. 2022; Jothimani et al. 2022; Dhatwalia et al. 2023). Phytochemicals and natural extracts from plants have a wide range of potential benefits for human health, including improving cognitive function, memory, and mood. More research is needed to fully understand the mechanisms by which these compounds work and to determine the optimal dosages and forms of supplementation.

## 10.7 Conclusion

Phytochemicals and natural extracts offer a promising avenue for exploring natural approaches to promote brain health and potentially reduce the risk of cognitive decline. These plant-based compounds demonstrate multifaceted potential, including antioxidant and anti-inflammatory properties, as well as the ability to influence brain cell signaling and growth. While further research is essential to fully understand the mechanisms and long-term effects of specific phytochemicals, it's clear that a diet rich in fruits, vegetables, herbs, and other plant-based sources can play a vital role in supporting cognitive well-being. Combining a phytochemical-rich diet with regular exercise, mental stimulation, and stress management is the most effective way to maintain optimal brain function throughout life.

## References

Dhatwalia J, Kumari A, Chauhan A, Batoo KM, Banerjee A, Radhakrishnan A, Thakur S, Guleria I, Lal S, Ghotekar S, Choi BH (2023) R ubus ellipticus fruits extract-mediated cuprous oxide nanoparticles: in vitro antioxidant, antimicrobial, and toxicity study. Chem Pap 77(3): 1377–1393

Esmeeta A, Adhikary S, Dharshnaa V, Swarnamughi P, Maqsummiya ZU, Banerjee A, Pathak S, Duttaroy AK (2022) Plant-derived bioactive compounds in colon cancer treatment: an updated review. Biomed Pharmacother 153:113384

Hornedo-Ortega R, Cerezo AB, de Pablos RM, Krisa S, Richard T, Garcia-Parrilla MC, Troncoso AM (2018) Phenolic compounds characteristic of the Mediterranean diet in mitigating microglia-mediated Neuroinflammation. Front Cell Neurosci 12:373

Jothimani G, Ganesan H, Pathak S, Banerjee A (2022) Molecular characterization of primary and metastatic colon cancer cells to identify therapeutic targets with natural compounds. Curr Top Med Chem 22(31):2598–2615

Kumar R, Arora R, Sarangi SC, Agarwal A, Kaleekal T, Gupta YK (2021) Pharmacodynamic and pharmacokinetic interactions of hydroalcoholic leaf extract of Centella asiatica with valproate and phenytoin in experimental models of epilepsy in rats. J Ethnopharmacol 270:113784

Lal S, Verma R, Chauhan A, Dhatwalia J, Guleria I, Ghotekar S, Thakur S, Mansi K, Kumar R, Kumari A, Kumar P (2022) Antioxidant, antimicrobial, and photocatalytic activity of green synthesized ZnO-NPs from Myrica esculenta fruits extract. Inorg Chem Commun 141:109518

Mehla J, Gupta P, Pahuja M, Diwan D, Diksha D (2020) Indian medicinal herbs and formulations for Alzheimer's disease, from traditional knowledge to scientific assessment. Brain Sci 10(12): 964. https://doi.org/10.3390/brainsci10120964. Published 2020 Dec 10

Nagy Z, Jobst KA, Esiri MM et al (1996) Hippocampal pathology reflects memory deficit and brain imaging measurements in Alzheimer's disease: clinicopathologic correlations using three sets of pathologic diagnostic criteria. Dementia 7(2):76–81

Paramita P, Sethu SN, Subhapradha N, Ragavan V, Ilangovan R, Balakrishnan A, Srinivasan N, Murugesan R, Moorthi A (2022) Neuro-protective effects of nano-formulated hesperetin in a traumatic brain injury model of Danio rerio. Drug Chem Toxicol 45(2):507–514

Sakina MR, Dandiya PC (1990) A psycho-neuropharmacological profileof *Centellaasiatica* extract. Fitoterapia 61:291–296

Vauzour D (2012) Dietary polyphenols as modulators of brain functions: biological actions and molecular mechanisms underpinning their beneficial effects. Oxidative Med Cell Longev 2012: 914273. https://doi.org/10.1155/2012/914273

# Flavonoids: A Promising Neuroprotectant and Its Salutary Effects on Age-Related Neurodegenerative Disorders

11

Sanjib Kumar Das, Koushik Sen, Tanmay Sanyal, Aloke Saha, and Nithar Ranjan Madhu

**Abstract**

Neurodegenerative disorders have become one of the most serious threats to global civilization in recent years. According to present data and future projections, the number of dementia cases will dramatically rise from 13.5 million cases in 2000 to 21.2 million in 2025 and 36.7 million in 2050. Medical advancements have increased human life expectancy and led to a rise in neurodegenerative disorders, which pose a growing social and economic burden. Unfortunately, the onset or advancement of these diseases cannot be stopped or slowed down by existing medication or therapy. Subsequently, there is an immediate need to develop alternative strategies that can prevent, delay, and even cure neurodegeneration. Recent research to understand the relationship between a healthy lifestyle, nutrition, and diseases indicates that nutritional interventions could be a possible and effective way to protect against age-related neurodegenerative disorders. Emerging evidence has revealed that dietary consumption of flavonoids significantly improves human memory and neuro-cognitive capabilities and inhibits or slows down the neuronal aging process and associated neurodegenerative diseases, including Alzheimer's disease. Flavonoids exert their beneficial effect via their ability to inhibit neuronal apoptosis, enhance

S. K. Das · K. Sen
Department of Zoology, Jhargram Raj College, Jhargram, West Bengal, India

T. Sanyal
Department of Zoology, Krishnagar Government College, Krishnagar, West Bengal, India

A. Saha
Department of Zoology, University of Kalyani, Kalyani, West Bengal, India

N. R. Madhu (✉)
Department of Zoology, Acharya Prafulla Chandra College, New Barrackpore, Kolkata, West Bengal, India



S. Pathak, A. Banerjee (eds.), *Neuroprotective Effects of Phytochemicals in Brain Ageing*, https://doi.org/10.1007/978-981-99-7269-2_11

neuronal function, and induce neuronal regeneration. Flavonoids induce neuronal proliferation and survival by stimulating various cellular signaling processes; it also reduces oxidative stress and promotes long-term potentiation in the brain's hippocampal region. All these neuroprotective actions of flavonoids thus impede the development and progression of neurodegeneration. They could be a promising medical armamentarium to fight against the development and progression of age-related neurodegenerative disorders.

**Keywords**

Flavonoids · Neurodegenerative disorders · Alzheimer's disease · Polyphenols · Dietary supplements · Phytochemicals

## 11.1 Introduction

Aging is a complex biological process and refers to the gradual and progressive deterioration of physiological functions over time, mainly characterized by the reduced capacity for stress management, failure to maintain homeostasis, and increased susceptibility to diseases such as cancer, cardiovascular disease, and neurological disorders (Park et al. 2020). Some changes are part of the normal aging process, while others are debilitating, referred to as pathological, and predisposing individuals to neurodegenerative disorders. As individuals age, changes occur at the cellular and molecular levels within the central nervous system (CNS) that can result in the abnormal aggregation of proteins, neuroinflammation, neuronal death, and decreased cognitive function. These age-associated changes in the brain sometimes lead to age-related neurodegenerative diseases such as cognitive impairment, amyotrophic lateral sclerosis (ALS), Parkinson's disease (PD), and Alzheimer's disease (AD) (Saxena and Caroni 2011; Mattson and Magnus 2006; Farkas and Luiten 2001). With increased life expectancy, the number of people over 60 years is expected to increase from 841 million in 2013 to more than 2 billion in 2050 (Department of Economic and Social Affairs Population Division 2013). As the global population ages, age-related neurodegenerative diseases such as AD and PD have become more common (Reeve et al. 2014; Saxena and Caroni 2011). From ancient times, scientists have been seeking remedies for various diseases by using natural herbals. Medicinal plants and their metabolites have been found to alleviate symptoms of several diseases through various mechanisms.

Flavonoids, a class of naturally occurring plant secondary metabolites, have a rich history of medicinal use for treating diverse ailments. Their ability to bind with a broad range of body proteins, alter transporters, enzymes, hormones, DNA, chelate heavy metals, and scavenge-free radicals positions flavonoids as vital players in drug design and discovery (Ayaz et al. 2019). Numerous research studies demonstrate the potential of flavonoids in managing various health conditions, including diabetes mellitus, cancer, cardiovascular diseases, neurological disorders, inflammation, and microbial diseases. Recent research suggests that flavonoids could have the potential to reverse neurodegenerative pathology and age-related deterioration of neurocognitive performance (Gomez-Pinilla and Nguyen 2012). Flavonoids have shown significant potential in managing neurological health by interacting with intracellular neuronal and glial signaling pathways. They modulate both the peripheral and cerebral vascular systems, offering protection against neuronal damage caused by various neurotoxic molecules. Additionally, flavonoids inhibit key enzymes that form amyloid plaques and other proteins associated with age-related neurodegeneration (Williams and Spencer 2012).

## 11.2 Distribution of Flavonoids in Nature

Flavonoids are plant secondary metabolites and an important component of the plant cell wall; the variety of flavonoids in the plant kingdom is thought to be more than 9000 (Williams and Grayer 2004; Andersen and Markham 2005). According to studies on flavonoids, they are abundant in angiosperms, gymnosperms, pteridophytes, and the rest of the plant kingdom (Hertog et al. 1993). Flavonoids are distinguished from one another by differences in the aglycon ring structure and state of oxidation/reduction. Furthermore, the extent of aglycon hydroxylation, positions of the hydroxyl groups, saturation of the pyran ring, and differences in the derivatization of the hydroxyl groups are major distinguishing features among the various classes of flavonoids. Extensive research regarding the source and structure of flavonoids enables us to determine which flavonoid subgroup is found in a specific plant subgroup. Fruits, juices, vegetables, tea, cereals, and wines are excellent sources of flavonoids (Manach et al. 2004). Flavonoids are a type of natural compound found in many plants. Some of the most commonly studied flavonoids are flavonols like quercetin and kaempferol, flavones such as luteolin and apigenin, and flavanols like epigallocatechin gallate (EGCG), catechin, epicatechin, and epigallocatechin. Other flavonoids include isoflavones in soy products and anthocyanidins in red wine and berries. Flavonoids can also be found in a wide range of foods and drinks like fruits, vegetables, tea, and chocolate. Recent research suggests that some flavonoids have protective effects on the brain, which makes them a promising therapy for neurodegenerative diseases.

## 11.3 Chemistry and Classification of Flavonoids

Flavonoids are natural compounds present in plants and are recognized by their diverse phenolic structures. In the 1930s, a substance extracted from oranges was initially believed to be a new type of vitamin called vitamin P. However, it was later identified as a flavonoid, specifically rutin. Since then, over 4000 types of flavonoids have been identified (Middleton et al. 2000).

The basic chemical structure of flavonoids consists of a 15-carbon skeleton with two benzene rings (A and C) connected by a heterocyclic pyran ring (B). One benzene ring (A) is attached to the pyran ring, while the other (C) is linked as a

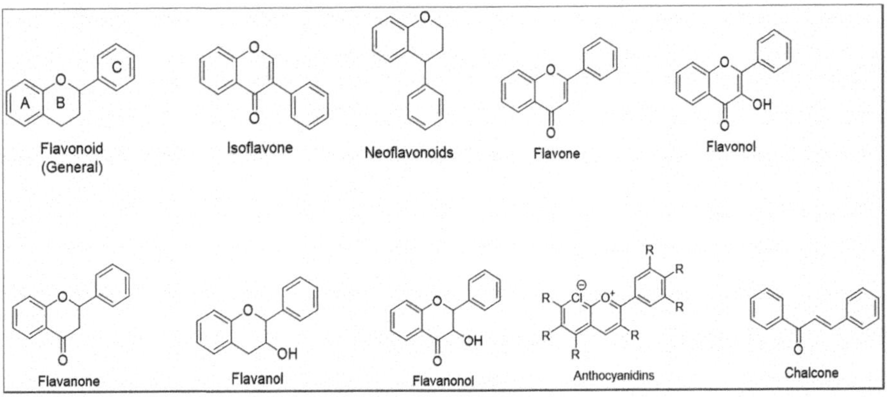


**Fig. 11.1** Basic chemical structure of major types of flavonoids

substituent to the pyran ring. Depending on the substitution pattern of the benzene rings and the substitution, oxidation, and saturation of the pyran ring, several derivatives of flavonoids can be produced. Flavonoids are categorized into different classes based on the differences in their carbon skeleton substitution and activity, such as flavones, flavonols, flavanones, isoflavones, and anthocyanins (Fig. 11.1). Additionally, chalcones, biflavones, coumarins, and aurones are also considered flavonoids (Ayaz et al. 2019).

### 11.3.1 Isoflavones

Isoflavones are a class of flavonoids in which the benzene ring (C) is attached to position 3 of the pyran ring. Isoflavones are majorly found in various natural products, especially soybean and other leguminous plants. Genistein and daidzein are popular isoflavonoids and well-accepted as phytoestrogens due to the estrogenic activity of isoflavones shown in animal models (Joshi et al. 2018); the anti-inflammatory effects of some medicinal plants are due to their richness in isoflavones (Dias et al. 2021).

### 11.3.2 Neoflavonoids

The neoflavonoids are a class of naturally occurring heterocyclic compounds characterized by a benzene ring (C) attached to position 4 of a pyran ring. Neoflavonoids are known for their antidiabetic activity and are composed of neoflavones and neoflavenes. While natural sources such as *Echinops niveus, Dalbergia odorifera, Nepalese propolis*, and *Polygonum perfoliatum* have been reported to contain neoflavonoids, various analogs have also been synthesized by researchers (Ayaz et al. 2019).

### 11.3.3 Flavones and Flavonols

Flavones are an abundant and widely distributed class of flavonoids that are derived from flavanones through dehydrogenation. These compounds are characterized by a 15-carbon skeleton consisting of two benzene rings (A and C) linked by a heterocyclic pyran ring (B). They are further classified based on their substitution pattern. Flavones contain a double bond on the pyran ring between positions 2 and 3 and a carbonyl (ketone) at position 4, and both aromatic rings have hydroxyl substituents. Flavones can exist as both O- and C-glycosides, and the two most common aglycones are apigenin and luteolin (Harborne 1994).

Flavonols, on the other hand, are the alcoholic derivatives of flavones and differ from flavones at position 3 of the pyran ring, where they have a hydroxyl group (Williams and Grayer 2004). These compounds are often referred to as 3-hydroxyflavones and are frequently synthesized. Quercetin is the most frequently found flavonol in nature, occurring in both aglycone and *O*-glycoside forms. Rutin, the most widespread flavonol glycoside in the plant kingdom, undergoes O-glycosylation at C-3 (Harborne 1994; Middleton et al. 2000; Williams and Grayer 2004).

### 11.3.4 Flavanones

Flavanones, commonly known as dihydroflavones, are a significant class of flavonoids with a unique saturated structure (Harborne 1994). The key difference between flavones and flavanones is that the latter lacks a double bond between positions 2 and 3 in the pyran ring. Citrus fruits such as grapefruits, oranges, mandarins, limes, and lemons are abundant sources of flavanones. Grapefruits, for instance, contain naringin, naringenin, and naringenin 7-*O*-neohesperidoside, whereas oranges, mandarins, limes, and lemons contain hesperidin, hesperetin, and hesperetin 7-*O*-rutinoside. In addition, lemons are also known for their high content of eriocitrin, eriodictyol, and diosmin (Manach et al. 2004).

### 11.3.5 Flavanols and Flavanonols

Flavanols are a group of flavonoids characterized by a saturated pyran ring with hydroxyl groups at positions 3 and 4 and a benzene ring attached at position 2. Due to the stereochemistry of the pyran ring, four possible diastereomers of flavanols differ in the orientation of the hydroxyl groups at positions 2, 3, and 4 relative to each other. Flavanols are widely present in the tea plant, Camellia sinensis, as epigallocatechin gallate, epicatechin gallate, epigallocatechin, and epicatechin. Various fruits, including apples, red grapes, peaches, mangoes, pears, plums, nectarines, and raspberries, also contain flavanols such as catechin, epicatechin, and epigallocatechin (Arts and Hollman 2005). Flavanonols, or dihydroflavonols, are another type of saturated pyran ring flavonoids with hydroxyl groups at positions 3 and 4 (Harborne 1994).

### 11.3.6 Anthocyanins

Anthocyanin pigments are responsible for flowers' and fruits' beautiful and lively colors. They are commonly found in the outer cell layers of berries like blueberries, cranberries, blackberries, strawberries, raspberries, and bilberries, as well as in red and merlot grapes, plums, pomegranates, and red wine (Gürler et al. 2020; Joshi et al. 2018). The coloring of these fruits is achieved through natural processes such as methylation or acylation of the hydroxyl groups on the A and B rings of anthocyanin. These unique flavonoids exist as ionizable chloride salts, known as 2-phenylchromenylium (flavylium) cation salt derivatives. In addition, nature has its collection of vibrant pigments known as Anthocyanidins, which are highly oxidized 2-aryl-3-hydroxychromenylium compounds. This group contains aurantinidin, capensinidin, cyaniding, delphinidin, europinidin, hirsutidin, malvidin, pelargonidin, peonidin, petunidin, pulchellidin, and rosinidin. All of them are different from each other based on the attached groups. The colors produced by the most common Anthocyanidin derivatives are quite remarkable. For instance, cyanidin generates shades of red to magenta, delphinidin creates a range of enchanting purple to blue hues, and pelargonidin imparts colors that span from orange to pink.

The introduction of a sugar moiety further amplifies the vividness of these colors. While glucose is the most commonly found sugar derivative with β-linkage that enhances color brightness, other derivatives such as galactose, rhamnose, and xylose have also been found to exhibit this attribute (Dias et al. 2021).

## 11.4 The Blood-Brain Barrier and Neuronal Availability of Flavonoids

The blood-brain barrier (BBB) is a critical component that protects the brain and distinguishes itself from other vascular barriers in both its physical properties and the type and number of transporters it contains. It is formed by interacting with highly specialized choroid plexus epithelial cells and brain microvessel endothelial cells. This intricate and dynamic system employs biomechanical and biochemical signals to communicate between the vascular system and the brain. The BBB is composed of three biological constituents: pericytes (PCs), astrocyte end-feet, and endothelial cells (Ballabh et al. 2004; Hawkins et al. 2006; Obermeier et al. 2013; Wong et al. 2013). Each cell type contributes vitally to the integrity of the BBB, disruption of which can result in or be a factor for developing the neurological disorder (Obermeier et al. 2013). The BBB protects the brain from harmful toxins and microorganisms, but it also inhibits around 95% of pharmaceuticals administered orally or intravenously. To treat their patients for neurodegenerative ailments like

Parkinson's, medical professionals frequently deliver medications through brain injections (Akinmoladun et al. 2019).

To enter the brain, flavonoids must first be required to traverse the BBB that governs the entry of foreign substances into the CNS. Growing evidence from numerous research indicates that flavonoids and some of their metabolites can effectively permeate the BBB (Youdim et al. 2004). The key difficulty with flavonoids' bioavailability to the brain is how they are carried over the BBB, whether they are conjugated or unconjugated. Although there have been observations of transporter-mediated and lipophilicity-dependent uptake of flavonoids, the exact mechanisms and extent of flavonoid penetration into the CNS are not yet fully understood. The lipophilicity level of flavonoids is a critical factor in their BBB penetration, with less polar O-methylated metabolites having a greater potential for brain uptake than more polar flavonoid glucuronides. However, it is possible for certain glucuronides to cross the BBB through a specific uptake mechanism in vivo (Aasmundstad et al. 1995). The ability of flavonoids to enter the brain may also be influenced by their interaction with specific efflux transporters, such as P-glycoprotein, expressed in the BBB. This transporter appears to be responsible for the differences in the flux of naringenin and quercetin into the brain in situ that it modulates their brain entry (Mitsunaga et al. 2000; Youdim et al. 2004; Weber et al. 2005). To improve flavonoids' water solubility in order to boost their bioavailability, several strategies have been developed, including methylation, lecithin complexation, cyclodextrin complexation, micro-emulsions, polyvinylpyrrolidone dispersion, and integration of flavonoids to borneol/methanol eutectic mixtures. Prenylation helps flavonoid tissue bioaccumulation by altering cell uptake and efflux, even though it does not directly increase bioavailability (Thilakarathna and Vasantha Rupasinghe 2013; Koirala et al. 2016).

An experimental study using the human brain endothelial cell line (HBMEC) found that, amongst three flavonoids tested [quercetin, EGCG, and cyanidin-3-glucoside (C3G)], EGCG is capable of crossing the BBB more rapidly than C3G. In contrast, quercetin failed to do so. Another study demonstrated that quercetin can cross the BBB when administered with α-tocopherol (Ferri et al. 2015). Although the exact brain distribution of flavonoids has not been extensively studied, several studies have found anthocyanins in different regions of rat and pig central nervous system following berry intervention, and O-methylated flavanone, tangerine in the hypothalamus, striatum, and hippocampus (Spencer 2009a, b). When radiolabeled epicatechin (EC) was administered to individuals, high levels of absorption and metabolism were observed, and the level of EC in the urine and plasma was undetectable. Recent studies have found that certain flavonoid metabolites, including quercetin-3-*O*-glucuronide (Q3GA), can cross the blood-brain barrier and accumulate in specific cells like the choroid plexus epithelial cells and macrophages. Moreover, after short-term and acute feeding with flavanols like (−) epicatechin and (+) catechin and their methylation and glucuronidated metabolites, they were also detected in the brain. These findings suggest that flavonoids and their metabolites may have the potential to exert beneficial effects on brain function (Schroeter et al. 2006; van Praag et al. 2014; Ottaviani et al. 2016). These findings demonstrate that flavonoids can be beneficial in treating neuronal disorders as they can penetrate the BBB and reach the site of damage to exert their therapeutic effects.

## 11.5 Role of Flavonoids in Neuroprotection

### 11.5.1 Anti-neuroinflammatory Effect of Flavonoids

Neuroinflammation refers to the immune response of the CNS, and it can be either acute or chronic. Acute neuroinflammation generally protects the nervous tissue, while chronic neuroinflammation has detrimental effects against peril signals. The acute neuroinflammatory response is attributed to increased rostaglandins (PGs) in the inflamed neuronal region (Ricciotti and Fitzgerald 2011). Conversely, chronic neuroinflammatory response leads to architectural and functional anomalies characterized by increased rostaglandin E2 (PGE2) (Zhang et al. 2009). The immune responses in the brain and spinal cord are mediated by two resident cell types: microglia and astrocytes. The microenvironment of the brain is monitored by ramified microglia, which are the macrophages of the CNS. These microglia employ different immune receptors, such as Toll-like receptors (TLRs), nucleotide-binding oligomerization domain-like receptors (NLRs), purinergic receptors (P2R), and receptors for advanced glycation end-products (RAGEs), expressed on their long protrusions to detect both foreign and self-harmful signals (Ransohoff 2016).

In the healthy brain, microglia display a 'deactivated' phenotype and produce anti-inflammatory and neurotrophic substances that may be crucial for maintaining tissue homeostasis through interactions with astrocytes and neurons, similar to the homeostatic functions of "alternatively activated" macrophages in other tissues. Microglia, when receptors detect danger or new signals of pathogen invasion or tissue injury, change to an 'activated' phenotype, which promotes an inflammatory response that further stimulates the immune system and commences the process of tissue repair. Such receptor-mediated activation of microglia is marked by histopathological alterations, leading to the overproduction of nitric oxide (NO) via upregulated inducible nitric oxide synthase (iNOS) and reactive oxygen species (ROS) generation, resulting in neuronal toxicity. Microgliosis followed by secretion of chemokines, pro-inflammatory cytokines such as interleukin (IL)-1b, IL-8, Tumor necrosis factor (TNF) $\alpha$, and matrix metalloproteinases (MMP) 2, 3, 9, suggests the microglial role in immune response modulation (Glass et al. 2010). Excessive or sustained microglia activation culminates into acute and chronic neuroinflammatory responses in the CNS, which can be corroborated with neurodegeneration in neurological disorders.

Astrocytes play crucial roles in the CNS, encompassing physical and metabolic support to neurons, assistance in neuronal migration, regulation of synaptic transmission, facilitation of synapse formation and synaptic plasticity, and maintenance of the BBB (Khakh and Sofroniew 2015). When activated, microglia release inflammatory factors such as cytokines, chemokines, and PGE2, which can stimulate astrocyte proliferation (Colombo and Farina 2016; Zhang et al. 2009). Finally, inflammatory signals get amplified in astrocytes via their autocrine secretion of pro-inflammatory factors (Liddelow and Barres 2015). Thus, several anti-inflammatory mechanisms, including inhibition of microglial activation, NO production, ROS generation, and astrocyte proliferation, can be the art of war against chronic neuroinflammation to retard or cease neurodegeneration (Amor et al. 2010; Farooqui 2010).

In vitro and in vivo studies with flavonoid-rich natural products over the last two decades unveiled the neuroprotective role of flavonoids through anti-inflammatory activity. Several flavonoids have been found to bring down astrogliosis, microgliosis, and the levels of pro-inflammatory mediators via down-regulating Nuclear Factor-kappa B (NF-κB) activation in different parts of CNS, thereby preventing neuroinflammation (Jaeger et al. 2018). In lipopolysaccharide (LPS)-stimulated brain-derived microglial cells, flavonoids such as Quercetin, Blueberry extract, and Nobiletin exhibit anti-inflammatory effects by suppressing the production of inflammatory mediators such as iNOS and cyclooxygenase-2 (COX-2) via inhibition of inhibitor of kappa B kinase (IKK), NF-κB and activator protein-1 (AP-1). They also decrease the level of NO, TNF-α, IL-1β, extracellular signal-regulated kinases (ERK), c-Jun NH(2)-terminal kinase (JNK), and p38 mitogen-activated protein kinases (MAPKs) and ROS (Chen et al. 2005; Lau et al. 2007; Cui et al. 2010). Catechin has been observed to exhibit a protective effect on murine microglia N9 cells against *tert*-butyl hydroperoxide-induced cell death. This protective effect is mediated by suppressing NF-kB, p53 activity, and ERK activation. Biochanin A and Naringenin have been shown to protect dopaminergic neurons against LPS-induced damage and cellular mortality by inhibiting microgliosis and reducing superoxide, TNF-α, NO, p38 signaling cascades, and signal transducer and activator of transcription (STAT)-1 transcription factor (Huang et al. 2005; Chen et al. 2007; Vafeiadou et al. 2009).

Naringenin activates the nuclear factor E2-related factor 2 (Nrf2), antioxidant response element (ARE) signaling pathway, and rapamycin complex-1 to suppress microgliosis in a murine model with neurotoxicity (Kim et al. 2016a, b; Lou et al. 2014). Treatment with luteolin reduces icrogliosis, which can be attributed to a compromised microglial transcriptome for genes involved in inflammation (Dirscherl et al. 2010; Chen et al. 2008).

Baicalein, another flavonoid, downregulates TNF-α and IL-1β in the substantia nigra and striatum of the 1-methyl-4-phenyl-1,2,3,6-tetrahydropyridine (MPTP) treated PD mice (Xue et al. 2014). Baicalein and apigenin can improve symptoms of Parkinson's disease in rats by reducing inflammation, promoting neuronal survival, and inhibiting cathepsin B production (Anusha et al. 2017; Hung et al. 2016). Apigenin can also inhibit IL-1β and caspase-1 and block the activation of inflammasomes via upregulation of peroxisome proliferator-activated receptor (PPAR)-γ, which are involved in the development of neurodegenerative diseases like Alzheimer's (Li et al. 2016). Experimental evidence suggests that during an inflammatory response, transcription factors NF-κB and AP-1, along with signaling molecules ERK1/2, p38, and JNK, regulate the expression of inflammatory genes through the MAPK and NF-kB signaling pathways. Activated microglia are recognized to express inflammatory markers, and studies have shown that flavonoids can effectively inhibit the activation of NF-kB, AP-1, and MAPK pathways, along with the expression of these markers. This suggests flavonoids can potentially prevent neuroinflammation (Chang and Karin 2001; Chen et al. 2005; Jang et al. 2008).

### 11.5.2 Anti-oxidative Effect of Flavonoids

Oxidative stress refers to the generation and accumulation of ROS, causing severe damage to important cellular components, which eventually jeopardizes cell viability. Many substances, including heavy metals, xenobiotics, herbicides, ultraviolet (UV) irradiation, and some other external factors, can markedly increase the production of ROS (Sinha et al. 2013). Cellular death due to oxidative stress could be associated with the human aging process (Orr and Sohal 1994) and might also be associated with age-related neurodegenerative disorders like AD, PD, and ALS (Halliwell and Gutteridge 2015). Baicalein, a flavonoid in Huangqin, protects against neuronal loss in PD mice by reducing oxidative stress, restoring dopamine (DA) and serotonin levels, and reducing inflammation in the brain. Baicalein has been shown to improve mitochondrial function, decrease ROS and lipid peroxidation, and protect against excitotoxicity in response to stroke in cellular studies. Additionally, Baicalein administration leads to increased production of mitochondrial DJ-1 protein and decreased mitochondrial lipid peroxidation. In astrocytes, Baicalein inhibits ROS production by preventing the degradation of the glutamine synthetase (GS) protein, which promotes glutamate disposal and protects against neuronal injury (Lee et al. 2014; Song et al. 2020).

Quercetin and luteolin and their metabolites are effective antioxidants tested in pheochromocytoma (PC12) neuronal cells. Luteolin has been shown to modulate stress response pathways in PC12 neuronal cells. In laboratory animal with PD, quercetin administration protects the striatum of the central nervous system from oxidative stress and prevent dopaminergic neuronal loss and significantly increases the function of sodium-potassium adenosine triphosphatase (Na+, K+-ATPase), glutathione peroxidase (GPx), superoxide dismutase (SOD), acetylcholinesterase (AchE), and DA content. Quercetin also contributes to neuroprotection against apoptosis by upregulating mitochondrial complex-I activity and increasing levels of catalase and SOD activity (Pavlica and Gebhardt 2010; Haleagrahara et al. 2011; Karuppagounder et al. 2013). According to research, kaempferol can improve behavior and decrease neuronal loss in the substantia nigra pars compacta (SNpc) of mice with Parkinson's disease by reducing oxidative stress. This is achieved by restoring the activity of key enzymes such as SOD and GPx, increasing dopamine levels in the striatum, and reducing levels of malondialdehyde in the substantia nigra (Fig. 11.2). Kaempferol promotes lipophagy, preventing lipid deposition and peroxidation that can lead to neuronal death due to mitochondrial damage in neurodegenerative diseases. Furthermore, Kaempferol protects SH-SY5Y cells and primary neurons from acute toxicity caused by rotenone that enhances mitochondrial

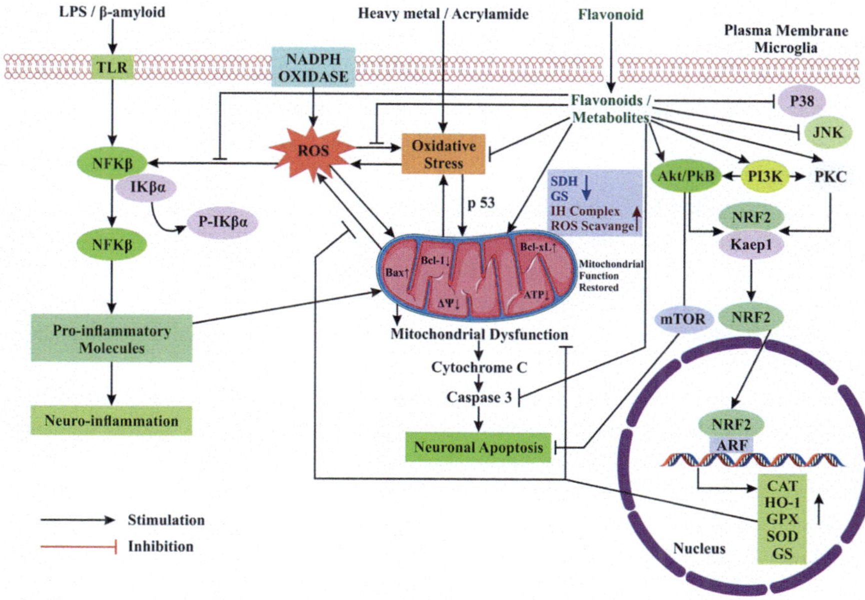


**Fig. 11.2** Anti-oxidative effects of flavonoids in neurodegenerative diseases

turnover through autophagy-mediated via antioxidative and antiapoptotic effects (Li and Pu 2011; Filomeni et al. 2012; Han et al. 2021).

In the 6-hydroxydopamine (6-OHDA) induced PD rat model, rutin protects dopaminergic neurons by improving antioxidant levels, increasing dopamine content and dopamine receptor expression, and enhancing neurobehavioral activity (Islam et al. 2012). Rutin pre-treatment also rescued PC-12 cells from cytotoxicity by activating the antioxidative enzyme profile (Magalingam et al. 2013). Nanoemulsion of rutin has an ameliorative effect against colistin-induced oxidative stress (Çelik et al. 2020). Hesperidin (Hsd) treatment in aged 6-OHDA induced PD mice showed significant protection against oxidative stress, with elevated DA content and improved neurobehavioral parameters (Antunes et al. 2014). In a *Drosophila* model of PD, Hsd reduced iron-induced oxidative stress in CNS, restored DA content and cholinergic activity, and improved motor activity (Poetini et al. 2018).

Phloretin and phlorizin, both flavonoids, protected human SH-SY5Y neuroblastoma cells from rotenone-induced toxicity by decreasing intracellular ROS levels, improving antioxidant enzymes, suppressing caspase-3 activity, preventing DNA damage, and normalizing mitochondrial transmembrane potential (Fig. 11.2) (Barreca et al. 2017). Similarly, quercetin, isoquercitrin, and afzelin extracted from *Actinidia okamotoanum* showed antioxidative properties by increasing cell survivability and reducing ROS production and LDH release. They also attenuated apoptosis and inflammation by suppressing caspase, iNOs, and COX-2 activation in the same cell line (Kim et al. 2019). Mulberry leaf-derived flavonoids, morachalcone D and E, were studied for their ability to protect HT22 cells from glutamate and erastin-induced oxidative damage. Morachalcone D was found to be more effective in preventing iron accumulation, ROS generation, and glutathione breakdown, along with inducing the expression of genes related to antioxidant mechanisms (Wen et al. 2020).

Morin, a bioflavonoid, has antioxidative, anti-inflammatory, and antiapoptotic properties. It protects against Ifosfamide-induced neurotoxicity in rats by inhibiting glutathione depletion and lipid peroxidation, increasing the activity of antioxidant enzymes, and suppressing the synthesis of various neurotoxic factors such as carbonic anhydrase, brain-derived neurotrophic factor (BDNF), glial fibrillary acidic protein (GFAP), Nrf-2, and expression of JNK, which are associated with neurotoxicity in rats (Çelik et al. 2020; Issac et al. 2021; Mohammadi et al. 2021). Silymarin, biochanin a, and dihydromyricetin have also shown protective roles against neuronal degeneration in rat models via various pathways (Haridevamuthu et al. 2022). Additionally, puerarin and blueberry anthocyanins attenuated oxidative stress-related neuronal mortality in rats induced by cadmium and acrylamide (Wen et al. 2021; Fang et al. 2022).

### 11.5.3 Impact on Memory and Cognition

Several experimental studies have found that flavonoid-rich foods are associated with better cognition and learning ability. The most commonly reported flavonoid-induced cognitive benefit is improving spatial learning or memory in elderly or otherwise impaired animals. Flavonoid-rich extracts from blueberries have been shown to improve spatial reference memory (Casadesus et al. 2004; Williams et al. 2008), spatial working memory (Rendeiro et al. 2013), and object recognition memory in elderly animals (Goyarzu et al. 2013). Chronic supplementation with blackberry (Shukitt-Hale et al. 2013), acai (Carey et al. 2015), or flavan-3-ols from green tea has also improved spatial working memory. Flavonoid supplementation with mulberry or green tea has improved avoidance learning in senescence-accelerated mice (Figueira et al. 2017). Flavonoids such as EGCG, epicatechin, and anthocyanins also improve the brain's cognitive function. In cell culture models, EGCG and other abundant catechins in the tea plant (Camellia sinensis) and grapevine (Vitis vinifera) have been shown to enhance cognitive function in Alzheimer's disease. Intragastrical administration of EGCG significantly reduced cognitive decline in normal aging rats and plaque formation in their brain (Wei et al. 2019). Epidemiological studies demonstrated that in healthy people, consumption of EGCG and catechin-rich green tea reduces cognitive aging, lowers the risk of dementia, and improves cognitive function (Cascella et al. 2017; Ide et al. 2014; Williams et al. 2010).

Bioflavonoid hesperidin found in citrus fruits also has neuroprotective and cogno-protective effects, as evident from both animal studies and clinical trials (Hajialyani et al. 2019). Administration of hesperidin ameliorates learning and memory impairments in AD mouse model and chemically induced mouse model of dementia (Wang et al. 2014a, b, c; Justin Thenmozhi et al. 2016). In humans, daily consumption of hesperidin-rich fruits sustains cognitive abilities in healthy older adults (Kean et al. 2015) and also reduces the incidence of dementia in elderly people (Zhang et al. 2016, 2017). Habitual chocolate eating, rich in cocoa flavanols, also augmented cognitive functions and lowered the risk of dementia (Crichton et al. 2013; Sokolov et al. 2013; Dominguez and Barbagallo 2018). Isoflavones, including genistein and daidzein, rich in soybean, have also been linked with better cognition (Zaw et al. 2017).

Experiments using a mouse model of AD reported that quercetin appears to be a promising learning and memory booster agent, particularly when administered as nanoparticles. Quercetin has been shown to enhance both short-term and long-term memory in mice that were subjected to chronic stress. In rodent studies, quercetin has also demonstrated the ability to improve spatial learning and memory in cases where impairments were induced by D-galactose administration or streptozotocin and enhance working memory (Chang et al. 2014; Chen et al. 2015; Nabavi et al. 2015). Luteolin, a flavonoid found in various plants, optimized spatial memory function in diabetic or obese rodents with cognitive decline. Additional evidence supports luteolin's beneficial effects in rodents with cognitive decline caused by scopolamine, streptozotocin, aging, excitotoxicity (kainic acid-KA), and chronic cerebral hypoperfusion. All of these studies indicate that luteolin has a positive effect on learning and memory functions, indicating that this compound may influence a variety of neurotransmitters in brain areas associated with cognition and behavior. Naringenin improved spatial learning and memory function in young rats and has also been shown to prevent cognitive decline in diabetic rats. Apigenin may also protect rats from the negative effects of Alzheimer's disease (transgenic model), isoflurane, or streptozotocin exposure. Some other compounds, such as liquiritigenin, phloretin, and acacetin astragalin, have improved memory and cognitive abilities in rodents and fruitflies (Bakoyiannis et al. 2019).

Flavonoids may exert beneficial effects on memory, cognition, and learning abilities via modulation of the cellular and molecular framework associated with the processes of memory. Flavonoids can traverse the blood-brain barrier and activate diverse signaling pathways involving receptors, kinases, and transcription factors. Although the precise loci of interaction are yet to be fully elucidated, experimental investigations have demonstrated that flavonoids can enhance memory and cognitive processes through diverse mechanisms, including ATP site binding to receptors and enzymes, modulation of kinase activity such as MAPK and/or MAPK kinase, modulation of phosphatase function, maintenance of $Ca^{2+}$ homeostasis, and inhibition of $Ca^{2+}$-dependent kinase activation in neurons, thereby modulating downstream signaling cascades. Flavonoid-binding sites within neurons may include adenosine, GABA, testosterone receptors, and a brain plasma membrane binding site specifically for polyphenols (Spencer et al. 2009).

Long-term potentiation (LTP) is a persistent enhancement of synaptic strength following high-frequency stimulation of a neuronal synapse. It is widely recognized as a major mechanism for information acquisition and storage in the brain, contributing to synaptic plasticity (Bliss and Collingridge 1993). LTP is characterized by an increase in the chemical strength of a synapse that can last from minutes to several days. This process is believed to underlie memory formation by increasing the strength of connections, or synapses, between two neurons (Malenka and Bear 2004). The activation of various pathways, including cAMP-dependent protein kinases such as protein kinase A, protein kinase B (PKB) or Akt, protein kinase C (PKC), calcium-calmodulin kinase IV, and ERK, contributes to LTP. These pathways converge to activate the cAMP-response element-binding protein (CREB), which initiates the expression of genes associated with synapse remodeling and synaptic plasticity. CREB also upregulates the actions of neurotrophins, such as BDNF, and enhances neuronal survival, differentiation, and synaptic function. BDNF also modifies the NMDA receptor composition in the hippocampus, with an upregulation in the NR2B subunit levels. The NR2B induces modification of NMDA receptor activation kinetics, leading to an increase in their ability to exhibit LTP (Rendeiro et al. 2015). The ingestion of flavonoids has been shown to activate the phosphoinositide 3-kinase (PI3K)/Akt and mammalian target of rapamycin (mTOR) pathways. Flavonoids can bind to the Tropomyosin receptor kinase B (TrkB) receptor and modulate PI3K activity by interacting with its ATP binding site. Certain flavonoids, such as quercetin, inhibit PI3K, while others, like hesperidin, activate Akt/PKB and inhibit pro-apoptotic proteins. Flavonoids promote Akt activation in hippocampal neurons, which triggers the activation of the mTOR pathway and increased translation of specific mRNA subpopulations, including Arc/Arg3.1. Increased Arc expression can facilitate LTP and synaptic strength and ultimately enhance cognitive ability (Spencer 2009a, b).

## 11.6 Therapeutic Potential of Flavonoids in Age-Associated Neurodegenerative Disorders

Flavonoids, a heterogeneous collection of bioactive chemicals derived from plants, have great potential for addressing agerelated neurodegenerative diseases. Their therapeutic efficacy arises from their capacity to combat oxidative stress and inflammation, both of which are significant factors in brain harm. Flavonoids can safeguard neurons by regulating cell signalingpathways, maintaining mitochondrial function, and preventing the accumulation of detrimental protein clumps such as amyloidbeta and tau, which are characteristic features of Alzheimer's disease. Studies indicate that flavonoids have the potential to not only decelerate the advancement of diseases but also improve cognitive function and memory. Flavonoids, due to their diverse advantages and proven safety, provide a natural and possibly effective method to maintain brain health and lower the likelihood of age-related neurodegeneration (Minocha et al. 2022).

### 11.6.1 Therapeutic Potential of Flavonoids in AD

Alzheimer's disease, primarily described by Alois Alzheimer almost a century ago, is an age-related neurodegenerative disorder characterized by gradual memory loss (dementia), progressive intellectual learning, and cognitive impairment, along with various behavioral and neuropsychiatric disturbances. The pathological hallmarks of AD are an accumulation of extracellular amyloid plaques (Aβ) and intracellular neurofibrillary tangles (NFTs) in the brain (Ayaz et al. 2019; Glass et al. 2010).

The amyloid precursor protein (APP) is a transmembrane protein that undergoes processing into the ER and is subsequently transported to the cell surface. Upon reaching the cell surface, processing of APP by α-secretase results in the production of the α-C-terminal fragment (α-CTF) and secretory APP (sAPPα), which follow the non-amyloidogenic pathway and do not generate amyloid-beta (Aβ). Alternatively, the pro-amyloidogenic pathway involves the cleavage of APP by β-secretase (BACE1) into β-CTF and sAPPβ, with β-CTF being further processed by γ-secretase to release Aβ. These intricate mechanisms are pivotal in understanding the pathogenesis of AD (Querfurth and LaFerla 2010). Self-aggregation of Aβ leads to oligomerization and subsequent production of amyloid plaques. Aβ monomers and oligomers promote microglial activation and proliferation, followed by subsequent secretion of pro-inflammatory cytokines. These pro-inflammatory cytokines cause hyperphosphorylation of axonal tau protein and result in faulty localization at the somatodendritic region due to phosphorylated-tau dissociation from the axonal microtubule. Hyperphosphorylated tau can now self-aggregate into toxic oligomers, generating intracellular neurofibrillary tangles (NFT). Incorrect localization of hyperphosphorylated tau to post-synaptic terminals along with Aβ leads to synaptic dysfunction and ultimately results into synaptic loss. In a nutshell, the whole mechanism can be threaded into the amyloid cascade hypothesis, according to which extracellular Aβ aggregation leads to neuroinflammation through microglial activation, followed by tau-induced toxicity that results in learning and memory impairment and dementia in AD due to synaptic dysfunction and neuronal loss (Hardy and Selkoe 2002).

Flavonoids stimulate non-amyloidogenesis, inhibit amyloidogenesis and Aβ aggregation, modulate tauopathy, inhibit (AchE), exert antioxidative response either via scavenging free radicals or preventing ROS production, interacts with different signaling pathways to modulate neuroinflammation and neuronal apoptosis, and promotes neurogenesis (Fig. 11.3). All these positive roles of flavonoids establish these phytochemicals as a potential therapeutic agent against AD. In the following section, we are discussing the therapeutic efficacy of different flavonoids in AD.

In the cerebral cortex of the AD mice model, apigenin treatment blocked the amyloidogenic process, inhibited oxidative stress, and upregulated the ERK/CREB/BDNF pathway (Zhao et al. 2013). Anthocyanins are also effective against ADs. Studies with HT22 and SH-SY5Y cells show that anthocyanin and anthocyanin-loaded nanoparticles reduce neuronal death from Aβ1–42 by scavenging free radicals, inhibiting ROS production, and stimulating p38-MAPK/JNK pathways. They also induce HO-1 and Nrf2, restoring mitochondrial membrane potential and

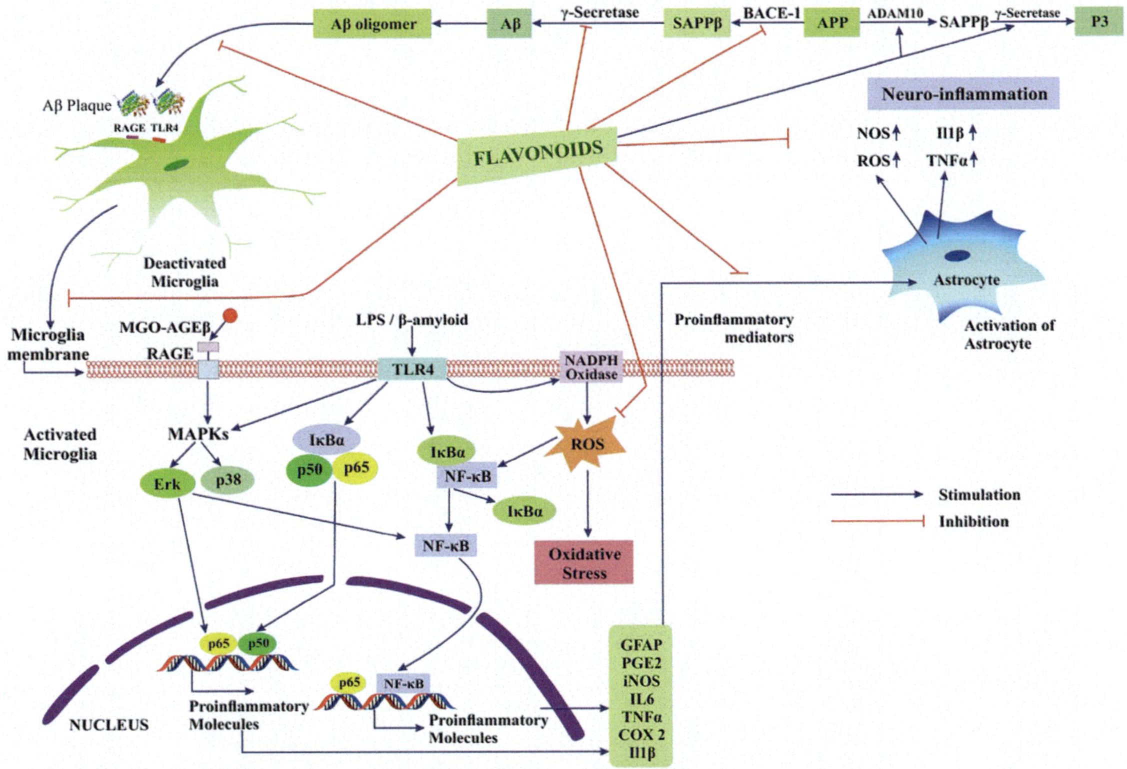


**Fig. 11.3** Anti-neuroinflammatory effects of flavonoids in neurodegenerative diseases with special reference to AD

intracellular $Ca^{2+}$ balance. Anthocyanins also reduce neuroinflammation, major Alzheimer's markers, and neuroapoptotic markers (Badshah et al. 2015; Amin et al. 2017; Ali et al. 2018), suggesting its neuroprotective role in AD. Both in vivo and in vitro investigations revealed that anthocyanins could modulate phosphorylated-phosphatidylinositol 3-kinase-Akt-glycogen synthase kinase 3 beta (p-PI3K/Akt/GSK3β) pathways. These modulated pathways counteract the effects of amyloid beta oligomer (AβO)-induced elevations in ROS levels and reduce oxidative stress (Ali et al. 2018). Treatment with anthocyanins in mice injected with LPS reduces GFAP and Iba-1 expression, indicating a decrease in astrogliosis and microgliosis. Anthocyanins also decrease p-JNK and inhibit inflammatory mediators such as IL-1β, TNF-α, and NF-kB, suggesting their potential to counteract LPS-induced ROS-mediated neuroinflammation (Fig. 11.3) (Khan et al. 2016).

In laboratory rats with AD, Baicalein improved behavior by enhancing histopathological and metabolic conditions. In a laboratory study, EGCG has shown anti-amyloidogenic properties when tested against both Aβ 42 and Aβ 42 fibrillation induced by aluminum (Al (III)). Nano EGCG transformed the insoluble Aβ 42 mature oligomeric fibrils into soluble Aβ 42 amorphous aggregates. This transformation resulted in reduced cellular toxicity. Additionally, EGCG demonstrated significant antioxidant and metal chelation potentiality (Singh et al. 2018a, b). EGCG prevents the phosphorylation of Tau proteins by binding to its phosphorylation region, masking it for kinase, and also restrains Tau from achieving proper conformation, preventing it from generating aggregates (Guéroux et al. 2017). EGCG prevented apoptosis, reduced oxidative stress, decreased TNF-α protein expression level, and improved working memory and cognition in a morphine-induced rat model with memory impairment (Saffar et al. 2020a, b). Hesperidin has anti-oxidative and anti-neuroinflammatory effects through Akt/Nrf2 signaling activation and RAGE/NF-κB signaling pathway inhibition, improving AD mouse models' learning and memory (Hong and An 2018). It reduces APP expression and Aβ peptide deposition, suppresses microglial activation, attenuates neuroinflammation, and improves non-cognitive nesting ability and social interaction deficit (Li et al. 2015).

Hesperidin improved cognitive performance in an AD mice model by reducing oxidative stress and mitochondrial dysfunction via suppressed Glycogen synthase kinase-3 activity. Hesperidin also prevented excessive extracellular glutamate release and neuronal death caused by KA-induced excitotoxicity in the hippocampus (Wang et al. 2014a, b, c; Chang et al. 2015). Phytochemical naringin increased memory in the unstressed and stressed Swiss albino mice model. In stressed mice, naringin reduced plasma corticosterone levels in addition to diminished AChE activity and iNOS, which is common with unstressed mice (Maratha and Mahadevan 2012). It also showed neuroprotective potentiality in neurodegenerative diseases by averting episodic memory deficiency (Ramalingayya et al. 2016). Naringin attenuated gp120-induced learning and memory dysfunction in rats (Qin et al. 2016). Naringenin inhibited Aβ1–42-induced microglial activation and polarization with upregulation of Aβ degradation enzymes, preventing neuroinflammation-induced neuronal death (Yang et al. 2019). In silico studies indicate that naringenin can prevent ROS production, maintain mitochondrial membrane potential, and lower toxic Aβ levels in neurons (Ahsan et al. 2020). Naringenin pretreatment res cued AD rat model from Aβ-induced learning and memory disability (Ghofrani et al. 2015). Naringenin improved spatial recognition in a dose-dependent manner and also reduced neuroinflammation by preventing oxidative stress, decreasing the activity of AChE, increasing Nrf2, and lowering the levels of NF-κB, TNFα, COX2, and iNOS (Khajevand-Khazaei et al. 2018).

Nobiletin, found in citrus fruit, exhibits neuroprotective properties and can serve as a potential preventive measure against dementia. Its administration in mice subjected to olfactory bulbectomy (OBX) surgery leads to a notable improvement in the levels of key enzymes, such as acetylcholinesterase and choline acetyltransferase, within the brain's hippocampal region. Nobiletin's neuroprotective ability following OBX surgery is attributed to the activation of the ERK / MAPK pathway, which triggers the phosphorylation of CREB protein and leads to the transcription of genes important for brain function and growth. This, in turn, stimulates the growth of new neurites from neurons in the brain (Nagase et al. 2005; Nakajima et al. 2007). Nobiletin can reverse the suppression of CREB phosphorylation in cultured hippocampal neurons and protect against memory impairment in AD model rats (Matsuzaki et al. 2006). In a separate study, nobiletin showed neuroprotective action against Aβ-treated mice with cognitive impairment via anti-amnesic effect and antiapoptotic activity that sharply downregulated the expression of Bax and split caspase-3 and upregulated the expression of Bcl-2 and Bcl-2/Bax in the cortex and hippocampus. Furthermore, nobiletin was reported to alleviate synaptic dysfunctions and neuronal loss in LPS-stimulated mice model with memory deficit and neuroinflammation by preventing microglial activation by modulating MAPKs, PI3K/AKT, and NF-κB signaling pathways, suppressing the secretion of proinflammatory cytokine and maintaining cellular oxidative balance (Fig. 11.3).

Quercetin, a pentahydroxyflavone, was found to have significant therapeutic effects in AD. It reduced Aβ plaque burden and oxidative stress while restoring mitochondrial function by activating AMP-activated protein kinase (AMPK) activity in the AD mouse model (Wang et al. 2014a, b, c). Additionally, Quercetin has been shown to improve memory and emotional behavior in mice by ameliorating tauopathy, extracellular β-amyloidosis, astrogliosis, and microgliosis in the amygdala and hippocampus. Furthermore, it decreased BACE1-mediated cleavage of APP, paired helical filament (PHF), Aβ1–40, and Aβ1–42 levels (Sabogal-Guáqueta et al. 2015). These findings suggest that Quercetin may be a promising therapeutic agent for treating AD.

In an AD rat model, rutin administration improves memory by reducing neurotoxicity mediated via the MAPK pathway, which led to increased levels of ERK1, cAMP response element-binding protein (CREB), and BDNF in the hippocampus. These findings suggest that rutin therapy may be a potential treatment option for improving memory in Alzheimer's disease (Moghbelinejad et al. 2014). Both in vitro and in vivo studies have shown that 7,8-Dihydroxyflavone (7,8-DHF), when administered orally or intraperitoneally to 5XFAD mice (mice with five familial Alzheimer's disease mutations), can downregulate the β-secretase enzyme (BACE1). This downregulation leads to lower levels of the β-secretase-cleaved C-terminal fragment of amyloid precursor protein (C99) and reduced production of Aβ40 and Aβ42, which are associated with Alzheimer's disease. As a result, this reduction in Aβ production has been observed to improve memory deficits associated with Alzheimer's disease (Devi and Ohno 2012).

Additionally, 7,8-DHF has been reported to prevent the loss of hippocampal synapses, restore synapse number and synaptic plasticity, and improve memory loss through the BDNF/TrkB signaling pathway (Zhang et al. 2014). A similar result has been reported when silibinin administration significantly reduced anxiety/depression-like behavior in an Aβ-induced rat model via ameliorating neuronal damage in the hippocampus by inhibiting autophagy through enhancing the activity of BDNF/TrkB pathway (Song et al. 2017). Silibinin treatment suppressed apoptosis via inhibition of activation of caspase-3 by suppressing Jun N-terminal kinase phosphorylation and the downstream hippocampal Bax/Bcl-2 ratio in APP/PS1 mice. It also inhibited oxidative stress via reducing malondialdehyde levels and increased glutathione and superoxide dismutase activity. Synaptophysin and PSD95 expression increase in association with memory improvement (Bai et al. 2017). Silibinin was found to attenuate decreased phosphorylation level of JNK and increased expression of ERα and Erβ following up-regulation of PI3K/Akt and MAPKs pathways in the hippocampus of Aβ1–42-treated AD rat model. It ameliorated learning and memory impairment in overtrained rats through reduced ROS generation, delayed cell senescence, and minimized neuronal loss by antiapoptosis (Liu et al. 2019). Silibinin also exerted neuroprotection by attenuating neuronal damage, tau hyperphosphorylation (ser404), and apoptosis in the hippocampus and cerebral cortex of streptozotocin (STZ)-treated rat model and upregulated insulin signaling pathway, suggesting its prospect in treating AD patients with type 2 diabetes mellitus (Liu et al. 2020).

### 11.6.2 Therapeutic Potential of Flavonoids in Parkinson's Disease

The prevalence of Parkinson's disease, the second-leading neurological condition after AD, is rising globally. Two key characteristics of PD are the development of intracellular Lewy bodies (inclusions containing α-synuclein) and Lewy neurites, mainly in the substantia nigra (SN) of the midbrain, and the chronic and gradual death of dopaminergic neurons. According to a recent study, aberrant iron buildup in the SNpc, reactive microglia, in association with melanin-containing dopamine (DA) neurons causes inert α-synuclein to assemble into lethal aggregates (Zhou et al. 2018). DA content decreases, and DA nerve activity is compromised due to SN lesions (Meng-Zhen et al. 2022). This upsets the equilibrium between dopaminergic neurons and cholinergic neurons. As a result, there may be motor symptoms like rigidity, resting tremors, bradykinesia, postural instability, and gait impairment, and these traits can be related to dopaminergic cell loss and the resulting malfunction of the basal ganglia, a cluster of deep nuclei that engage in the initiation and execution of movements. Nonmotor-related symptoms include neuropsychiatric manifestations (depression, dementia, and hallucination), cognitive deficits, olfactory deficits, autonomic dysfunction, and sleep disorders, and all these features are probably caused by the pathology spreading past the basal ganglia while oxidative stress is still present (Glass et al. 2010; Antunes et al. 2014). Studies have shown that cellular oxidative stress, including mitochondrial activities, neuroinflammation, autophagy, and dopamine homeostasis, plays a key role in PD, even though the exact origins of SN lesions and the ensuing damage to dopaminergic neurons are still unknown. The pharmaceutical industry and researchers have recently focused their attention on flavonoids due to their excellent antioxidative properties and other therapeutic qualities since natural antioxidants are thought to lower the incidence of PD (Han et al. 2021; Meng-Zhen et al. 2022).

Baicalein's efficacy in treating intracerebral hemorrhage-induced brain injury in rat models is mainly due to its ability to reduce apoptosis, oxidative stress, and neuroinflammation (Wei et al. 2017). Particularly, its oxidized forms, at lower concentrations, not only inhibit the synthesis of α-synuclein fibrils but are also able to disaggregate existing α-synuclein fibrils in a soluble oligomer of α-synuclein, in which the protein molecules have been covalently changed by baicalein quinone to create a Schiff base with a lysine side chain, that becomes the main byproduct of the inhibitory reaction (Zhu et al. 2004).

Baicalein can reduce the formation of fibrilization caused by a specific genetic mutation (E46K) in the α-synuclein protein, which is associated with PD. This effect depends on the concentration of Baicalein. Baicalein also prevents the harmful effects of E46K on two important cellular processes: proteasome inhibition and mitochondrial depolarization. Moreover, Baicalein can protect cells from Parkinson's disease by reversing mitochondrial dysfunction and reducing the production of ROS (Jiang et al. 2010; Wang et al. 2013). Baicalein also reduces monoamine oxidase-B (MAO-B) and upregulates tyrosine hydroxylase (TH), which improves motor function and protects against neurotoxicity. In mice models, Baicalein prevents MPTP-induced build-up of iron and lipid peroxides. In SH-SY5Y cells treated with MPP+, Baicalein reduces malondialdehyde and ROS levels while increasing glutathione (GSH) levels, showing an antioxidative impact (Song et al. 2021). Icaritin improves symptoms of PD in a mouse model induced by MPTP by attenuating NOD-like receptor family pyrin domain-containing protein 3 (NLRP3) inflammasome and the secretion of IL-1β. Icaritin also improves mitochondrial function by stabilizing important mitochondrial proteins, such as voltage-dependent anion channel (VDAC) and ATP synthase function, increasing levels of ATP, ADP, and antioxidants (Wu et al. 2021).

Kaempferol protects neurons against rotenone-induced toxicity by enhancing autophagy-mediated mitochondrial turnover, stabilizing the striatal glutamatergic response, and exhibiting anti-apoptotic and antioxidative effects. In a mouse model of PD induced by MPTP/p, Kaempferol reduces behavioral deficits by promoting autophagy, mainly autophagy-mediated degradation of lipid droplets, which prevents lipid peroxidation-related stress and alleviates mitochondrial damage in the substantia nigra pars compacta (Filomeni et al. 2012; Han et al. 2021).

Kaempferol treatment modulates the expressions of several proteins, including p-Akt, p-GSK-3β, Nrf-2, p-NF-κB, TNF-α, IL-1β, and IL-6, which suggests an immune-modulatory, anti-inflammatory, anti-oxidative, and anti-apoptotic role. This treatment improves neurons' morphology, arrangement, and distribution in Sprague Dawley rats with cerebral ischemia-reperfusion injury (Wang et al. 2020).

Tilianin pretreatment in a dopaminergic neuron PD cell model improved cell viability by inhibiting the MAPK signaling pathway, apoptosis, oxidative stress, and inflammatory cytokine secretion, ameliorating dopaminergic neurons (Li and Xu 2022). In a rotenone-induced PD rat model, hibiscetin markedly improved endogenous antioxidant levels, mitigating nitrative and oxidative stress, neuroinflammation, and behavioral abnormalities; it also ameliorated metabolic abnormalities and restored dopamine levels (Alzarea et al. 2022).

Pretreatment with fisetin suppressed several inflammatory and apoptotic pathways in PD cells, rescuing dopaminergic neurons from cytotoxicity-related death by decreasing α-synuclein expression and improving mitochondrial membrane potential, ion transporter activities, programmed cell death, and neuroinflammation in aged rat brains. In traumatic brain injury (TBI) models, fisetin ameliorated cerebral edema, brain lesion, and BBB disruption while improving neurological function via activation of the Nrf2-ARE pathway (Patel et al. 2012; Singh et al. 2018a, b; Zhang et al. 2018). Silibinin treatment has been shown to improve motor behavior and minimize dopamine depletion in aging animals, suggesting a neuroprotective role. In addition, it attenuated MPP+-induced neurotoxicity and increased the expression of proinflammatory molecules in the SN, indicating an ameliorative effect on the degeneration of the nigrostriatal dopaminergic pathway in PD. Similarly, hesperidin administration improved behavioral parameters and memory impairment in a 6-OHDA-induced PD mouse model through its anti-oxidative activity. Furthermore, it augmented synaptogenesis in the cortex by promoting cortical astrocyte's ability via higher expression of TGF-β1. Naringin has also been found to promote GDNF production in DA neurons and prevent degeneration of DA neurons in MPP+-induced neurotoxicity in the SN of PD rat brains, with anti-inflammatory effects (Lu et al. 2010; Jung et al. 2014; Antunes et al. 2014; Leem et al. 2014; Matias et al. 2017).

Similarly, naringenin treatment ameliorated learning ability and memory retention in an isoflurane-induced neurodegenerative model. It prevented neuroapoptosis via modulating PI3/Akt/PTEN and NF-kB signaling pathways (Hua et al. 2016) and can protect DA from oxidative stress and glutamate-induced excitotoxicity by activating Nrf2 expression for increased astroglial neurotrophic effect (Wang et al. 2019). Naringenin treatment in SH-SY5Y cells resulted in increased cell survival, enhanced mitochondrial membrane potential, greater cellular ATP levels, and decreased oxidative stress. In vivo studies showed that naringenin provided significant neuroprotection against various impairments, such as oxidative stress, mitochondrial dysfunction, and astrocytosis, in paraquat-induced rats with neurotoxicity and neurodegeneration. Naringenin administration also significantly altered the mRNA expression of specific genes such as SNCA, LRRK2, β-catenin, and caspase-3 associated with neurotoxicity and neurodegeneration (Ahmad et al. 2021).

Myricetin has been demonstrated to have antioxidative and anti-apoptotic effects by suppressing ROS production, restoring mitochondrial transmembrane potential, improving the Bcl-2/Bax ratio, lowering caspase-3 activation, and suppressing MKK4 and JNK activation in MPP+-treated MES23.5 cells (Zhang et al. 2011). Additionally, myricitrin was found to inhibit the production of advanced glycation end products, control RAGE expression and NF-kB activation and prevent mitochondrial dysfunction and promote cell survival through DJ-1 antioxidation in SH-SY5Y and SN4741 cells (Wang et al. 2014a, b, c; Cai et al. 2015). Myricitrin also protects against 6-OHDA-induced neurodegeneration by retaining TH enzyme activity and mTORC1 activation in nigral DA neurons in mice and preventing the expression of tumor necrosis factor in activated microglia (Kim et al. 2016a, b).

EGCG prevented iron buildup in the SNpc of the MPTP-induced neurodegenerative model of PD, revived movement behavior and shielded TH+ cells via modulating the peripheral immune response, and decreased Il-6 and TNF-α levels (Mandel et al. 2004; Zhou et al. 2018). EGCG reduced cell mortality induced by 6-OHDA and boosted the proliferation of neuronal cells by counteracting the drop in STAT3 activity in SH-SY5Y cells (Wang et al. 2009). EGCG promoted neurogenesis and synaptic plasticity, exerted protection against mitochondrial oxidative stress-induced neuronal apoptosis, and regulated the expression of antioxidant enzymes CAT and GST in response to Cd exposure (Ding et al. 2017; Nan et al. 2018; Halder et al. 2018). Quercetin (QN) alone and in combination with piperine reduced 6-OHDA-induced rotating behavior and motor impairments in rats (Haleagrahara et al. 2013).

Tangeretin has neuroprotective and anti-inflammatory properties, preserving striatonigral integrity and improving dopamine content in 6-OHDA lesioned rat model with PD, reducing NO, TNF-α, IL-6, and IL-1 in microglia, and inhibiting ROS production. It also activates sirtuin 1 (SIRT1) and adenosine monophosphate-activated protein kinase (AMPK), inhibits NF-kB, and increases the expression of heme oxygenase-1 (HO-1) and tissue inhibitor of metalloproteinase-2 while inhibiting MMP-3 and MMP-8 (Datla et al. 2001; Singh and Kumar 2018).

### 11.6.3 Therapeutic Potential of Flavonoids in Amyotrophic Lateral Sclerosis

Amyotrophic lateral sclerosis (ALS), a globally fatal neurodegenerative disease, is characterized by muscle wasting and weakness, increased spasticity, and hyperreflexia, ultimately leading to paralysis and death within 2–5 years of diagnosis, mainly caused by the loss of motor neurons that control voluntary muscle movements (Glass et al. 2010; Riancho et al. 2019). Neuroinflammation, characterized by overexpression of ROS, COX2, IL-1b, TNF-α, and IL-6, is observed in both ALS mouse models and human patients (McGeer and McGeer 2002). Familial ALS, which accounts for only 10% of total cases, is caused by heritable gene mutations, whereas sporadic ALS develops due to complex gene-gene and gene-environment interactions (Glass et al. 2010; Lutz 2018).

Few investigations have been made to unveil the effectiveness of flavonoids in the ALS model (Maher 2019). In the ALS mice model, EGCG treatment prolonged the onset of symptoms and increased life span through improved motor neuronal count by preserving survival signals and attenuating death signals via caspase-3 cleavage. It also reduces NOS and NFκB levels, which were found to be associated with decreased microglial activation in the spinal cord, indicating a potential therapeutic role of EGCG in amyotrophic lateral sclerosis (ALS) by targeting neuroinflammation (Xu et al. 2006; Koh et al. 2006). Intraperitoneal administration of 7,8-DHF to the SOD1-G93A mice model was also found to ameliorate total motor neuron count, dendritic spine density on motor neurons, and motor performance (Korkmaz et al. 2014). Orally administered Fiestin exhibited antioxidative and neuroprotective effects in multiple mutant SOD1 models of ALS through ERK and HO-1 activation, with extended lifespan (Wang et al. 2018). Together, all these reports suggested that few flavonoids have therapeutic potential in ALS treatment.

## 11.7 Conclusion

Existing research in the field of neurodegenerative disorders suggests the need for a potential therapeutic target to restrain age-related neurodegeneration. Regular consumption of flavonoid-enriched food, such as berries and cocoa, may potentially restrict or cure age-related dwindling in cognition and memory and also restrict the onset of dementia. The mechanisms by which flavonoids modulate memory and cognition are still hidden behind the curtain. However, significant evidence suggests that flavonoids stimulate salutary effects in cognitive outcomes in both aging and healthy animal models and humans through both direct and indirect actions. The beneficial effect of flavonoids to protect against neurodegeneration is mainly attributed to their antioxidant and anti-inflammatory properties, which can alleviate oxidative stress and inflammation, two key culprits of neurodegeneration.

Despite promising preclinical data and encouraging results, several flavonoids, including resveratrol, epicatechin, and quercetin, have shown neuroprotective effects in preclinical studies on AD, PD, and ALS. These flavonoids are also known to mitigate neurodegenerative pathophysiology, such as amyloid-beta plaques, tau hyperphosphorylation, neuroinflammation, and oxidative stress. However, clinical trials on flavonoids are currently limited and often yield conflicting results. While some research findings suggest a positive relationship between flavonoid consumption and cognitive performance, others find no such correlation. Future research should prioritize understanding the mechanisms of action of different flavonoids and their interactions with the gut microbiome. Recent studies indicate a connection between the gut-brain axis and neurodegeneration.

Further research in this field, particularly clinical trials, is necessary to validate the beneficial impact of flavonoids in neurodegenerative disorders. Additionally, attributes such as bioavailability, absorption, metabolism, potential toxicities, and accumulation sites must be taken into account for future preclinical and clinical trials. Furthermore, exploring modified flavonoid molecules and innovative delivery systems, such as nano or microencapsulation, is crucial to evaluating their role and effectiveness as therapeutics in neurological disorders. In summary, while flavonoids show promise as a pivotal molecule with potential elixir-like qualities for enhancing memory and alleviating age-related neurodegeneration, it is imperative that future investigations, including comprehensive clinical trials, are undertaken to strengthen our understanding in this field.

## References

Aasmundstad TA, Mørland J, Paulsen RE (1995) Distribution of morphine 6-glucuronide and morphine across the blood-brain barrier in awake, freely moving rats investigated by in vivo microdialysis sampling. J Pharmacol Exp Ther 275(1):435. https://bpspubs.onlinelibrary.wiley.com/doi/pdfdirect/10.1038/sj.bjp.0704406

Ahmad MH, Fatima M, Ali M, Rizvi MA, Mondal AC (2021) Naringenin alleviates paraquat-induced dopaminergic neuronal loss in SH-SY5Y cells and a rat model of Parkinson's disease. Neuropharmacology 201:108831. https://doi.org/10.1016/j.neuropharm.2021.108831

Ahsan AU, Sharma VL, Wani A, Chopra M (2020) Naringenin upregulates AMPK-mediated autophagy to rescue neuronal cells from β-amyloid (1–42) evoked neurotoxicity. Mol Neurobiol 57(8):3589–3602. https://doi.org/10.1007/s12035-020-01969-4

Akinmoladun AC, Farombi T, Farombi EO (2019) Food for brain health: flavonoids. In: Elsevier eBooks, pp 370–386. https://doi.org/10.1016/b978-0-08-100596-5.21752-6

Ali T, Kim T, Rehman SU, Khan MS, Amin FU, Khan M et al (2018) Natural dietary supplementation of anthocyanins via PI3K/Akt/Nrf2/HO-1 pathways mitigate oxidative stress, neurodegeneration, and memory impairment in a mouse model of Alzheimer's disease. Mol Neurobiol 55:6076–6093. https://doi.org/10.1007/s12035-017-0798-6

Alzarea SI, Afzal M, Alharbi KS, Alzarea AI, Alenezi SK, Alshammari MS, Alquraini A, Kazmi I (2022) Hibiscetin attenuates oxidative, nitrative stress and neuroinflammation via suppression of TNF-α signaling in rotenone induced parkinsonism in rats. Saudi Pharm J 30(12):1710–1717. https://doi.org/10.1016/j.jsps.2022.09.016

Amin FU, Shah SA, Badshah H, Khan M, Kim MO (2017) Anthocyanins encapsulated by PLGA@PEG nanoparticles potentially improved its free radical scavenging capabilities via p38/JNK pathway against Aβ 1–42-induced oxidative stress. J Nanobiotechnol 15:1–16. https://doi.org/10.1186/s12951-016-0227-4

Amor S, Puentes F, Baker D, Van Der Valk P (2010) Inflammation in neurodegenerative diseases. Immunology 129(2):154–169. https://doi.org/10.1111/J.1365-2567.2009.03225.X

Andersen OM, Markham KR (Eds) (2005) Flavonoids: chemistry, biochemistry and applications. CRC Press. https://doi.org/10.1201/9781420039443

Antunes MS, Goes ATR, Boeira SP, Prigol M, Jesse CR (2014) Protective effect of hesperidin in a model of Parkinson's disease induced by 6-hydroxydopamine in aged mice. Nutrition 30(11–12):1415–1422. https://doi.org/10.1016/J.NUT.2014.03.024

Anusha C, Sumathi T, Joseph LD (2017) Protective role of apigenin on rotenone induced rat model of Parkinson's disease: suppression of neuroinflammation and oxidative stress mediated apoptosis. Chem Biol Interact 269:67–79. https://doi.org/10.1016/J.CBI.2017.03.016

Arts ICW, Hollman PCH (2005) Polyphenols and disease risk in epidemiologic studies. Am J Clin Nutr 81(1 suppl):317S. https://doi.org/10.1093/AJCN/81.1.317S

Ayaz M, Sadiq A, Junaid M, Ullah F, Ovais M, Ullah I, Ahmed J, Shahid M (2019) Flavonoids as prospective neuroprotectants and their therapeutic propensity in aging associated neurological disorders. Front Aging Neurosci 11(JUN):155–155. https://doi.org/10.3389/FNAGI.2019.00155

Badshah H, Kim TH, Kim MO (2015) Protective effects of anthocyanins against amyloid beta-induced neurotoxicity in vivo and in vitro. Neurochem Int 80:51–59. https://doi.org/10.1016/j.neuint.2014.10.009

Bai D, Jin G, Yin S, Zou D, Zhu Q, Yang Z et al (2017) Antioxidative and anti-apoptotic roles of silibinin in reversing learning and memory deficits in APP/PS1 mice. Neurochem Res 42(12): 3439–3445. https://doi.org/10.1007/s11064-017-2389-3

Bakoyiannis I, Daskalopoulou A, Pergialiotis V, Perrea D (2019) Phytochemicals and cognitive health: are flavonoids doing the trick? Biomed Pharmacother 109:1488–1497. https://doi.org/10.1016/J.BIOPHA.2018.10.086

Ballabh P, Braun A, Nedergaard M (2004) The blood-brain barrier: an overview: structure, regulation, and clinical implications. Neurobiol Dis 16(1):1–13. https://doi.org/10.1016/J.NBD.2003.12.016

Barreca D, Currò M, Bellocco E, Ficarra S, Laganà G, Tellone E, Laura Giunta M, Visalli G, Caccamo D, Galtieri A, Ientile R (2017) Neuroprotective effects of phloretin and its glycosylated derivative on rotenone-induced toxicity in human SH-SY5Y neuronal-like cells. Biofactors 43(4):549–557. https://doi.org/10.1002/BIOF.1358

Bliss TVP, Collingridge GL (1993) A synaptic model of memory: long-term potentiation in the hippocampus. Nature 361(6407):31–39. https://doi.org/10.1038/361031A0

Cai Z, Zeng W, Tao K, Lu F, Gao G, Yang Q (2015) Myricitrin alleviates MPP+-induced mitochondrial dysfunction in a DJ-1-dependent manner in SN4741 cells. Biochem Biophys Res Commun 458(2):227–233. https://doi.org/10.1016/j.bbrc.2015.01.060

Carey AN, Miller MG, Fisher DR, Bielinski DF, Gilman CK, Poulose SM, Shukitt-Hale B (2015) Dietary supplementation with the polyphenol-rich açaí pulps (Euterpe oleracea Mart. and Euterpe precatoria Mart.) improves cognition in aged rats and attenuates inflammatory signaling in BV-2 microglial cells. Nutr Neurosci 20(4):238–245. https://doi.org/10.1080/1028415X.2015.1115213

Casadesus G, Shukitt-Hale B, Stellwagen HM, Zhu X, Lee HG, Smith MA, Joseph JA (2004) Modulation of hippocampal plasticity and cognitive behavior by short-term blueberry supplementation in aged rats. Nutr Neurosci 7(5–6):309–316. https://doi.org/10.1080/10284150400020482

Cascella M, Bimonte S, Muzio MR, Schiavone V, Cuomo A (2017) The efficacy of Epigallocatechin-3-gallate (green tea) in the treatment of Alzheimer's disease: an overview of pre-clinical studies and translational perspectives in clinical practice. Infect Agents Cancer 12(1):1–7. https://doi.org/10.1186/S13027-017-0145-6/TABLES/2

Çelik H, Kandemir FM, Caglayan C, Özdemir S, Çomaklı S, Kucukler S, Yardım A (2020) Neuroprotective effect of rutin against colistin-induced oxidative stress, inflammation and apoptosis in rat brain associated with the CREB/BDNF expressions. Mol Biol Rep 47(3): 2023–2034. https://doi.org/10.1007/S11033-020-05302-Z/METRICS

Chang L, Karin M (2001) Mammalian MAP kinase signalling cascades. Nature 410(6824):37–40. https://doi.org/10.1038/35065000

Chang HC, Yang YR, Wang PS, Wang RY (2014) Quercetin enhances exercise-mediated neuroprotective effects in brain ischemic rats. Med Sci Sports Exerc 46(10):1908–1916. https://doi.org/10.1249/MSS.0000000000000310

Chang CY, Lin TY, Lu CW, Huang SK, Wang YC, Chou SSP, Wang SJ (2015) Hesperidin inhibits glutamate release and exerts neuroprotection against excitotoxicity induced by kainic acid in the hippocampus of rats. Neurotoxicology 50:157–169. https://doi.org/10.1016/j.neuro.2015.08.014

Chen JC, Ho FM, Chao PDL, Chen CP, Jeng KCG, Hsu HB, Lee ST, Wen TW, Lin WW (2005) Inhibition of iNOS gene expression by quercetin is mediated by the inhibition of IκB kinase, nuclear factor-kappa B and STAT1, and depends on heme oxygenase-1 induction in mouse BV-2 microglia. Eur J Pharmacol 521(1–3):9–20. https://doi.org/10.1016/J.EJPHAR.2005.08.005

Chen HQ, Jin ZY, Li GH (2007) Biochanin A protects dopaminergic neurons against lipopolysaccharide-induced damage through inhibition of microglia activation and proinflammatory factors generation. Neurosci Lett 417(2):112–117. https://doi.org/10.1016/J.NEULET.2006.11.045

Chen HQ, Jin ZY, Wang XJ, Xu XM, Deng L, Zhao JW (2008) Luteolin protects dopaminergic neurons from inflammation-induced injury through inhibition of microglial activation. Neurosci Lett 448(2):175–179. https://doi.org/10.1016/J.NEULET.2008.10.046

Chen FY, Cao LF, Wan HX, Zhang MY, Cai JY, Shen LJ, Zhong JH, Zhong H (2015) Quercetin enhances adriamycin cytotoxicity through induction of apoptosis and regulation of mitogen-activated protein kinase/extracellular signal-regulated kinase/c-Jun N-terminal kinase signaling in multidrug-resistant leukemia K562 cells. Mol Med Rep 11(1):341–348. https://doi.org/10.3892/MMR.2014.2734

Colombo E, Farina C (2016) Astrocytes: key regulators of neuroinflammation. Trends Immunol 37(9):608–620. https://doi.org/10.1016/J.IT.2016.06.006

Crichton GE, Bryan J, Murphy KJ (2013) Dietary antioxidants, cognitive function and dementia - a systematic review. Plant Foods Hum Nutr 68(3):279–292. https://doi.org/10.1007/S11130-013-0370-0/METRICS

Cui Y, Wu J, Jung SC, Park DB, Maeng YH, Hong JY, Kim SJ, Lee SR, Kim SJ, Kim SJ, Eun SY (2010) Anti-neuroinflammatory activity of nobiletin on suppression of microglial activation. Biol Pharm Bull 33(11):1814–1821. https://doi.org/10.1248/BPB.33.1814

Datla KP, Christidou M, Widmer WW, Rooprai HK, Dexter DT (2001) Tissue distribution and neuroprotective effects of citrus flavonoid tangeretin in a rat model of Parkinson's disease. Neuroreport 12(17):3871–3875

Department of Economic and Social Affairs Population Division (2013) World population ageing 2013. United Nations, New York

Devi L, Ohno M (2012) 7, 8-dihydroxyflavone, a small-molecule TrkB agonist, reverses memory deficits and BACE1 elevation in a mouse model of Alzheimer's disease. Neuropsychopharmacology 37(2):434–444. https://doi.org/10.1038/npp.2011.191

Dias MC, Pinto DCGA, Silva AMS (2021) Plant flavonoids: chemical characteristics and biological activity. Molecules 26(17):5377. https://doi.org/10.3390/MOLECULES26175377

Ding ML, Ma H, Man YG, Lv HY (2017) Protective effects of a green tea polyphenol, epigallocatechin-3-gallate, against sevoflurane-induced neuronal apoptosis involve regulation of CREB/BDNF/TrkB and PI3K/Akt/mTOR signalling pathways in neonatal mice. Can J Physiol Pharmacol 95(12):1396–1405. https://doi.org/10.1139/cjpp-2016-0333

Dirscherl K, Karlstetter M, Ebert S, Kraus D, Hlawatsch J, Walczak Y, Moehle C, Fuchshofer R, Langmann T (2010) Luteolin triggers global changes in the microglial transcriptome leading to a unique anti-inflammatory and neuroprotective phenotype. J Neuroinflammation 7(1):1–16. https://doi.org/10.1186/1742-2094-7-3/FIGURES/7

Dominguez LJ, Barbagallo M (2018) Nutritional prevention of cognitive decline and dementia. Acta Biomed 89(2):276. https://doi.org/10.23750/ABM.V89I2.7401

Fang Z, Luo Y, Ma C, Dong L, Chen F (2022) Blueberry anthocyanins extract attenuates acrylamide-induced oxidative stress and neuroinflammation in rats. Oxid Med Cell Longev 2022:7340881. https://doi.org/10.1155/2022/7340881

Farkas E, Luiten PGM (2001) Cerebral microvascular pathology in aging and Alzheimer's disease. Prog Neurobiol 64(6):575–611. https://doi.org/10.1016/S0301-0082(00)00068-X

Farooqui AA (2010) Neurochemical aspects of neurodegenerative diseases. In: Neurochemical aspects of neurotraumatic and neurodegenerative diseases. Springer, New York, pp 249–324. https://doi.org/10.1007/978-1-4419-6652-0_8

Ferri P, Angelino D, Gennari L, Benedetti S, Ambrogini P, Del Grande P, Ninfali P (2015) Enhancement of flavonoid ability to cross the blood–brain barrier of rats by co-administration with α-tocopherol. Food Funct 6(2):394–400. https://doi.org/10.1039/C4FO00817K

Figueira I, Garcia G, Pimpão RC, Terrasso AP, Costa I, Almeida AF, Tavares L, Pais TF, Pinto P, Ventura MR, Filipe A, McDougall GJ, Stewart D, Kim KS, Palmela I, Brites D, Brito MA,

Brito C, Santos CN (2017) Polyphenols journey through blood-brain barrier towards neuronal protection. Sci Rep 7(1):11456. https://doi.org/10.1038/S41598-017-11512-6

Filomeni G, Graziani I, de Zio D, Dini L, Centonze D, Rotilio G, Ciriolo MR (2012) Neuroprotection of kaempferol by autophagy in models of rotenone-mediated acute toxicity: possible implications for Parkinson's disease. Neurobiol Aging 33(4):767–785. https://doi.org/10.1016/J.NEUROBIOLAGING.2010.05.021

Ghofrani S, Joghataei MT, Mohseni S, Baluchnejadmojarad T, Bagheri M, Khamse S, Roghani M (2015) Naringenin improves learning and memory in an Alzheimer's disease rat model: insights into the underlying mechanisms. Eur J Pharmacol 764:195–201. https://doi.org/10.1016/j.ejphar.2015.07.001

Glass CK, Saijo K, Winner B, Marchetto MC, Gage FH (2010) Mechanisms underlying inflammation in neurodegeneration. Cell 140(6):918–934. https://doi.org/10.1016/J.CELL.2010.02.016

Gomez-Pinilla F, Nguyen TTJ (2012) Natural mood foods: the actions of polyphenols against psychiatric and cognitive disorders. Nutr Neurosci 15(3):127–133. https://doi.org/10.1179/1476830511Y.0000000035

Goyarzu P, Malin DH, Lau FC, Taglialatela G, Moon WD, Jennings R, Moy E, Moy D, Lippold S, Shukitt-Hale B, Joseph JA (2013) Blueberry supplemented diet: effects on object recognition memory and nuclear factor-kappa B levels in aged rats. Nutr Neurosci 7(2):75–83. https://doi.org/10.1080/10284150410001710410

Guéroux M, Fleau C, Slozeck M, Laguerre M, Pianet I (2017) Epigallocatechin 3-gallate as an inhibitor of tau phosphorylation and aggregation: a molecular and structural insight. J Prev Alzheimers Dis 4(4):218–225

Gürler SB, Kiraz Y, Baran Y (2020) Flavonoids in cancer therapy: current and future trends. In: Biodiversity and biomedicine. Academic Press, San Diego, pp 403–440. https://doi.org/10.1016/B978-0-12-819541-3.00021-9

Hajialyani M, Farzaei MH, Echeverría J, Nabavi SM, Uriarte E, Eduardo SS (2019) Hesperidin as a neuroprotective agent: a review of animal and clinical evidence. Molecules 24(3):648. https://doi.org/10.3390/MOLECULES24030648

Halder S, Kar R, Chandra N, Nimesh A, Mehta AK, Bhattacharya SK et al (2018) Alteration in cognitive behaviour, brain antioxidant enzyme activity and their gene expression in F1 generation mice, following Cd exposure during the late gestation period: modulation by quercetin. Metab Brain Dis 33:1935–1943. https://doi.org/10.1007/s11011-018-0299-y

Haleagrahara N, Siew CJ, Mitra NK, Kumari M (2011) Neuroprotective effect of bioflavonoid quercetin in 6-hydroxydopamine-induced oxidative stress biomarkers in the rat striatum. Neurosci Lett 500(2):139–143. https://doi.org/10.1016/J.NEULET.2011.06.021

Haleagrahara N, Siew CJ, Ponnusamy K (2013) Effect of quercetin and desferrioxamine on 6-hydroxydopamine (6-OHDA) induced neurotoxicity in striatum of rats. J Toxicol Sci 38(1): 25–33. https://doi.org/10.2131/jts.38.25

Halliwell B, Gutteridge JM (2015) Free radicals in biology and medicine. Oxford university press, USA. https://doi.org/10.1093/acprof:oso/9780198717478.001.0001

Han X, Zhao S, Song H, Xu T, Fang Q, Hu G, Sun L (2021) Kaempferol alleviates LD-mitochondrial damage by promoting autophagy: implications in Parkinson's disease. Redox Biol 41:101911. https://doi.org/10.1016/J.REDOX.2021.101911

Harborne JB (1994) The flavonoids: advances in research since 1986. Chapman & Hall, London. https://link.springer.com/book/9780412480706

Hardy J, Selkoe DJ (2002) The amyloid hypothesis of Alzheimer's disease: progress and problems on the road to therapeutics. Science 297(5580):353–356. https://doi.org/10.1126/SCIENCE.1072994

Haridevamuthu B, Guru A, Murugan R, Sudhakaran G, Pachaiappan R, Almutairi MH, Almutairi BO, Juliet A, Arockiaraj J (2022) Neuroprotective effect of Biochanin a against Bisphenol A-induced prenatal neurotoxicity in zebrafish by modulating oxidative stress and locomotory defects. Neurosci Lett 790:136889. https://doi.org/10.1016/J.NEULET.2022.136889

Hawkins RA, O'Kane RL, Simpson IA, Viña JR (2006) Structure of the blood-brain barrier and its role in the transport of amino acids. J Nutr 136(1 suppl):218S. https://doi.org/10.1093/JN/136.1.218S

Hertog MGL, Hollman PCH, van de Putte B (1993) Content of potentially anticarcinogenic flavonoids of tea infusions, wines, and fruit juices. J Agric Food Chem 41(8):1242–1246. https://doi.org/10.1021/JF00032A015/ASSET/JF00032A015.FP.PNG_V03

Hong Y, An Z (2018) Hesperidin attenuates learning and memory deficits in APP/PS1 mice through activation of Akt/Nrf2 signaling and inhibition of RAGE/NF-κB signaling. Arch Pharm Res 41: 655–663. https://doi.org/10.1007/s12272-015-0662-z

Hua FZ, Ying J, Zhang J, Wang XF, Hu YH, Liang YP et al (2016) Naringenin pre-treatment inhibits neuroapoptosis and ameliorates cognitive impairment in rats exposed to isoflurane anesthesia by regulating the PI3/Akt/PTEN signalling pathway and suppressing NF-κB-mediated inflammation. Int J Mol Med 38(4):1271–1280. https://doi.org/10.3892/ijmm.2016.2715

Huang Q, Wu LJ, Tashiro SI, Gao HY, Onodera S, Ikejima T (2005) (+)-Catechin, an ingredient of green tea, protects murine microglia from oxidative stress-induced DNA damage and cell cycle arrest. J Pharmacol Sci 98(1):16–24. https://doi.org/10.1254/JPHS.FPJ04053X

Hung KC, Huang HJ, Wang YT, Lin AMY (2016) Baicalein attenuates α-synuclein aggregation, inflammasome activation and autophagy in the MPP+-treated nigrostriatal dopaminergic system in vivo. J Ethnopharmacol 194:522–529. https://doi.org/10.1016/J.JEP.2016.10.040

Ide K, Yamada H, Takuma N, Park M, Wakamiya N, Nakase J, Ukawa Y, Sagesaka YM (2014) Green tea consumption affects cognitive dysfunction in the elderly: a pilot study. Nutrients 6(10):4032–4042. https://doi.org/10.3390/NU6104032

Islam F, Khan MM, Raza SS, Javed H, Ahmad A, Khan A, Islam F, Safhi MM (2012) Rutin protects dopaminergic neurons from oxidative stress in an animal model of Parkinson's disease. Neurotox Res 22(1):1–15. https://doi.org/10.1007/S12640-011-9295-2/METRICS

Issac PK, Guru A, Velayutham M, Pachaiappan R, Arasu MV, Al-Dhabi NA, Choi KC, Harikrishnan R, Arockiaraj J (2021) Oxidative stress induced antioxidant and neurotoxicity demonstrated in vivo zebrafish embryo or larval model and their normalization due to morin showing therapeutic implications. Life Sci 283:119864. https://doi.org/10.1016/J.LFS.2021.119864

Jaeger BN, Parylak SL, Gage FH (2018) Mechanisms of dietary flavonoid action in neuronal function and neuroinflammation. Mol Aspects Med 61:50–62. https://doi.org/10.1016/J.MAM.2017.11.003

Jang S, Kelley KW, Johnson RW (2008) Luteolin reduces IL-6 production in microglia by inhibiting JNK phosphorylation and activation of AP-1. Proc Natl Acad Sci U S A 105(21): 7534–7539. https://doi.org/10.1073/PNAS.0802865105/ASSET/DA66273E-76BE-4F64-87ED-AB55C61493CB/ASSETS/GRAPHIC/ZPQ0170803930006.JPEG

Jiang M, Porat-Shliom Y, Pei Z, Cheng Y, Xiang L, Sommers K et al (2010) Baicalein reduces E46K α-synuclein aggregation in vitro and protects cells against E46K α-synuclein toxicity in cell models of familiar Parkinsonism. J Neurochem 114(2):419–429. https://doi.org/10.1111/j.1471-4159.2010.06752.x

Joshi R, Kulkarni YA, Wairkar S (2018) Pharmacokinetic, pharmacodynamic and formulations aspects of Naringenin: an update. Life Sci 215:43–56. https://doi.org/10.1016/J.LFS.2018.10.066

Jung UJ, Jeon MT, Choi MS, Kim SR (2014) Silibinin attenuates MPP+-induced neurotoxicity in the substantia nigra in vivo. J Med Food 17(5):599–605. https://doi.org/10.1089/jmf.2013.2926

Justin Thenmozhi A, William Raja TR, Manivasagam T, Janakiraman U, Essa MM (2016) Hesperidin ameliorates cognitive dysfunction, oxidative stress and apoptosis against aluminium chloride induced rat model of Alzheimer's disease. Nurt Neurosci 20(6):360–368. https://doi.org/10.1080/1028415X.2016.1144846

Karuppagounder SS, Madathil SK, Pandey M, Haobam R, Rajamma U, Mohanakumar KP (2013) Quercetin up-regulates mitochondrial complex-I activity to protect against programmed cell death in rotenone model of Parkinson's disease in rats. Neuroscience 236:136–148. https://doi.org/10.1016/J.NEUROSCIENCE.2013.01.032

Kean RJ, Lamport DJ, Dodd GF, Freeman JE, Williams CM, Ellis JA, Butler LT, Spencer JPE (2015) Chronic consumption of flavanone-rich orange juice is associated with cognitive benefits: an 8-wk, randomized, double-blind, placebo-controlled trial in healthy older adults. Am J Clin Nutr 101(3):506–514. https://doi.org/10.3945/AJCN.114.088518

Khajevand-Khazaei MR, Ziaee P, Motevalizadeh SA, Rohani M, Afshin-Majd S, Baluchnejadmojarad T, Roghani M (2018) Naringenin ameliorates learning and memory impairment following systemic lipopolysaccharide challenge in the rat. Eur J Pharmacol 826: 114–122. https://doi.org/10.1016/j.ejphar.2018.03.001

Khakh BS, Sofroniew MV (2015) Diversity of astrocyte functions and phenotypes in neural circuits. Nat Neurosci 18(7):942–952. https://doi.org/10.1038/NN.4043

Khan MS, Ali T, Kim MW, Jo MH, Jo MG, Badshah H, Kim MO (2016) Anthocyanins protect against LPS-induced oxidative stress-mediated neuroinflammation and neurodegeneration in the adult mouse cortex. Neurochem Int 100:1–10. https://doi.org/10.1016/j.neuint.2016.08.005

Kim HD, Jeong KH, Jung UJ, Kim SR (2016a) Myricitrin ameliorates 6-hydroxydopamine-induced dopaminergic neuronal loss in the substantia nigra of mouse brain. J Med Food 19(4):374–382. https://doi.org/10.1089/jmf.2015.3581

Kim HD, Jeong KH, Jung UJ, Kim SR (2016b) Naringin treatment induces neuroprotective effects in a mouse model of Parkinson's disease in vivo, but not enough to restore the lesioned dopaminergic system. J Nutr Biochem 28:140–146. https://doi.org/10.1016/J.JNUTBIO.2015.10.013

Kim JH, Quilantang NG, Kim HY, Lee S, Cho EJ (2019) Attenuation of hydrogen peroxide-induced oxidative stress in SH-SY5Y cells by three flavonoids from Acer okamotoanum. Chem Pap 73(5):1135–1144. https://doi.org/10.1007/S11696-018-0664-7/METRICS

Koh SH, Lee SM, Kim HY, Lee KY, Lee YJ, Kim HT et al (2006) The effect of epigallocatechin gallate on suppressing disease progression of ALS model mice. Neurosci Lett 395(2):103–107. https://doi.org/10.1016/j.neulet.2005.10.056

Koirala N, Thuan NH, Ghimire GP, Van Thang D, Sohng JK (2016) Methylation of flavonoids: chemical structures, bioactivities, progress and perspectives for biotechnological production. Enzyme Microb Technol 86:103–116. https://doi.org/10.1016/J.ENZMICTEC.2016.02.003

Korkmaz OT, Aytan N, Carreras I, Choi JK, Kowall NW, Jenkins BG, Dedeoglu A (2014) 7, 8-Dihydroxyflavone improves motor performance and enhances lower motor neuronal survival in a mouse model of amyotrophic lateral sclerosis. Neurosci Lett 566:286–291. https://doi.org/10.1016/j.neulet.2014.02.058

Lau FC, Bielinski DF, Joseph JA (2007) Inhibitory effects of blueberry extract on the production of inflammatory mediators in lipopolysaccharide-activated BV2 microglia. J Neurosci Res 85(5): 1010–1017. https://doi.org/10.1002/JNR.21205

Lee E, Park HR, Ji ST, Lee Y, Lee J (2014) Baicalein attenuates astroglial activation in the 1-methyl-4-phenyl-1,2,3,4-tetrahydropyridine-induced Parkinson's disease model by downregulating the activations of nuclear factor-κB, ERK, and JNK. J Neurosci Res 92(1): 130–139. https://doi.org/10.1002/JNR.23307

Leem E, Nam JH, Jeon MT, Shin WH, Won SY, Park SJ et al (2014) Naringin protects the nigrostriatal dopaminergic projection through induction of GDNF in a neurotoxin model of Parkinson's disease. J Nutr Biochem 25(7):801–806. https://doi.org/10.1016/j.jnutbio.2014.03.006

Li S, Pu XP (2011) Neuroprotective effect of kaempferol against a 1-methyl-4-phenyl-1,2,3,6-tetrahydropyridine-induced mouse model of Parkinson's Disease. Biol Pharm Bull 34(8): 1291–1296. https://doi.org/10.1248/BPB.34.1291

Li J, Xu S (2022) Tilianin attenuates MPP+ induced oxidative stress and apoptosis of dopaminergic neurons in a cellular model of Parkinson's disease. Exp Ther Med 23(4):1–7. https://doi.org/10.3892/etm.2022.11223

Li C, Zug C, Qu H, Schluesener H, Zhang Z (2015) Hesperidin ameliorates behavioral impairments and neuropathology of transgenic APP/PS1 mice. Behav Brain Res 281:32–42. https://doi.org/10.1016/j.bbr.2014.12.012

Li R, Wang X, Qin T, Qu R, Ma S (2016) Apigenin ameliorates chronic mild stress-induced depressive behavior by inhibiting interleukin-1β production and NLRP3 inflammasome activation in the rat brain. Behav Brain Res 296:318–325. https://doi.org/10.1016/J.BBR.2015.09.031

Liddelow S, Barres B (2015) SnapShot: astrocytes in health and disease. Cell 162(5):1170–1170.e1. https://doi.org/10.1016/j.cell.2015.08.029

Liu B, Liu W, Liu P, Liu X, Song X, Hayashi T et al (2019) Silibinin alleviates the learning and memory defects in overtrained rats accompanying reduced neuronal apoptosis and senescence. Neurochem Res 44:1818–1829. https://doi.org/10.1007/s11064-019-02816-2

Liu P, Cui L, Liu B, Liu W, Hayashi T, Mizuno K et al (2020) Silibinin ameliorates STZ-induced impairment of memory and learning by up-regulating insulin signaling pathway and attenuating apoptosis. Physiol Behav 213:112689. https://doi.org/10.1016/j.physbeh.2019.112689

Lou H, Jing X, Wei X, Shi H, Ren D, Zhang X (2014) Naringenin protects against 6-OHDA-induced neurotoxicity via activation of the Nrf2/ARE signaling pathway. Neuropharmacology 79:380–388. https://doi.org/10.1016/J.NEUROPHARM.2013.11.026

Lu P, Mamiya T, Lu L, Mouri A, Niwa M, Kim HC et al (2010) Silibinin attenuates cognitive deficits and decreases of dopamine and serotonin induced by repeated methamphetamine treatment. Behav Brain Res 207(2):387–393. https://doi.org/10.1016/j.bbr.2009.10.024

Lutz C (2018) Mouse models of ALS: past, present and future. Brain Res 1693:1–10. https://doi.org/10.1016/j.brainres.2018.03.024

Magalingam KB, Radhakrishnan A, Haleagrahara N (2013) Rutin, a bioflavonoid antioxidant protects rat pheochromocytoma (PC-12) cells against 6-hydroxydopamine (6-OHDA)-induced neurotoxicity. Int J Mol Med 32(1):235–240. https://doi.org/10.3892/IJMM.2013.1375/HTML

Maher P (2019) The potential of flavonoids for the treatment of neurodegenerative diseases. Int J Mol Sci 20(12):3056. https://doi.org/10.3390/ijms20123056

Malenka RC, Bear MF (2004) LTP and LTD: an embarrassment of riches. Neuron 44(1):5–21. https://doi.org/10.1016/j.neuron.2004.09.012

Manach C, Scalbert A, Morand C, Rémésy C, Jiménez L (2004) Polyphenols: food sources and bioavailability. Am J Clin Nutr 79(5):727–747. https://doi.org/10.1093/AJCN/79.5.727

Mandel S, Maor G, Youdim MB (2004) Iron and α-synuclein in the substantia nigra of MPTP-treated mice: effect of neuroprotective drugs R-apomorphine and green tea polyphenol (−)-epigallocatechin-3-gallate. J Mol Neurosci 24:401–416. https://doi.org/10.1385/JMN:24:3:401

Maratha SR, Mahadevan N (2012) Memory enhancing activity of naringin in unstressed and stressed mice: possible cholinergic and nitriergic modulation. Neurochem Res 37(10): 2206–2212. https://doi.org/10.1007/s11064-012-0844-8

Matias I, Diniz LP, Buosi A, Neves G, Stipursky J, Gomes FCA (2017) Flavonoid hesperidin induces synapse formation and improves memory performance through the astrocytic TGF-β1. Front Aging Neurosci 9:184. https://doi.org/10.3389/fnagi.2017.00184

Matsuzaki K, Yamakuni T, Hashimoto M, Haque AM, Shido O, Mimaki Y et al (2006) Nobiletin restoring β-amyloid-impaired CREB phosphorylation rescues memory deterioration in Alzheimer's disease model rats. Neurosci Lett 400(3):230–234. https://doi.org/10.1016/j.neulet.2006.02.077

Mattson MP, Magnus T (2006) Ageing and neuronal vulnerability. Nat Rev Neurosci 7(4): 278–294. https://doi.org/10.1038/NRN1886

McGeer PL, McGeer E (2002) Inflammatory processes in amyotrophic lateral sclerosis. Muscle Nerve 26(4):459–470. https://doi.org/10.1002/mus.10191

Meng-Zhen S, Ju L, Lan-Chun Z, Cai-Feng D, Shu-da Y, Hao-Fei Y, Wei-Yan H (2022) Potential therapeutic use of plant flavonoids in AD and PD. Heliyon 8:e11440. https://doi.org/10.1016/j.heliyon.2022.e11440

Middleton E, Kandaswami C, Theoharides T (2000) The effects of plant flavonoids on mammalian cells: implications for inflammation, heart disease, and cancer. Pharmacol Rev 52:673

Minocha T, Birla H, Obaid AA, et al. (2022) Flavonoids as Promising Neuroprotectants and Their Therapeutic Potential against Alzheimer's Disease. Oxidative Med Cell Longev. 2022 Aug 28:2022:6038996. https://doi.org/10.1155/2022/6038996

Mitsunaga Y, Takanaga H, Matsuo H, Naito M, Tsuruo T, Ohtani H, Sawada Y (2000) Effect of bioflavonoids on vincristine transport across blood–brain barrier. Eur J Pharmacol 395(3): 193–201. https://doi.org/10.1016/S0014-2999(00)00180-1

Moghbelinejad S, Nassiri-Asl M, Farivar TN, Abbasi E, Sheikhi M, Taghiloo M et al (2014) Rutin activates the MAPK pathway and BDNF gene expression on beta-amyloid induced neurotoxicity in rats. Toxicol Lett 224(1):108–113. https://doi.org/10.1016/j.toxlet.2013.10.010

Mohammadi N, Asle-Rousta M, Rahnema M, Amini R (2021) Morin attenuates memory deficits in a rat model of Alzheimer's disease by ameliorating oxidative stress and neuroinflammation. Eur J Pharmacol 910:174506. https://doi.org/10.1016/J.EJPHAR.2021.174506

Nabavi SF, Braidy N, Gortzi O, Sobarzo-Sanchez E, Daglia M, Skalicka-Woźniak K, Nabavi SM (2015) Luteolin as an anti-inflammatory and neuroprotective agent: a brief review. Brain Res Bull 119(pt A):1–11. https://doi.org/10.1016/J.BRAINRESBULL.2015.09.002

Nagase H, Omae N, Omori A, Nakagawasai O, Tadano T, Yokosuka A et al (2005) Nobiletin and its related flavonoids with CRE-dependent transcription-stimulating and neuritegenic activities. Biochem Biophys Res Commun 337(4):1330–1336. https://doi.org/10.1016/j.bbrc.2005.10.001

Nakajima A, Yamakuni T, Haraguchi M, Omae N, Song SY, Kato C et al (2007) Nobiletin, a citrus flavonoid that improves memory impairment, rescues bulbectomy-induced cholinergic neurodegeneration in mice. J Pharmacol Sci 105(1):122–126. https://doi.org/10.1254/jphs.SC0070155

Nan W, Zhonghang X, Keyan C, Tongtong L, Wanshu G, Zhongxin X (2018) Epigallocatechin-3-gallate reduces neuronal apoptosis in rats after middle cerebral artery occlusion injury via PI3K/AKT/eNOS signaling pathway. Biomed Res Int 2018:6473580. https://doi.org/10.1155/2018/6473580

Obermeier B, Daneman R, Ransohoff RM (2013) Development, maintenance and disruption of the blood-brain barrier. Nat Med 19(12):1584–1596. https://doi.org/10.1038/nm.3407

Orr WC, Sohal RS (1994) Extension of life-span by overexpression of superoxide dismutase and catalase in Drosophila melanogaster. Science 263(5150):1128–1130. https://doi.org/10.1126/SCIENCE.8108730

Ottaviani JI, Borges G, Momma TY, Spencer JPE, Keen CL, Crozier A, Schroeter H (2016) The metabolome of [2-14C](−)-epicatechin in humans: implications for the assessment of efficacy, safety and mechanisms of action of polyphenolic bioactives. Sci Rep 6(1):1–10. https://doi.org/10.1038/srep29034

Park MH, Jin HK, Bae J, sung. (2020) Potential therapeutic target for aging and age-related neurodegenerative diseases: the role of acid sphingomyelinase. Exp Mol Med 52(3):380. https://doi.org/10.1038/S12276-020-0399-8

Patel MY, Panchal HV, Ghribi O, Benzeroual KE (2012) The neuroprotective effect of fisetin in the MPTP model of Parkinson's disease. J Parkinsons Dis 2(4):287–302. https://doi.org/10.3233/JPD-012110

Pavlica S, Gebhardt R (2010) Protective effects of flavonoids and two metabolites against oxidative stress in neuronal PC12 cells. Life Sci 86(3–4):79–86. https://doi.org/10.1016/J.LFS.2009.10.017

Poetini MR, Araujo SM, Trindade de Paula M, Bortolotto VC, Meichtry LB, Polet de Almeida F, Jesse CR, Kunz SN, Prigol M (2018) Hesperidin attenuates iron-induced oxidative damage and dopamine depletion in Drosophila melanogaster model of Parkinson's disease. Chem Biol Interact 279:177–186. https://doi.org/10.1016/J.CBI.2017.11.018

Qin S, Chen Q, Wu H, Liu C, Hu J, Zhang D, Xu C (2016) Effects of naringin on learning and memory dysfunction induced by gp120 in rats. Brain Res Bull 124:164–171. https://doi.org/10.1016/j.brainresbull.2016.04.020

Querfurth HW, LaFerla FM (2010) Alzheimer's disease. N Engl J Med 362(4):329–344. https://doi.org/10.1056/NEJMRA0909142

Ramalingayya GV, Nampoothiri M, Nayak PG, Kishore A, Shenoy RR, Rao CM, Nandakumar K (2016) Naringin and rutin alleviates episodic memory deficits in two differentially challenged object recognition tasks. Pharmacogn Mag 12(suppl 1):S63

Ransohoff RM (2016) How neuroinflammation contributes to neurodegeneration. Science 353(6301):777–783. https://doi.org/10.1126/SCIENCE.AAG2590

Reeve A, Simcox E, Turnbull D (2014) Ageing and Parkinson's disease: why is advancing age the biggest risk factor? Ageing Res Rev 14(100):19–30. https://doi.org/10.1016/J.ARR.2014.01.004

Rendeiro C, Vauzour D, Rattray M, Waffo-Téguo P, Mérillon JM, Butler LT, Williams CM, Spencer JPE (2013) Dietary levels of pure flavonoids improve spatial memory performance and increase hippocampal brain-derived neurotrophic factor. PloS One 8(5):e63535. https://doi.org/10.1371/JOURNAL.PONE.0063535

Rendeiro C, Rhodes JS, Spencer JPE (2015) The mechanisms of action of flavonoids in the brain: direct versus indirect effects. Neurochem Int 89:126–139. https://doi.org/10.1016/J.NEUINT.2015.08.002

Riancho J, Gil-Bea FJ, Santurtun A, de Munaín AL (2019) Amyotrophic lateral sclerosis: a complex syndrome that needs an integrated research approach. Neural Regen Res 14(2):193. https://doi.org/10.4103/1673-5374.244783

Ricciotti E, Fitzgerald GA (2011) Prostaglandins and inflammation. Arterioscler Thromb Vasc Biol 31(5):986–1000. https://doi.org/10.1161/ATVBAHA.110.207449/FORMAT/EPUB

Sabogal-Guáqueta AM, Munoz-Manco JI, Ramirez-Pineda JR, Lamprea-Rodriguez M, Osorio E, Cardona-Gomez GP, Davila-Ortiz G (2015) The flavonoid quercetin ameliorates Alzheimer's disease pathology and protects cognitive and emotional function in aged triple transgenic Alzheimer's disease model mice. Neuropharmacology 93:134–145. https://doi.org/10.1016/j.neuropharm.2015.01.028

Saffar S, Fatemi I, Rahmani M, Hassanshahi J, Sahamsizadeh A, Allahtavakoli M et al (2020a) The effect of epigallocatechin-3-gallate on morphine-induced memory impairments in rat: EGCG effects on morphine neurotoxicity. Hum Exp Toxicol 39(7):994–1002. https://doi.org/10.1177/0960327120909540

Saffar S, Fatemi I, Rahmani MR, Hassanshahi J, Pak-Hashemi M, Kaeidi A (2020b) The effect of epigallocatechin gallate on morphine induced working memory defects in rats: an experimental study. J Rafsanjan Univ Med Sci 19(2):115–124. https://doi.org/10.29252/jrums.19.2.115

Saxena S, Caroni P (2011) Selective neuronal vulnerability in neurodegenerative diseases: from stressor thresholds to degeneration. Neuron 71(1):35–48. https://doi.org/10.1016/J.NEURON.2011.06.031

Schroeter H, Heiss C, Balzer J, Kleinbongard P, Keen CL, Hollenberg NK, Sies H, Kwik-Uribe C, Schmitz HH, Kelm M (2006) (-)-Epicatechin mediates beneficial effects of flavanol-rich cocoa on vascular function in humans. https://www.pnas.org.

Shukitt-Hale B, Cheng V, Joseph JA (2013) Effects of blackberries on motor and cognitive function in aged rats. Nutr Neurosci 12(3):135–140. https://doi.org/10.1179/147683009X423292

Singh S, Kumar P (2018) Piperine in combination with quercetin halt 6-OHDA induced neurodegeneration in experimental rats: biochemical and neurochemical evidences. Neurosci Res 133:38–47. https://doi.org/10.1016/j.neures.2017.10.006

Singh NA, Mandal AKA, Khan ZA (2018a) Inhibition of Al (III)-induced Aβ 42 fibrillation and reduction of neurotoxicity by epigallocatechin-3-gallate nanoparticles. J Biomed Nanotechnol 14(6):1147–1158. https://doi.org/10.1166/jbn.2018.2552

Singh S, Singh AK, Garg G, Rizvi SI (2018b) Fisetin as a caloric restriction mimetic protects rat brain against aging induced oxidative stress, apoptosis and neurodegeneration. Life Sci 193: 171–179. https://doi.org/10.1016/j.lfs.2017.11.004

Sinha K, Das J, Pal PB, Sil PC (2013) Oxidative stress: the mitochondria-dependent and mitochondria-independent pathways of apoptosis. Arch Toxicol 87(7):1157–1180. https://doi.org/10.1007/S00204-013-1034-4

Sokolov AN, Pavlova MA, Klosterhalfen S, Enck P (2013) Chocolate and the brain: neurobiological impact of cocoa flavanols on cognition and behavior. Neurosci Biobehav Rev 37(10): 2445–2453. https://doi.org/10.1016/J.NEUBIOREV.2013.06.013

Song X, Liu B, Cui L, Zhou B, Liu W, Xu F et al (2017) Silibinin ameliorates anxiety/depression-like behaviors in amyloid β-treated rats by upregulating BDNF/TrkB pathway and attenuating autophagy in hippocampus. Physiol Behav 179:487–493. https://doi.org/10.1016/j.physbeh.2017.07.023

Song X, Gong Z, Liu K, Kou J, Liu B, Liu K (2020) Baicalin combats glutamate excitotoxicity via protecting glutamine synthetase from ROS-induced 20S proteasomal degradation. Redox Biol 34:101559. https://doi.org/10.1016/J.REDOX.2020.101559

Song Q, Peng S, Zhu X (2021) Baicalein protects against MPP+/MPTP-induced neurotoxicity by ameliorating oxidative stress in SH-SY5Y cells and mouse model of Parkinson's disease. Neurotoxicology 87:188–194. https://doi.org/10.1016/j.neuro.2021.10.003

Spencer JPE (2009a) Flavonoids and brain health: multiple effects underpinned by common mechanisms. Genes Nutr 4(4):243–250. https://doi.org/10.1007/S12263-009-0136-3

Spencer JPE (2009b) The impact of flavonoids on memory: physiological and molecular considerations. Chem Soc Rev 38(4):1152–1161. https://doi.org/10.1039/b800422f

Spencer JPE, Vauzour D, Rendeiro C (2009) Flavonoids and cognition: the molecular mechanisms underlying their behavioural effects. Arch Biochem Biophys 492(1–2):1–9. https://doi.org/10.1016/J.ABB.2009.10.003

Thilakarathna SH, Vasantha Rupasinghe HP (2013) Flavonoid bioavailability and attempts for bioavailability enhancement. Nutrients 5(9):3367–3387. https://doi.org/10.3390/NU5093367

Vafeiadou K, Vauzour D, Lee HY, Rodriguez-Mateos A, Williams RJ, Spencer JPE (2009) The citrus flavanone naringenin inhibits inflammatory signalling in glial cells and protects against neuroinflammatory injury. Arch Biochem Biophys 484(1):100–109. https://doi.org/10.1016/J.ABB.2009.01.016

van Praag X, Fleshner M, Schwartz MW, Mattson MP (2014) Exercise, energy intake, glucose homeostasis, and the brain. J Neurosci 34(46):15139. https://doi.org/10.1523/JNEUROSCI.2814-14.2014

Wang L, Xu S, Xu X, Chan P (2009) (-)-Epigallocatechin-3-gallate protects SH-SY5Y cells against 6-OHDA-induced cell death through STAT3 activation. J Alzheimers Dis 17(2):295–304. https://doi.org/10.3233/JAD-2009-1048

Wang YH, Yu HT, Pu XP, Du GH (2013) Baicalein prevents 6-hydroxydopamine-induced mitochondrial dysfunction in SH-SY5Y cells via inhibition of mitochondrial oxidation and up-regulation of DJ-1 protein expression. Molecules 18(12):14726–14738. https://doi.org/10.3390/molecules181214726

Wang D, Liu L, Zhu X, Wu W, Wang Y (2014a) Hesperidin alleviates cognitive impairment, mitochondrial dysfunction and oxidative stress in a mouse model of Alzheimer's disease. Cell Mol Neurobiol 34:1209–1221. https://doi.org/10.1007/s10571-014-0098-x

Wang D, Liu L, Zhu X, Wu W, Wang Y (2014b) Hesperidin alleviates cognitive impairment, mitochondrial dysfunction and oxidative stress in a mouse model of Alzheimer's disease. Cell Mol Neurobiol 34(8):1209–1221. https://doi.org/10.1007/S10571-014-0098-X/METRICS

Wang YH, Yu HT, Pu XP, Du GH (2014c) Myricitrin alleviates methylglyoxal-induced mitochondrial dysfunction and AGEs/RAGE/NF-κB pathway activation in SH-SY5Y cells. J Mol Neurosci 53:562–570. https://doi.org/10.1007/s12031-013-0222-2

Wang TH, Wang SY, Wang XD, Jiang HQ, Yang YQ, Wang Y et al (2018) Fisetin exerts antioxidant and neuroprotective effects in multiple mutant hSOD1 models of amyotrophic lateral sclerosis by activating ERK. Neuroscience 379:152–166. https://doi.org/10.1016/j.neuroscience.2018.03.008

Wang GQ, Zhang B, He XM, Li DD, Shi JS, Zhang F (2019) Naringenin targets on astroglial Nrf2 to support dopaminergic neurons. Pharmacol Res 139:452–459. https://doi.org/10.1016/j.phrs.2018.11.043

Wang J, Mao J, Wang R, Li S, Wu B, Yuan Y (2020) Kaempferol protects against cerebral ischemia reperfusion injury through intervening oxidative and inflammatory stress induced apoptosis. Front Pharmacol 11:424. https://doi.org/10.3389/fphar.2020.00424

Weber CC, Kressmann S, Ott M, Fricker G, Müller WE (2005) Inhibition of P-glycoprotein function by several antidepressants may not contribute to clinical efficacy. Pharmacopsychiatry 38(6):293–300. https://doi.org/10.1055/S-2005-916184

Wei N, Wei Y, Li B, Pang L (2017) Baicalein promotes neuronal and behavioral recovery after intracerebral hemorrhage via suppressing apoptosis, oxidative stress and neuroinflammation. Neurochem Res 42:1345–1353. https://doi.org/10.1007/s11064-017-2179-y

Wei BB, Liu MY, Zhong X, Yao WF, Wei MJ (2019) Increased BBB permeability contributes to EGCG-caused cognitive function improvement in natural aging rats: pharmacokinetic and distribution analyses. Acta Pharmacol Sin 40(11):1490–1500. https://doi.org/10.1038/s41401-019-0243-7

Wen L, Shi D, Zhou T, Tu J, He M, Jiang Y, Yang B (2020) Identification of two novel prenylated flavonoids in mulberry leaf and their bioactivities. Food Chem 315:126236. https://doi.org/10.1016/J.FOODCHEM.2020.126236

Wen S, Wang L, Zou H, Gu J, Song R, Bian J, Yuan Y, Liu Z (2021) Puerarin attenuates cadmium-induced neuronal injury via stimulating cadmium excretion, inhibiting oxidative stress and apoptosis. Biomolecules 11(7):978. https://doi.org/10.3390/BIOM11070978

Williams CA, Grayer RJ (2004) Anthocyanins and other flavonoids. Nat Prod Rep 21(4):539–573. https://doi.org/10.1039/B311404J

Williams RJ, Spencer JPE (2012) Flavonoids, cognition, and dementia: actions, mechanisms, and potential therapeutic utility for Alzheimer disease. Free Radic Biol Med 52(1):35–45. https://doi.org/10.1016/J.FREERADBIOMED.2011.09.010

Williams CM, El Mohsen MA, Vauzour D, Rendeiro C, Butler LT, Ellis JA, Whiteman M, Spencer JPE (2008) Blueberry-induced changes in spatial working memory correlate with changes in hippocampal CREB phosphorylation and brain-derived neurotrophic factor (BDNF) levels. Free Radic Biol Med 45(3):295–305. https://doi.org/10.1016/J.FREERADBIOMED.2008.04.008

Williams P, Sorribas A, Howes MJR (2010) Natural products as a source of Alzheimer's drug leads. Nat Prod Rep 28(1):48–77. https://doi.org/10.1039/C0NP00027B

Wong AD, Ye M, Levy AF, Rothstein JD, Bergles DE, Searson PC (2013) The blood-brain barrier: an engineering perspective. Front Neuroeng 6:7. https://doi.org/10.3389/FNENG.2013.00007/ABSTRACT

Wu H, Liu X, Gao ZY, Lin M, Zhao X, Sun Y, Pu XP (2021) Icaritin provides neuroprotection in Parkinson's disease by attenuating neuroinflammation, oxidative stress, and energy deficiency. Antioxidants (Basel, Switzerland) 10(4):529. https://doi.org/10.3390/antiox10040529

Xu Z, Chen S, Li X, Luo G, Li L, Le W (2006) Neuroprotective effects of (-)-epigallocatechin-3-gallate in a transgenic mouse model of amyotrophic lateral sclerosis. Neurochem Res 31:1263–1269. https://doi.org/10.1007/s11064-006-9166-z

Xue X, Liu H, Qi L, Li X, Guo C, Gong D, Qu H (2014) Baicalein ameliorated the upregulation of striatal glutamatergic transmission in the mice model of Parkinson's disease. Brain Res Bull 103:54–59. https://doi.org/10.1016/J.BRAINRESBULL.2014.02.004

Yang Z, Kuboyama T, Tohda C (2019) Naringenin promotes microglial M2 polarization and Aβ degradation enzyme expression. Phytother Res 33(4):1114–1121. https://doi.org/10.1002/ptr.6305

Youdim KA, Shukitt-Hale B, Joseph JA (2004) Flavonoids and the brain: interactions at the blood-brain barrier and their physiological effects on the central nervous system. Free Radic Biol Med 37(11):1683–1693. https://doi.org/10.1016/j.freeradbiomed.2004.08.002

Zaw JJT, Howe PRC, Wong RHX (2017) Does phytoestrogen supplementation improve cognition in humans? A systematic review. Ann N Y Acad Sci 1403(1):150–163. https://doi.org/10.1111/NYAS.13459

Zhang D, Hu X, Qian L, Wilson B, Lee C, Flood P, Langenbach R, Hong JS (2009) Prostaglandin E2 released from activated microglia enhances astrocyte proliferation in vitro. Toxicol Appl Pharmacol 238(1):64–70. https://doi.org/10.1016/J.TAAP.2009.04.015

Zhang K, Ma Z, Wang J, Xie A, Xie J (2011) Myricetin attenuated MPP+-induced cytotoxicity by anti-oxidation and inhibition of MKK4 and JNK activation in MES23. 5 cells. Neuropharmacology 61(1-2):329–335. https://doi.org/10.1016/j.neuropharm.2011.04.021

Zhang Z, Liu X, Schroeder JP, Chan CB, Song M, Yu SP et al (2014) 7, 8-Dihydroxyflavone prevents synaptic loss and memory deficits in a mouse model of Alzheimer's disease. Neuropsychopharmacology 39(3):638–650. https://doi.org/10.1038/npp.2013.243

Zhang DM, Ye JX, Mu JS, Cui XP (2016) Efficacy of vitamin B supplementation on cognition in elderly patients with cognitive-related diseases. J Geriatr Psychiatry Neurol 30(1):50–59. https://doi.org/10.1177/0891988716673466

Zhang S, Tomata Y, Sugiyama K, Sugawara Y, Tsuji I (2017) Citrus consumption and incident dementia in elderly Japanese: the Ohsaki Cohort 2006 Study. Br J Nutr 117(8):1174–1180. https://doi.org/10.1017/S000711451700109X

Zhang L, Wang H, Zhou Y, Zhu Y, Fei M (2018) Fisetin alleviates oxidative stress after traumatic brain injury via the Nrf2-ARE pathway. Neurochem Int 118:304–313. https://doi.org/10.1016/j.neuint.2018.05.011

Zhao L, Wang JL, Liu R, Li XX, Li JF, Zhang L (2013) Neuroprotective, anti-amyloidogenic and neurotrophic effects of apigenin in an Alzheimer's disease mouse model. Molecules 18(8): 9949–9965. https://doi.org/10.3390/molecules18089949

Zhou T, Zhu M, Liang Z (2018) (-)-Epigallocatechin-3-gallate modulates peripheral immunity in the MPTP-induced mouse model of Parkinson's disease. Mol Med Rep 17(4):4883–4888. https://doi.org/10.3892/mmr.2018.8470

Zhu M, Rajamani S, Kaylor J, Han S, Zhou F, Fink AL (2004) The flavonoid baicalein inhibits fibrillation of α-synuclein and disaggregates existing fibrils. J Biol Chem 279(26): 26846–26857. https://doi.org/10.1074/jbc.M403129200

Minocha T, Birla H, Obaid AA, et al. Flavonoids as Promising Neuroprotectants and Their Therapeutic Potential against Alzheimer's Disease. Oxid Med Cell Longev. 2022;2022: 6038996. Published 2022 Aug 28. doi:10.1155/2022/6038996

# Bioactive Compounds of Algae: Potential Neuroprotective Agents in Neurodegenerative Disorders

12

Sripriya Paranthaman and Paheshwari Palraj

**Abstract**

Neurodegenerative diseases are diverse in nature and are characterized by the continuous loss of neurons, glial cells, and neural networks in the brain and spinal cord, as well as selective malfunctioning of these tissues. Dementia pathologies such as Alzheimer's disease are reaching epidemic proportions, yet they are not successfully managed by effective symptomatic treatments. Even with a high prevalence, neurogenerative diseases are treated with drugs that alleviate the symptoms of patients, but which develop adverse events and still do not inhibit the progression of the disease. Only five drugs have been developed to alleviate cognitive symptoms and behavioral and psychological symptoms of dementia. As two of these licensed drugs, cholinesterase inhibitors are natural derived galantamine and rivastigmine, the potential for plants to yield new therapeutic agents has stimulated extensive research to discover new cholinesterase inhibitors together with plant extracts, phytochemicals and their derivatives with other mechanistic effects relevant to dementia treatment. Consequently, algae pave way for exploring potential compounds with therapeutic uses. In the search of neuroprotective natural product, algae have been recognised as an alternative to explore advanced bioactive natural materials because algae contain a wide range of bioactive compounds. Thus, within a new pharmacological perspective, this chapter aimed to discuss the therapeutic potential of natural compounds of algae against neurogenerative diseases.

**Keywords**

Bioactive compounds · Algae · Neuroprotection · Neurodegenerative disorders

S. Paranthaman (✉) · P. Palraj
Department of Plant Biology and Plant Biotechnology, Presidency College (Autonomous), Chennai, Tamil Nadu, India





S. Pathak, A. Banerjee (eds.), *Neuroprotective Effects of Phytochemicals in Brain Ageing*, https://doi.org/10.1007/978-981-99-7269-2_12

## 12.1 Introduction

Neurodegeneration refers to a phenomenon that results in neuron death due to the progressive loss of function and structure over time. Indeed, neurons do not normally reproduce and are incapable of being substituted by the body when damaged or killed. Therefore, continuous loss of neurons creates serious neurological disorders. During physiological or pathological aging, ischemic stroke and brain injuries, the neurons and supporting cells of the brain undergo degenerative changes (Chi et al. 2018). Degenerative brain diseases such as Alzheimer's disease (AD) and Parkinson's disease (PD) are the result of pathological brain aging characterized by region-specific loss of neurons (Vasili et al. 2019). Worldwide, these diseases are the leading cause of dementia in the elderly. Some common pathophysiologies such as oxidative stress, neuroinflammation, mitochondrial dysfunction, protein misfolding and defective protein purification system, which in turn complicate these diseases (Ganguly et al. 2017; Verri et al. 2012), while ischemic, traumatic and other brain injuries, if not fate, cause secondary damage. and are important causes of cognitive impairment in patients. Regardless of the types of dementia conditions, available treatment options only help reduce symptoms, rather than slow the progression of the disease. In addition, there are several negative effects associated with modern drugs. Researchers are working to identify potential therapeutic compounds that can target the etiology of the disease without harming the health of patients, given the enormous social and economic impact of these diseases. Although synthetic drugs have certain advantages, such as ease of development, naturally derived agents have gained popularity because they are generally well tolerated. The natural compounds are said to have anti-inflammatory, antioxidant and immunomodulatory effects (Yahfoufi et al. 2018). Compounds have been claimed to show pharmacological effects offer a better solution for the remedy of neurological disorders with complex pathomechanisms (León et al. 2013). A significant quantity of natural products has been reported to show neuroprotective activity against a wide range of toxic insults (Angeloni and David 2019; Rehman et al. 2019). There are investigations reporting therapeutic promise in preclinical studies (Kim et al. 2010) and clinical traits (Salehi et al. 2019; Turner et al. 2005).

Algae are a great source of bioactive molecules of diverse therapeutic value and have broad commercial applications. Both microalgae and seaweeds are a rich source of primary and secondary metabolites such as carbohydrates, fatty acids, carotenoids, vitamins, lectins, mycosporin-like amino acids, phenolics, polyphenols, phycobiliproteins, alginic acid, agar carrageenan, etc. Microalgae, in recent years, have been in the limelight for their novel and unique bioactive compounds, which contribute to various fields such as the pharmaceutical and food industries. They are being used in the development of new healthier food products and are also considered as an alternative for protein-rich sources (Batista et al. 2012; Becker 2013). Microalgae are capable of synthesizing rich metabolites viz. proteins (carotenoids), lipids, carbohydrates, and dietary fiber (Cardoso et al. 2017). The carotenoids are linked with health-promoting effects such as neuroprotective in addition to others such as cardioprotective, anti-inflammatory, anti-diabetes and anti-cancer activities (Sathasivam et al. 2018).

The marine ecosystem is a source for unlimited novel secondary metabolites with varied chemical structures, and they are paving for the discovery of novel drugs for the treatment of many diseases, including NDD (Molinski et al. 2008; Yang and Song 2018). Among the diverse marine organisms, seaweeds are a valuable source of bioactives that are not seen in other organisms. They exhibit various biological activities viz. antitumor, antioxidant, antimicrobial, neuroprotective, antiviral, anti-inflammatory, and antidiabetic. There are still many algae that have not been explored. Many algal species have gained significant attention in recent years because of their beneficial health impact. Phyco-medicine denotes the use of algal sources as herbal medicine with therapeutic potential. There is an increased curiosity to develop algae-based medicine and its commercial application.

## 12.2 Natural Products as Neuroprotective Agents

Natural substances derived from edible plants and animals are reported as potential neuroprotective agents, as these ingredients could be developed as cheap and low-risk medicinal food and yet have promising effects on the prevention of neurodegenerative diseases (Sairazi et al. 2020). Neural problems have traditionally been treated with herbal medication. Despite the fact that the exact mechanisms of action of herbal medications are still unknown, several of them have demonstrated anti-inflammatory and/or antioxidant effects in a range of peripheral systems. Currently, anti-inflammatory herbal medicine and its constituents are being demonstrated to be a potent neuroprotector against various therapeutic targets that are newly discovered by genomics, proteomics, and high-throughput screening, indicating that neuroglia-derived chronic inflammatory responses play a pathological role in the central nervous system (CNS) (Kumar and Khanum 2012). Numerous plants have been used to treat cognitive disorders, including neurodegenerative diseases like Alzheimer's disease and other memory-related disorders, in traditional medical practices. In Asian nations, more than 120 traditional medicines are utilized to treat problems of the central nervous system (CNS) (Uriarte et al. 2009).

Research in animal models of neurodegenerative illnesses and epidemiological studies of human populations have shown that phytochemicals in fruits and vegetables can shield the brain system from diseases. For example, resveratrol, quercetin, and catechins reduced oxidative stress and protected cultured hippocampal neurons against nitric-oxide-mediated cell death (Bastianetto et al. 2000). Interestingly, epidemiological studies pointed to the possible capacity of moderate consumption of red wine, an alcoholic beverage that contains elevated amounts of polyphenols (German and Walzem 2000) and flavonoids to reduce the incidence of certain age-related neurological disorders including macular degeneration (Obisesan et al. 1998) and dementia (Commenges et al. 2000; Orgogozo et al. 1997). To support this hypothesis, in vitro studies reported that red wine polyphenols such as quercetin, catechin, and resveratrol displayed protective abilities in various models of toxicity (Moosmann and Behl 2002; Oyama et al. 1996; Subirade et al. 1995). Phenolic compounds such as catechins and epicatechins are reported to protect neurons against a range of oxidative and metabolic insults in the case of Parkinson's and Huntington's disease models and are also known to suppress the pathogenesis of AD and to protect neurons against the processes of AD (Ehrnhoefer et al. 2006).

## 12.3 Neuroprotective Potentials of Algal Metabolites

The sterols from the microalgae exhibit neuroprotective abilities on the nerves scavenge free radicals and chelate metals thus inhibiting metal-induced peroxidation in the brain (Koller et al. 2014). Recent studies have demonstrated that astaxanthin, a carotenoid found in marine organisms, has a controlling effect on neurodegenerative diseases that are induced by oxidative stress (Gandla et al. 2023). Astaxanthin, a red carotenoid with potent antioxidant activity, can be 10 times greater than that of other carotenoids and 500 times greater than that of α-tocopherol (Zhao et al. 2019). The most common astaxanthin from microalgae is *Haemotococcus pluvialis* (Oslan et al. 2021a). Zeaxanthin and carotene from algae such as *Dunaliella saline*, *D. tertiolecta*, *Chlorella saccharophila*, *C. zofingiensis*, *Odontella aurita*, and *Haemotococcus pluvialis* have been reported to exhibit neuropathic inhibition in PD and amyotrophic lateral sclerosis enzyme activation. They were also known for the activation of antioxidant and anti-inflammatory properties (Gallego et al. 2022; Isonaka et al. 2012; Repine et al. 1997). Table 12.1 summarizes the metabolites derived from various micro and macroalgae with potential biological activities.

For decades, phycocyanin has received attention in the field of pharmaceuticals as it exhibits diverse pharmacological traits. Phycocyanin is one of the vital pigment components of *Spirulina* and possesses many biological activities, viz. antioxidant, anti-inflammatory, neuroprotective, and hepatoprotective. Romay et al. (2003) reported the neuroprotective effect of phycocyanin in rat cerebellar granule cell cultures and in kainite-induced injury in rats. Heteropolysaccharides of six species of *Sargassum* were reported to possess a neuroprotective effect. The neuroprotective activities of the six heteropolysaccharides were not corelating with the hydroxyl-radical scavenging activity, DDPH-radical scavenging activity, and reducing power they exhibited. Thus, it was concluded that the neuroprotective effects of the family of compounds of *Sargassum* could be a result of the combination of multiple factors (Jin et al. 2014). The investigated evidences strongly support the hypothesis that these compounds possessing activities, such as antioxidant and inhibiting neuronal death through scavenging of free radicals, indicate that they might be considered for the treatment of neurodegenerative disorders. Silva et al. (2018) reported the neuroprotective effect of various seaweeds such as *Saccorhiza polyshides*, *Padina pavonica*, *Sargassum muticum*, *Codium tomentosum*, *Ulva compressa*, and *Bifurcaria bifurcata* to possess neuroprotective through the mechanism of mitochondrial membrane potential, $H_2O_2$ production, Caspase-3 activity, and DNA fragmentation (Silva et al. 2018). Neuroprotective effect and antioxidant activity in the gerbil model were observed by treating with *Halimeda incrassata* aqueous extracts (Rivero

**Table 12.1** Metabolites derived from various micro and macroalgae with potential biological activities

| S. no. | Algae | Extract/compounds | Biological activity-neuroprotection | References |
|---|---|---|---|---|
| 1 | *Amphiroa ephedraea*, *Agarum clathratum* subsp. *yakishiriense*, *Amphiroa bowerbankii*, *Asaragopsis armata*, *Ecklonia bicyclis*, *E. maxima*, *Gelidiella acerosa*, *Gelidium amansii*, *Gloiopeltis furcata*, *Gracilaria*, *Gracilariopsis chorda* *Hydropuntia edulis*, *Hypnea valentine*, *Kappaphycus alvarezii*, *Caulerpa racemose*, *Sargassum polycystum*, *Rhodomelopsis africana*, *Nannochloropsis oceanica*, *Neopyropia yezoensis*, *Myagropsis myagroides*, *Porphyra yezoensis*, *Ochtodes secundiramea*, *Oscillatoria* sp., *Padina australis*, *P. gymnospora*, *P. tetratromatica Papenfussiella lutea*, *Scytothamnus australis*, *Splachnidium rugosum*, *Phormidium* sp., *Porphyra*, *Rhodomela confervoides*, *Myagropsis myagroides Caulerpa racemose*, *S. fulvellum*, *S. fusiforme*, *S. polycystum*, *S. sagamianum*, *S. swartzii*, *S. wightii*, *Ulva* | Acetone/alkaline/benzene/chloroform/dichloromethane/ethanol/hexane/methanol/n-butanol extract/petroleum ether/polyphenolic fraction | Cholinesterase inhibitory activity, anti-oxidative activity: INHIBIT superoxide radicals, hydroxyl radicals, and lipid peroxidation, restoration of cellular viability, neuronal protection from ischemic injury, improvement of behavioral deficits and prevention of dopaminergic neuron loss, improvement of lipopolysaccharide (LPS)-induced memory impairment, increased neurite outgrowth, anti-inflammatory, neuronal cell viability and cell cytotoxicity, neuroprotection and anti-Alzheimer's, significant anticonvulsant activity | Gany et al. (2015), Murugan et al. (2015), De Souza et al. (2007), Ghannadi et al. (2013), Kawee-ai et al. (2013), Hannan et al. (2012, 2014a, b), Jin et al. (2006), Syad et al. (2012), Syad et al. (2013), Machado et al. (2015), Kim et al. (2013c, 2014a, b), Zhou et al. (2015), Cui et al. (2012), Choi et al. (2017), Decheng et al. (2018), Mohibbullah et al. (2016a), Tirtawijaya et al. (2016), Li et al. (2012), Ryan et al. (2011), Aisya et al. (2014), Stirk et al. (2007), Natarajan et al. (2009), Fang et al. (2010), Yoon et al. (2008, 2009), Suganthy et al. (2010), Ryu et al. (2003), Sarithakumari et al. (2013), Zhang et al. (2014b), Bhuiyan et al. (2015), Gany et al. (2014), Shanmuganathan et al. (2015), Yende et al. (2016), Custódio et al. (2016), Pereira and Valado (2021), Giacobini (2010), Aluko (2021), Yang et al. (2015) |

(continued)

**Table 12.1** (continued)

| S. no. | Algae | Extract/compounds | Biological activity-neuroprotection | References |
|---|---|---|---|---|
| | *conglobate*, *Undaria pinnatifida* | | | |
| 2 | *S. vulgare* | Alginate | Anti-inflammation, microglial phagocytosis of Aβ | Zhou et al. (2015), Chen et al. (2014) |
| 3 | *Chlorella vulgaris*, *Diacronema lutheri*, *Haematococcus lacuscris*, *Leptolyngbya* sp., *Phaeodactylum tricornutum*, *Porphyridium purpureum*, *Tetraselmis suecica* | Apigenin | Norepinephrine-elevating action, anti-depressive effect, sleep cycle regulation, sedative effect | Goiris et al. (2014), Tayab et al. (2022) |
| 4 | *Lithothamnion corallioides* | Aquamin | Anti-inflammation | Ryan et al. (2011) |
| 5 | *Haematococcus pluvialis* | *Astaxanthin*, lutein, zeaxanthin, canthaxanthin, β- carotein, oleic acid | Antioxidant, amelioration in cerebral edema, reversed LPS-induced depression-like behaviors, blood-brain barriers disruption, neurological dysfunction, and neuronal degeneration after the induction of subarachnoid hemorrhage | Santhosh et al. (2016) |
| 6 | *Chlorella* sp., *S. macrocarpum*, *Dunaliella saline*, *D. tertiolecta*, *Chlorella saccharophila*, *C. zofingiensis*, *Haemotococcus pluvialis Odontella aurita* | Carotenoids, sulphated polysaccharides, sterols | Antioxidant, anti-inflammatory, promote neurite outgrowth activity | Santhosh et al. (2016), Tsang et al. (2001), Gallego et al. (2022), Repine et al. (1997), Isonaka et al. (2012) |
| 7 | *Chondracanthus acicularis*, *Eucheuma denticulatum*, *H. musciformis* | Carrageenan and derivatives | Anti-apoptotic, anti-inflammation, antioxidative, inhibition of lipid peroxidation | Souza et al. (2018), Yao et al. (2014), De Souza et al. (2009) |

| | | | | |
|---|---|---|---|---|
| 8 | *Caulerpa prolifera and C. racemosa* | Caulerpenyne, caulerpin | Inhibitory activity against LOX; antinociceptive and anti-inflammatory | Cengiz et al. (2011), De Medeiros et al. (2021) |
| 9 | *Spirulina* sp., *S. platensis, Ecklonia cava, Aphanizomenon flos-aquae, Arthrospira platensis* | c-phycocyanin | improves the neurologic deficit score, protects neurons, improves the functional outcome (locomotor behavior) and promotes survival, antioxidant, neuroprotection, anti-inflammation activity, neuroinflammation, oxidative stress, and brain mitochondrial, and synaptic alterations | Pentón-Rol et al. (2011b), Romay et al. (2003), Mitra et al. (2015), Cho et al. (2012), Wang et al. (2021), Scoglio et al (2014), Chen et al. (2012) |
| 10 | *Amphiroa bowerbanii, A. ephearaca, Cladophora fasicularis, Codium capitatum, Dictyota humifusa, Gelidiella acerosa, Gracilaria edulis, G. gracilis, Halimeda cuneata, Hypnea musciformi, H. valentiae, Laurencia translucida, Porphyra perforate, Sargassum* sp., *S. boveanum, S. oligocystum, S. sagamianum, Spirogyra gratiana, Ulva fasciata, U. reticulata, Capsosiphon fulvescens* | Crude extract | Cholinesterase inhibitory activities,<br>Improves mood, counteracts anxiety, and enhances attention and learning | Choi et al. (2007, 2015b), Fang et al. (2012), Syad et al. (2016), Kartal et al. (2009), Natarajan et al. (2009) |
| 11 | *Sargassum fulvellum* | Dimethyl sulfonio propionate | Protects against neurotoxicity; attenuates stress response and mitochondrial damages | Wichmann et al. (2016) |
| 12 | *Nannochloropsis* sp. | Eicosapentaenoic acid | Antioxidant | Santhosh et al. (2016) |

(continued)

**Table 12.1** (continued)

| S. no. | Algae | Extract/compounds | Biological activity-neuroprotection | References |
|---|---|---|---|---|
| 13 | *Arthrospira* sp., *Chlorella* sp., *Desmodesmus* sp., *Diaeronema* sp., *Dunaliella salina*, *Haematococcus* sp., *Nannochloropsis* sp., *Phaeodactylum* sp., *Porphyridium* sp., *Tetraselmis* sp. | Ferulic acid | Mood regulation, anti-depressive action, stress lowering activity, anti-inflammatory action, increased norepinephrine concentration | Goiris et al. (2014), Ferdous et al. (2021), Tayab et al. (2022), Bellahcen et al. (2020) |
| 14 | *Chondrus cripus*, *Laurencia undulate* | Floridoside, glycerol glycosides, d-isofloridoside | Anti-inflammation extract mediated protection against Parkinson's disease pathology | Kim et al. (2013b), Liu et al. (2015) |
| 15 | *Alaria esculenta*, *Brythamnion triquetrum*, *Cystoseira humilis*, *Dictyota humifusa*, *Ericaria selaginoides*, *Gongolaria nodicaulis* | Fractions | Inhibit the formation of amyloid aggregates by α-synuclein, AChE BuChE inhibitory capacity | Giffin et al. (2017), Fallarero et al. (2003, 2006), Custódio et al. (2016) |
| 16 | *Chlorella pyrenoidosa*, *Dictyota ciliolate*, *Chrysoderma* sp. *Chrysomeris* sp., *Chyrsowaernella* sp., *Diacronema lutheri*, *E. maxima*, *E. stolonifera*, *Ecklonia cava*, *Fucus vesiculosus*, *Ishige okamurae*, *Isochrysis galbana*, *Laminaria japonica*, *Nannochloropsis* sp. BR2, *Olisthodiscus luteus*, *Phaedactylum tricornutum*, *S. crassifolium*, *S. fulvellum*, | Fucoidan; (sulphated polysaccharide) Fucose- rich sulfated | Anti-inflammation, antioxidant, neuroprotective, glial activation, mitochondria biogenesis, inhibits beta amyloid induced microglial clustering, ameliorating learning, and memory deficiencies, neuroprotective effects in Parkinson's disease in vivo anticholinergic action, elevated serotonin and norepinephrine concentrations, improvement of cognitive function, promotion of | Cui et al. (2010), Han et al. (2019), Chale-Dzul et al. (2017), Kim et al. (2014d, 2019), Jhamandas et al. (2005), Park et al. (2011), Alghazwi et al. (2019), Li et al. (2004), De Souza et al. (2007), Lee et al. (2007, 2020), Do et al. (2009), Ryu et al. (2016), Che et al. (2017), Huang et al. (2018), Hu et al. (2016), Gao et al. (2012b), Gao et al. (2012a), Yang et al. (2017), Zhang et al. (2014b), Meenakshi et al. (2016), Wei et al. |

| | | | | |
|---|---|---|---|---|
| | *S. hemiphyllum*, *S. horneri*, *S. polycystum*; *Saccharina japonica*; *Turbinaria durren*, *Tetraselmis* sp. M8 *Undaria pinnatifida* | | cell longevity, inhibition of amyloid beta-induced toxicity | (2017), Luo et al. (2009), Liang et al. (2019), Wu et al. (2019), Wong et al. (2018), Park et al. (2019), Sanjeewa et al. (2018), Barreira et al. (2015), Yehuda et al. (1998), Reigada et al. (2021), Zhang et al. (2014a) |
| 17 | *Ecklomia bicyclis*, *E. stolonifera* *U. pinnatifita*, *Eisenia bicycylis*, *Hizikia fusiformis*, *Padina australis*, *Sargassum binderi*, *S. horridum*, *S. siliquatrum*, *U. pinnatifida*, *Enteromorpha prolifera*, *Padina australis*, *Papenfussiella lutea*, *Pelvetia siliquosa*, *S. fulvellum*, *Ulva lactuca*, *Undaria* sp. | Fucosterol 3,6,17-trihydroxy-stigmasta-4,7,24(28)-triene, 14,15,18,20-diepoxyturbinarin (sterols), fucofuroeckols; and 24-hydroperoxy24-vinylcholesterol | Reduced inflammatory response, Attenuates Aβ1-42 induced neurotoxicity, reduced apoptosis in a β-induced ameliorates aβ1–42- induced memory impairment, protection against oxidative stress, reduced inflammatory response, extended lifespan, inhibiting AChE and BChE activities, antioxidant, neuronal survival against oxidative stress, memory enhancement, improvement in the neurologic deficit score, anti-apoptosis, induction of autophagy, reduced lipid peroxidation, neurite outgrowth | Jung et al. (2013), Alghazwi et al. (2019), Oh et al. (2018), Gan et al. (2019), Lee et al. (2003), Sun et al. (2015), Liu et al. (2019), Oktaviani et al. (2019), Wong et al. (2018), Fernando et al. (2019), Choi et al. (2015c), Castro-Silva et al. (2019), Rengasamy et al. (2015), Rafiquzzaman et al. (2015), Pangestuti et al. (2013), Yoon et al. (2008, 2012), Mohibbullah et al. (2018), Lin et al. (2017), Hu et al. (2018), Zhang et al. (2017), Jung et al. (2016), Yang et al. (2017), Sangeetha et al. (2009), Kawee-ai et al. (2013), Aisya et al. (2014), Hoong et al. (2018), Stirk et al. (2007), Custódio et al. (2016), Syad et al. (2012), Shrestha et al. (2022) |
| 18 | *Phaeodactylum*, *Pinnatifida*, *Sargassum* sp., *S. horneri*, *S. siliquatrum*, *Hijikia fusiformis* | Fucoxanthin, xanthophyll | Antioxidant, anti-inflammation, neuronal survival against oxidative stress, secondary injury, cell survival, attenuation of Aβ42, | Liu et al. (2019), Pangestuti et al. (2013), Zhao et al. (2017), Ikeda et al. (2003), Zhang et al. (2017), Lin et al. (2017), Hu et al. (2018), |

(continued)

**Table 12.1** (continued)

| S. no. | Algae | Extract/compounds | Biological activity-neuroprotection | References |
|---|---|---|---|---|
| | | | cholinesterase inhibitory activities, prevention of neuronal apoptosis | Alghazwi et al. (2019), Heo et al. (2008), Natarajan et al. (2009), Jung et al. (2008), Kim et al. (2013b), Yu et al. (2017), Yan et al. (1999), Liu et al. (2017) |
| 19 | *Arthrospira platensis*, *Anabaena* sp., *Chlamydomonas*, *Chondrophycus undalatus*, *Chlorella vulgaris*, *Undaria pinnatifida* | Glycerol glycosides: floridoside, glycoprotein | Suppress pro-inflammatory responses in microglia, antidepressant, improved memory and learning ability, attenuates oxidative stress, neurogenesis, neuroprotection, promotion of cell survival and neurite extension, reduced anxiety | Rafiquzzaman et al. (2015), Shanmuganathan et al. (2015, 2018), Fernando et al. (2019), Abdul et al. (2016) |
| 20 | *Tetraselmis* | Hexadecatrienoic acid | Antioxidant | Custódio et al. (2014) |
| 21 | *E. cava* | Laminarian | Antioxidant, neuro-inflammation, reduced reactive gliosis (M1 microglia), analgesic activity | Park et al. (2020a, b), Lee et al. (2020), Kim et al. (2014c) |
| 22 | *Isochrysis* sp. *Phacodactylum tricornutum Phacodactylum tricornutum*, *Cryptomonas* sp., *Dunaliella* sp., *Nephroselmis* sp., *Rhodomonas* sp., *Tetraselmis viridis* | Lipids, fatty acid, carotenoids lipids, phytol lipids, phytol, polysaturated fatty acids including omega-3 fatty acid | Antioxidant, neuroprotective, neuroprotective action, sleep promoting action, improvement of cognitive ability, mood regulation and anti-depressive action, anxiety lowering effects | Santhosh et al. (2016), Singh et al. (2017), Custódio et al. (2014), Huang et al. (2009), Barreira et al. (2015), Yehuda et al. (1998), Reigada et al. (2021) |
| 23 | *Sargassum fulvellum*, *S. swartzii*, *S. micracanthum* | Meroterpenoid, meroditerpenoids, sargachromenol, sargaquinoic acid | Antioxidant, neurite outgrowth | Natarajan et al. (2009), Tsang et al. (2005), Kamei et al. (2003) |

| | | | | |
|---|---|---|---|---|
| 24 | *Pyropia yezoensis* | Oligo-porphyran | Antiapoptotic, antioxidation, anti-inflammation, ameliorates behavioral deficits | Liu et al. (2018, 2019) |
| 25 | *Amphiroa beauvoisii* *Gelidium foliaceum* | Phenolic and flavonoid compound | Inhibiting AChE | Hannan et al. (2014b), Rengasamy et al. (2015) |
| 26 | *Spirulina* sp. | Phenols, vitamins E, docosahexaenoic acid, linolenic | Antioxidant, anti-inflammatory | Santhosh et al. (2016), Singh et al. (2017) |
| 27 | *Sargassum fulvellum* | Pheophytin A | NGF-independent neurite outgrowth | Ina et al. (2007) |
| 28 | *Ecklonia* sp., *E. bicyclis*, *E. cava*, *E. kurome*, *E. maxima*, *E. radiata*, *E. stolonifera*, *Eisenia*, *Fucus guiryi*, *F. serratus*, *F. spiralis*, *F. vesiculosus*, *Ishige okamurae*, *Marginariella boryana* *Padina australis*, *Pelvetia canaliculata* *Undaria pinnatifida*, *Zonaria spiralis* | Phlorotannins, phlorglucinol, eckol, triphloroethol A, eckstolonol, bieckol, dieckol; phlorofucofuroeckol, phlorotannin- rich fucoidan | Cholinesterase inhibitory activities (AChE and BuChE), potential ability to enhance memory in neurodegenerative disorders; neuroprotective activity, anti- apoptotic properties, improvement of cognitive impairments, aggregation of b-amyloid, protection from retinal neuronal death, protects against neuronal cells against Aß1–42 evoked toxicity, suppression of inflammatory responses | Kang et al. (2012), Kim et al. (2012, 2013a, 2014d, 2016a), Zhang et al. (2015), Jung et al. (2009), Kannan et al. (2013), Myung et al. (2005), Choi et al. (2015c), Liu et al. (2019), Lee et al. (2019), Yoon et al. (2008, 2009), Heo et al. (2012), Yu et al. (2015), Kang et al. (2005), Ahn et al. (2012), Rafiquzzaman et al. (2015), Jung et al. (2013), Zhang et al. (2012), Choi et al. (2015c), Jung et al. (2010), Liu et al. (2012), Yang et al. (2014), Li et al. (2009), Cho et al. (2012), Shrestha et al. (2020), Wang et al. (2018b), Yotsu-Yamashita et al. (2013), Wang et al (2018a), Kim et al. (2016b) |
| 29 | *Gelidiella acerosa* | Phytol, acyclic diterpene alcohol | Antioxidant activities and cholinesterase inhibitory potential for AChE and BuChE | Syad et al. (2016), Fang et al. (2010) |

(continued)

**Table 12.1** (continued)

| S. no. | Algae | Extract/compounds | Biological activity-neuroprotection | References |
|---|---|---|---|---|
| 30 | *S. macrocarpum* | Pigment: pheophytin A | Produce neurite out growth | Ina et al. (2007) |
| 31 | *S. fusiforme*, *S. sagamianum* | Plastoquinones, sargaquinoic acid and sargachromenol | Anti-oxidative activity, lipid peroxidation, inhibiting AChE inhibiting BuChE | Mori et al. (2005), Heo et al. (2008) |
| 32 | *Spirulina platensis*, *Porphyridium* sp., *Porphyra* sp. | Polysaccharide, sulphated polysaccharides, phycobiliproteins | Neuroprotective, antioxidant, anti-inflammatory | Zhang et al. (2015), Santhosh et al. (2016), Yabuta et al. (2010) |
| 33 | *Porphyra capensis*, *Pyropia haitanensis* | Porphyrin | Anti-apoptotic, improved learning and memory deficits | Liu et al. (2018), Zhang et al. (2020) |
| 34 | *Arthorospira platensis*, *Calothrix brevissima*, *Diacronema lutheri*, *Haematococcus pluvialis*, *Hapalosiphon fontinalis*, *Limnothrix obliqueaeumina* *Microchaete tenue*, *Nostoc ellipsosporum* *Porphyridium pureum*, *Scenedesmus quadricauda*, *Westiellopsis prolifica* | Quercetin | Improvement in cognitive functions, locomotor activity and mood regulation | Tayab et al. (2022), Fu et al. (2019), De Medeiros et al. (2021) |
| 35 | *Caulerpa racemose* | Recemosins A | Increased cell viability, reduced apoptosis | Liu et al. (2013) |
| 36 | *Sargassum micracanthum* *S. siliquastrum*, *S. macrocarpum* *S. macrocarpum* *S. sagamianum* *S. fusiforme* | Sarachromenol, sargaquinoic acid sargachromenol saringosterol | Anti-inflammation, neurite outgrowth and survival, cell differentiation and ChE activity | Yang et al. (2013), Kang et al. (2013), Tsang et al. (2005), Kamei et al. (2003), Myung et al. (2005), Hoang et al. (2012) |
| 37 | *E. stolonifera*, *S. sagamianum* | Sterol (fucosterol), total sterols and β-sitosterol | Inhibiting AChE, antidepressant effects | Yoon et al. (2008), Zhao et al. (2016) |

| | | | | |
|---|---|---|---|---|
| 38 | *Ulva lactuca*, *E. prolifera*, *Codium fraglie Chlamydomonas reinhardtii Gracilaria cornea*, *P. gymnospora*, *Turbinaria decurrens*, *Undaria pinnatifida* | Sulphate agaran, fucans, oligosaccharides, and polysaccharide | Anti-inflammatory, antioxidant activity, anti-PD activity, restores behavioral deficits and locomotors performance, prevents the accumulation of Aβ, anti-Parkinson's activity | Wozniak et al. (2015), Liu et al. (2019), Wang et al. (2020b), Panigrahi et al. (2019), Souza et al. (2017, 2018), Seong et al. (2019), Wang et al. (2020a) |
| 39 | *Ulva lactuca*; *U. intentinalis* | Ulvan | Antioxidant, inhibition of AchE and BuchE activity, immunostimulation | Olasehinde et al. (2019), Guo et al. (2018) |
| 40 | *Sargassum fulvellum* | Vitamin B12 (chlorophyll-related analog to pheophytin) | Cell differentiation | Ina et al. (2006) |
| 41 | *Arthrospira platensis*, *Chlamydomonas reinhardii*, *Chlorella*, *Euglena gracilis* | Whole biomass | Antioxidant, anti-inflammation activity, neuroprotection, norepinephrine elevating action, anti-depressive effect, sleep cycle regulation, sedative effect oxygen generation in CNS | Voss et al. (2021), Nuzzo et al. (2018), Goiris et al. (2014), Tayab et al. (2022), Nakashima et al. (2009) |
| 42 | *Dictyopteris undulate*, *Padina gymnospora Eckolnia bicyclis*, *S. siliquastrum*, *Saccharina japonica* | Zonarol (p-hydroquinone sesquiterpene), α-bisabolol, monocyclic sesquiterpene alcohol | Neuronal survival against oxidative stress, anti-apoptosis, protection against Aβ-induced proteotoxicity, reduced AChE and BuChE activity | Shimizu et al. (2015), Shanmuganathan et al. (2018, 2019), Lin et al. (2017), Jung et al. (2016), Gany et al. (2015), Choi et al. (2007), Wozniak et al. (2015) |
| 43 | *Aphanizomenon flos-aquae* | β-phenylethylamine | Sleep cycle regulation, anti-depressive effect, neuron regeneration, antioxidative and anti-inflammatory action | Goiris et al. (2014), Ferdous et al. (2021) |

et al. 2003). *Dicyota*, a brown alga with a rich source of diterpenes, was reported to produce hydroazylene diterpenes, exhibiting neuroprotective effects against Cerebral Ischemia-reperfusion injury (Wu et al. 2021). Gold nanoparticles of *Dictyopteris*, when examined for the neuroprotective effects in neuroblastoma cells and treated with oxygen and glucose deprivation, it was found that their neuroprotection was accompanied by the regulation of apoptosis-related proteins (Park et al. 2020b). A recent report on phlorotannins, for the first time, reported the cell wall phlorotannins from two species of the order Fucales, *Fucus vesiculosis* and *Pelvetia canaliculate*, were exhibiting neuroprotective, cytotoxic, antioxidant, and antibacterial properties (Meshalkina et al. 2023).

### 12.3.1 Anti-oxidant Activity

In normal conditions, cells are able to balance the production of oxidants and antioxidants. Any disproportion in the natural derivative of oxidation, i.e., reactive oxygen species (ROS), causes oxidative damage of biomolecules, viz. RNA and DNA (Moncada et al. 1991; Pizzino et al. 2017). *Sargassum wightii* was considered the best natural source of phytochemicals with excellent antioxidant properties that improve antibacterial activity too (Rajivgandhi et al. 2021). *Spirulina* includes various species of primitive unicellular blue-green algae, most commonly *S. platensis* and *S. maxima*. One of the common compounds, spirulina maxima, which is used as a food additive, is a microalga rich in protein and other essential nutrients. It has been reported that *Spirulina maxima* has antioxidant and anti-inflammatory effects. Recently, many reports showed the neurotoxicity effects on central nervous system (CNS). Trimethyltin as a compound used a model of cognitive disorders and induced neurotoxicity in CNS. *S. maxima* extract elevated protein related with neuroprotective effects against trimethyltin-induced cognitive disorders in HT-22 cells (Koh et al. 2016). The potential of phlorotannins from brown alga *Ecklonia cava* was reported to inhibit peroxidation and protection of neuronal cells against $H_2O_2$-induced neurotoxicity using murine hippocampal HT22 cells (Kang et al. 2012). Cytoprotective properties of hydroazulene diterpenes of brown alga *Dictyota* against oxidative stress was reported (Wu et al. 2021). As such, hydroazulene diterpenes could serve as lead structures for the development of novel neuroprotective agents against cerebral ischemia-reperfusion injury. Neuroprotective effects of sulfated agaran from marine alga *Gracilaria cornea* exhibited behavioral, neurochemical, and transcriptional alteration in rat 6-hydroxydopamine Parkinson's disease model (Souza et al. 2017). Seaweeds from *Halimeda* genus showed highest antioxidant activities, but *Galaxaura* genus, as well as *Dictyopteris justis*, *Bryothamnium triquetrum*, and *Caulerpa racemosa*, exhibited effectiveness in the inhibition of spontaneous lipoperoxidation of rat brain homogenates. *Halimeda incrassata* was reported to exhibit possible existence of several bioactive water soluble compounds conferring in vitro antioxidant properties and in vivo neuroprotective effect (Rivero et al. 2003).

### 12.3.2 Anti-inflammatory Activity

Inflammation is a reaction of tissues toward infection, irritation, or foreign substance linked with the pathogenesis of many diseases such as cancer, neurodegenerative diseases, arthritis, and other metabolic disorders. Parkinson's disease is a multifactorial disease associated with the degeneration of dopaminergic neurons and behavioral alterations. New pharmacological alternatives to neurological disorders, such as Parkinson's disease, can be found in natural bioactive compounds. The marine alga *Gracilaria cornea* has sulfated polysaccharide (SA-Gc) was reported to demonstrate anti-parkinsonian activity against behavioral and weight gain alternations induced by 6-OHDA. In addition, SA-Gc showed transcriptional modulation of the NFKb/IL-1β pathway. Their study also revealed the neuroprotective effects that might be associated with the upregulation of the GSH levels and BDNF gene transcription (Souza et al. 2017). *Sargussum homeri* possessed sulfated polysaccharide, and the purified compounds have been reported as an anti-inflammatory agent (Wen et al. 2016). Raman et al. (2014) reported the maximum anti-inflammatory activity that it prevents prostaglandin biosynthesis by inhibiting cyclooxygenease (COX) enzymes. Astaxanthin is one of the most potent carotenoid compounds on the market. In recent decades, they have gained interests due to their powerful antioxidant, anti-inflammatory, mending, antiproliferative, and possible antiaging properties (Oslan et al. 2021b).

### 12.3.3 Anti-aging Activity

Aging is a factor in NDD, suggesting that the accumulation of neuronal stresses over time causes cell death. However, during normal aging, there is relatively little neuronal loss, which is in contrast to high degree of neuronal death in neurodegenerative diseases (Park et al. 2020b). The lavish use of synthetic components in cosmetics can lead to harmful effects, and hence, to mitigate the consequences, over the past decade, there is a preference toward nature-derived skincare products. Rhodophyceae members are considered the primitive group of algae (Qiu et al. 2022), and they have been consumed for decades as food supplements toward meeting the nutritional requirement for a healthy living (Leandro et al. 2020). They are abundant producers of sulfates, carrageenan, polysaccharides, mycosporin-like amino acids, and phenolic compounds, which scavenge free radicals and antioxidants, thereby preventing damage to humans (Zhang and Duan 2018). Carrageenan aids in enhancing the elasticity and appearance of the skin by moisturizing and hence prevents anti-aging. *Kappaphycus alvarezii* and *Chondrus crispus* are the potentially red algae from which top three marine components are obtained (Resende et al. 2021). The potential anti-aging effects of protocatechuic acid were reported to promote type-1 collagen (Shin et al. 2020). Flavonoids are favored in anti-aging therapies due to their antioxidant and anti-aging potentialities. Anthocyanins are reported to slow down skin aging by stopping inflammation and peroxidation of lipids (Dixit and Reddy 2017). Seaweeds such as *Ecklonia cava*,

*Ericaria selaginoides*, *Gongolaria nodicaulis*, *G. usneoides* and *Cystoseira usneoides* are known for their anti-wrinkling, anti-aging, anti-photoaging, and antioxidant properties (Ferreres et al. 2012; Wang et al. 2020b). *Ericaria amentacea*, a brown alga, was reported to slow down the formation of wrinkles and uneven pigmentation and was recommended ideal component toward the preparation of sunburn symptoms and anti-aging lotions (Mirata et al. 2023). The application of microalgae in cosmeceutics have recently gained interest as the extensive studies were limited to *Chlorella* and *Arthrospira* strains (Ioannou and Labrou 2022; Morais et al. 2021; Savvidou et al. 2023). A recent study by Georgakis et al. (2023) reported the potential cosmeceutical effect of yellow-green microalgae, *Trachydiscus minutus*.

### 12.3.4 Cholesterol Homeostasis, Anti-amyloidogenic, and Aggregation Inhibition Activity

One of the leading causes of death worldwide is atherosclerosis, with an estimated number of more than 72 million deaths every year. Phytosterol consumption is suggested as a therapeutic option to reduce plasma cholesterol levels and atherosclerosis. The most copious sterol in marine algae, fucosterol, has been reported to inhibit the absorption of cholesterol from the intestine (Tapiero et al. 2003). In addition, another derivative of fucosterol, 4-methylsterol from the seaweed *Crypthecodinium cohnii* was showing an effect on cholesterol metabolism in rats (Kritchevsky et al. 1999). Various fractions of *Gelidiella acerosa* exhibited antioxidant and anticholinesterase activities and the pooled fractions verified the neuroprotective activity. The LC-MS analysis revealed phytol and the assessment of phytol for neuroprotective effect revealed very promising results and further supported with molecular docking studies for the interaction of phytol with cholinesterase (Syad et al. 2016). Beta-amyloid (Aβ) can form aggregates through self-assembly and produce neurotoxicity in the early stage of AD. Therefore, the inhibition of Aβ assembly is considered the primary target for AD therapy. Xiang et al. (2017) reported that fucoxanthin, a marine carotenoid, potently reduced the formation of Aβ fibrils and oligomers. Likewise, fucoxanthin significantly inhibited oxidative stress, enhanced the expression of brain-derived neurotrophic factor, and increased ChAT-positive regions in the hippocampus of mice (Xiang et al. 2017).

### 12.3.5 Neurotrophic Activity

In addition to support neuronal development, neurotrophic factors are crucial to protect mature neurons from atrophy in neurodegenerative diseases. Despite the natural sources that are known to exhibit neurotrophic activity, not much are focused on natural substances from marine source. Seaweeds, one of the major marine inhabitants with potential sources for such substances are in need of an in-depth investigation. Hannan et al. (2013) reported seaweed species, *Gelidium amansii*, followed by *Undaria pinnatifida* and *Sargassum fulvellum* extracts to possess potent activity. *Porphyra yezoensis* is the most popular edible red alga known for its economic and medicinal importance. An investigation by Mohibullah et al. (2016b) demonstrated the neurotrophic and neuroprotective effects of the ethanol extracts of *P. yezoensis* in the cultures of hippocampal neurons. One of natural carotenoids, fucoxanthin, is abundant in edible brown seaweeds, which has been shown to possess, for the first time, effective protection against scopolamine-induced cognitive impairments in mice (Lin et al. 2016).

## 12.4 Concluding Remarks

As a promising strategy for the treatment or prevention of neurodegenerative disease, algae could be used to create novel functional ingredients in pharmaceuticals and functional foods. Algae are a valuable source of neuroprotective agents and can be used in pharmaceuticals and functional foods as an alternative to synthetic ingredients that can contribute to neuroprotection. Although a variety of algae has been found to contain neuroactive compounds, seaweed species belonging to the phaeophyceae group produce the most compounds. However, some promising metabolites may also be available from species from other groups, such as *Gelidium amansii*, which belongs to the rhodophyceae and has significant neuromodulatory effects. Algal natural compounds, such as phlorotannins, alginates, fucoidan, sargaquinoic acid, SPs, and carotenoids, have a wide range of biological activities, Despite a sizeable collection of algae-based natural products with distinct neuroprotective functions, only sodium oligomannate has emerged as successful drug for AD. Marine algae have uncovered a few compounds with potential for neurite outgrowth, in contrast to neuroprotective compounds that could help neuronal survival. Therefore, it is necessary to screen compounds reported to exhibit neuroprotective ability as well as the compounds that have not been explored for their potential to promote neurite extension. Moreover, most of the studies on the biological activities of the algal metabolites are performed in vitro and in mouse model organisms. Hence, there is a necessity for further research on humans to investigate the neuroprotective activities to increase the potential to expand neuroprotective effects and health-beneficial value of algae in the pharmaceutical industry. Prioritizing the investigation of less-explored algal compounds with potential for neurite outgrowth promotion could lead to breakthroughs in neuronal regeneration. Furthermore, transitioning from promising in vitro and animal studies to rigorous human trials will be crucial for establishing the safety, efficacy, and translational potential of these algal compounds. By addressing these challenges and collaborating with the pharmaceutical industry, we can pave the way for the development of novel algae-derived therapeutics for neurodegenerative diseases.

## References


Abdul QA, Choi RJ, Jung HA et al (2016) Health benefit of fucosterol from marine algae: a review. J Sci Food Agric 96:1856–1866. https://doi.org/10.1002/jsfa.7489

Ahn BR, Moon HE, Kim HR et al (2012) Neuroprotective effect of edible brown alga *Eisenia bicyclis* on amyloid beta peptide-induced toxicity in PC12 cells. Arch Pharm Res 35:1989–1998. https://doi.org/10.1007/s12272-012-1116-5.59

Aisya GS, Ching TS, Yee GS (2014) Antioxidative, anticholinesterase and anti-neuroinflammatory properties of malaysian brown and green seaweeds. Int J Ind Manuf Eng 8:895–906. https://doi.org/10.5281/zenodo.1099988

Alghazwi M, Smid S, Karpiniec S et al (2019) Comparative study on neuroprotective activities of fucoidans from *Fucus vesiculosus* and *Undaria pinnatifida*. Int J Biol Macromol 122:255–264. https://doi.org/10.1016/j.ijbiomac.2018.10.168

Aluko RE (2021) Food-derived acetylcholinesterase inhibitors as potential agents against Alzheimer's disease. eFood 22:49–58. https://doi.org/10.2991/efood.k.210318.001

Angeloni C, David V (2019) Natural products and neuroprotection. Int J Mol Sci 22:5570. https://doi.org/10.3390/ijms20225570

Barreira L, Pereira H, Gangadhar KN et al (2015) Medicinal effects of microalgae-derived fatty acids. In: Kim SK (ed) Handbook of marine microalgae. Elsevier, London, pp 209–231. https://doi.org/10.1016/B978-0-12-800776-1.00013-3

Bastianetto S, Zheng WH, Quirion R (2000) Neuroprotective abilities of resveratrol and other red wine constituents against nitric oxide-related toxicity in cultured hippocampal neurons. Br J Pharmacol 131(4):711–720. https://doi.org/10.1038/sj.bjp.0703626

Batista AP, Nunes MC, Fradinho P (2012) Novel foods with microalgal ingredients–effect of gel setting conditions on the linear viscoelasticity of *Spirulina* and *Haematococcus* gels. J Food Eng 110(2):182–189. https://doi.org/10.1016/j.jfoodeng.2011.05.044

Becker EW (2013) Microalgae for human and animal nutrition. In: Handbook of microalgal culture: applied phycology and biotechnology, pp 461–503. https://doi.org/10.1002/9781118567166.ch25

Bellahcen TO, AAmiri A, Touam I et al (2020) Evaluation of Moroccan microalgae: *Spirulina platensis* as a potential source of natural antioxidants. J Complement Integr Med 17:20190036. https://doi.org/10.1515/jcim-2019-0036

Bhuiyan MMH, Mohibbullah M, Hannan MA et al (2015) *Undaria pinnatifida* promotes spinogenesis and synaptogenesis and potentiates functional presynaptic plasticity in hippocampal neurons. Am J Chin Med 43:529542. https://doi.org/10.1142/S0192415X15500330

Cardoso DCLA, Kanno KYF, Karp SG (2017) Microbial production of carotenoids a review. Afr J Biotechnol 16(4):139–146. https://doi.org/10.5897/AJB2016.15763

Castro-Silva ES, Bello M, Hernandez-Rodriguez M et al (2019) *In vitro* and *in silico* evaluation of fucosterol from *Sargassum horridum* as potential human acetylcholinesterase inhibitor. J Biomol Struct Dyn 37:3259–3268. https://doi.org/10.1080/07391102.2018.1505551

Cengiz S, Cavas L, Yurdakoc K et al (2011) The sesquiterpene *Caulerpenyne* from *Caulerpa spp.* is a lipoxygenase inhibitor. Mar Biotechnol 13:321–326. https://doi.org/10.1007/s10126-010-9303-1

Chale-Dzul J, Freile-Pelegrín Y, Robledo D et al (2017) Protective effect of fucoidans from tropical seaweeds against oxidative stress in HepG2 cells. J Appl Phycol 29:2229–2238. https://doi.org/10.1007/s10811-017-1194-3

Che N, Ma Y, Xin Y et al (2017) Protective role of fucoidan in cerebral ischemia-reperfusion injury through inhibition of MAPK signalling pathway. Biomol Ther 25:272–278. https://doi.org/10.4062/biomolther.2016.098

Chen J-C, Liu KS, Yang T-J et al (2012) *Spirulina* and *C-phycocyanin* reduce cytotoxicity and inflammation-related genes expression of microglial cells. Nutr Neurosci 15(6):252–256. https://doi.org/10.1179/1476830512Y.0000000020

Chen PB, Wang HC, Liu YW et al (2014) Immunomodulatory activities of polysaccharides from *Chlorella pyrenoidosa* in a mouse model of Parkinson's disease. J Funct Foods 11:103–113

Chi H, Chang HY, Sang TK (2018) Neuronal cell death mechanisms in major neurodegenerative diseases. Int J Mol Sci 19(10):3082. https://doi.org/10.3390/ijms19103082

Cho S, Han D, Kim SB et al (2012) Depressive effects on the central nervous system and underlying mechanism of the enzymatic extract and its phlorotannin-rich fraction from *Ecklonia cava* edible brown seaweed. Biosci Biotechnol Biochem 76:163–168. https://doi.org/10.1271/bbb.110702

Choi JS, Han YR, Byeon JS et al (2015b) Protective effect of fucosterol isolated from the edible brown algae, *Ecklonia stolonifera* and *Eisenia bicyclis*, on tert-butyl hydroperoxide- and tacrine-induced HepG2 cell injury. J Pharm Pharmacol 67:1170–1178. https://doi.org/10.1111/jphp.12404

Choi JS, Haulader S, Karki S et al (2015a) Acetyl- and butyryl-cholinesterase inhibitory activities of the edible brown alga *Eisenia bicyclis*. Arch Pharm Res 38:1477–1487. https://doi.org/10.1007/s12272-014-0515-1

Choi JY, Hwang CJ, Lee HP et al (2017) Inhibitory effect of ethanol extract of *Nannochloropsis oceanica* on lipopolysaccharide-induced neuroinflammation, oxidative stress, amyloidogenesis and memory impairment. Oncotarget 8:45517–45530. https://doi.org/10.18632/oncotarget.17268

Choi BW, Lee HS, Shin HC et al (2015c) Multifunctional activity of polyphenolic compounds associated with a potential for Alzheimer's disease therapy from *Ecklonia cava*. Phytother Res 29:549–553. https://doi.org/10.1002/ptr.5282

Choi BW, Rhu G, Park SH et al (2007) Anticholinesterase activity of plastoquinones from *Sargassum sagamianum*, lead compounds for Alzheimer's disease therapy. Phytother Res 21:423–426. https://doi.org/10.1002/ptr.2090

Commenges D, Scotet V, Renaud S et al (2000) Intake of flavonoids and risk of dementia. Eur J Epidemiol 16(4):357–363. https://doi.org/10.1023/a:1007614613771

Cui YQ, Jia YJ, Zhang T et al (2012) Fucoidan protects against lipopolysaccharide-induced rat neuronal damage and inhibits the production of proinflammatory mediators in primary microglia. CNS Neurosci Ther 18:827–833. https://doi.org/10.1111/j.1755-5949.2012.00372.x

Cui YQ, Zhang LJ, Zhang T et al (2010) Inhibitory effect of fucoidan on nitric oxide production in lipopolysaccharide activated primary microglia. Clin Exp Pharmacol Physiol 37:422–428. https://doi.org/10.1111/j.1440-1681.2009.05314.x

Custódio L, Silvestre L, Rocha MI et al (2016) Methanol extracts from *Cystoseira tamariscifolia* and *Cystoseira nodicaulis* are able to inhibit cholinesterases and protect a human dopaminergic cell line from hydrogen peroxide-induced cytotoxicity. Pharm Biol 54:1687–1696. https://doi.org/10.3109/13880209.2015.1123278

Custódio L, Soares F, Pereira H et al (2014) Fatty acid composition and biological activities of *Isochrysis galbana T-ISO*, *Tetraselmis sp.* and *Scenedesmus sp.*: possible application in the pharmaceutical and functional food industries. J Appl Phycol 26(1):151–161. https://doi.org/10.1007/s10811-013-0098-0

De Medeiros VPB, de Souza EL, de Albuquerque TMR et al (2021) Freshwater microalgae biomasses exert a prebiotic effect on human colonic microbiota. Algal Res 60:102547. https://doi.org/10.1016/j.algal.2021.102547

De Souza ET, de Lira DP, de Queiroz AC et al (2009) The antinociceptive and anti-inflammatory activities of caulerpin, a bisindole alkaloid isolated from seaweeds of the genus *Caulerpa*. Mar Drugs 7(4):689–704. https://doi.org/10.3390/md7040689

De Souza MCR, Marques CT, Dore CMG et al (2007) Antioxidant activities of sulfated polysaccharides from brown and red seaweeds. J Appl Phycol 19:153–160. https://doi.org/10.1007/s10811-006-9121-z

Decheng B, Qiuxian L, Qingguo H et al (2018) Seleno-polymannuronate attenuates neuroinflammation by suppressing microglial and astrocytic activation. J Funct Foods 51:113–120. https://doi.org/10.1016/j.jff.2018.10.010

Dixit D, Reddy CRK (2017) Non-targeted secondary metabolite profile study for deciphering the cosmeceutical potential of red marine macro alga *Jania rubens*—an LCMS-based approach. Cosmetics 4(4):45. https://doi.org/10.3390/cosmetics4040045

Do H, Pyo S, Sohn EH et al (2009) Suppression of iNOS expression by fucoidan is mediated by regulation of p38 MAPK, JAK/STAT, AP-1 and IRF-1, and depends on up-regulation of scavenger receptor B1 expression in TNF-alpha- and IFN-gamma-stimulated C6 glioma cells. J Nutr Biochem 21:671–679. https://doi.org/10.1016/j.jnutbio.2009.03.013

Ehrnhoefer DE, Duennwald M, Markovic P et al (2006) Green tea (−)-epigallocatechin-gallate modulates early events in huntingtin misfolding and reduces toxicity in Huntington's disease models. Hum Mol Genet 15(18):2743–2751. https://doi.org/10.1093/hmg/ddl210

Fallarero A, Loikkanen JJ, Männistö PT et al (2003) Effects of aqueous extracts of *Halimeda incrassata* (Ellis) *Lamouroux and Bryothamnion triquetrum* (S.G.Gmelim) howe on hydrogen peroxide and methyl mercury-induced oxidative stress in GT1–7 mouse hypothalamic immortalized cells. Phytomedicine 10:39–47. https://doi.org/10.1078/094471103321648647

Fallarero A, Peltoketo A, Loikkanen J et al (2006) Effects of the aqueous extract of *Bryothamnion triquetrum* on chemical hypoxia and aglycemia-induced damage in GT1-7 mouse hypothalamic immortalized cells. Phytomedicine 13:240–245. https://doi.org/10.1016/j.phymed.2003.10.009

Fang Z, Jeong SY, Jung HA et al (2010) Anticholinesterase and antioxidant constituents from *Gloiopeltis furcata*. Chem Pharm Bull 58:1236–1239. https://doi.org/10.1248/cpb.58.1236

Fang Z, Jeong SY, Jung HA et al (2012) Capsofulvesins A–C, cholinesterase inhibitors from *Capsosiphon fulvescens*. Chem Pharm Bull 60(11):1351–1358. https://doi.org/10.1248/cpb.c12-00268

Ferdous UT, Balia Yusof ZN et al (2021) Insight into potential anticancer activity of algal flavonoids: current status and challenges. Molecules 26(22):6844. https://doi.org/10.3390/molecules26226844

Fernando IPS, Jayawardena TU, Kim HS et al (2019) Beijing urban particulate matter-induced injury and inflammation in human lung epithelial cells and the protective effects of fucosterol from *Sargassum binderi* (sonder ex J. Agardh). Environ Res 172:150–158. https://doi.org/10.1016/j.envres.2019.02.016

Ferreres F, Lopes G, Gil-Izquierdo A et al (2012) Phlorotannin extracts from fucales characterized by HPLC-DAD-ESI-MS$^n$: approaches to hyaluronidase inhibitory capacity and antioxidant properties. Mar Drugs 10(12):2766–2781. https://doi.org/10.3390/md10122766

Fu H, Liu L, Tong Y et al (2019) The antidepressant effects of hesperidin on chronic unpredictable mild stress-induced mice. Eur J Pharmacol 853:236–246. https://doi.org/10.1016/j.ejphar.2019.03.035

Gallego R, Valdes A, Suárez-Montenegro ZJ et al (2022) Anti-inflammatory and neuroprotective evaluation of diverse microalgae extracts enriched in carotenoids. Algal Res 67:830. https://doi.org/10.1016/j.algal.2022.102830

Gan SY, Wong LZ, Wong JW et al (2019) Fucosterol exerts protection against amyloid β-induced neurotoxicity, reduces intracellular levels of amyloid β and enhances the mRNA expression of neuroglobin in amyloid β-induced SH-SY5Y cells. Int J Biol Macromol 121:207–213. https://doi.org/10.1016/j.ijbiomac.2018.10.021

Gandla K, Babu AK, Unnisa A et al (2023) Carotenoids: role in neurodegenerative diseases remediation. Brain Sci 13(3):457. https://doi.org/10.3390/brainsci13030457

Ganguly G, Chakrabarti S, Chatterjee U (2017) Proteinopathy, oxidative stress and mitochondrial dysfunction: cross talk in Alzheimer's disease and Parkinson's disease. Drug Des Devel Ther 11:797–810. https://doi.org/10.2147/DDDT.S130514

Gany SA, Tan SC, Gan SY et al (2014) Antioxidative, anticholinesterase and anti-neuroinflammatory properties of Malaysian brown and green seaweeds. Int J Biol Biomol Agric Food Biotechnol Eng 8:1251–1257. https://doi.org/10.5281/zenodo.1099988

Gany SA, Tan SC, Gan SY et al (2015) Antioxidative, anticholinesterase and anti-neuroinflammatory properties of Malaysian brown and green seaweeds. World Acad Sci Eng Technol 8:1269–1275. https://doi.org/10.5281/zenodo.1099988

Gao Y, Dong C, Yin J et al (2012a) Neuroprotective effect of fucoidan on $H_2O_2$-induced apoptosis in PC12 cells via activation of PI3K/Akt pathway. Cell Mol Neurobiol 32:523–529. https://doi.org/10.1007/s10571-011-9792-0

Gao Y, Li C, Yin J et al (2012b) Fucoidan, a sulfated polysaccharide from brown algae, improves cognitive impairment induced by infusion of Aβ peptide in rats. Environ Toxicol Pharmacol 33: 304–311. https://doi.org/10.1016/j.etap.2011.12.022

Georgakis ND, Ioannou E, Chatzikonstantinou M et al (2023) The cosmeceutical potential of the yellow-green alga *Trachydiscus minutus* aqueous extract: preparation of a natural-based dermal formula as a proof of concept. Cosmetics 10(3):75. https://doi.org/10.3390/cosmetics10030075

German JB, Walzem RL (2000) The health benefits of wine. Annu Rev Nutr 20:561–593. https://doi.org/10.1146/annurev.nutr.20.1.561

Ghannadi A, Plubrukarn A, Zandi K et al (2013) Screening for antimalarial and acetylcholinesterase inhibitory activities of some Iranian seaweeds. Res Pharm Sci 8(2):113

Giacobini E (2010) Cholinesterases new roles in brain function and in Alzheimer's disease. Neurochem Res 28(3–4):515–522. https://doi.org/10.1023/A:1022869222652

Giffin JC, Richards RC, Craft C et al (2017) An extract of the marine alga *Alaria esculenta* modulates α-synuclein folding and amyloid formation. Neurosci Lett 644:87–93. https://doi.org/10.1016/j.neulet.2017.02.055

Goiris K, Muylaert K, Voorspoels S et al (2014) Detection of flavonoids in microalgae from different evolutionary lineages. J Phycol 50:483–492. https://doi.org/10.1111/jpy.12180

Guo L, Ma R, Sun H et al (2018) Anti-inflammatory activities and related mechanism of polysaccharides isolated from *Sargentodoxa cuneata*. Chem Biodivers 15:e1800343. https://doi.org/10.1002/cbdv.201800343

Han YS, Lee JH, Lee SH et al (2019) Fucoidan suppresses mitochondrial dysfunction and cell death against 1-methyl-4-phenylpyridinuminduced neuronal cytotoxicity via regulation of PGC-1α expression. Mar Drugs 17:518. https://doi.org/10.3390/md17090518

Hannan MA, Kang JY, Hong YK et al (2012) A brown alga *Sargassum fulvellum* facilitates neuronal maturation and synaptogenesis. In Vitro Cell Dev Biol Anim 48:535–544. https://doi.org/10.1007/s11626-012-9537-5

Hannan MA, Kang JY, Hong YK et al (2013) The marine alga *Gelidium amansii* promotes the development and complexity of neuronal cytoarchitecture. Phytother Res 27(1):21–29. https://doi.org/10.1002/ptr.4684

Hannan MA, Mohibbullah M, Hong YK et al (2014b) *Gelidium amansii* promotes dendritic spine morphology and synaptogenesis, and modulates NMDA receptor-mediated postsynaptic current. In Vitro Cell Dev Biol Anim 50:445–452. https://doi.org/10.1007/s11626-013-9721-2

Hannan MA, Mohibbullah M, Hwang SY et al (2014a) Differential neuritogenic activities of two edible brown macroalgae, *Undaria pinnatifida* and *Saccharina japonica*. Am J Chin Med 42: 1371–1384. https://doi.org/10.1142/S0192415X14500864.91

Heo SJ, Cha SH, Kim KN et al (2012) Neuroprotective effect of phlorotannin isolated from *Ishige okamurae* against $H_2O_2$-induced oxidative stress in murine hippocampal neuronal cells, HT22. Appl Biochem Biotechnol 166:1520–1532. https://doi.org/10.1007/s12010-012-9545-7

Heo SJ, Ko SC, Kang SM et al (2008) Cytoprotective effect of fucoxanthin isolated from brown algae *Sargassum siliquastrum* against $H_2O_2$-induced cell damage. Eur Food Res Technol 228: 145–151. https://doi.org/10.1007/s00217-008-0918-7

Hoang MH, Jia Y, Jun HJ et al (2012) Fucosterol is a selective liver X receptor modulator that regulates the expression of key genes in cholesterol homeostasis in macrophages, hepatocytes, and intestinal cells. J Agric Food Chem 60:11567–11575. https://doi.org/10.1021/jf3019084

Hoong WC, John IGS, Ling CEW et al (2018) Fucosterol inhibits the cholinesterase activities and reduces the release of pro-inflammatory mediators in lipopolysaccharide and amyloid-induced microglial cells. J Appl Phycol 30:3261–3270. https://doi.org/10.1007/s10811-018-1495-1

Hu L, Chen W, Tian F et al (2018) Neuroprotective role of fucoxanthin against cerebral ischemic/reperfusion injury through activation of Nrf2/HO-1 signaling. Biomed Pharmacother 106:1484–1489. https://doi.org/10.1016/j.biopha.2018.07.088

Hu P, Li Z, Chen M et al (2016) Structural elucidation and protective role of a polysaccharide from *Sargassum fusiforme* on ameliorating learning and memory deficiencies in mice. Carbohydr Polym 139:150–158. https://doi.org/10.1016/j.carbpol.2015.12.019

Huang CY, Kuo CH, Chen PW et al (2018) Compressional-puffing pretreatment enhances neuroprotective effects of fucoidans from the brown seaweed *Sargassum hemiphyllum* on 6-hydroxydopamine-induced apoptosis in SH-SY5Y cells. Molecules 23:78. https://doi.org/10.3390/molecules23010078

Huang WC, Yen FC, Shiao YJ et al (2009) Enlargement of A-beta aggregates through chemokine-dependent microglial clustering. Neurosci Res 63:280–287. https://doi.org/10.1016/j.neures.2009.01.001

Ikeda K, Kitamura A, Machida H et al (2003) Effect of *Undaria pinnatifida* (wakame) on the development of cerebrovascular diseases in stroke-prone spontaneously hypertensive rats. Clin Exp Pharmacol Physiol 30:44–48. https://doi.org/10.1046/j.1440-1681.2003.03786.x

Ina A, Hayashi K, Nozaki H et al (2007) Pheophytin a, a low molecular weight compound found in the marine brown alga *Sargassum fulvellum*, promotes the differentiation of PC12 cells. Int J Dev Neurosci 25:63–68. https://doi.org/10.1016/j.ijdevneu.2006.09.323

Ina A, Kamei Y et al (2006) Vitamin B(12), a chlorophyll-related analog to pheophytin a from marine brown algae, promotes neurite outgrowth and stimulates differentiation in PC12 cells. Cytotechnology 52:181–187. https://doi.org/10.1007/s10616-006-9038-z

Ioannou E, Labrou NE (2022) Development of enzyme-based cosmeceuticals: studies on the proteolytic activity of *Arthrospira platensis* and its efficient incorporation in a hydrogel formulation. Cosmetics 9(5):106. https://doi.org/10.3390/cosmetics9050106

Isonaka R, Katakura T, Kawakami T (2012) Effect of inhibition of superoxide dismutase on motor neurons during growth: comparison of phosphorylated and non-phosphorylated neurofilament-containing spinal neurons by histogram distribution. Brain Res 1470:11–16. https://doi.org/10.1016/j.brainres.2012.06.014

Jhamandas JH, Wie MB, Harris K et al (2005) Fucoidan inhibits cellular and neurotoxic effects of beta-amyloid (a beta) in rat cholinergic basal forebrain neurons. Eur J Neurosci 21:2649–2659. https://doi.org/10.1111/j.1460-9568.2005.04111.x

Jin DQ, Lim CS, Sung JY (2006) *Ulva conglobata*, a marine algae, has neuroprotective and anti-inflammatory effects in murine hippocampal and microglial cells. Neurosci Lett 402:154–158. https://doi.org/10.1016/j.neulet.2006.03.068

Jin W, Zhang W, Wang J et al (2014) A study of neuroprotective and antioxidant activities of heteropolysaccharides from six *Sargassum* species. Int J of Bio Macromol 67:336–342. https://doi.org/10.1016/j.ijbiomac.2014.03.031

Jung HA, Ali MY, Choi RJ et al (2016) Kinetics and molecular docking studies of fucosterol and fucoxanthin, BACE1 inhibitors from brown algae *Undaria pinnatifida* and *Ecklonia stolonifera*. Food Chem Toxicol 89:104–111. https://doi.org/10.1016/j.fct.2016.01.014

Jung WK, Heo SJ, Jeon YJ et al (2009) Inhibitory effects and molecular mechanism of dieckol isolated from marine brown alga on COX-2 and iNOS in microglial cells. J Agric Food Chem 57:4439–4446. https://doi.org/10.1021/jf9003913

Jung M, Jang KH, Kim B et al (2008) Meroditerpenoids from the brown alga *Sargassum siliquastrum*. J Nat Prod 71:1714–1719. https://doi.org/10.1021/np800321y

Jung HA, Jin SE, Ahn BR et al (2013) Anti-inflammatory activity of edible brown alga *Eisenia bicyclis* and its constituents fucosterol and phlorotannins in LPS-stimulated RAW264.7 macrophages. Food Chem Toxicol 59:199–206. https://doi.org/10.1016/j.fct.2013.05.061

Jung HA, Oh SH, Choi JS et al (2010) Molecular docking studies of phlorotannins from *Eisenia bicyclis* with BACE1 inhibitory activity. Bioorg Med Chem Lett 20:3211–3215. https://doi.org/10.1016/j.bmcl.2010.04.093.58

Kamei Y, Tsang CK et al (2003) Sargaquinoic acid promotes neurite outgrowth via protein kinase a and MAP kinases-mediated signaling pathways in PC12D cells. Int J Dev Neurosci 21:255–262. https://doi.org/10.1016/s0736-5748(03)00068-6

Kang SM, Cha SH, Ko JY et al (2012) Neuroprotective effects of phlorotannins isolated from a brown alga, *Ecklonia cava*, against $H_2O_2$-induced oxidative stress in murine hippocampal HT22 cells. Environ Toxicol Pharmacol 34:96–105. https://doi.org/10.1016/j.etap.2012.03.006

Kang GJ, Han SC, Yoon WJ et al (2013) Sargaquinoic acid isolated from *Sargassum siliquastrum* inhibits lipopolysaccharide-induced nitric oxide production in macrophages via modulation of nuclear factor-κB and c-Jun N-terminal kinase pathways. Immunopharmacol Immunotoxicol 35:80–87. https://doi.org/10.3109/08923973.2012.698622

Kang KA, Lee KH, Chae S et al (2005) Triphlorethol-a from *Ecklonia cava* protects V79-4 lung fibroblast against hydrogen peroxide induced cell damage. Free Radic Res 39:883–892. https://doi.org/10.1080/10715760500161165

Kannan RRR, Aderogba MA, Ndhlala AR et al (2013) Acetylcholinesterase inhibitory activity of phlorotannins isolated from the brown alga, *Ecklonia maxima* (Osbeck) Papenfuss. Food Res Int 54:1250–1254. https://doi.org/10.1016/j.foodres.2012.11.017

Kartal M, Orhan I, Abu-Asaker M et al (2009) Antioxidant and anticholinesterase assets and liquid chromatography-mass spectrometry preface of various fresh-water and marine macroalgae. Pharmacogn Mag 5(20):29. https://doi.org/10.4103/0973-1296.58147

Kawee-ai A, Kuntiya A, Kim SM et al (2013) Anticholinesterase and antioxidant activities of fucoxanthin purified from the microalga *Phaeodactylum tricornutum*. Nat Prod Commun 8(10): 1934578X1300801010

Kim H, Ahn JH, Song M et al (2019) Pretreated fucoidan confers neuroprotection against transient global cerebral ischemic injury in the gerbil hippocampal CA1 area via reducing of glial cell activation and oxidative stress. Biomed Pharmacother 109:1718–1727. https://doi.org/10.1016/j.biopha.2018.11.015

Kim JJ, Kang YJ, Shine SA et al (2016a) Phlorofucofuroeckol improves glutamate-induced neurotoxicity through modulation of oxidative stress-mediated mitochondrial dysfunction in PC12 cells. PLoS One 11:e0163433. https://doi.org/10.1371/journal.pone.0163433

Kim S, Kim JI, Choi JW (2013c) Anti-inflammatory effect of hexane fraction from *Myagropsis myagroides* ethanolic extract in lipopolysaccharide-stimulated BV-2 microglial cells. J Pharm Pharmcol 65:895–906. https://doi.org/10.1111/jphp.12049

Kim KA, Kim SM, Kang SW et al (2012) Edible seaweed, *Eisenia bicyclis*, protects retinal ganglion cells death caused by oxidative stress. Mar Biotechnol 14:383–395. https://doi.org/10.1007/s10126-012-9459-y

Kim AR, Lee MS, Choi JW et al (2013b) Phlorofucofuroeckol a suppresses expression of inducible nitric oxide synthase, cyclooxygenase-2, and pro-inflammatory cytokines via inhibition of nuclear factor-κB, c-Jun NH2-terminal kinases, and Akt in microglial cells. Inflammation 36: 259–271. https://doi.org/10.1007/s10753-012-9542-6

Kim R, Lee B, Joung EJ et al (2016b) 6,60 -Bieckol suppresses inflammatory responses by down-regulating nuclear factor-κB activation via Akt, JNK, and p38 MAPK in LPS-stimulated microglial cells. Immunopharmacol Immunotoxicol 38:244–252. https://doi.org/10.3109/08923973.2016.1173060

Kim EA, Lee SH, Ko C et al (2014d) Protective effect of fucoidan against AAPH-induced oxidative stress in zebrafish model. Carbohydr Polym 102:185–191. https://doi.org/10.1016/j.carbpol.2013.11.022

Kim J, Lee HJ, Lee KW (2010) Naturally occurring phytochemicals for the prevention of Alzheimer's disease. J Neurochem 112(6):1415–1430. https://doi.org/10.1111/j.1471-4159.2009.06562.x

Kim S, Lee MS, Lee B et al (2014b) Anti-inflammatory effects of sargachromenol-rich ethanolic extract of *Myagropsis myagroides* on lipopolysaccharide-stimulated BV-2 cells. BMC Complement Altern Med 14:231. https://doi.org/10.1186/1472-6882-14-231

Kim M, Li YX, Dewapriya P et al (2013a) Floridoside suppresses pro-inflammatory responses by blocking MAPK signaling in activated microglia. BMB Rep 46:398–403. https://doi.org/10.5483/bmbrep.2013.46.8.237

Kim JG, Lim DW, Cho S et al (2014c) The edible brown seaweed *Ecklonia cava* reduces hypersensitivity in postoperative and neuropathic pain models in rats. Molecules 19:7669–7678. https://doi.org/10.3390/molecules19067669

Kim IH, Yoo KY, Park JH et al (2014a) Comparison of neuroprotective effects of extract and fractions from *Agarum clathratum* against experimentally induced transient cerebral ischemic damage. Pharm Biol 52:335–343. https://doi.org/10.3109/13880209.2013.837074

Koh EJ, Seo YJ, Choi J (2016) Neuroprotective effect of *Spirulina maxima* extract against trimethyltin-induced cognitive disorders in HT-22 cells. 444. KFN symposium and annual meeting

Koller M, Muhr A, Braunegg G (2014) Microalgae as versatile cellular factories for valued products. Algal Res 6:52–63. https://doi.org/10.1016/j.algal.2014.09.002

Kritchevsky D, Tepper SA, Czarnecki SK (1999) Effects of 4-methylsterols from algae and of β sitosterol on cholesterol metabolism in rats. Nutr Res 19(11):1649–1654. https://doi.org/10.1016/S0271-5317(99)00120-7

Kumar GP, Khanum F (2012) Neuroprotective potential of phytochemicals. Pharmacogn Rev 12: 81–90. https://doi.org/10.4103/0973-7847.99898

Leandro A, Pereira L, Gonçalves AMM (2020) Diverse applications of marine macroalgae. Mar Drugs 18(1):17. https://doi.org/10.3390/md18010017

Lee TK, Ahn JH, Park CW et al (2020) Pre-treatment with *laminarin* protects hippocampal CA1 pyramidal neurons and attenuates reactive gliosis following transient forebrain ischemia in gerbils. Mar Drugs 18:52. https://doi.org/10.3390/md18010052

Lee HR, Do H, Lee SR et al (2007) Effects of fucoidan on neuronal cell proliferation: association with NO production through the iNOS pathway. J Food Nutr Sci 12:74–78. https://doi.org/10.3746/jfn.2007.12.2.074

Lee S, Lee YS, Jung SH et al (2003) Anti-oxidant activities of fucosterol from the marine algae Pelvetia siliquosa. Arch Pharm Res 26:719–722. https://doi.org/10.1007/BF02976680

Lee S, Youn K, Kim DH et al (2019) Anti-neuroinflammatory property of phlorotannins from *Ecklonia cava* on Aβ25-35-induced damage in PC12 cells. Mar Drugs 17(1):7. https://doi.org/10.3390/md17010007

León R, Garcia AG, Marco-Contelles J (2013) Recent advances in the multitarget-directed ligands approach for the treatment of Alzheimer's disease. Med Res Rev 33(1):139–189. https://doi.org/10.1002/med.20248

Li K, Li XM, Gloer JB et al (2012) New nitrogen-containing bromophenols from the marine red alga *Rhodomela confervoides* and their radical scavenging activity. Food Chem 135:868–872. https://doi.org/10.1016/j.foodchem.2012.05.117

Li Y, Liu L, Liu D et al (2004) Microglial activation by uptake of fDNA via a scavenger receptor. J Neuroimmunol 147:50–55. https://doi.org/10.1016/j.jneuroim.2003.10.043

Li Y, Qian ZJ, Ryu B et al (2009) Chemical components and its antioxidant properties *in vitro*: an edible marine brown alga, *Ecklonia cava*. Bioorg Med Chem 17:1963–1973. https://doi.org/10.1016/j.bmc.2009.01.031

Liang Z, Liu Z, Sun X et al (2019) The effect of fucoidan on cellular oxidative stress and the CatD-Bax signaling axis in MN9D cells damaged by 1-Methyl-4-Phenypyridinium. Front Aging Neurosci 10:429. https://doi.org/10.3389/fnagi.2018.00429

Lin J, Huang L, Yu J et al (2016) Fucoxanthin, a marine carotenoid, reverses scopolamine-induced cognitive impairments in mice and inhibits acetylcholinesterase *in vitro*. Mar Drugs 14(4):67. https://doi.org/10.3390/md14040067

Lin J, Yu J, Zhao J et al (2017) Fucoxanthin, a marine carotenoid, attenuates β-amyloid oligomer-induced neurotoxicity possibly via regulating the PI3K/Akt and the ERK pathways in SH-SY5Y cells. Oxidative Med Cell Longev 2017:6792543. https://doi.org/10.1155/2017/6792543

Liu J, Banskota AH, Critchley AT et al (2015) Neuroprotective effects of the cultivated *Chondrus crispus* in a *C. elegans* model of Parkinson's disease. Mar Drugs 13:2250–2266. https://doi.org/10.3390/md13042250

Liu Y, Geng L, Zhang J et al (2018) Oligo-porphyran ameliorates neurobehavioral deficits in parkinsonian mice by regulating the PI3K/Akt/Bcl-2 pathway. Mar Drugs 16:82. https://doi.org/10.3390/md16030082

Liu H, Gu L et al (2012) Phlorotannins from brown algae (*Fucus vesiculosus*) inhibited the formation of advanced glycation end products by scavenging reactive carbonyls. J Agric Food Chem 60:1326–1334. https://doi.org/10.1021/jf204112f

Liu Y, Jiang L, Li X et al (2017) κ-Carrageenan-derived pentasaccharide attenuates Aβ25-35-induced apoptosis in SH-SY5Y cells via suppression of the JNK signaling pathway. Mol Med Rep 15:285–290. https://doi.org/10.3892/mmr.2016.6006

Liu XY, Liu D, Lin GP et al (2019) Anti-ageing and antioxidant effects of sulfate oligosaccharides from green algae *Ulva lactuca* and *Enteromorpha prolifera* in SAMP8 mice. Int J Biol Macromol 139:342–351. https://doi.org/10.1016/j.ijbiomac.2019.07.195

Liu DQ, Mao SC, Zhang HY et al (2013) Racemosins a and B, two novel bisindole alkaloids from the green alga *Caulerpa racemosa*. Fitoterapia 91:15–20. https://doi.org/10.1016/j.fitote.2013.08.014.126

Luo D, Zhang Q, Wang H et al (2009) Fucoidan protects against dopaminergic neuron death *in vivo* and *in vitro*. Eur J Pharmacol 617:33–40. https://doi.org/10.1016/j.ejphar.2009.06.015

Machado LP, Carvalho LR, Young MCM et al (2015) Evaluation of acetylcholinesterase inhibitory activity of Brazilian red macroalgae organic extracts. Rev Bras 25:657–662. https://doi.org/10.1016/j.bjp.2015.09.003

Meenakshi S, Umayaparvathi S, Saravanan R et al (2016) Neuroprotective effect of fucoidan from *Turbinaria decurrens* in MPTP intoxicated Parkinsonic mice. Int J Biol Macromol 86:425–433. https://doi.org/10.1016/j.ijbiomac.2015.12.025

Meshalkina D, Tsvetkova E, Orlova A (2023) First insight into the neuroprotective and antibacterial effects of Phlorotannins isolated from the cell walls of Brown algae *Fucus vesiculosus* and *Pelvetia canaliculata*. Antioxidants 12(3):696. https://doi.org/10.3390/antiox12030696

Mirata S, Asnaghi V, Chiantore M (2023) Photoprotective and anti-aging properties of the apical frond extracts from the Mediterranean seaweed *Ericaria amentacea*. Mar Drugs 21(5):306. https://doi.org/10.3390/md21050306

Mitra S, Siddiqu WA, Khandelwal S (2015) C-phycocyanin protects against acute tributyltin chloride neurotoxicity by modulating glial cell activity along with its anti-oxidant and anti-inflammatory property: a comparative efficacy evaluation with N-acetyl cysteine in adult rat brain. Chem Biol Interact 238:138–150. https://doi.org/10.1016/j.cbi.2015.06.016

Mohibbullah M, Bhuiyan MMH, Hannan MA (2016b) The edible red alga *Porphyra yezoensis* promotes neuronal survival and cytoarchitecture in primary hippocampal neurons. Cell Mol Neurobiol 36:669–682. https://doi.org/10.1007/s10571-015-0247-x

Mohibbullah M, Hannan MA, Park IS et al (2016a) The edible red seaweed *Gracilariopsis chorda* promotes axodendritic architectural complexity in hippocampal neurons. J Med Food 19:638–644. https://doi.org/10.1089/jmf.2016.3694.118

Mohibbullah M, Haque MN, Khan MNA et al (2018) Neuroprotective effects of fucoxanthin and its derivative fucoxanthinol from the phaeophyte *Undaria pinnatifida* attenuate oxidative stress in hippocampal neurons. J Appl Phycol 30:3243–3252. https://doi.org/10.1007/s10811-018-1458-6

Molinski TF, Dalisay DS, Lievens SL (2008) Drug development from marine natural products. Nat Rev Drug Discov 8(1):69–85. https://doi.org/10.1038/nrd2487

Moncada S, Palmer RM, Higgs EA (1991) Nitric oxide: physiology, pathophysiology, and pharmacology. Pharmacol Rev 43(2):109–142

Moosmann B, Behl C (2002) Antioxidants as treatment for neurodegenerative disorders. Expert Opin Investig Drugs 11:1407–1435. https://doi.org/10.1517/13543784.11.10.1407

Morais T, Cotas J, Pacheco D et al (2021) Seaweeds compounds: an ecosustainable source of cosmetic ingredients? Cosmetics 8(1):8. https://doi.org/10.3390/cosmetics8010008

Mori J, Iwashima M, Wakasugi H et al (2005) New plastoquinones isolated from the brown alga, *Sargassum micracanthum*. Chem Pharm Bull 53:1159–1163. https://doi.org/10.1248/cpb.53.1159

Murugan AC, Vallal D, Karim MR (2015) *In vitro* antiradical and neuroprotective activity of polyphenolic extract from marine algae *Padina autralis*. J Chem Pharm Res 7:355–362

Myung CS, Shin HC, Bao HY et al (2005) Improvement of memory by dieckol and phlorofucofuroeckol in ethanol-treated mice: possible involvement of the inhibition of acetylcholinesterase. Arch Pharm Res 28:691–698. https://doi.org/10.1007/BF02969360

Nakashima Y, Ohsawa KF et al (2009) Preventive effects of *chlorella* on cognitive decline in age-dependent dementia model mice. Neurosci Lett 464:193–198. https://doi.org/10.1016/j.neulet.2009.08.044

Natarajan S, Shanmugiahthevar KP, Kasi PD et al (2009) Cholinesterase inhibitors from *Sargassum* and *Gracilaria gracilis*: seaweeds inhabiting south Indian coastal areas (Hare Island, gulf of Mannar). Nat Prod Res 23:355–369. https://doi.org/10.1080/14786410802156036

Nuzzo D, Presti G, Picone P et al (2018) Effects of the *Aphanizomenon flos-aquae* extract (Klamin®) on a neurodegeneration cellular model. Oxidative Med Cell Longev 2018: 9089016. https://doi.org/10.1155/2018/9089016

Obisesan TO, Hirsch R, Kosoko O (1998) Moderate wine consumption is associated with decreased odds of developing age-related macular degeneration in NHANES-1. J Am Geriatr Soc 46(1): 1–7. https://doi.org/10.1111/j.1532-5415.1998.tb01005.x

Oh JH, Choi JS, Nam TJ et al (2018) Fucosterol from an edible brown alga *Ecklonia stolonifera* prevents soluble amyloid beta-induced cognitive dysfunction in aging rats. Mar Drugs 16:368. https://doi.org/10.3390/md16100368

Oktaviani DF, Bae YS, Meinita MDN et al (2019) An ethanol extract of the brown seaweed *Hizikia fusiformis* and its active constituent, fucosterol, extend the lifespan of the nematode *Caenorhabditis elegans*. J Life Sci 29:1120–1125. https://doi.org/10.5352/JLS.2019.29.10.1120

Olasehinde TA, Mabinya LV, Olaniran AO et al (2019) Chemical characterization of sulfated polysaccharides from *Gracilaria gracilis* and *Ulva lactuca* and their radical scavenging, metal chelating, and cholinesterase inhibitory activities. Int J Food Prop 22:100–110. https://doi.org/10.1080/10942912.2019.1573831

Orgogozo JM, Dartigues JF, Lafont S (1997) Wine consumption and dementia in the elderly: a prospective community study in the Bordeaux area. Rev Neurol (Paris) 153(3):185–192

Oslan SNH, Shoparwe NF, Yusoff AH (2021a) A review on *Haematococcus pluvialis* bioprocess optimization of green and red stage culture conditions for the production of natural astaxanthin. Biomol Ther 11(2):256. https://doi.org/10.3390/biom11020256

Oslan SNH, Tan JS, Oslan SN (2021b) N. *Haematococcus pluvialis* as a potential source of astaxanthin with diverse applications in industrial sectors: current research and future directions. Molecules 26(21):6470. https://doi.org/10.3390/molecules26216470

Oyama Y, Chikahisa L, Ueha T et al (1996) Ginkgo biloba extract protects brain neurons against oxidative stress induced by hydrogen peroxide. Brain Res 712(2):349–352. https://doi.org/10.1016/0006-8993(95)01440-3

Pangestuti R, Vo TS, Ngo DH et al (2013) Fucoxanthin ameliorates inflammation and oxidative reponses in microglia. J Agric Food Chem 61:3876–3883. https://doi.org/10.1021/jf400015k

Panigrahi GP, Rane AR, Vavilala SL et al (2019) Deciphering the anti-Parkinson's activity of sulphated polysaccharides from Chlamydomonas reinhardtii on the α-synuclein mutants A30P, A53T, E46K, E57K and E35K. J Biochem 166(6):463–474. https://doi.org/10.1093/jb/mvz064

Park JH, Ahn JH, Lee TK et al (2020a) *Laminarin* pretreatment provides neuroprotection against forebrain ischemia/reperfusion injury by reducing oxidative stress and neuroinflammation in aged gerbils. Mar Drugs 18:213. https://doi.org/10.3390/md18040213

Park HY, Han MH, Ark C et al (2011) Anti-inflammatory effects of fucoidan through inhibition of NF-κB, MAPK and Akt activation in lipopolysaccharide-induced BV2 microglia cells. Food Chem Toxicol 2011(49):1745–1752. https://doi.org/10.1016/j.fct.2011.04.020

Park MH, Jin HK, Bae JS (2020b) Potential therapeutic target for aging and age-related neurodegenerative diseases: the role of acid sphingomyelinase. Exp Mol Med 52(3):380–389. https://doi.org/10.1038/s12276-020-0399-8

Park SK, Kang JY, Kim JM et al (2019) Fucoidan-rich substances from *Ecklonia cava* improve trimethyltin-induced cognitive dysfunction via down-regulation of amyloid β production/tau hyperphosphorylation. Mar Drugs 17:591. https://doi.org/10.3390/md17100591

Pentón-Rol G, Marín-Prida J, Pardo-Andreu G et al (2011b) C-phycocyanin is neuroprotective against global cerebral ischemia/reperfusion injury in gerbils. Brain Res Bull 86:42–52. https://doi.org/10.1016/j.brainresbull.2011.05.016

Pentón-Rol G, Martínez-Sánchez G, Cervantes-Llanos M et al (2011a) C-phycocyanin ameliorates experimental autoimmune encephalomyelitis and induces regulatory T cells. Int Immunopharmacol 11(1):29–38. https://doi.org/10.1016/j.intimp.2010.10.001

Pereira L, Valado A (2021) The seaweed diet in prevention and treatment of the neurodegenerative diseases. Mar Drugs 19(3):128. https://doi.org/10.3390/md19030128

Pizzino G, Irrera N, Cucinotta MD (2017) Oxidative stress: harms and benefits for human health. Oxidative Med Cell Longev 2017:8416763. https://doi.org/10.1155/2017/8416763

Qiu SM, Aweya JJ, Liu X (2022) Bioactive polysaccharides from red seaweed as potent food supplements: a systematic review of their extraction, purification, and biological activities. Carbohydr Polym 275:118696. https://doi.org/10.1016/j.carbpol.2021.118696

Rafiquzzaman SM, Kim EY, Lee JM et al (2015) Anti-Alzheimers and anti-inflammatory activities of a glycoprotein purified from the edible brown alga *Undaria pinnatifida*. Food Res Int 77:118–124. https://doi.org/10.1016/j.foodres.2015.08.021

Rajivgandhi GN, Kanisha G, Ramachandran N et al (2021) Phytochemical screening and anti-oxidant activity of *Sargassum wightii* enhances the anti-bacterial activity against *Pseudomonas aeruginosa*. Saudi J Biol Sci 28:1763–1769. https://doi.org/10.1016/j.sjbs.2020.12.018

Raman RS, Parthiban S, Karthikeyan MS et al (2014) Antimicrobial and anti-inflammatory studies on *Sargassum wightii* extracts. Int J Pharm Pharm Sci 6:611–614

Rehman MU, Wali AF, Ahmed A et al (2019) Neuroprotective strategies for neurological disorders by natural products: an update. Curr Neuropharmacol 17(3):247–267. https://doi.org/10.2174/1570159X16666180911124605

Reigada LC, Buchanan EM, Hazeltine DB et al (2021) A pilot randomized controlled trial testing supplements of omega-3 fatty acids, probiotics, combination or placebo on symptoms of depression, anxiety and stress. J Affect Disord Rep 5:100141

Rengasamy KRR, Amoo SO, Aremu AO et al (2015) Phenolic profiles, antioxidant capacity, and acetylcholinesterase inhibitory activity of eight south African seaweeds. J Appl Phycol 27: 1599–1605. https://doi.org/10.1007/s10811-014-0438-8

Repine JE, Bast A, Lankhorst I (1997) Oxidative stress in chronic obstructive pulmonary disease. Am J Respir Crit Care Med 156:341–357. https://doi.org/10.1164/ajrccm.156.2.9611013

Resende DI, Ferreira M, Magalhães C et al (2021) Trends in the use of marine ingredients in anti-aging cosmetics. Algal Res 55:102273. https://doi.org/10.1016/j.algal.2021.102273

Rivero F, Fallarero A, Castañeda O et al (2003) Antioxidant activity *in vivo* and *in vitro* of *Halimeda incrassata* aqueous extracts. Food Sci Technol 23:256–263. https://doi.org/10.1590/S0101-20612003000200026

Romay CH, Gonzaléz R, Ledón N et al (2003) C-phycocyanin: a biliprotein with antioxidant, anti-inflammatory and neuroprotective effects. Curr Protein Pept Sci 4(3):207. https://doi.org/10.2174/1389203033487216

Ryan S, O'Gorman DM, Nolan YM et al (2011) Evidence that the marine-derived multi-mineral aquamin has anti-inflammatory effects on cortical glial-enriched cultures. Phytother Res 25: 765–767. https://doi.org/10.1002/ptr.3309

Ryu MJ, Chung HS et al (2016) Fucoidan reduces oxidative stress by regulating the gene expression of HO-1 and SOD-1 through the Nrf2/ERK Signaling pathway in HaCaT cells. Mol Med Rep 14:3255–3260. https://doi.org/10.3892/mmr.2016.5623

Ryu G, Park SH, Kim ES et al (2003) Cholinesterase inhibitory activity of two farnesylacetone derivatives from the brown alga *Sargassum sagamianum*. Arch Pharm Res 26(10):796–799. https://doi.org/10.1007/BF02980022

Sairazi MNS, Sirajudeen KNS et al (2020) Natural products and their bioactive compounds neuroprotective potentials against neurodegenerative diseases. Evid Based Complement Alternat Med 2020:6565396. https://doi.org/10.1155/2020/6565396

Salehi B, Sharifi-Rad J et al (2019) Current trends on seaweeds looking at chemical composition. Phytopharmacol and cosmetic applications. Molecules 24(22):4182. https://doi.org/10.3390/molecules24224182

Sangeetha R, Bhaskar N, Baskaran V et al (2009) Comparative effects of β-carotene and fucoxanthin on retinol deficiency induced oxidative stress in rats. Mol Cell Biochem 331:59–67. https://doi.org/10.1007/s11010-009-0145-y

Sanjeewa KKA, Fernando IPS, Kim SY et al (2018) *In vitro* and *in vivo* anti-inflammatory activities of high molecular weight sulfated polysaccharide containing fucose separated from *Sargassum horneri*. Short communication. Int J Biol Macromol 107:803–807. https://doi.org/10.1016/j.ijbiomac.2017.09.050

Santhosh S, Dhandapani R, Hemalatha N et al (2016) Bioactive compounds from microalgae and its different applications-a review. Pelagia Res Libr Adv Appl Sci Res 7(4):153–158

Sarithakumari CH, Renju GL, Kurup GM et al (2013) Anti-inflammatory and antioxidant potential of alginic acid isolated from the marine algae *Sargassum wightii* on adjuvant-induced arthritic rats. Inflammopharmacology 21:261–268. https://doi.org/10.1007/s10787-012-0159-z

Sathasivam R, Ki JS et al (2018) A review of the biological activities of microalgal carotenoids and their potential use in healthcare and cosmetic industries. Mar Drugs 16(1):26. https://doi.org/10.3390/md16010026

Savvidou MG, Georgiopoulou I, Antoniou N et al (2023) Extracts from *Chlorella vulgaris* protect mesenchymal stromal cells from oxidative stress induced by hydrogen peroxide. Plan Theory 12(2):361. https://doi.org/10.3390/plants12020361

Scoglio S, Benedetti Y, Benvenuti F et al (2014) Selective monoamine oxidase B inhibition by an Aphanizomenon flos-aquae extract and by its constitutive active principles phycocyanin and mycosporine-like amino acids. Phytomedicine 21(7):992–997. https://doi.org/10.1016/j.phymed.2014.03.006

Seong SH, Paudel P, Choi JW et al (2019) Probing multi-target action of phlorotannins as new monoamine oxidase inhibitors and dopaminergic receptor modulators with the potential for treatment of neuronal disorders. Mar Drugs 17:377. https://doi.org/10.3390/md17060377

Shanmuganathan B, Malar DS, Sathya S et al (2015) Antiaggregation potential of *Padina gymnospora* against the toxic Alzheimer's beta-amyloid peptide 25–35 and cholinesterase inhibitory property of its bioactive compounds. PLoS One 10:1–14. https://doi.org/10.1371/journal.pone.0141708

Shanmuganathan B, Sathya S, Balasubramaniam B et al (2019) Amyloid-β induced neuropathological actions are suppressed by *Padina gymnospora (Phaeophyceae*) and its active constituent α-bisabolol in Neuro2a cells and transgenic *Caenorhabditis elegans* Alzheimer's model. Nitric Oxide 91:52–66. https://doi.org/10.1016/j.niox.2019.07.009

Shanmuganathan B, Suryanarayanan V, Sathya S et al (2018) Anti-amyloidogenic and anti-apoptotic effect of α-bisabolol against Aβ induced neurotoxicity in PC12 cells. Eur J Med Chem 143:1196–1207. https://doi.org/10.1016/j.ejmech.2017.10.017

Shimizu H, Koyama T, Yamada S et al (2015) A sesquiterpene from the brown algae *Dictyopteris undulata*, provides neuroprotection by activating the Nrf2/ARE pathway. Biochem Biophys Res Commun 457:718–722. https://doi.org/10.1016/j.bbrc.2015.01.059.56

Shin S, Cho SH, Park D et al (2020) Anti-skin aging properties of protocatechuic acid *in vitro* and *in vivo*. J Cosmet Dermatol 19(4):977–984. https://doi.org/10.1111/jocd.13086

Shrestha S, Choi JS, Zhang W et al (2022) Neuroprotective activity of macroalgal fucofuroeckols against amyloid β peptide-induced cell death and oxidative stress. Int J Food Sci Technol 57:4286–4295. https://doi.org/10.1111/ijfs.15753

Shrestha S, Zhang W, Begbie AJ et al (2020) *Ecklonia radiata* extract containing eckol protects neuronal cells against aß1-42evoked toxicity and reduces aggregate density. Food Funct 11: 6509–6516. https://doi.org/10.1039/D0FO01438A

Silva J, Alves C, Pinteus S et al (2018) Neuroprotective effects of seaweeds against 6-hydroxidopamine-induced cell death on an *in vitro* human neuroblastoma model. BMC Complement Altern Med 18(1):1–10. https://doi.org/10.1186/s12906-018-2103-2

Singh R, Parihar P, Singh M et al (2017) Uncovering potential applications of cyanobacteria and algal metabolites in biology agriculture and medicine current status and future prospects. Front Microbiol 8:1–37. https://doi.org/10.3389/fmicb.2017.00515

Souza RB, Frota AF, Silva J et al (2018) *In vitro* activities of kappa-carrageenan isolated from red marine alga *Hypnea musciformis*: antimicrobial, anticancer and neuroprotective potential. Int J Biol Macromol 112:1248–1256. https://doi.org/10.1016/j.ijbiomac.2018.02.029

Souza RB, Frota AF, Sousa RS et al (2017) Neuroprotective effects of sulphated agaran from marine alga *Gracilaria cornea* in rat 6-hydroxydopamine Parkinson's disease model behavioural neurochemical and transcriptional alterations. Basic Clin Pharmacol Toxicol 120: 159–170. https://doi.org/10.1111/bcpt.12669

Stirk WA, Reinecke DL, Staden JV et al (2007) Seasonal variation in antifungal antibacterial and acetylcholinesterase activity in seven south African seaweeds. J Appl Phycol 19:271–276. https://doi.org/10.1007/s10811-006-9134-7.57

Subirade I, Fernandez Y, Periquet A et al (1995) Catechin protection of 3T3 Swiss fibroblasts in culture under oxidative stress. Biol Trace Elem Res 47(1–3):313–319. https://doi.org/10.1007/BF02790132

Suganthy N, Pandian SK, Devi KP et al (2010) Neuroprotective effect of seaweeds inhabiting south Indian coastal area (Hare Island gulf of Mannar marine biosphere reserve) cholinesterase inhibitory effect of *Hypnea valentiae* and *Ulva reticulate*. Neurosci Lett 468:216–219. https://doi.org/10.1016/j.neulet.2009.11.001

Sun Z, Mohamed MAA, Park SY et al (2015) Fucosterol protects cobalt chloride induced inflammation by the inhibition of hypoxia-inducible factor through PI3K/Akt pathway. Int Immunopharmacol 29:642–647. https://doi.org/10.1016/j.intimp.2015.09.016

Syad AN, Rajamohamed BS, Shunmugaiah KP et al (2016) Neuroprotective effect of the marine macroalga *Gelidiella acerosa* identification of active compounds through bioactivity-guided fractionation. Pharm Biol 54:2073–2081. https://doi.org/10.3109/13880209.2016.1145700

Syad AN, Shunmugiah KP, Kasi PD (2013) Antioxidant and anti-cholinesterase activity of *Sargassum wightii*. Pharm Biol 51(11):1401–1410. https://doi.org/10.3109/13880209.2013.793721

Syad AN, Shunmugiah KP, Kasi PD et al (2012) Assessment of anticholinesterase activity of *Gelidiella acerosa* implications for its therapeutic potential against Alzheimer's disease. Evid Based Complement Alternat Med 2012:497242. https://doi.org/10.1155/2012/497242

Tapiero H, Townsend DM, Tew KD (2003) Phytosterols in the prevention of human pathologies. Biomed Pharmacother 57(8):321–325. https://doi.org/10.1016/S0753-3322(03)00104-5

Tayab MA, Islam MN, Chowdhury KAA et al (2022) Targeting neuroinflammation by polyphenols a promising therapeutic approach against inflammation-associated depression. Biomed Pharmacother 147:112668. https://doi.org/10.1016/j.biopha.2022.112668

Tirtawijaya G, Mohibbullah M, Meinita MDN et al (2016) The ethanol extract of the rhodophyte *Kappaphycus alvarezii* promotes neurite outgrowth in hippocampal neurons. J Appl Phycol 28: 2515–2522. https://doi.org/10.1007/s10811-016-0795-6

Tsang C, Ina A, Goto T et al (2005) A novel nerve growth factor-potentiating substance isolated from *Sargassum macrocarpum* promotes neurite outgrowth and survival via distinct signaling pathways in PC12D cells. Neuroscience 132:633–643. https://doi.org/10.1016/j.neuroscience.2005.01.028

Tsang C, Sagara A, Kamei Y et al (2001) Structure-activity relationship of a neurite outgrowth-promoting substance purified from the brown alga *Sargassum macrocarpum* and its analogues on PC12D cells. J Appl Phycol 13:349–357. https://doi.org/10.1023/A:1017540620106

Turner SR, Thomas RG et al (2005) For the Alzheimer's disease cooperative study a randomized double-blind placebo-controlled trial of resveratrol for Alzheimer's disease. Neurology 85(16): 1383–1391

Uriarte PI, Calvo MI et al (2009) Phytochemical study and evaluation of antioxidant, neuroprotective and acetylcholinesterase inhibitor activities of *Galeopsis ladanum* L extracts. Pharmacogn Mag 5:287–290. https://doi.org/10.4103/0973-1296.58146

Vasili E, Dominguez-Meijide A, Outeiro TF et al (2019) Spreading of α-synuclein and tau a systematic comparison of the mechanisms involved. Front Mol Neurosci 12:107. https://doi.org/10.3389/fnmol.2019.00107

Verri M, Pastoris O, Dossena M et al (2012) Mitochondrial alterations oxidative stress and neuroinflammation in Alzheimer's disease. Int J Immunopathol Pharmacol 25(2):345–353. https://doi.org/10.1177/039463201202500204

Voss LJ, Plouviez M, Whittle N et al (2021) Microalgae-based photosynthetic strategy for oxygenating avascularised mouse brain tissue—an *in vitro* proof of concept study. Brain Res 1768:147585. https://doi.org/10.1016/j.brainres.2021.147585

Wang L, Jayawardena TU, Yang HW et al (2020a) The potential of sulfated polysaccharides isolated from the brown seaweed *Ecklonia maxima* in cosmetics antioxidant anti-melanogenesis and photoprotective activities. Antioxidants 9(8):724. https://doi.org/10.3390/antiox9080724

Wang L, Oh JY, Je JG et al (2020b) Protective effects of sulfated polysaccharides isolated from the enzymatic digest of *Codium fragile* against hydrogen peroxide-induced oxidative stress in *in vitro* and *in vivo* models. Algal Res 48:101891. https://doi.org/10.1016/j.algal.2020.101891

Wang X, Yi K, Zhao Y et al (2018b) Fucoidan inhibits amyloid-β-induced toxicity in transgenic *Caenorhabditis elegans* by reducing the accumulation of amyloid-β and decreasing the production of reactive oxygen species. Food Funct 9:552–560. https://doi.org/10.1039/C7FO00662D

Wang C, Zhao Y, Wang L et al (2021) C-phycocyanin mitigates cognitive impairment in doxorubicin-induced chemobrain impact on neuroinflammation, oxidative stress and brain mitochondrial and synaptic alterations. Neurochem Res 46:149–158. https://doi.org/10.1007/s11064-020-03164-2

Wang J, Zheng J, Huang C et al (2018a) Eckmaxol a phlorotannin extracted from *Ecklonia maxima* produces anti-β-amyloid oligomer neuroprotective effects possibly via directly acting on glycogen synthase kinase 3β. ACS Chem Neurosci 9:1349–1356. https://doi.org/10.1021/acschemneuro.7b00527

Wei H, Gao Z, Zheng L et al (2017) Protective effects of fucoidan on Aβ25-35 and d-gal-induced neurotoxicity in PC12 cells and d-gal-induced cognitive dysfunction in mice. Mar Drugs 15:77. https://doi.org/10.3390/md15030077

Wen ZS, Xiang XW, Jin HX et al (2016) Composition and anti-inflammatory effect of polysaccharides from *Sargassum horneri* in RAW264. 7 macrophages. Int J Biol Macromol 88:403–413. https://doi.org/10.1016/j.ijbiomac.2016.02.025

Wichmann H, Brinkhoff T, Simon M et al (2016) Dimethylsulfoniopropionate promotes process outgrowth in neural cells and exerts protective effects against tropodithietic acid. Mar Drugs 14: 89. https://doi.org/10.3390/md14050089

Wong CH, Gan SY, Tan SC et al (2018) Fucosterol inhibits the cholinesterase activities and reduces the release of pro-inflammatory mediators in lipopolysaccharide and amyloid-induced microglial cells. J Appl Phycol 30:3261–3270. https://doi.org/10.1007/s10811-018-1495-1

Wozniak M, Bell T, Dénes Á et al (2015) Anti-HSV1 activity of brown algal polysaccharides and possible relevance to the treatment of Alzheimer's disease. Int J Biol Macromol 74:530–540. https://doi.org/10.1016/j.ijbiomac.2015.01.003

Wu H, Gao S, Terakawa S et al (2019) Inhibitory effects of fucoidan on NMDA receptors and l-type ca(2+) channels regulating the ca(2+) responses in rat neurons. Pharm Biol 57:1–7. https://doi.org/10.1080/13880209.2018.1548626

Wu J, Xi Y, Li G et al (2021) Hydroazulene diterpenes from a *Dictyota* brown alga and their antioxidant and neuroprotective effects against cerebral ischemia-reperfusion injury. J Nat Prod 84(4):1306–1315. https://doi.org/10.1021/acs.jnatprod.1c00027

Xiang S, Liu F, Lin J et al (2017) Fucoxanthin inhibits β-amyloid assembly and attenuates β-amyloid oligomer-induced cognitive impairments. J Agri Food Chem 65(20):4092–4102. https://doi.org/10.1021/acs.jafc.7b00805

Yabuta Y, Fujimura H, Kwak CS et al (2010) Antioxidant activity of the phycoerythrobilin compound formed from a dried Korean purple laver (*Porphyra sp)* during *in vitro* digestion. Food Sci Technol 16:347–352. https://doi.org/10.3136/fstr.16.347

Yahfoufi N, Alsadi N, Jambi M et al (2018) The immunomodulatory and anti-inflammatory role of polyphenols. Nutrients 10(11):1618. https://doi.org/10.3390/nu10111618

Yan XJ, Chuda Y, Suzuki M et al (1999) Fucoxanthin as the major antioxidant in *Hijikia fusiformis*, a common edible seaweed. Biosci Biotechnol Biochem 63:605–607. https://doi.org/10.1271/bbb.63.605

Yang WN, Chen PW, Huang CY et al (2017) Compositional characteristics and *in vitro* evaluations of antioxidant and neuroprotective properties of crude extracts of fucoidan prepared from compressional puffing-pretreated *Sargassum crassifolium*. Mar Drugs 15:183. https://doi.org/10.3390/md15060183

Yang EJ, Ham YM, Yang KW et al (2013) Sargachromenol from *Sargassum micracanthum* inhibits the lipopolysaccharide-induced production of inflammatory mediators in RAW 264. 7 macrophages. Sci World J 2013:712303. https://doi.org/10.1155/2013/712303

Yang YI, Jung SH, Lee KT et al (2014) 8, 8'-Bieckol, isolated from edible brown algae exerts its anti-inflammatory effects through inhibition of NF-kB signaling and ROS production in LPS-stimulated macrophages. Int Immunopharmacol 23:460–468. https://doi.org/10.1016/j.intimp.2014.09.019

Yang P, Liu DQ, Liang TJ et al (2015) Bioactive constituents from the green alga *Caulerpa racemosa*. Bioorg Med Chem 23:38–45. https://doi.org/10.1016/j.bmc.2014.11.031

Yang N, Song F (2018) Bioprospecting of novel and bioactive compounds from marine actinomycetes isolated from South China Sea sediments. Curr Microbiol 75(2):142–149. https://doi.org/10.1007/s00284-017-1358-z

Yao ZA, Xu L, Wu HG et al (2014) Immunomodulatory function of κ-carrageenan oligosaccharides acting on LPS-activated microglial cells. Neurochem Res 39:333–343. https://doi.org/10.1007/s11064-013-1228-4

Yehuda S, Rabinovitz S, Mostofsk DI et al (1998) Essential fatty acids and sleep mini-review and hypothesis. Med Hypotheses 50:139–145. https://doi.org/10.1016/S0306-9877(98)90200-6

Yende SR, Harle UN, Ittadwar AM et al (2016) Insignificant anticonvulsant activity of *Padina tetrastromatica* (*Brown macroalgae*) in mice. J Pharm Negat Results 7:33–36. https://doi.org/10.1016/0006-8993(91)90822-d

Yoon NY, Chung HY, Kim HR et al (2008) Acetyl- and butyrylcholinesterase inhibitory activities of sterols and phlorotannins from *Ecklonia stolonifera*. Fish Sci 74:200–207. https://doi.org/10.1111/j.1444-2906.2007.01511.x

Yoon NY, Lee SH, Kim SKP et al (2009) Phlorotannins from *Ishige okamurae* and their acetyl-and butyrylcholinesterase inhibitory effects. J Funct Foods 1(4):331–3359. https://doi.org/10.1016/j.jff.2009.07.002

Yoon MS, Shin JS, Choi HE et al (2012) Fucosterol isolated from *Undaria pinnatifida* inhibits lipopolysaccharide-induced production of nitric oxide and pro-inflammatory cytokines via the inactivation of nuclear factor-kappaB and p38 mitogen-activated protein kinase in RAW264.7 macrophages. Food Chem 135:967–975. https://doi.org/10.1016/j.foodchem.2012.05.039

Yotsu-Yamashita M, Kondo S et al (2013) Isolation and structural determination of two novel phlorotannins from the brown alga *Ecklonia kurome Okamura* and their radical scavenging activities. Mar Drugs 11:165–183. https://doi.org/10.3390/md11010165

Yu DK, Lee B, Kwon M et al (2015) Phlorofucofuroeckol B suppresses inflammatory responses by down-regulating nuclear factor κB activation via Akt ERK and JNK in LPS-stimulated microglial cells. Int Immunopharmacol 28:1068–1075. https://doi.org/10.1016/j.intimp.2015.08.028

Yu J, Lin JJ, Yu R et al (2017) Fucoxanthin prevents $H_2O_2$-induced neuronal apoptosis via concurrently activating the PI3-K/Akt cascade and inhibiting the ERK pathway. Food Nutr Res 61:1304678. https://doi.org/10.1080/16546628.2017.1304678

Zhang S, Duan E (2018) Fighting against skin aging the way from bench to bedside. Cell Transplant 27(5):729–738. https://doi.org/10.1177/0963689717725755

Zhang FL, He Y, Zheng Y et al (2014a) Therapeutic effects of fucoidan in 6-hydroxydopamine-lesioned rat model of Parkinson's disease role of NADPH oxidase-1. CNS Neurosci Ther 20: 1036–1044. https://doi.org/10.1111/cns.12340

Zhang F, Lu J, Zhang J et al (2015) Protective effects of a polysaccharide from *Spirulina platensis* on dopaminergic neurons in an MPTP-induced Parkinson's disease model in C57BL/6J mice. Neural Regen Res 10:308. https://doi.org/10.4103/1673-5374.152387

Zhang L, Wang H, Fan Y et al (2017) Fucoxanthin provides neuroprotection in models of traumatic brain injury via the Nrf2-ARE and Nrf2-autophagy pathways. Sci Rep 7:46763. https://doi.org/10.1038/srep46763

Zhang Z, Wang X, Pan Y et al (2020) The degraded polysaccharide from Pyropia haitanensis represses amyloid beta peptide-induced neurotoxicity and memory *in vivo*. Int J Biol Macromol 146:725–729. https://doi.org/10.1016/j.ijbiomac.2019.09.243

Zhang H, Xiao X, Conte MM et al (2012) Spiralisones A–D Acylphloroglucinol hemiketals from an Australian marine brown alga, *Zonaria spiralis*. Org Biomol Chem 10:9671–9676. https://doi.org/10.1039/C2OB26988K

Zhang XS, Zhang X, Wu Q et al (2014b) Astaxanthin offers neuroprotection and reduces neuroinflammation in experimental subarachnoid hemorrhage. J Surg Res 192:206–213. https://doi.org/10.1016/j.jss.2014.05.029

Zhao D, Kwon SH, Chun YS (2017) Anti-neuroinflammatory effects of fucoxanthin via inhibition of Akt/NF-κB and MAPKs/AP-1 pathways and activation of PKA/CREB pathway in lipopolysaccharide-activated BV-2 microglial cells. Neurochem Res 42:667–677. https://doi.org/10.1007/s11064-016-2123-6

Zhao X, Zhang X, Liu H et al (2019) Enzyme-assisted extraction of astaxanthin from *Haematococcus pluvialis* and its stability and antioxidant activity. Food Sci Biotechnol 28: 1637–1647. https://doi.org/10.1007/s10068-019-00608-6

Zhao D, Zheng L, Qi L et al (2016) Structural features and potent antidepressant effects of total sterols and β-sitosterol extracted from *Sargassum horneri*. Mar Drugs 14:123. https://doi.org/10.3390/md14070123

Zhou R, Shi XY, Bi DC et al (2015) Alginate-derived oligosaccharide inhibits neuroinflammation and promotes *Microglial phagocytosis* of β-amyloid. Mar Drugs 13:5828–5846. https://doi.org/10.3390/md13095828

# Neurotrophins and Phytochemicals: Potential Partners to Combat Neurodegenerative Diseases

# 13

Megala Jayaraman, Annesha Dutta, Swathi Mahanta, Diveyaa Sivakumar, and Hanumanth Rao Balaji Raghavendran

**Abstract**

Neurotrophins and phytochemicals have emerged as promising areas of research in the field of neuroscience and human health. Neurotrophins are a group of proteins that are essential for the growth and development of neurons in the nervous system. Four basic types of neurotrophins have been discovered in mammals: nerve growth factor (NGF), brain-derived neurotrophic factor (BDNF), neurotrophin-3 (NT-3), and neurotrophin-4 (NT-4), and these proteins bind to specific receptors on neurons, initiating various signaling cascades that regulate neuronal processes. Dysregulation of neurotrophin expression and signaling have been implicated in various neurodegenerative diseases, such as Alzheimer's disease (AD), Parkinson's disease (PD), Huntington's disease (HD), and amyotrophic lateral sclerosis (ALS). In the search for treatment of these diseases, phytochemicals have gained attention due to their antioxidant, anti-inflammatory, and neuroprotective properties. These plant-derived bioactive compounds have been found to protect against neurodegeneration by neutralizing reactive oxygen species, reducing inflammation, and modulating cellular signaling pathways involved in the diseases. Phytochemicals have been found to offer a

Megala Jayaraman, Annesha Dutta, Swathi Mahanta, Diveyaa Sivakumar and Hanumanth Rao Balaji Raghavendran contributed equally with all other contributors.

M. Jayaraman (✉) · A. Dutta · S. Mahanta
Department of Genetic Engineering, Faculty of Engineering and Technology, SRM Institute of Science and Technology, Kattankulathur, Chennai, Tamil Nadu, India

D. Sivakumar
School of Dental Sciences, Universiti Sains Malaysia, Kubang Kerian, Kelantan, Malaysia

H. R. B. Raghavendran
Biomaterial Laboratory, Faculty of Clinical Research, Sri Ramachandra Institute of Higher Education and Research, Central Research Facility, Porur, Chennai, Tamil Nadu, India



S. Pathak, A. Banerjee (eds.), *Neuroprotective Effects of Phytochemicals in Brain Ageing*, https://doi.org/10.1007/978-981-99-7269-2_13

promising avenue for targeting the dysregulation of neurotrophins observed in neurodegenerative diseases, either by modulating neurotrophin expression or the activity of neurotrophin receptors and downstream signaling molecules. In this chapter, we have provided an overview of neurotrophins, their receptors, and signaling pathways, as well as discussed the potential of phytochemicals in targeting and modulating neurotrophin expression in different neurodegenerative disorders.

## 13.1 Neurotrophins

Neurotrophins are a family of proteins essential for the development, survival, and function of neurons (nerve cells) throughout the nervous system.

### 13.1.1 Types of Neurotrophins

The growth, development, and maturation of neurons in the vertebrate nervous system are mediated by a class of structurally and functionally similar proteins called neurotrophins. They play key roles in neuronal differentiation, axonal and dendritic growth, synapse formation, and synaptic plasticity (Huang and Reichardt 2001). In mammals, four different neurotrophins have been identified: nerve growth factor (NGF), brain-derived neurotrophic factor (BDNF), neurotrophin-3 (NT-3), and neurotrophin-4 (NT-4). Neurotrophins are first produced as proneurotrophins, the precursor forms of mature ligands. These proneurotrophins would then undergo cleavage by proteases, either within the cell or outside, to generate the biologically active and mature forms of the neurotrophins (Lu et al. 2005).

NGF was the first neurotrophic factor to be discovered and characterized. It has been found to play a role in both the peripheral nervous system (PNS) and the central nervous system (CNS). In the PNS, NGF plays a key role in the development and maintenance of sympathetic neurons, while in the CNS, NFG contributes to the maintenance and development and cholinergic neurons that are predominantly involved in the research of aging and neural degradation (Aloe et al. 2012). NGF has also shown significant repair capacity for nerve injuries in various animal models and clinical trials (Apfel and Kessler 1995; McMahon and Priestley 1995). NGF exerts its effect by binding to two classes of NGF receptors, TrkA and p75 neurotrophin receptor ($p75^{NTR}$) (Micera et al. 2007).

BDNF also stands as an important member of the neurotrophin family, which significantly contributes to the regulation of pronociceptive signaling in inflammatory and neuropathic pain conditions. Dorsal root ganglion (DRG) BDNF expression levels have been reported to be increased in animal models of inflammation and nerve injury, thus indicating its role in pain signaling (Obata and Noguchi 2006). BDNF was found predominantly involved in the modulation of neurotransmitters and regulates neural plasticity, which is important for learning and memory. It also helps in the growth and maintenance of new neurons from neural stem cells. Upon synthesis in the DRG, BDNF is carried to the central terminals of the primary afferents and subsequently released at the dorsal horn (Li et al. 1999), where it interacts with TrkB receptors present on second-order neurons, playing a role as a neuromodulator in synaptic transmission and spinal nociception (Garraway et al. 2003).

NT-3 and NT-4 were then later discovered due to the DNA sequence homology present in NGF and BDNF. While working on designing primers for amplifying both NGF and BDNF genes, several research groups were able to identify other members of the NGF gene family. This was due to the conserved regions that are present within these genes. By utilizing these primers, NT-3 and NT-4 were discovered and characterized (Maisonpierre et al. 1990; Hallböök et al. 1991). High expression of these proteins was reported during early development. Specifically, the expression of NT-3 increases significantly in the brains of mice embryos by post-natal day 4 and then slowly decreases over time and is limited to the hippocampal dentate gyrus in adults (Thompson et al. 2014). In adults, NT-3 functions to regulate neurogenesis, facilitate hippocampal plasticity, and enhance learning and memory (Shimazu et al. 2006). NT-4 has been reported to exhibit high expression in the trigeminal ganglion (TRG), DRG, and sympathetic nodose ganglion neurons in the PNS of the adult rat, as well as in various regions in the CNS, including the forebrain and medulla (Holtzman and Mobley 1994).

### 13.1.2 Receptors and Signaling Pathways

The neurotrophins predominantly exert their effects by binding to four different receptors: $p75^{NTR}$, TrkA, TrkB, and TrkC. While $p75^{NTR}$ has the ability to interact with all neurotrophins, albeit with varying affinities, the TrkA, TrkB, and TrkC receptors have distinct binding preferences based on their respective affinities for specific neurotrophins. TrkA has a higher affinity for NGF, TrkB preferentially binds to BDNF and NT-4, and TrkC shows a greater affinity for NT-3 (Fig. 13.1). These binding preferences allow for specific signaling cascades and biological responses to be activated in response to the respective neurotrophin-receptor interactions (Huang and Reichardt 2003).

The tyrosine kinase (Trk) family of receptors is a class of receptors extensively used by neurotrophins for biological activities. Trk receptors are transmembrane proteins with heavily glycosylated extracellular domains that contain a cysteine-rich cluster, three leucine-rich repeats, followed by another cysteine-rich repeat, and two immunoglobulin-like domains. The intracellular domain of the Trk receptors consists of a tyrosine kinase domain (Huang and Reichardt 2003). The Trk receptors go through a process of dimerization upon binding to dimeric neurotrophin ligands, which causes the phosphorylation of particular tyrosine residues located inside the intracellular domain of the receptors. These phosphorylated tyrosine residues subsequently function as adaptor protein docking sites, making it easier to recruit other kinases for extra downstream signaling. The well-known signaling pathways

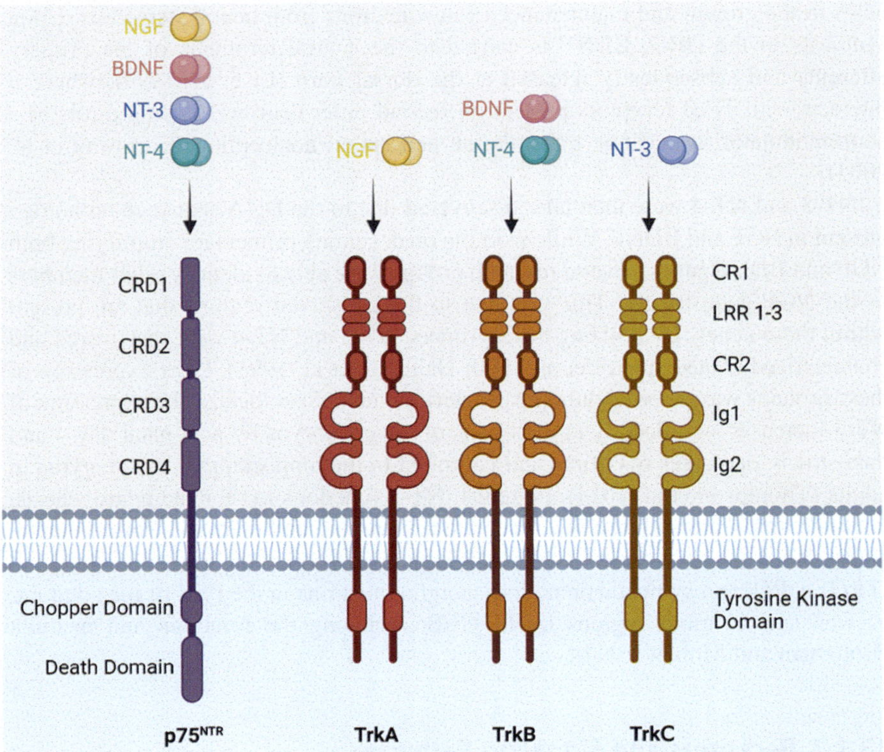


**Fig. 13.1** Neurotrophins and their different affinities for their receptors. While all neurotrophins bind to and activate the p75$^{NTR}$, Trk receptors have specific affinities for different neurotrophins. As seen in the figure, TrkA has a higher affinity for NGF, TrkB preferentially binds to BDNF and NT-4, while TrkC shows a greater affinity for NT-3. *NGF* nerve growth factor, *BDNF* brain-derived neurotrophic factor, *NT-3* neurotrophin-3, *NT-4* neurotrophin-4, *CRD* cysteine-rich domains, *CR* cysteine-rich clusters, *LRR* leucine-rich repeats, *Ig* immunoglobulin-like domains, *p75NTR* p75 neurotrophin receptor, *Trk* tyrosine kinase

activated by Trk receptors are the ERK/MAPK, PI3K/Akt, and PLCγ/PKCγ pathways (Conroy and Coulson 2022). Through the activation of these pathways, neurotrophins regulate a variety of pathways involved in crucial neuronal processes, including cell proliferation, growth, and survival, as well as the formation and maintenance of synapses (Deinhardt and Chao 2014).

The p75 neurotrophin receptor is a member of the TNF receptor/Fas/CD40 superfamily of receptors. It is a transmembrane protein with four cysteine-rich extracellular domains, an α-helical transmembrane domain, and an intracellular domain that contains a chopper domain and a death domain (Conroy and Coulson 2022). Upon binding of neurotrophin ligands, p75$^{NTR}$ initiates various downstream signaling events. The intracellular domain of the p75$^{NTR}$ interacts with different signaling molecules, including tumor necrosis factor receptor-associated factors (TRAFs) and neurotrophin receptor-interacting factors (NRIFs), activating different intracellular signaling pathways, such as the NF-κB pathway and the JNK pathway. It is known to undergo interactions with proteins that play roles in the transmission of signals important for the survival, differentiation, and synaptic plasticity of neuronal cells (Dechant and Neumann 2002).

By forming heteromeric complexes with them, $p75^{NTR}$ can modulate the activity of Trk receptors, thereby influencing the specificity and potency of neurotrophin signaling (Lee et al. 2001). The affinity with which the $p75^{NTR}$ binds to the neurotrophins also depends on the cell context. For example, the co-expression of TrkA and p75 forms a high-affinity binding site for NGF ligand (Hempstead et al. 1991). Additionally, it has been discovered that $p75^{NTR}$ enhances the selectivity of TrkA for its preferred ligand, NGF, and reduces the sensitivity of the receptor to NT-3 (Mischel et al. 2001). These findings suggest that the presence of $p75^{NTR}$ can fine-tune the responsiveness of Trk receptors to different neurotrophins, ultimately influencing the specificity of ligand binding and downstream signaling pathways.

## 13.2 Neuro-nutraceuticals

### 13.2.1 Overview of Nutraceuticals

The term "nutraceutical" was created in 1989 by Stephen Defelice, the creator and executive director of the Foundation for Innovation in Medicine, an American organization that promotes medical health (DeFelice 1995). One of the predominant factors affecting the growth and maintenance of the human body remains to be food components. Apart from the aspect of nutrition, they are seen to be contributing toward medical benefits in the body. Often different plant-derived foods are deemed to have beneficial effects on human health. Nutraceuticals are a subset of functional foods, which are defined as foods that have a positive effect on health beyond their basic nutritional content, including the prevention and treatment of diseases, that are related to dietary habits such as type II diabetes, neurodegenerative diseases, etc. (Espín et al. 2007). The capacity of some plant-derived meals to reduce the risk of chronic diseases has been associated, at least in part, with the existence of non-nutrient secondary metabolites (phytochemicals), which have been proved to exhibit a wide range of biological functions. Glucosinolates, sulfur-containing Alliaceae compounds, terpenoids (carotenoids, monoterpenes, and phytosterols), and various groups of polyphenols (anthocyanins, flavones, flavan-3-ols, isoflavones, stilbenoids, ellagic acid, etc.) are phytochemicals that are present in the diet and have been linked to health benefits. Food products used in the form of nutraceuticals occur in different forms such as probiotics, omega-3 fatty acids, prebiotics, antioxidants, and dietary fibers. Polyphenols, including anthocyanins, stilbenes, proanthocyanidins, flavonols, hydroxycinnamates, coumarins, isoflavones, lignans, etc., are some of the most prevalent phytochemicals found in the nutraceutical industry. Nutraceuticals are dietary supplements that provide a concentrated form of a supposedly food-based bioactive element in a non-food matrix. They are used to boost health at levels greater than those that might be obtained from everyday foods (Espín et al. 2007).

### 13.2.2 Characterization of Nutraceuticals

Nutraceuticals can be categorized based on the type of bioactive compounds they contain. Some common types of nutraceuticals include the following: (1) Polyphenols: plant-derived compounds found in foods such as fruits, vegetables, tea, and wine. Polyphenols are known for their antioxidant properties and are thought to have a variety of health benefits (Piccolella et al. 2019). Polyphenols have been shown to reduce inflammation, improve heart health, and may help protect against certain types of cancer. (2) Carotenoids: natural pigments found in plants, algae, and some bacteria. They are responsible for the colors of many fruits and vegetables. Carotenoids have been found to have anti-inflammatory and antioxidant qualities, and they may help prevent several diseases (Jaswir et al. 2011). (3) Omega-3 fatty acids: a type of polyunsaturated fat found in oily fish, such as salmon and sardines, as well as in nuts and seeds. Omega-3 fatty acids are known for their anti-inflammatory properties and have been linked to a reduced risk of heart disease, among other health benefits (Gupta and Gupta 2020). (4) Probiotics: live microorganisms that are found in certain foods, such as yogurt and fermented vegetables. Probiotics are believed to provide a number of health advantages, including enhancing immune function and enhancing digestive health (Penner et al. 2005). (5) Prebiotics: non-digestible fibers that are found in certain foods, such as bananas, onions, and garlic. Prebiotics help promote the growth of beneficial bacteria in the gut and may have a variety of health benefits (Kumar et al. 2022). (6) Vitamins and minerals: essential nutrients that the body needs in small amounts to function effectively. Nutraceuticals that contain vitamins and minerals are often marketed as supplements to a healthy diet and may be beneficial for individuals who are not getting enough of these nutrients from their regular diet (Houston 2005).

## 13.3 Phytochemicals

Derived from the Greek word "phyto" that stands for plants, phytochemicals are found to be produced by all the members of the botanical world. Phytochemicals are, as suggested by the word, chemicals produced within plants to fight against diseases—such as infections and illnesses caused by bacteria, fungi, and viruses. With a low molecular weight, these chemical compounds may be the result of primary and/or secondary metabolic pathways in the plant. While extensive research has been conducted over the past few decades, and phytochemicals have been associated with various beneficial effects due to their antioxidant, anti-inflammatory, and anticancer properties, a clear link between the onset of human diseases and the effect of phytochemicals has yet to be established (Egbuna and Ifemeje 2015).

Based on biosynthetic origins, phytochemicals can be broadly classified into five categories: phenolic compounds (e.g., phenolic acids, flavonoids, stilbenes, and tannins), alkaloids (e.g., isoquinolines, indoles, and pyridines), carotenoids (e.g., carotenes, zeaxanthin, and lycopene), and terpenoids (e.g., monoterpenes, sesquiterpenes, and diterpenes), and organosulfur compounds (e.g., isothiocyanates and allyl sulfur compounds) (Bellik et al. 2013).

**Table 13.1** List of plant compounds and their neurological effects

| Plant source | Compound | Neurological effects | Reference |
|---|---|---|---|
| *Astragalus* sp. | Astragaloside IV | Anti-inflammatory, antioxidant, and antiapoptotic properties have therapeutic impact on preventing neurodegenerative diseases. | Abd Elrahim Abd Elkader et al. (2022) |
| *Ginkgo biloba* | EGb761 | Scavenger for many types of free radicals; protects against neuronal death induced by various neurotoxic agents; activates natural antioxidant defenses. | Christen (2004) |
| *Panax ginseng* | Ginsenoside | Decreases ROS generation; elevates bax/bcl-2 ratio, release of cytochrome C, and activation of caspase-3. | Cho (2012) |
| *Bacopa monnieri* | Bacoside A | Showed anti-epileptogenic properties. | Anand et al. (2011) |
| *Gardenia jasminoides* | Geniposide | Attenuates insulin-deficiency-induced acceleration of b-amyloidosis in an APP/PS1 transgenic model of Alzheimer's disease. | Zhang et al. (2005) |
| *Sophora flavescens* | Oxymatrine | Anti-inflammatory activities boost antioxidant defenses, which reduces the degree of damage seen following hypoxic-ischemic brain injury. | Williams et al. (2015) |
| *Picrorhiza kurroa* | Apocynin | Reduces the creation of reactive oxygen species and glutathione depletion, which inhibits the cholesterol oxidation product's ability to cause neuronal cells to undergo programmed cell death. | Williams et al. (2015) |
| *Curcuma longa* | Curcumin | Reduced glial activation and suppressed the activation of cytokines and chemokines in patients with persistent epilepsy. | Kaur et al. (2015) |
| *Allium sativum* | S-allyl cysteine | Variety of anti-inflammatory properties in addition to its neuroprotective effects, which are presumably mediated by antioxidant actions, possibly via Nrf2. | Colín-González et al. (2012) |
| *Crocus sativus* | Crocin | Attenuated endoplasmic reticulum stress in a cellular model of Parkinson's disease. | Zhang et al. (2015) |

Phytochemicals are naturally occurring plant compounds that can improve brain health. They work in the human body by undergoing different mechanisms to nurture it in different ways. There are many phytochemicals upon which active studies are being conducted, many of which are discussed in Table 13.1. Multiple studies on these low molecular-weight chemical compounds have opened new doors to finding potential phytochemicals for curing a multitude of illnesses in humans. A higher intake of flavonoids, such as flavonols, anthocyanidins, and proanthocyanidins, has been linked to a lower risk of cardiovascular disease. For every 10 mg/day increase in flavonol consumption, a 5% reduction in risk was noted

(Wang et al. 2014). Studies have also shown that the consumption of α-carotene, β-carotene, lycopene, and carotenoids is associated with a decreased risk of breast cancer by up to 28% (Eliassen et al. 2015). According to a meta-analysis of future cohort research, a higher intake of total flavonoids was associated with a 9% lower relative risk of type 2 diabetes, and each increase in consumption of 500 mg/day was associated with a 5% lower risk (Liu et al. 2014). A separate study showed that diets high in carotenoids, especially β-carotene and α-carotene, were associated with a reduced risk of type 2 diabetes (Sluijs et al. 2015). Turmeric contains a polyphenolic substance called curcumin, which has been demonstrated to enhance memory and cognitive performance. It accomplishes this by lowering brain inflammation, oxidative stress, and beta-amyloid buildup, all of which are associated with cognitive loss. Additionally, it has been discovered that curcumin raises levels of brain neurotrophic factor (BDNF), a protein needed for the growth and survival of brain neurons (Mishra and Palanivelu 2008). Resveratrol is a polyphenol found in grapes, red wine, and some fruits, and it has been found to have anti-inflammatory properties that can protect the brain from damage. In addition, resveratrol has been found to increase blood flow to the brain and promote vascular growth and survival (Poulose et al. 2015). Quercetin is a flavonoid found in many fruits and vegetables such as apples, berries, and onions. It has been found to have anti-inflammatory properties that can protect the brain from damage by increasing blood flow and neural connectivity. In addition, animal studies have shown that quercetin improves cognitive function and memory (Amanzadeh et al. 2019). Epigallocatechin gallate (EGCG) is a catechin found in green tea. It has been found to have anti-inflammatory properties that can protect the brain from damage. In addition, EGCG has been reported to promote neurogenesis and improve cognitive function in animal studies (Unno et al. 2017). Luteolin is a plant pigment known as a flavonoid, which is part of the wider class of polyphenols. Numerous plants, including celery, peppers, parsley, thyme, and chamomile tea, contain this pigment. In performing various animal studies, it was found that luteolin plays a significant role in the reduction of microglial activity of the brain, which accelerates during the process of aging, hence increasing cognitive function and memory (Burton et al. 2016). Additionally, it has been discovered that luteolin has therapeutic potential for the treatment of mood disorders. Its effects on anxiety and depression have been demonstrated in animal models, and it may be able to control the brain's levels of neurotransmitters like dopamine and serotonin. Apigenin is a bioactive flavonoid that is a member of the group of plant chemicals known as flavones. It is frequently discovered in a variety of fruits and vegetables, including celery, parsley, thyme, and chamomile tea. A major method by which apigenin exerts its neuroprotective benefits is via modifying important signaling pathways in the brain. Apigenin, for instance, has been demonstrated to activate the Nrf2/ARE pathway, which is essential for preventing oxidative stress and inflammation. Additionally, it has been discovered that apigenin reduces the activity of enzymes responsible for creating amyloid-beta peptides, which are thought to contribute to the onset of AD (Salehi et al. 2019). Beta-sitosterol is a plant sterol found in various plants, including fruits, vegetables, nuts, and seeds. In tests on animals, beta-sitosterol was proven to enhance memory and cognitive function, in terms of brain nutrition. Additionally, it has been demonstrated to have neuroprotective effects by lowering brain inflammation and oxidative stress, which can help stop or slow the progression of neurodegenerative diseases like Parkinson's and Alzheimer's. Additionally, it has been discovered that beta-sitosterol enhances the effects of other phytochemicals, such as quercetin and resveratrol, on brain function. This shows that it might work in concert with other substances to support brain health (Chandra et al. 2022). *Ginkgo biloba* is an herb that has been used in traditional Chinese medicine for centuries. It has favorable benefits on the brain because it includes a number of bioactive substances, such as flavonoids, terpenoids, and other phytochemicals. Increased blood flow to the brain is one of *Ginkgo biloba*'s most well-known advantages for maintaining brain health. This is assumed to be because *Ginkgo biloba* contains terpenoids, which have been found to widen blood vessels and enhance circulation. More oxygen and nutrients can be delivered to brain cells through improved blood flow, which helps support their growth and function (Gohil 2002).

### 13.3.1 Neuroprotective Potential of Phytochemicals

The brain is a network of nearly 100 billion neurons and glial cells, which sets up the foundation of the layout of every chemical, biological process, and function in the body. Even after being a mere 2% of the body weight, it requires a staggering amount of energy to function properly. Found to take up about 15% of the cardiac output and 30% of the resting metabolic output alone, the brain dysfunctions without the obligatory energy supply (Youdim and Joseph 2001). Neurodegeneration is the word used to refer to the aging and deterioration of the neurons in the brain. This gradual degradation is collectively found to be linked to genetic disorders with various speeds of brain aging. While brain aging can be attributed to various factors, some notable ones include chronic inflammation, exposure to UV light, and infections resulting from foreign organisms, as well as endogenous factors such as mitochondrial dysfunction, genetic modifications, and abnormal enzymatic activities (Wang et al. 2018). Neurodegenerative diseases are associated with high fatality rates globally and can have devastating effects on individuals and their families. While there are treatments available to manage symptoms and slow disease progression to some extent, a complete cure for most neurodegenerative diseases is still elusive.

It is not a secret that traditional and herbal medicines were used to treat multiple diseases in the past, including neural symptoms. It is well known that plant extracts contain compounds that have been found to possess potent neuroprotective properties (Jayaraman et al. 2022). These compounds are now being explored as potential targets for gene therapy and specific mechanisms to address genetic disorders, making plant extracts a valuable source for potential therapeutic interventions (Kumar and Khanum 2012). As individuals age, cognitive abilities in the brain tend to decline, leading to challenges in memory. This is particularly evident in elderly individuals and patients with neurodegenerative disorders, who commonly experience issues related to memory impairment. To address cognitive issues and support brain health, a class of drugs known as nootropic drugs is often prescribed. Nootropic drugs function by modulating the dopaminergic, glutamatergic, and serotonergic systems, aiming to slow down the rapid deterioration of cognitive functions (Onaolapo et al. 2019).

This brought a new interest in looking further into these plant compounds as pharmacological agents. Researchers have found ways to slow down the progressive degradation of the brain by incorporating phytochemicals into the dietary needs of patients. Several dietary or phytobioactive compounds that have been observed to have anti-inflammatory and antioxidant effects are able to play major roles in the inhibition of oxidative mechanisms that can hinder the progression of neurodegenerative diseases (Wang et al. 2016). Recent studies have discovered that resveratrol, quercetin, and catechins have the ability to reduce oxidative stress and protect against nitric oxide-mediated cell death in cultured hippocampal neurons. These findings demonstrate the potential of these phytochemicals in the treatment of neurodegenerative diseases (Kumar and Khanum 2012).

Surprisingly, phenolic compounds found in apples, grapes, citrus fruits, and green tea have been found to have a higher neuroprotective capability than most antioxidant vitamins due to their ability to protect neurons from oxidative decay and dopaminergic neurons from the damage dealt by 6-hydroxydopamine (Guo et al. 2007). Resveratrol found in berries, peanuts, and grapes is a polyphenol compound that is well known for its antioxidant, anti-carcinogenic, and anti-inflammatory effects on the brain (Bernier et al. 2016). In part through the activation of protein kinase C, catechins, a kind of polyphenol, display neuroprotective properties. They also have transcription factors that stimulate the expression of genes for cell survival (Ehrnhoefer et al. 2006). Another polyphenol found in green tea, called EGCG, also contains anti-inflammatory properties to combat neurodegeneration. Several compounds of the phenol family such as curcumin have been used to potentially slow down the progression of AD. Flavonoid-rich plants have been found to possess the ability to enhance memory and protect vulnerable neurons by promoting the initiation and stimulation of new neuron growth. Among the flavonoids, quercetin has been identified as a neuroprotective agent with beneficial effects on neuronal health and function (Ganesan et al. 2015). Multiple plant species such as *Panax ginseng, Bacopa monniera, Centella asiatica, Ginkgo biloba, and Dinh lang* are species known for medicinal properties and were used as traditional medicines to treat symptoms associated with various diseases and are even thought to have neuroprotective properties (Youdim and Joseph 2001; Kumar and Khanum 2012). Needless to say, the phytochemicals contained in these plants are now being studied for their therapeutic properties.

## 13.4 Neurotrophin-Mediated Actions of Phytochemicals in Neurodegenerative Disorders

Neurodegenerative disorders share a common characteristic: the premature degradation of the neuronal system. By modulating pathways and systems that control and have a direct effect on the neuronal system, a cure for neurodegenerative diseases can be developed. However, highly effective drugs for such diseases would have side effects and can impact overall bodily functions. On the other hand, phytochemicals and their derivatives are considered much less toxic when compared to other novel synthetic drugs (Mathur et al. 2023). By utilizing phytochemicals, it is possible to explore treatment options that are potentially more favorable in terms of safety and tolerability. These natural compounds hold promise in providing neuroprotective effects and supporting the health and function of the neuronal system. Thus, directing therapeutic efforts toward identifying and developing phytochemicals that are able to target neurotrophins and their factors could lead to novel approaches for treating neurodegenerative diseases while minimizing the risk of adverse effects on the body (Sharifi-Rad et al. 2020).

Phytochemicals have the ability to bind to receptors on neurotrophic factors, initiating a cascade of signaling pathways that directly influence brain functions. In recent studies on neurodegenerative disorders, NGF and BDNF have been targeted by various means to study the pathology of diseases and their symptoms (Venkatesan et al. 2015). Flavonoids, deoxygedunin, and diosmetin have been reported to bind to TrkB and initiate the activation of P13K-Akt-extracellular down-signaling pathways to increase the synthesis of BDNF. Curcumin was found to activate TrkB-MAPK and P-13 K as seen in rat models, where it plays an essential role in protecting the cerebral cortical neurons from glutamate cytotoxicity (Naoi et al. 2017). Even coffee fruit extracts, when consumed in a regulated and controlled manner, are observed to increase the production of BDNF. Studying the effects of neurotrophin mediated by phytochemicals in specific neurodegenerative disorders can help understand the effectiveness of these compounds.

### 13.4.1 Alzheimer's Disease (AD)

Alzheimer's disease (AD) is considered to be the most widespread neurodegenerative disorder that accounts for the most deaths occurring due to neurological issues. It affects a significant proportion of the elderly population, with approximately 30% of individuals over the age of 50 experiencing symptoms associated with dementia (Scheltens et al. 2016). AD is responsible for one in three seniors' deaths, highlighting its significant impact on public health. Being a neurodegenerative disorder, the main focus of the disease falls on the misfolded amyloid-beta (Aβ) and tau proteins causing amyloid plaque and neurofibrillary tangles to form. The buildup of these amyloid plaque and neurofibrillary tangles can lead to severe neuron damage, leading to neuronal death. The exact factors that contribute to the onset of AD are still not fully understood. However, research has focused on studying the sequence and concentration of Aβ peptides, as well as the conditions under which they form, as potential destabilizing factors in the synthesis of mature neurons (Scheltens et al. 2016). Mutations in the hereditary genes of APP, PSEN1, and PSEN2 are an important factor in affecting the Aβ peptide cleavage and aggregation, and their specific display becomes the cause of early or late onset (Scheltens et al. 2016). Symptoms of AD could range from short-term memory loss in day-to-day life to trouble understanding visual images and spatial awareness. As with other neurological diseases, AD is an incurable disorder, and medications, such as galantamine and rivastigmine, can only work to control some of the cognitive impairment.

Phytochemicals were proven to work against AD in experiments where the administration of curcumin was directly related to the decreased oxidative stress in the mouse due to the increase of the cytoprotective enzymes such as catalase, superoxide dismutase, glutathione peroxidase, and glutathione-S-transferase in mouse models of AD (D'Onofrio et al. 2016). Diterpenes, ginkgolides, sesquiterpene bilobalide, and numerous other flavonoids are phytochemicals that have been extracted from the medicinal plant *Ginkgo biloba*. These are said to directly affect the down-regulation of amyloid-β neurotoxicity by increasing the blood flow in the cerebrum (Wightman 2017). A list of phytochemicals that have been found to be associated with the effective treatment of AD have been included in the table below (Table 13.2).

In the context of AD, targeting the NGF and BDNF signaling pathways is an area of active research. Neurons require mature NGF for essential processes such as cell proliferation, development, and division. However, the upregulation of metalloproteinase 9 (MMP9), an enzyme that degrades NGF, has been observed in AD (Mitra et al. 2019). This degradation of NGF can negatively impact neuronal function and survival. Curcumin was found to affect the nuclear content of glutathione by interception with certain transcriptional factors such as 12-tetradecanoate 13-acetate (TPA) as well as promote the production of NGF to combat the skewed TrkA/p75 receptor ratio, thus helping in the altered maturation of NGF in AD models (Mitra et al. 2019). In individuals affected by neurodegenerative disorders like AD, BDNF expression and signaling in the brain is altered. Specifically, there is a decrease in BDNF mRNA levels, which indicates a disruption in the production of BDNF. It is also observed that the amyloid protein, which is the key pathological feature of AD, directly hinders the maturation of BDNF, thus causing rapid and premature degradation of neurons (Tanila 2017). Phytochemicals such as curcumin and flavonoid have been found to play a role in activating the NMDA-CREB pathway to increase the synthesis of BDNF in a rat model for AD (Naoi et al. 2017).

### 13.4.2 Parkinson's Disease (PD)

Parkinson's disease (PD), the second most prevalent neurodegenerative disorder, affects around 0.5–1% of individuals aged 65–69, and its occurrence increases to 1–3% in people aged 80 and above. While PD is a multifactorial neurodegenerative disorder influenced by both genetic and environmental factors, the most significant

**Table 13.2** List of phytochemicals that are associated with the effective treatment of Alzheimer's disease

| Phytochemicals | Chemical compound | Plant source | Effect of compound | References |
|---|---|---|---|---|
| Quercetin | 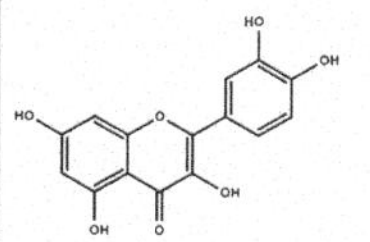 | *Morus alba* | Inhibition of AChE and anti-inflammatory | Khan et al. (2019) |
| Resveratrol | 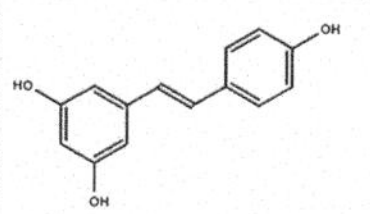 | *Vitis vinifera* | Exhibits antioxidant and neuroprotective properties and delays Aβ-induced toxicity | Vingtdeux et al. (2008) |
| Ginsenosides | 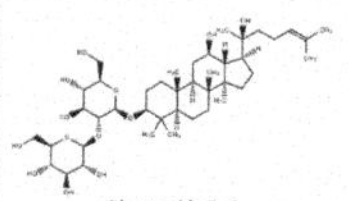 Ginsenoside Rg3 | *Panax ginseng* | Promotes Aβ degradation by enhancing neprilysin gene expression | Yang et al. (2010) |
| | 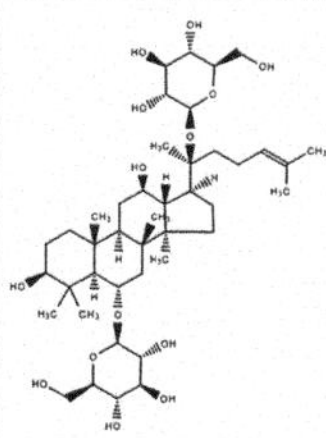 Ginsenoside Rg1 | | Attenuates the amount of accumulated Aβ by activating the protein kinase A/cAMP response element binding protein signaling pathway | Fang et al. (2012) |
| | 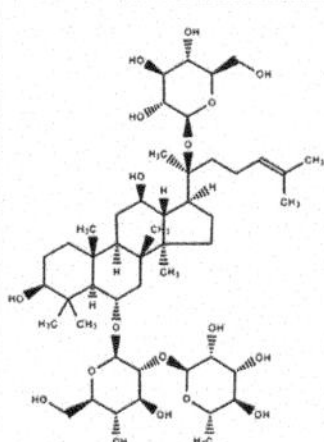 Ginsenoside Re | | Attenuates neuroinflammation markers in the hippocampus | Wang et al. (2011) |
| Ginkgolides | 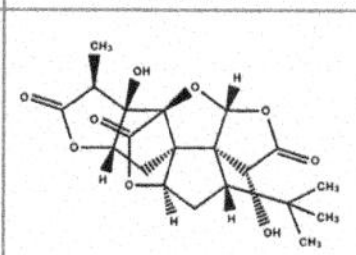 Ginkgolide A | *Gingko biloba* | Protects neuronal cells from synaptic damage and increases neuronal survival against Aβ-induced toxicity | Bate et al. (2008) |
| | 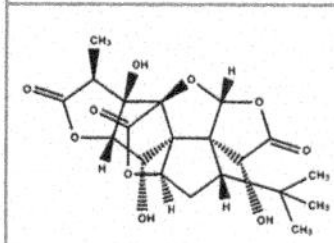 Ginkgolide B | | Rescues hippocampal neurons from Aβ-induced apoptosis by increasing the production of brain-derived neurotrophic factor (BDNF) | Xiao et al. (2010) |
| | 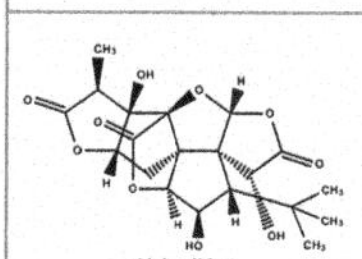 Ginkgolide J | | Inhibits Aβ-induced hippocampal neuronal cell death | Vitolo et al. (2009) |
| Curcumin | 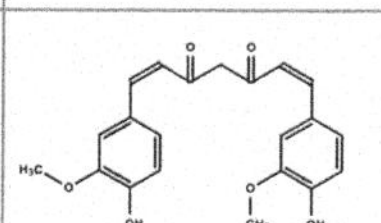 | *Curcuma longa* | Exhibits antioxidant, anti-inflammatory properties and inhibits the formation of Aβ 42 oligomers | Hamaguchi et al. (2010) |

risk factor for PD is age, and the average age at which symptoms typically appear is around 60. (15) While PD is commonly considered an idiopathic condition, there is a small proportion of cases (approximately 10–15%) that have reported family history, and 5% may exhibit Mendelian inheritance patterns (Deng et al. 2018). Certain genes associated with PD have been identified, and mutations in these genes can indicate autosomal dominant inheritance (SCNA, LRRK2, and VPS32) or autosomal recessive inheritance (PRKN, PINK1, and DJ-1) (Schulte and Gasser 2011). PD is characterized by neuronal degeneration as well as the degradation of dopamine-producing neurons in the substantia nigra in the midbrain. The neuronal system in this region acts as the bridge between the neurons of the basal ganglia by releasing dopamine, which is known to be responsible for motor movement. This loss of dopamine leads to motor symptoms such as tremors, rigidity, and bradykinesia. However, these symptoms only tend to appear when 70–80% of the dopaminergic neurons have already been damaged (El-Agnaf et al. 2006). The accumulation of Lewy bodies and neurofibrillary tangles, due to the misfolding of tau protein, in neuronal cell bodies is also seen in PD patients (Schulz-Schaeffer 2010). Other mechanisms such as abnormal protein clearance, mitochondrial dysfunction, and neuroinflammation have also been implicated in the progression of PD; however, the pathways remain uncertain (Kouli et al. 2018). Current treatment mainly focuses on symptomatic relief via drugs that help regulate the levels of dopamine, thus helping alleviate the motor symptoms. Recent studies are exploring deep brain stimulation for PD where electrodes are implanted in various regions of the brain and are triggered to deliver specific electrical impulses. This has been found to help regulate abnormal brain activity and thus reduce motor symptoms experienced by the patients (Mahlknecht et al. 2022).

The effect of various natural compounds has been explored and studied for their neuroprotective properties in PD. For example, curcumin was found to reduce the levels of reactive oxygen species (ROS) and protect against α-synuclein cytotoxicity by inhibiting α-synuclein aggregation as well as destabilizing preformed aggregates (Pandey et al. 2008). EGCG and N-methyl-4-phenyl-1,2,3,6-tetrahydropyridine (MPTP) were found interacting with the neurons and preventing the degradation of striatal dopamine production due to the catechol-like structure of the constituents (Wightman 2017). EGCG was reported to reduce oxidative stress and partially restore dopamine concentrations. At higher concentrations, EGCG was also effective against α-synuclein oligomerization and fibrillization (Schulte and Gasser 2011). In a PD mouse model, resveratrol was shown to exhibit neuroprotective potential by inhibiting α-synuclein aggregation and reducing α-synuclein-induced cytotoxicity, thus improving motor and cognitive performance (Zhang et al. 2018).

There are various studies that provide evidence of a connection between the decline of BDNF levels and the progression of PD. Patients with depression or PD have been observed to exhibit lower blood and brain BDNF levels (Scalzo et al. 2010). A separate study reported that this decrease in BDNF concentration was found to coincide with an increase in dopaminergic neuron degeneration (Huang et al. 2018). BDNF was also found to play a role in regulating tyrosine hydroxylase (TH), which is also diminished in PD, ultimately leading to motor disorders

(Baydyuk et al. 2011). Considering the effect that BDNF seems to have on PD progression, exploring the potential therapeutic implications of modulating BDNF levels in the context of PD holds promise for the development of novel treatment strategies.

A study on the MPTP-induced PD mouse model reported that caffeic acid treatment resulted in the restored expression of BDNF, GDNF, and TH. As these proteins were found to be decreased in PD, these findings support that this compound was effective in preventing the development of PD (Tsai et al. 2011). Curcumin was found to play a neuroprotective role in a PD rat model by activating the BDNF/TrkB pathway and promoting the regeneration of neurons in the hippocampal tissue (Yang et al. 2014). Asiatic acid, found in plants such as *Centella asiatica* and *Olea europaea* (olive), had similar effects as caffeic acid on an MPTP rat model. Asiatic acid treatment was shown to increase striatal levels of dopamine, BDNF, and GDNF. Additionally, treated rats also displayed decreased striatal expression of α-synuclein (Chao et al. 2016). These novel findings show the therapeutic potential of phytochemicals in modulating BDNF levels for the development of treatment strategies for PD.

### 13.4.3 Huntington's Disease (HD)

Huntington's disease (HD) is a genetic neurodegenerative ailment that is characterized by a steady decline in motor coordination, mental capacity, and psychiatric symptoms. The creation of a mutant huntingtin protein (mHTT), which clumps together in neurons and impairs their normal function, is what causes it. This mutant protein is the result of an abnormal number of CAG repeats in the Huntingtin (HTT) gene. A greater number of CAG repeats is linked to an earlier onset of the disease and a faster development of symptoms (MacDonald et al. 1993). In most cases, HD is inherited in an autosomal dominant pattern, meaning that there is a 50% chance of an affected individual passing a copy of the mutated gene to the next generation. The pathogenesis of HD revolves around the toxic effects of mHTT aggregates, which leads to neuronal dysregulation as well as the damage and loss of striatal and cortical neurons. These aggregates impair intracellular transport, disrupt protein degradation pathways, promote oxidative stress, and trigger inflammatory responses, ultimately leading to the progressive loss of neurons in specific brain regions (Md and Frcp 2010). While the exact mechanisms underlying HD pathogenesis are complex and not fully understood, several key factors, including oxidative stress excitotoxicity, mitochondrial dysfunction, and impaired protein clearance, contribute to the progressive neurodegeneration observed in HD (Choudhary et al. 2013). Unfortunately, there is currently no cure for HD and treatment mainly focuses on managing symptoms and providing support to improve the quality of life for affected individuals. Medications such as neuroleptics can be prescribed to help control movement problems, manage psychiatric symptoms, and address other complications associated with the disease. Physical therapy, occupational therapy, and speech therapy have also been shown to be beneficial in maintaining mobility, function, and communication abilities (Novak and Tabrizi 2010).

While research on phytochemical therapy for HD is still in its early stages, some studies have shown promising results regarding the potential benefits of phytochemicals in HD animal models. Curcumin has been well studied for its neuroprotective properties in HD. In a study of HD mouse models where the diet was supplemented with curcumin, a decrease in HTT aggregates was observed and the behavioral deficits that were seen in the non-treated mice were restored (Hickey et al. 2012). In another study, curcumin treatment was found to significantly increase mitochondrial and superoxide dismutase activity while restoring glutathione levels in HD rat models. Additionally, treated rats demonstrated significant improvements in neuromotor coordination, indicating that administering curcumin holds promise as a potential therapeutic intervention for HD (Sandhir et al. 2014). Extensive research has also been conducted on *Ginkgo biloba* for its anti-inflammatory and antioxidant characteristics, particularly focusing on the impact of its extract on the CNS (DeFeudis and Drieu 2000). Findings indicate that the extract EGb 761 derived from *Ginkgo biloba* can inhibit the aggregation of polyQ proteins and enhance their breakdown by promoting proteasome activity (Stark and Behl 2014). Other than these compounds, many other compounds are also being explored for their potential therapeutic benefits in the treatment of HD, and further research is being conducted to evaluate their efficacy and safety (Dey et al. 2020).

In HD, BDNF levels have been reported to be reduced in the striata, brainstem regions, and prefrontal cortexes of HD patients and animal models (Azman and Zakaria 2022). Plasma BDNF and NFG levels were also found to be decreased in HD patients (Tasset et al. 2012). As the decrease in BDNF levels has been widely considered a significant characteristic of HD, various therapeutic strategies focused on regulating the production of BDNF have been explored. When considering phytochemical therapy, curcumin, as discussed above, has been shown to elevate BDNF levels in preclinical studies of various diseases including HD. In an HD animal model, dietary curcumin supplementation was found to preserve the normal levels of BDNF in the cortex and striatum. Curcumin also reduced the aggregation of mHTT and typical motor impairment linked to the progression of the disease in these mice (Elifani et al. 2019). Studies have also shown that resveratrol, a polyphenol found in grapes and red wine, activated the ERK1/2/CREB-mediated signaling pathway and thus induced BDNF and GDNF secretion in rat primary astroglia-enriched cultures (Zuccato and Cattaneo 2009). In a similar study, [6]-Shogaol, a chemical found in ginger, was reported to increase the BDNF expression and attenuate the cell death of lipopolysaccharide (LPS)-treated astrocytes, thus inferring that the neuroprotective effects exhibited by [6]-shogaol treatments are related to the upregulation of BDNF (Shim et al. 2012). While these studies were not done in HD models, the increase in BDNF and GDNF expressions provides evidence of the potential therapeutic properties of certain phytochemicals as novel treatment strategies for neurological disorders.

### 13.4.4 Amyotrophic Lateral Sclerosis (ALS)

Amyotrophic lateral sclerosis (ALS) is a progressive neurodegenerative disorder characterized by the degeneration and death of motor neurons in the brain and spinal cord. It is considered the most common adult-onset motor neuron disease. Studies have shown that environmental and genetic variables are linked to the development of the disease, even if the exact origins of the condition are still not entirely understood. The majority of ALS cases (90–95%) occur sporadically without a clear genetic link, while the remaining 5–10% of cases have a familial component, suggesting a genetic basis for the disease. Several mutations associated with ALS have been discovered, including mutations in the C9orf72, SOD1, TARDBP, FUS, and UBQLN2 genes, among others (Vildan et al. 2019). The hallmark feature of ALS is the progressive degeneration of motor neurons, which are essential for the communication between the brain and the muscles. As these neurons get damaged and begin to die, this results in the loss of muscle control and voluntary movement. The initial symptoms often involve muscle weakness, twitching, and muscle stiffness. As the disease progresses, individuals may experience difficulty speaking, swallowing, and breathing, leading to respiratory failure, which is a common cause of death in ALS patients (Keon et al. 2021). Numerous possible factors have been attributed to the development and advancement of ALS, including oxidative stress due to superoxide dismutase 1 (SOD1) mutations, glutamate excitotoxicity, mitochondrial dysfunction, and unregulated apoptosis (Rossi et al. 2013). However, it is still necessary to explore whether these mechanisms interact with each other, operate simultaneously, or follow a specific sequence in inducing the death of neurons. Similar to the other diseases discussed above, currently, there is no curative drug for ALS. However, the FDA had approved three drugs: riluzole, edaravone, and relyvrio, which have been found to be effective in slowing down disease progression and prolonging patient survival (Aschenbrenner 2023; Neupane et al. 2023). Additionally, drugs to manage symptoms and physical and speech therapy are commonly prescribed to improve the quality of life for individuals with ALS.

The effect of *Withania somnifera* (Ashwagandha) root extract was studied in mice with A315T mutation in the TDP-43 gene. After 8 weeks of treatment, researchers observed improved innervation at neuromuscular junctions and better cognitive performance. They also reported that the treatment attenuated neuroinflammation in the spinal cord and caused reduced hTDP-43 aggregation in the motor neurons (Dutta et al. 2017). Another study on *Withania somnifera* root extract also showed that treatment of mice with the G93A mutant of the SOD1 gene resulted in improved motor performance and delayed disease progression (Dutta et al. 2018). The neuroprotective effects of genistein, an isoflavone from soybean, were studied using the same ALS mice model. Genistein administration was found to increase the viability of spinal motor neurons, thus delaying the onset of symptoms as well as slightly increasing the lifespan of the mice (Zhao et al. 2019).

As BDNF is a protein that has a crucial role in the survival and growth of neurons, it is no surprise that studies have shown that BDNF dysregulation and signaling may impact the degeneration of motor neurons in ALS. In the ALS model with the G93A mutant of the SOD1 gene, BDNF and NT-4 levels were shown to be significantly increased as compared to the wild-type mouse (Just-Borràs et al. 2020). In humans, while higher BDNF levels were observed in the serum of ALS patients, lower Pro-BDNF levels were seen. This study also reported that significantly higher cerebrospinal fluid (CSF) BDNF levels were associated with faster disease progression and lower serum Pro-BDNF levels were associated with a shorter survival (Riolo et al. 2022). Apart from PBNF, NGF levels were also observed to be increased in the spinal cord of mice with the G93A SOD1 mutation. In the same study, spinal cord astrocytes were cocultured with motor neurons, and when astrocytes were triggered to the secretion of NGF, a significant reduction in motor neuron survival was observed. Further experiments also reported that NGF only induced apoptosis in $p75^{NTR}$-positive motor neurons but failed to have any effect on $p75^{NTR}$-negative motor neurons (Pehar et al. 2004). These findings indicate that NGF not only plays an important role in promoting neuronal survival and differentiation by activating TrkA, but it can also induce neuronal death through the activation of $p75^{NTR}$. Similar results were seen in another study where glycation, a post-translational oxidative modification, was able to regulate the ability of NGF to induce cell death. The modified neurotrophin was found to exert its effects by binding to receptors for advanced glycation end products (RAGEs) and $p75^{NTR}$ to induce motor neuron death. Additionally, it was reported that this altered form of NGF was detected in the spinal cord of an ALS mouse model. It was found that through signaling pathways involving RAGE and $p75^{NTR}$, this modified NGF played a role in astrocyte-mediated toxicity toward motor neurons, which is a pathological characteristic of ALS (Kim et al. 2018).

While there have been many studies done exploring the neuroprotective properties of phytochemicals in the context of ALS, no studies have reported specifically targeting neurotrophins or their signaling in ALS animal models. However, various studies have shown the effectiveness of phytochemicals in modulating and regulating BNDF and NGF levels; it can be assumed that these properties would also be seen in experiments involving ALS models (Venkatesan et al. 2015; Naoi et al. 2017). Moreover, further research is still required to determine the complete therapeutic potential of these phytochemicals in ALS and to translate these findings into clinical applications.

## 13.5 Future Aspects of Neuro-nutraceuticals

Commonly known as nootropics or mood enhancers, neuro-nutraceuticals are compounds that can improve memory, cognition, creativity, and overall brain health. In today's world, the neuro-nutraceutical market is rapidly expanding due to the need for greater intellectual development. A report from Bonafide Research projects that the global market for nutraceuticals would be worth USD 650 billion by 2022 and increase of 8.13% through 2027. It looks like future neuro-nutrition will be interesting, including improving cognitive performance. The future of neuro-nutraceuticals looks promising, with the potential to develop new products that can enhance cognitive function and promote brain health. The future of neuro-nutraceuticals is in personalized nutrition, where foods will be chosen to meet each person's unique requirements in terms of cognition. It will be feasible to determine a person's genetic propensity for cognitive decline and create personalized neuro-nutraceuticals that can stop or slow down cognitive decline thanks to advancements in genetic testing and analysis (Bhattacharya et al. 2022).

In the market for neuro-nutraceuticals, natural and herbal products are becoming more popular. In traditional medicine, natural substances like *Ginkgo biloba, Bacopa monnieri,* and turmeric have been used for generations to improve cognitive abilities. There is growing interest in the creation of novel natural and herbal neuro-nutraceuticals that can offer secure and efficient cognitive enhancement due to the rising demand for natural products (Kumar and Khanum 2012). Neuro-nutraceuticals may be used in combination with other treatment modalities to enhance their efficacy. This could involve combining neuro-nutraceuticals with pharmaceutical drugs, non-invasive brain stimulation techniques, or cognitive training programs. Synergistic effects may be achieved by addressing multiple aspects of brain health simultaneously.

Brain-computer interfaces (BCIs) are devices that can interpret brain signals and translate them into actions. BCIs have the potential to revolutionize the neuro-nutraceutical market by providing real-time feedback on cognitive function and enabling personalized neuro-nutraceuticals to be developed. For example, a BCI can be used to measure an individual's attention span and provide a neuro-nutraceutical that can improve attention span in real-time.

The gut-brain axis is a sophisticated network of neural connections that connects the gut with the brain. The collection of bacteria in the stomach known as the microbiome is essential to the gut-brain axis. It has been demonstrated that altering the microbiome with the use of prebiotics, probiotics, and other microbiome-modulating substances improves mental performance and brain health. Neuro-nutraceuticals that regulate the microbiome will be more sought after in the future (Spisni et al. 2022).

With the increasing use of digital health technologies, there is a growing interest in the development of digital neuro-nutraceuticals that can provide cognitive enhancement through mobile apps, wearables, and other digital devices. For example, a mobile app can provide personalized cognitive training exercises that can enhance cognitive function (Georgiev et al. 2021).

In conclusion, the future of neuro-nutraceuticals looks promising, with the potential to develop new products that can enhance cognitive function, memory, focus, creativity, and overall brain health. Personalized nutrition, natural and herbal products, combination products, brain-computer interfaces, microbiome modulation, and digital health are some of the future aspects of neuro-nutraceuticals. As the demand for cognitive enhancement continues to grow, the neuro-nutraceuticals market will continue to evolve and innovate.

## 13.6 Conclusion

Neurodegenerative diseases pose a significant global health challenge, and the search for effective treatments remains a priority in medical research. While a definitive cure may still elude us, the use of phytochemicals presents a promising avenue for neuroprotection. Phytochemicals, derived from plants, have shown potential in delaying or preventing the onset of numerous neurodegenerative diseases (Jayaraman et al. 2022). Importantly, these natural compounds have exhibited non-cytotoxic properties, which is the major side effect seen when synthetic compounds are used as treatment. This non-cytotoxic nature of phytochemicals is an essential attribute that supports their potential therapeutic utility as this allows for the long-term administration of these compounds, potentially providing sustained neuroprotection without significant toxicities.

These compounds exhibit pleiotropic effects, meaning they can act on various molecular targets simultaneously. This characteristic allows for phytochemicals to exert their neuroprotective effects through different mechanisms, enhancing their potential therapeutic value (Limanaqi et al. 2020). Additionally, these compounds possess antioxidant and anti-inflammatory properties, which can combat oxidative stress and neuroinflammation, both of which are implicated in the development of neurodegenerative illnesses (Forni et al. 2019). By reducing oxidative damage and inflammation, phytochemicals create an environment conducive to neuronal health, thereby potentially preserving cognitive function and delaying disease progression.

The neurodegenerative disorders have been discussed in this chapter, where the ability of phytochemicals to regulate neurotrophins in neurodegenerative diseases is of particular significance. As dysregulation of neurotrophin signaling has been linked to the progressive degeneration of neurons observed in various neurodegenerative disorders, targeting neurotrophin regulation offers a promising approach to counteract the underlying pathophysiological processes contributing to these diseases (Chen et al. 2018). By modulating neurotrophin expression, mimicking their functions, and activating their receptors, phytochemicals have the potential to counteract neurodegenerative diseases by enhancing the availability and activity of these crucial proteins. Studies have shown that increased levels of these neurotrophins have been associated with neuroprotection and enhanced neuronal survival (El Ouaamari et al. 2023). By targeting pathways involved in neurotrophin synthesis, phytochemicals can potentially support the growth and functionality of neurons, thereby mitigating the development and progression of neurodegenerative diseases. By focusing on the interaction between phytochemicals and neurotrophins, researchers can contribute to advancing the development of therapeutic interventions that promote the maintenance and survival of neurons, potentially leading to improved outcomes for individuals affected by these debilitating conditions.

## References

Abd Elrahim Abd Elkader H-T, Essawy AE, Al-Shami AS (2022) Astragalus species: phytochemistry, biological actions and molecular mechanisms underlying their potential neuroprotective effects on neurological diseases. Phytochemistry 202:113293. https://doi.org/10.1016/j.phytochem.2022.113293

Aloe L, Rocco ML, Bianchi P, Manni L (2012) Nerve growth factor: from the early discoveries to the potential clinical use. J Transl Med 10(1):1–15. https://doi.org/10.1186/1479-5876-10-239/TABLES/4

Amanzadeh E et al (2019) Application of quercetin in neurological disorders: from nutrition to nanomedicine. Rev Neurosci 30(5):555–572. https://doi.org/10.1515/revneuro-2018-0080

Anand T, Naika M, Swamy MSL, Khanum F (2011) Antioxidant and DNA damage preventive properties of Bacopa Monniera (L) Wettst. Free Radic Antioxid 1(1):84–90. https://doi.org/10.5530/ax.2011.1.13

Apfel SC, Kessler JA (1995) Neurotrophic factors in the therapy of peripheral neuropathy. Baillieres Clin Neurol 4(3):593–606. Available at: https://europepmc.org/article/med/8599726?utm_medium=email&utm_source=transaction (Accessed: 14 May 2023)

Aschenbrenner DS (2023) New drug approved for ALS. AJN, Am J Nurs 123(1):22–23. https://doi.org/10.1097/01.NAJ.0000911516.31267.67

Azman KF, Zakaria R (2022) Recent advances on the role of brain-derived neurotrophic factor (BDNF) in neurodegenerative diseases. Int J Mol Sci 23:12. https://doi.org/10.3390/ijms23126827

Bate C, Tayebi M, Williams A (2008) Ginkgolides protect against amyloid-β1–42-mediated synapse damage in vitro. Mol Neurodegener 3:1. https://doi.org/10.1186/1750-1326-3-1

Baydyuk M, Nguyen MT, Xu B (2011) Chronic deprivation of TrkB signaling leads to selective late-onset nigrostriatal dopaminergic degeneration. Exp Neurol 228(1):118–125. https://doi.org/10.1016/j.expneurol.2010.12.018

Bellik Y, Boukraâ L, Alzahrani HA et al (2013) Molecular mechanism underlying anti-inflammatory and anti-allergic activities of phytochemicals: an update. Molecules 18(1):322–353. https://doi.org/10.3390/molecules18010322

Bernier M, Wahl D, Ali A et al (2016) Resveratrol supplementation confers neuroprotection in cortical brain tissue of nonhuman primates fed a high-fat/sucrose diet. Aging 8(5):899–916. https://doi.org/10.18632/aging.100942

Bhattacharya T et al (2022) Applications of Phyto-nanotechnology for the treatment of neurodegenerative disorders. Materials 15(3):804. https://doi.org/10.3390/ma15030804

Burton MD et al (2016) Dietary luteolin reduces proinflammatory microglia in the brain of senescent mice. Rejuvenation Res 19(4):286–292. https://doi.org/10.1089/rej.2015.1708

Chandra R, Singh S, Ganguly C (2022) β-Sitosterol & quercetin enhances brain development in iodine deficient rat models. Nutr Health:026010602211222. https://doi.org/10.1177/02601060221122209

Chao P, Lee H, Yin M (2016) Asiatic acid attenuated apoptotic and inflammatory stress in the striatum of MPTP-treated mice. Food Funct 7(4):1999–2005. https://doi.org/10.1039/C6FO00041J

Chen X-Q, Sawa M, Mobley WC (2018) Dysregulation of neurotrophin signaling in the pathogenesis of Alzheimer disease and of Alzheimer disease in down syndrome. Free Radic Biol Med 114:52–61. https://doi.org/10.1016/j.freeradbiomed.2017.10.341

Cho I-H (2012) Effects of Panax ginseng in neurodegenerative diseases. J Ginseng Res 36(4):342–353. https://doi.org/10.5142/jgr.2012.36.4.342

Choudhary S, Kumar P, Malik J (2013) Plants and phytochemicals for Huntington′s disease. Pharmacogn Rev 7(14):81–91. https://doi.org/10.4103/0973-7847.120505

Christen Y (2004) Ginkgo biloba and neurodegenerative disorders. Front Biosci 9(1–3):3091. https://doi.org/10.2741/1462

Colín-González AL, Santana RA, Silva-Islas CA et al (2012) The antioxidant mechanisms underlying the aged garlic extract- and S-Allylcysteine-induced protection. Oxidative Med Cell Longev 2012:1–16. https://doi.org/10.1155/2012/907162

Conroy JN, Coulson EJ (2022) High-affinity TrkA and p75 neurotrophin receptor complexes: a twisted affair. J Biol Chem 298(3):101568. https://doi.org/10.1016/j.jbc.2022.101568

D'Onofrio G, Sancarlo D, Ruan Q et al (2016) Phytochemicals in the treatment of Alzheimer's disease: a systematic review. Curr Drug Targets 18:13. https://doi.org/10.2174/1389450117666161102121553

Dechant G, Neumann H (2002) Neurotrophins. Adv Exp Med Biol 513:303–334. https://doi.org/10.5578/kvj.7239

DeFelice SL (1995) The nutraceutical revolution: its impact on food industry R&D. Trends Food Sci Technol 6(2):59–61. https://doi.org/10.1016/S0924-2244(00)88944-X

DeFeudis F, Drieu K (2000) Ginkgo biloba extract (EGb 761) and CNS functions basic studies and clinical applications. Curr Drug Targets 1(1):25–58. https://doi.org/10.2174/1389450003349380

Deinhardt K, Chao MV (2014) Trk receptors. Handb Exp Pharmacol 220:103–119. https://doi.org/10.1007/978-3-642-45106-5_5

Deng H, Wang P, Jankovic J (2018) The genetics of Parkinson disease. Ageing Res Rev 42:72–85. https://doi.org/10.1016/j.arr.2017.12.007

Dey A, Nanda B, Nandy S et al (2020) Implications of phytochemicals as disease-modifying agents against Huntington's disease (HD): bioactivity, animal models and transgenics, synergism and structure–activity studies. pp 27–79. https://doi.org/10.1016/B978-0-12-819483-6.00002-3

Dutta K, Patel P, Julien JP (2018) Protective effects of Withania somnifera extract in SOD1 G93A mouse model of amyotrophic lateral sclerosis. Exp Neurol 309:193–204. https://doi.org/10.1016/j.expneurol.2018.08.008

Dutta K, Patel P, Rahimian R et al (2017) Withania somnifera reverses transactive response DNA binding protein 43 Proteinopathy in a mouse model of amyotrophic lateral sclerosis/frontotemporal lobar degeneration. Neurotherapeutics 14(2):447–462. https://doi.org/10.1007/s13311-016-0499-2

Egbuna C, Ifemeje J (2015) Biological functions and anti-nutritional effects of phytochemicals in living system. IOSR J Pharm Biol Sci (IOSR-JPBS) [Preprint]. Available at: http://www.iosrjournals.org/iosr-jpbs/papers/Vol10-issue2/Version-3/C010231019.pdf (Accessed: 14 May 2023)

Ehrnhoefer DE, Duennwald M, Markovic P et al (2006) Green tea (−)-epigallocatechin-gallate modulates early events in huntingtin misfolding and reduces toxicity in Huntington's disease models. Hum Mol Genet 15(18):2743–2751. https://doi.org/10.1093/hmg/ddl210

El Ouaamari Y, Van den Bos J, Willekens B et al (2023) Neurotrophic factors as regenerative therapy for neurodegenerative diseases: current status, challenges and future perspectives. Int J Mol Sci 24(4):3866. https://doi.org/10.3390/ijms24043866

El-Agnaf OMA, Salem SA, Paleologou KE et al (2006) Detection of oligomeric forms of α-synuclein protein in human plasma as a potential biomarker for Parkinson's disease. FASEB J 20(3):419–425. https://doi.org/10.1096/fj.03-1449com

Eliassen AH, Liao X, Rosner B et al (2015) Plasma carotenoids and risk of breast cancer over 20 y of follow-up. Am J Clin Nutr 101(6):1197–1205. https://doi.org/10.3945/ajcn.114.105080

Elifani F, Amico E, Pepe G et al (2019) Curcumin dietary supplementation ameliorates disease phenotype in an animal model of Huntington's disease. Hum Mol Genet 28(23):4012–4021. https://doi.org/10.1093/hmg/ddz247

Espín JC, García-Conesa MT, Tomás-Barberán FA (2007) Nutraceuticals: facts and fiction. Phytochemistry 68(22–24):2986–3008. https://doi.org/10.1016/j.phytochem.2007.09.014

Fang F, Chen X, Huang T et al (2012) Multi-faced neuroprotective effects of Ginsenoside Rg1 in an Alzheimer mouse model. Biochim Biophys Acta 1822:286. https://doi.org/10.1016/J.BBADIS.2011.10.004

Forni C, Facchiano F, Bartoli M et al (2019) Beneficial role of phytochemicals on oxidative stress and age-related diseases. Biomed Res Int 2019:1–16. https://doi.org/10.1155/2019/8748253

Ganesan P, Ko HM, Kim IS, Choi DK (2015) Recent trends in the development of nanophytobioactive compounds and delivery systems for their possible role in reducing oxidative stress in Parkinson's disease models. Int J Nanomedicine 10:6757–6772. https://doi.org/10.2147/IJN.S93918

Garraway SM, Petruska JC, Mendell LM (2003) BDNF sensitizes the response of lamina II neurons to high threshold primary afferent inputs. Eur J Neurosci 18(9):2467–2476. https://doi.org/10.1046/J.1460-9568.2003.02982.X

Georgiev D et al (2021) Virtual reality for neurorehabilitation and cognitive enhancement. Brain Sci 11(2):221. https://doi.org/10.3390/brainsci11020221

Gohil K (2002) Genomic responses to herbal extracts: lessons from in vitro and in vivo studies with an extract of Ginkgo biloba. Biochem Pharmacol 64(5–6):913–917. https://doi.org/10.1016/S0006-2952(02)01163-2

Guo S, Yan J, Yang T et al (2007) Protective effects of green tea polyphenols in the 6-OHDA rat model of Parkinson's disease through inhibition of ROS-NO pathway. Biol Psychiatry 62(12):1353–1362. https://doi.org/10.1016/j.biopsych.2007.04.020

Gupta J, Gupta R (2020) Nutraceutical status and scientific strategies for enhancing production of Omega-3 fatty acids from microalgae and their role in healthcare. Curr Pharm Biotechnol 21(15):1616–1631. https://doi.org/10.2174/1389201021666200703201014

Hallböök F, Ibáñez CF, Persson H (1991) Evolutionary studies of the nerve growth factor family reveal a novel member abundantly expressed in xenopus ovary. Neuron 6(5):845–858. https://doi.org/10.1016/0896-6273(91)90180-8

Hamaguchi T, Ono K, Yamada M (2010) Curcumin and Alzheimer's disease. CNS Neurosci Ther 16(5):285–297. https://doi.org/10.1111/j.1755-5949.2010.00147.x

Hempstead BL, Martin-Zanca D, Kaplan DR et al (1991) High-affinity NGF binding requires coexpression of the trk proto-oncogene and the low-affinity NGF receptor. Nature 350(6320):678–683. https://doi.org/10.1038/350678a0

Hickey MA, Zhu C, Medvedeva V et al (2012) Improvement of neuropathology and transcriptional deficits in CAG 140 knock-in mice supports a beneficial effect of dietary curcumin in Huntington's disease. Mol Neurodegener 7(1):12. https://doi.org/10.1186/1750-1326-7-12

Holtzman DM, Mobley WC (1994) Neurotrophic factors and neurologic disease. West J Med 161(3):246–254

Houston MC (2005) Nutraceuticals, vitamins, antioxidants, and minerals in the prevention and treatment of hypertension. Prog Cardiovasc Dis 47(6):396–449. https://doi.org/10.1016/j.pcad.2005.01.004

Huang EJ, Reichardt LF (2001) Neurotrophins: roles in neuronal development and function. Annu Rev Neurosci 24:677–736. https://doi.org/10.1146/annurev.neuro.24.1.677

Huang EJ, Reichardt LF (2003) Trk receptors: roles in neuronal signal transduction. Annu Rev Biochem 72:609–642. https://doi.org/10.1146/annurev.biochem.72.121801.161629

Huang Y, Yun W, Zhang M et al (2018) Serum concentration and clinical significance of brain-derived neurotrophic factor in patients with Parkinson's disease or essential tremor. J Int Med Res 46(4):1477–1485. https://doi.org/10.1177/0300060517748843

Jaswir I et al (2011) Carotenoids: sources, medicinal properties and their application in food and nutraceutical industry. J Med Plant Res 5(33):7119–7131. https://doi.org/10.5897/JMPRx11.011

Jayaraman M, Dutta P, Krishnan S et al (2022) Emerging promise of phytochemicals in ameliorating neurological disorders. CNS Neurol Disord Drug Targets 21. https://doi.org/10.2174/1871527321666220701153926

Just-Borràs L, Hurtado E, Cilleros-Mañé V et al (2020) Running and swimming prevent the deregulation of the BDNF/TrkB neurotrophic signalling at the neuromuscular junction in mice with amyotrophic lateral sclerosis. Cell Mol Life Sci 77(15):3027–3040. https://doi.org/10.1007/s00018-019-03337-5

Kaur H, Patro I, Tikoo K, Sandhir R (2015) Curcumin attenuates inflammatory response and cognitive deficits in experimental model of chronic epilepsy. Neurochem Int 89:40–50. https://doi.org/10.1016/j.neuint.2015.07.009

Keon M, Musrie B, Dinger M et al (2021) Destination amyotrophic lateral sclerosis. Front Neurol 12:392. https://doi.org/10.3389/FNEUR.2021.596006/BIBTEX

Khan H, Ullah H, Aschner M et al (2019) Neuroprotective effects of quercetin in Alzheimer's disease. Biomol Ther 10(1):59. https://doi.org/10.3390/biom10010059

Kim MJ, Vargas MR, Harlan BA et al (2018) Nitration and glycation turn mature NGF into a toxic factor for motor neurons: a role for p75 NTR and RAGE signaling in ALS. Antioxid Redox Signal 28(18):1587–1602. https://doi.org/10.1089/ars.2016.6966

Kouli A, Torsney KM, Kuan W-L (2018) Parkinson's disease: etiology, neuropathology, and pathogenesis. Parkinson's Disease: Pathogenesis and Clinical Aspects 3–26. https://doi.org/10.15586/codonpublications.parkinsonsdisease.2018.ch1

Kumar D et al (2022) Functional fermented probiotics, prebiotics, and synbiotics from non-dairy products: a perspective from nutraceutical. Mol Nutr Food Res 66(14):2101059. https://doi.org/10.1002/mnfr.202101059

Kumar GP, Khanum F (2012) Neuroprotective potential of phytochemicals. Pharmacogn Rev 6(12):81–90. https://doi.org/10.4103/0973-7847.99898

Lee FS, Kim AH, Khursigara G, Chao MV (2001) The uniqueness of being a neurotrophin receptor. Curr Opin Neurobiol 11(3):281–286. https://doi.org/10.1016/S0959-4388(00)00209-9

Li WP, Xian C, Rush RA, Zhou XF (1999) Upregulation of brain-derived neurotrophic factor and neuropeptide Y in the dorsal ascending sensory pathway following sciatic nerve injury in rat. Neurosci Lett 260(1):49–52. https://doi.org/10.1016/S0304-3940(98)00958-6

Limanaqi F, Biagioni F, Mastroiacovo F et al (2020) Merging the multi-target effects of phytochemicals in neurodegeneration: from oxidative stress to protein aggregation and inflammation. Antioxidants 9(10):1022. https://doi.org/10.3390/antiox9101022

Liu YJ, Zhan J, Liu XL et al (2014) Dietary flavonoids intake and risk of type 2 diabetes: a meta-analysis of prospective cohort studies. Clin Nutr 33(1):59–63. https://doi.org/10.1016/j.clnu.2013.03.011

Lu B, Pang PT, Woo NH (2005) The yin and yang of neurotrophin action. Nat Rev Neurosci 6(8): 603–614. https://doi.org/10.1038/nrn1726

MacDonald ME, Ambrose CM, Duyao MP et al (1993) A novel gene containing a trinucleotide repeat that is expanded and unstable on Huntington's disease chromosomes. Cell 72:971–983

Mahlknecht P, Foltynie T, Limousin P, Poewe W (2022) How does deep brain stimulation change the course of Parkinson's disease? Mov Disord 37(8):1581–1592. https://doi.org/10.1002/mds.29052

Maisonpierre P, Belluscio L, Squinto S et al (1990) Neurotrophin-3: a neurotrophic factor related to NGF and BDNF. Science 247(4949):1446–1451. https://doi.org/10.1126/science.247.4949.1446

Mathur S, Gawas C, Ahmad IZ et al (2023) Neurodegenerative disorders: assessing the impact of natural vs drug-induced treatment options. Aging Med 6(1):82–97. https://doi.org/10.1002/agm2.12243

McMahon SB, Priestley JV (1995) Peripheral neuropathies and neurotrophic factors: animal models and clinical perspectives. Curr Opin Neurobiol 5(5):616–624. https://doi.org/10.1016/0959-4388(95)80067-0

Md DCAR, Frcp SJT (2010) Huntington's disease: from molecular pathogenesis to clinical treatment. Lancet Neurol 10(1):83–98. https://doi.org/10.1016/S1474-4422(10)70245-3%5Cnpapers2://publication/doi/10.1016/S1474-4422(10)70245-3

Micera A, Lambiase A, Stampachiacchiere B et al (2007) Nerve growth factor and tissue repair remodeling: trkANGFR and p75NTR, two receptors one fate. Cytokine Growth Factor Rev 18(3–4):245–256. https://doi.org/10.1016/j.cytogfr.2007.04.004

Mischel PS, Smith SG, Vining ER et al (2001) The extracellular domain of p75NTR is necessary to inhibit Neurotrophin-3 signaling through TrkA. J Biol Chem 276(14):11294–11301. https://doi.org/10.1074/jbc.M005132200

Mishra S, Palanivelu K (2008) The effect of curcumin (turmeric) on Alzheimer's disease: an overview. Ann Indian Acad Neurol 11(1):13–19. https://doi.org/10.4103/0972-2327.40220

Mitra S, Behbahani H, Eriksdotter M (2019) Innovative therapy for Alzheimer's disease-with focus on biodelivery of NGF. Front Neurosci 13(FEB). https://doi.org/10.3389/fnins.2019.00038

Naoi M, Inaba-Hasegawa K, Shamoto-Nagai M, Maruyama W (2017) Neurotrophic function of phytochemicals for neuroprotection in aging and neurodegenerative disorders: modulation of intracellular signaling and gene expression. J Neural Transm 124(12):1515–1527. https://doi.org/10.1007/s00702-017-1797-5

Neupane P, Thada PK, Singh P et al (2023) Investigating edaravone use for management of amyotrophic lateral sclerosis (ALS): a narrative review. Cureus 15:1. https://doi.org/10.7759/CUREUS.33746

Novak MJU, Tabrizi SJ (2010) Huntington's disease. BMJ 340(jun30 4):c3109. https://doi.org/10.1136/bmj.c3109

Obata K, Noguchi K (2006) BDNF in sensory neurons and chronic pain. Neurosci Res 55(1):1–10. https://doi.org/10.1016/j.neures.2006.01.005

Onaolapo AY, Obelawo AY, Onaolapo OJ (2019) Brain ageing, cognition and diet: a review of the emerging roles of food-based nootropics in mitigating age-related memory decline. Curr Aging Sci 12(1):2–14. https://doi.org/10.2174/1874609812666190311160754

Pandey N, Strider J, Nolan WC et al (2008) Curcumin inhibits aggregation of α-synuclein. Acta Neuropathol 115(4):479–489. https://doi.org/10.1007/s00401-007-0332-4

Pehar M, Cassina P, Vargas MR et al (2004) Astrocytic production of nerve growth factor in motor neuron apoptosis: implications for amyotrophic lateral sclerosis. J Neurochem 89(2):464–473. https://doi.org/10.1111/j.1471-4159.2004.02357.x

Penner R, Fedorak RN, Madsen KL (2005) Probiotics and nutraceuticals: non-medicinal treatments of gastrointestinal diseases. Curr Opin Pharmacol 5(6 SPEC. ISS):596–603. https://doi.org/10.1016/j.coph.2005.06.009

Piccolella S et al (2019) Nutraceutical polyphenols: new analytical challenges and opportunities. J Pharm Biomed Anal 175. https://doi.org/10.1016/j.jpba.2019.07.022

Poulose SM et al (2015) Effects of pterostilbene and resveratrol on brain and behavior. Neurochem Int 89:227–233. https://doi.org/10.1016/j.neuint.2015.07.017

Riolo G, Ricci C, De Angelis N et al (2022) BDNF and pro-BDNF in amyotrophic lateral sclerosis: a new perspective for biomarkers of neurodegeneration. Brain Sci 12(5):617. https://doi.org/10.3390/brainsci12050617

Rossi F, Franco M, Estevez A (2013) Pathophysiology of amyotrophic lateral sclerosis. In: Current Advances in Amyotrophic Lateral Sclerosis. InTech. https://doi.org/10.5772/56562.

Salehi B et al (2019) The therapeutic potential of apigenin. Int J Mol Sci 20(6):1305. https://doi.org/10.3390/ijms20061305

Sandhir R, Yadav A, Mehrotra A et al (2014) Curcumin nanoparticles attenuate neurochemical and neurobehavioral deficits in experimental model of Huntington's disease. NeuroMolecular Med 16(1):106–118. https://doi.org/10.1007/s12017-013-8261-y

Scalzo P, Kümmer A, Bretas TL et al (2010) Serum levels of brain-derived neurotrophic factor correlate with motor impairment in Parkinson's disease. J Neurol 257(4):540–545. https://doi.org/10.1007/s00415-009-5357-2

Scheltens P, Blennow K, Breteler MMB et al (2016) Alzheimer's disease. Lancet 388(10043):505–517. https://doi.org/10.1016/S0140-6736(15)01124-1

Schulte C, Gasser T (2011) Genetic basis of Parkinson's disease: inheritance, penetrance, and expression. Appl Clin Genet 4:67–80. https://doi.org/10.2147/TACG.S11639

Schulz-Schaeffer WJ (2010) The synaptic pathology of α-synuclein aggregation in dementia with Lewy bodies, Parkinson's disease and Parkinson's disease dementia. Acta Neuropathol 120(2):131–143. https://doi.org/10.1007/s00401-010-0711-0

Sharifi-Rad M, Lankatillake C, Dias DA et al (2020) Impact of natural compounds on neurodegenerative disorders: from preclinical to pharmacotherapeutics. J Clin Med 9:4. https://doi.org/10.3390/jcm9041061

Shim S, Kim S, Kwon Y-B, Kwon J (2012) Protection by [6]-shogaol against lipopolysaccharide-induced toxicity in murine astrocytes is related to production of brain-derived neurotrophic factor. Food Chem Toxicol 50(3–4):597–602. https://doi.org/10.1016/j.fct.2011.11.042

Shimazu K, Zhao M, Sakata K et al (2006) NT-3 facilitates hippocampal plasticity and learning and memory by regulating neurogenesis. Learn Mem 13(3):307–315. https://doi.org/10.1101/lm.76006

Sluijs I, Cadier E, Beulens JWJ et al (2015) Dietary intake of carotenoids and risk of type 2 diabetes. Nutr Metab Cardiovasc Dis 25(4):376–381. https://doi.org/10.1016/j.numecd.2014.12.008

Spisni E et al (2022) Nutraceuticals in the modulation of the intestinal microbiota: current status and future directions. Front Pharmacol 13. https://doi.org/10.3389/fphar.2022.841782

Stark M, Behl C (2014) The Ginkgo biloba extract EGb 761 modulates proteasome activity and polyglutamine protein aggregation. Evid Based Complement Alternat Med 2014. https://doi.org/10.1155/2014/940186

Tanila H (2017) The role of BDNF in Alzheimer's disease. Neurobiol Dis 97:114–118. https://doi.org/10.1016/j.nbd.2016.05.008

Tasset I, Sánchez-López F, Agüera E et al (2012) NGF and nitrosative stress in patients with Huntington's disease. J Neurol Sci 315(1–2):133–136. https://doi.org/10.1016/j.jns.2011.12.014

Thompson CL, Ng L, Menon V et al (2014) A high-resolution spatiotemporal atlas of gene expression of the developing mouse brain. Neuron 83(2):309–323. https://doi.org/10.1016/j.neuron.2014.05.033

Tsai S, Chao C, Yin M (2011) Preventive and therapeutic effects of caffeic acid against inflammatory injury in striatum of MPTP-treated mice. Eur J Pharmacol 670(2–3):441–447. https://doi.org/10.1016/j.ejphar.2011.09.171

Unno K et al (2017) Blood–brain barrier permeability of green tea catechin metabolites and their neuritogenic activity in human neuroblastoma SH-SY5Y cells. Mol Nutr Food Res 61(12):1700294. https://doi.org/10.1002/mnfr.201700294

Venkatesan R, Ji E, Kim SY (2015) Phytochemicals that regulate neurodegenerative disease by targeting neurotrophins: a comprehensive review. Biomed Res Int 2015. https://doi.org/10.1155/2015/814068

Vildan C, Sule D, Turker B et al (2019) Genetic alterations of C9orf72, SOD1, TARDBP, FUS, and UBQLN2 genes in patients with amyotrophic lateral sclerosis. Cogent Med. Edited by A. Spurkland 6(1):1582400. https://doi.org/10.1080/2331205X.2019.1582400

Vingtdeux V, Dreses-Werringloer U, Zhao H et al (2008) Therapeutic potential of resveratrol in Alzheimer's disease. BMC Neurosci 9(S2):S6. https://doi.org/10.1186/1471-2202-9-S2-S6

Vitolo O, Gong B, Cao Z et al (2009) Protection against β-amyloid induced abnormal synaptic function and cell death by Ginkgolide. J Neurobiol Aging 30:257. https://doi.org/10.1016/J.NEUROBIOLAGING.2007.05.025

Wang Y, Liu J, Zhang Z et al (2011) Anti-neuroinflammation effect of ginsenoside Rbl in a rat model of Alzheimer disease. Neurosci Lett 487:70–72. https://doi.org/10.1016/J.NEULET.2010.09.076

Wang X, Ouyang YY, Liu J, Zhao G (2014) Flavonoid intake and risk of CVD: a systematic review and meta-analysis of prospective cohort studies. Br J Nutr 111(1):1–11. https://doi.org/10.1017/S000711451300278X

Wang J, Song Y, Gao M et al (2016) Neuroprotective effect of several phytochemicals and its potential application in the prevention of neurodegenerative diseases. Geriatrics (Switzerland) 1:4. https://doi.org/10.3390/geriatrics1040029

Wang J, Song Y, Chen Z, Leng SX (2018) Connection between systemic inflammation and neuroinflammation underlies neuroprotective mechanism of several phytochemicals in neurodegenerative diseases. Oxidative Med Cell Longev 2018. https://doi.org/10.1155/2018/1972714

Wightman EL (2017) Potential benefits of phytochemicals against Alzheimer's disease. Proc Nutr Soc 76(2):106–112. https://doi.org/10.1017/S0029665116002962

Williams RJ, Mohanakumar KP, Beart PM (2015) Neuro-nutraceuticals: the path to brain health via nourishment is not so distant. Neurochem Int 89:1–6. https://doi.org/10.1016/j.neuint.2015.08.012

Xiao Q, Wang C, Li J et al (2010) Ginkgolide B protects hippocampal neurons from apoptosis induced by beta-amyloid 25-35 partly via up-regulation of brain-derived neurotrophic factor. Eur J Pharmacol 647:48–54. https://doi.org/10.1016/J.EJPHAR.2010.08.002

Yang L, Hao J, Zhang J et al (2010) Ginsenoside Rg3 promotes beta-amyloid peptide degradation by enhancing gene expression of neprilysin. J Pharm Pharmacol 61:375–380. https://doi.org/10.1211/JPP.61.03.0013

Yang J, Song S, Li J, Liang T (2014) Neuroprotective effect of curcumin on hippocampal injury in 6-OHDA-induced Parkinson's disease rat. Pathol Res Pract 210(6):357–362. https://doi.org/10.1016/j.prp.2014.02.005

Youdim KA, Joseph JA (2001) A possible emerging role of phytochemicals in improving age-related neurological dysfunctions: a multiplicity of effects. Free Radic Biol Med 30(6):583–594. https://doi.org/10.1016/S0891-5849(00)00510-4

Zhang L-M, He K, Ren Z, Pu Q (2005) Biochemical mechanism of hepatic protective effect of geniposide on liver-injured mice induced by CCl4. Chin J Appl Environ Biol 11(6):669–672

Zhang G-F, Zhang Y, Zhao G (2015) Crocin protects PC12 cells against MPP+-induced injury through inhibition of mitochondrial dysfunction and ER stress. Neurochem Int 89:101–110. https://doi.org/10.1016/j.neuint.2015.07.011

Zhang LF, Yu XL, Ji M et al (2018) Resveratrol alleviates motor and cognitive deficits and neuropathology in the A53T α-synuclein mouse model of Parkinson's disease. Food Funct 9(12):6414–6426. https://doi.org/10.1039/C8FO00964C

Zhao Z, Fu J, Li S, Li Z (2019) Neuroprotective effects of genistein in a SOD1-G93A transgenic mouse model of amyotrophic lateral sclerosis. J Neuroimmune Pharmacol 14(4):688–696. https://doi.org/10.1007/s11481-019-09866-x

Zuccato C, Cattaneo E (2009) Brain-derived neurotrophic factor in neurodegenerative diseases. Nat Rev Neurol 5(6):311–322. https://doi.org/10.1038/nrneurol.2009.54

# Recent Advancement in the Use of Medicinal Plants in Brain Aging

14

Sandip Paul, Anjoy Majhi, and Pinki Saha Sardar

**Abstract**

Humans now have a wide range of treatment options, thanks to plant-derived phytochemicals. In the twenty-first century, cognitive dysfunction is a significant health concern, and many neuropsychiatric and neurodegenerative disorders, including Parkinsonism, schizophrenia, depression, Alzheimer's disease, dementia, cerebrovascular impairment, and seizure disorders, can be severely functionally disabling. Over time, several neurotransmitters and signaling molecules have been discovered that have been thought of as potential therapeutic targets. The action of the principal inhibitory neurotransmitter receptors is influenced by phytochemicals from medicinal plants, which are essential for maintaining the chemical balance of the brain. The beneficial effects of dietary phytochemicals on general health and longevity are well documented. Dietary phytochemicals are a substantial array of nonnutritive substances generated from numerous plant-derived foods and chemical subgroups. Remarkable progress has been achieved in the last 10 years in understanding the key risk factors for brain aging, chronic, low-grade inflammation, and oxidative and nitrosative stress (OS & NS). Increasing evidence indicates that phytochemicals from fruits, vegetables, herbs, and spices may have anti-O&NS and/or immunoregulatory effects that are relevant to the aging brain. Despite the translational gap between basic and clinical research, the current knowledge of the molecular interactions between phytochemicals and

S. Paul · P. S. Sardar
Department of Chemistry, The Bhawanipur Education Society College, Kolkata, West Bengal, India
e-mail: pinki.sahasardar@thebges.edu.in

A. Majhi (✉)
Department of Chemistry, Presidency University, Kolkata, West Bengal, India
e-mail: anjoy.chem@presiuniv.ac.in


S. Pathak, A. Banerjee (eds.), *Neuroprotective Effects of Phytochemicals in Brain Ageing*, https://doi.org/10.1007/978-981-99-7269-2_14

immune-inflammatory and O&NS (IO&NS) pathways may aid in the development of successful dietary interventions to slow down brain aging and enhance cognitive function.

**Keywords**

Phytochemicals · Blood–brain barrier · Signaling pathways · Brain aging · Drugs · Herbs · Neurodegenerative diseases

## 14.1 Introduction

A lifelong buildup of molecular and cellular damage defines aging, which is a complex biological process (Lopez-Otin et al. 2013; Kirkwood 2005). As a result of this, the physiological functions are subsequently lost due to the incapacity of the human body to repair this damage (Lopez-Otin et al. 2013). Age is a significant risk factor for a number of fatal illnesses, such as cancer, heart disease, and neurological disorders (Lopez-Otin et al. 2013). Chronological age, however, is not necessarily a valid indicator of age-related risk because the course of aging differs among the population. Among the population, genetic and environmental factors vary and have different effects on the aging processes that take place in individual cells and different tissue types (Kirkwood 2005). The impacts of aging are particularly noticeable in the brain, which manifests as changes to its structure and cognitive function (Anderton 2002; Damoiseaux 2017). These alterations can be observed in vivo, thanks to neuroimaging technologies like magnetic resonance imaging (MRI). The most frequent alterations brought on by aging are aberrant functional connectivity, reduced white matter integrity and volume, and brain atrophy (loss of grey matter volume and cortical thinning) (Bennett and Madden 2014; Fjell and Walhovd 2010; Gunning-Dixon et al. 2009). The search for specific age-related brain alteration, biomarkers is one of these advancements (Cole et al. 2019). Neuroimaging features are used by so-called "brain age" algorithms to record the changes in the brain that are frequently brought on by aging (Cole et al. 2019). In order to forecast individual brain ages in a group of interest, a multivariate statistical model must typically be trained to understand normative patterns of brain aging. The discrepancy between anticipated biological age and actual chronological age denotes a departure from the expected course of aging and may be used to detect disease, track the effectiveness of treatment, or pinpoint lifestyle choices that are either good for or bad for brain health (Cole et al. 2018, 2019). When chronological age could be predicted from neuroimaging features obtained from any imaging modality, estimates of brain age were deemed to be admissible (e.g., MRI). Studies that looked at brain aging as the difference between brain age and chronological age were considered eligible. The group differences in models of brain features as a function of age or studies utilizing different methods for measuring brain aging, such as the slope between chronological age and brain age, were disregarded (Ronan et al. 2016).

The development of well-liked pharmaceutical compounds by the modern pharmaceutical industry was made possible by natural phytochemicals. The term "phytochemical" refers to a broad category of naturally occurring substances originating from plants (phyto). Humans first learned about the medicinal properties of plants by purposeful or unintentional ingestion, which led to the advancement of herbology into the field of medicine. Plant-derived molecules have been successfully used in medication translation through contemporary pharmaceutical engineering. In general, it begins with the isolation and purification of target phytochemicals, followed by pharmacological and pharmacodynamic testing in a lab, before moving on to druggability testing and the drug development process. Some phytochemicals undergo structural optimization and derivatization for greater potency. Phytochemicals have been included in a number of small-molecule libraries for computer-based virtual drug screening and high-throughput drug screening. Additionally, their rather complicated structures offer a fresh source for therapeutic development and library growth. For high-throughput drug screening and virtual drug screening carried out on computers, phytochemicals have also been added to a number of small-molecule libraries. They offer a fresh source for drug design and the development of current libraries, thanks to their relatively complicated structures. Modern organic chemistry also places a lot of emphasis on the entire synthesis of naturally occurring phytochemicals with high economic values, with goals including lowering synthesis costs, enhancing yield and purity, and obtaining precursors that can be derivatized in a variety of ways. Since many phytochemicals were found to have medicinal qualities through food sources, they typically have relatively high gastric acid tolerance and intestinal absorption, making oral administration of these substances possible (Homayun et al. 2019). However, the bioavailability of orally administered phytochemicals is often poor due to their low targeted organ accumulation and the presence of the blood–brain barrier (BBB). Due to their unsatisfactory direct therapeutic effects, a vast number of phytochemical candidates either continued to be evaluated in laboratories or were turned into healthcare items (Homayun et al. 2019). Drug concentrations in the blood are increased by intravenous delivery, which prevents drug loss in the gastrointestinal tract, although it is still only suitable for brain targeting. Innovative brain delivery platforms with promising outcomes in animal trials have been made available by the expanding field of nanomedicine and few even reached clinical acceptance (Zheng et al. 2018; Guo et al. 2021). Additionally, nasal delivery is regarded as a possible method to promote drug accumulation in the brain. It would be ideal to limit the amount of medicine lost in blood circulation and make it easier for the drug to enter the brain through the olfactory bulb by combining nasal administration with specific drug formulations, such as lipid-soluble solvents. However, many medications are still ineffective for human use because of the stark contrast between the olfactory bulbs of mice and humans in terms of their size in relation to the brain (Khan et al. 2017). In light of this, nanodelivery formulations are also anticipated to increase the effectiveness of nasogastric administration's BBB translocation.

Current clinical applications of phytochemicals encounter difficulties with adequate targeted organ accumulation, which requires future research into much more definite processes of their functional pathogenic mechanisms. Numerous phytochemicals were discovered to be involved in a variety of treatments, although it is still unclear how they act at the molecular level. A clinical drug, as opposed to healthcare goods, needs to have a defined dose, be effective, and have specific indications in addition to a thorough understanding of its pathological interaction and metabolism. Additional research on its mechanisms of action, particularly in conjunction with structural biology and computational drug design, may lead to the development of more novel drug candidates derived from phytochemical sources, helping to overcome the current pharmaceutical screening banks' relatively simple structures and limited screening libraries.

## 14.2 Physical Changes

According to numerous studies, the volume and/or weight of the brain decreases with age at a rate of around 5% each decade beyond the age of 40, with the real rate of loss likely rising with age, especially over the age of 70 (Svennerholm et al. 1997). This happens in a less obvious way than how. It is usually claimed that neuronal cell death causes the grey matter to decrease, although it is unclear if this is the only cause or even the main result (Anderton 2002; Trollor 2001). The alterations in an aging brain may be related to sex, with various regions most impacted in men and women, and it has been hypothesized that a loss in neuronal volume rather than number leads to these changes (Murphy et al. 1996). Additionally, the dentritic arbor, spines, and synapses might alter. In order to retain the same number of synapses and make up for any cell death, dendritic sprouting may take place (Anderton 2002). On the other hand, a reduction in dendritic synapses or loss of synaptic plasticity has also been reported (Barnes 2003). Functional organizational change and compensatory mechanisms could happen and resemble those seen in patients recovering from moderate traumatic brain damage (Levine et al. 2002). However, there are not many cases studied in the latter field. White matter's function in the aging brain must also be taken into account (Raz 2004). Although not all studies agree, it has been claimed that the late myelinating regions of the frontal lobes are the most impacted by white matter lesions (WML). White matter may deteriorate with age, the myelin sheath eroding after roughly the age of 40 even in normal aging (Bartzokis et al. 2003). All areas of the brain do not undergo the same degree of change (Trollor 2001). Using two MRI images separated by about 1 or 2 years, a longitudinal investigation supports the idea that these brain alterations are not uniform (Scahill et al. 2003). The latter, in contrast to the majority of the other work in this field, only included research that compared younger (aged less than 30) and older (aged greater than 60) populations. Based on a review of more than seven research, the striatum came in second. With between 8 and 18 investigations, the volume of the prefrontal white matter as well as the temporal lobe, cerebellar vermis, cerebellar hemispheres, and hippocampus decreased (five studies) (Peters 2006). The least impacted area was the occipital cortex (five studies) (Peters 2006). The results, which show that the prefrontal cortex is most damaged and the occipital is least, are consistent with the cognitive changes associated with aging, while some research also implies that the hippocampus is the area of the brain most affected by aging (Anderton 2002; Barnes 2003). Another difference between men and women is that the frontal and temporal lobes are more likely to be affected in males than the hippocampal and parietal lobes are in women (Murphy et al. 1996). A correlation was discovered between aging, a decline in prefrontal cortical volume, an increase in subcortical white matter lesions, and a rise in perseverative behavior in a study that examined cortical volume and white matter hyperintensity volume in 140 people aged 50 to 81 who had undergone dementia and depression screenings (decreased executive function) (Gunning-Dixon and Raz 2003).

## 14.3 Mechanisms of Change

Dopamine and serotonin are the neurotransmitters that are most frequently mentioned in relation to aging. From early adulthood, dopamine levels decrease by about 10% every 10 years, which has been linked to deficits in cognitive and motor function (Nyberg and Bäckman 2004). It's possible that as people age, the dopaminergic connections between the frontal brain and the striatum become less effective. Dopamine levels may also decrease, and receptor binding or synaptic density may decrease (Nyberg and Bäckman 2004). Age-related declines in serotonin and brain-derived neurotrophic factor levels may be related to the regulation of synaptic plasticity and neurogenesis in the adult brain (Mattson et al. 2004). Age-associated increases in monoamine oxidase, a molecule connected to neurotransmitter levels, may allow for the release of free radicals from processes beyond the capacity of the body's own antioxidants (Peters 2006). Additionally, mitochondrial failure, reactive oxygen species generation, and calcium dysregulation have all been linked to an aging brain (Peters 2006). Hormonal effect is another issue to take into account when thinking about the aging brain and its cognitive function. It is well-recognized that sex hormones can influence cognitive functions in adults and that these hormones vary as people age, especially in women going through menopause. Even with a greater life expectancy, women still have a higher incidence of AD (Peters 2006). Memory loss is a hallmark of AD, and there has been speculation that estrogen therapy may boost dopaminergic responsiveness and provide protection against AD (Peters 2006). However, it is important to keep in mind that HRT use has been linked to a higher risk of developing cancer recently. Growth hormone levels may be related to cognitive function and decline with age, although the evidence is not entirely clear (Peters 2006). As cerebrovascular efficiency declines, the aging brain may also experience impaired glucose metabolism or a decreased input of glucose or oxygen. However, the loss in glucose may partially be caused by shrinkage rather than any alteration in glucose metabolism. Vascular changes are significant, and white matter lesions and stroke are other prevalent ischemia-related findings in older people's brains (Peters 2006).

## 14.4 Brain Diseases and the Blood–Brain Barrier (BBB)

The BBB is a dynamic multicellular layer that separates the central nervous system (CNS) from the peripheral blood circulation. It is made up of brain microvessel endothelial cells (ECs), which are supported by astrocytes, pericytes, and extracellular matrix elements (Abbott et al. 2006). Limiting the entry of toxins, infections, or cells from the blood into the brain and maintaining the right environment, nutrition, and oxygen for neuronal functioning are all important roles of the blood–brain barrier (BBB). Compared to ECs in other tissues, brain ECs (BECs) have a complicated molecular transportation system and distinct characteristics. The paracellular and transcellular molecular exchanges of nearly 98% of generic molecules are severely constrained by the built-in continuous intercellular tight junctions (TJs) of BECs, which lack fenestrations and have a very low transcytosis rate, leaving only a few lipid-soluble small molecules (i.e., molecules with a molecular weight of less than 400 Da or fewer than eight hydrogen bonds) that can cross (Abbott et al. 2006). Therefore, the BBB can only be crossed by a restricted number of unique transporters that allow certain payloads, such as fluid, ions, a few peptides, and proteins, to pass through (Sweeney et al. 2018). For instance, aquaporin (AQP) and the abluminal sodium pump (Na+-, K+-ATPase), respectively, play a key role in the transfer of water and ions across the blood–brain barrier (BBB). Alcohol and other tiny lipophilic substances can easily diffuse over the BBB. Peptides and proteins are mostly transported through receptor-mediated transcytosis (RMT), but metabolites, nucleotides, and other substrates can be carried by solute carrier-mediated transport with concentration gradient (Bellettato and Scarpa 2018). The BECs' comparatively modest expression of immune cell recognition indicators, such as leukocyte adhesion molecules, also contributes to a few immune cells' ability to enter the healthy CNS across the BBB (Muldoon et al. 2013). As a result, the immunological defense system in the CNS is likewise distinctive. Together, these discrepancies prevented us from understanding brain illnesses more fully.

New research has shown that the BBB and pathological alterations in the brain are closely related. By disrupting the microenvironment and/or transport pathways, the intact BBB is broken down by tissue damage, injury, pathogens, or pathological molecular events, which can result in a number of different brain illnesses. Immune cells must also remove debris from the damaged brain. Additionally, when the BBB ruptures, there is an increase in immune cell invasion and chemical flux dysregulations, and immune surveillance and effector responses to brain infections may result in inflammation and a number of illnesses (Ransohoff and Engelhardt 2012). A number of pathological changes, including increased BBB permeability, microbleeds, impaired p-glycoprotein function, impaired glucose transport, CNS leukocyte infiltration, capillary leakages, pericyte degeneration, endothelial degeneration, abnormal angiogenesis, and other molecular pathological changes, are associated with the development of neurodegenerative diseases (Sweeney et al. 2018).

## 14.5 Glioblastoma Therapy and BBB Permeability

The most prevalent and dangerous primary malignant brain tumor of the central nervous system in adults is glioblastoma (Liu et al. 2022). Patients' median survival time was fewer than 15 months due to its extremely dismal prognosis (Liu et al. 2022). Chemotherapy, radiation, and surgery are being used to treat GBM. However, there are very few chemotherapeutic agents available for GBM treatment, with temozolomide (TMZ) being the only oral medication authorized by the US Food and Drug Administration (FDA) (Liu et al. 2022). The DNA double-strand is broken down by the DNA-alkylating chemical TMZ, which ultimately causes cell death. TMZ, the sole frontline treatment for GBM, was given via several ways to GBM xenograft mice models. In mouse GBM xenograft models, TMZ treatment given orally once at a dose of 600 mg/kg or three intraperitoneal injections at a dose of 50 mg/kg over the course of 2 weeks both produced the best results (Liu et al. 2022). Extensive preclinical (animal models) and clinical experiments are required for the formulation of an informative dosage of chemotherapeutic medications like TMZ (reported in Medscape and Mayo Clinic). More significantly, for GBM therapy, the number of active medicines at the site (i.e., brain tumor tissue) matters more than the prescribed dosage. The main interest was focused on phytochemicals because they are natural substances that are in the hunt. Clinical trials for several cancer types have indicated the use of phytochemicals in cancer therapy. Examples include the use of Allium sativum to treat pancreatic, liver, or colorectal cancer that is incurable (Hosseini and Ghorbani 2015). Despite this, the BBB blocked most exogenous chemicals' entry to the brain, as was already mentioned. To be more precise, the in vivo BBB transport of a solute is reported to decrease by 10 times for each pair of hydrogen bonds on the solute, and the BBB permeability reduces 100-fold when a drug's molecular weight goes from 150 to 450 Da (Liu et al. 2022). Generally, more complicated chemical structures of phytochemicals are likely to be impermeable to the BBB and restrict their subsequent therapeutic usage toward brain illnesses, leaving uncommon examples for their use of phytochemicals to treat GBM.

## 14.6 Limited Numbers of Phytochemicals Are BBB Permeable for Brain Diseases

The permeability of phytochemicals through the BBB is a key factor in their current therapeutic applications for brain illnesses, and their corresponding therapeutic dosage is still debatable. As was previously said, one of the directions for establishing whether a medicine has great therapeutic efficacy is the provided dosage. Despite the extremely little information available, we work to learn more, and in this chapter, we provide instances of phytochemicals whose administration at doses lower or higher than 10 mg/kg/day, respectively, increases or decreases the likelihood that they will have BBB permeability. Given that all current administration methods suffer from significant loss during blood circulation or gastrointestinal consumption, this dosage measurement simply serves as a suggestion and is independent of the routes used to provide the medication. The most researched phytochemical subgroups, polyphenols and flavonoids, were thought to have neuroprotective effects and broad-spectrum disease-treating properties, including the ability to treat cancer, cardiovascular disorders, and metabolic syndrome (Liu et al. 2022). Unfortunately, the mechanism by which these identified phytochemicals cross the BBB more easily is still unknown. Furthermore, knowledge of the underlying functional mechanisms of bioactive phytochemicals for the treatment of brain diseases is very restricted. In both in vivo and in vitro settings, luteolin encourages nuclear erythroid 3-related factor 2 (Nrf2) translocation to the nucleus. It has been proposed that luteolin's neuroprotective effects on traumatic brain injury are caused by activation of the Nrf2-ARE pathway (Liu et al. 2022). In a mouse model of Parkinson's disease, the natural substance icariin (ICA), which is derived from the herb HerbaEpimedii, had neuroprotective effects on dopaminergic neurons. The potential mechanisms may involve the phosphatidylinositol-3-kinase (PI3K)/protein kinase B (Akt) and mitogen-activated protein kinase (MEK)/extracellular signal-regulated kinase (Chen et al. 2017). Administration of 4-O-methylhonokiol, isolated from Magnolia officinalis, increased lipopolysaccharide-induced neuroinflammation, amyloidogenesis, and memory impairment via inhibition of nuclear factor-kappaB (NF-kB) signaling pathway in in vitro and in vivo models in a dose-dependent manner (Liu et al. 2022).

Phytochemicals frequently have considerable therapeutic effects that have been extensively described, but their underlying mechanisms are still unknown. The injection of obovatol with mice expressing mutant human amyloid precursor proteins greatly improved cognitive capacities. Obovatol is a biphenolic substance from Magnolia obovate that has neuroprotective properties against neuroinflammation (Liu et al. 2022).

## 14.7 BBB Is a Challenging Boundary for Most Phytochemical Drugs with Great Brain Therapeutic Potential

The BBB's existence severely limits the therapeutic potential of given phytochemicals and prevents most of them from being used to treat brain diseases. The extremely high dose administration thus continues to be a biocompatibility concern in order to acquire significant brain accumulation of the phytochemical medicines. Amyloid precursor protein and presenilin 1 (PS1) double-transgenic (APP/PS1) mice were given an intraperitoneal injection of 150 mg/kg/day for 4 weeks to reduce cytokine production, inhibit the NF-kB signaling pathway, and suppress neuroinflammation. Curcumin is a hydrophobic polyphenol extracted from the dried rhizomes of Curcuma longa L (Liu et al. 2016). Other studies have shown how curcumin prevents the production of Ab42 oligomers and the disaggregation of the fibrils that are already produced, both of which are advantageous for AD outcomes (Liu et al. 2022). Curcumin has important neuroprotective properties, but its use as a treatment for neurodegenerative disorders and brain tumors is still constrained by the poor brain bioavailability it exhibits due to poor absorption and stability at physiological pH, a high rate of metabolism, rapid systemic elimination, and restricted BBB permeation (Askarizadeh et al. 2020). By controlling the production of thioredoxin, thioredoxin-interacting proteins, and NF-kB pathway proteins, salidroside from Rhodiolarosea is able to cross the BBB and improve cognitive recovery in AD rats (Liu et al. 2022). In a senescence-accelerated mouse model, salidroside exhibits therapeutic effects at a dose of 50 mg/kg/day for 3 months with decreased Ab1-42 accumulation and microglial activation (Liu et al. 2022). The hippocampal CA1 area of aging mice's brains saw less neuronal degeneration after receiving a gavaging therapy of 25 mg/kg/day for 8 weeks. Salidroside stimulates telomerase reverse transcriptase expression via the PI3K/Akt pathway, according to further analysis of the treated animals (Zhu et al. 2021). Currently, geniposide, an iridoidglucoside isolated from Gardenia jasminoides Ellis, has been shown to be able to inhibit the generation of tumor necrosis factor-a (TNF-a) cerebral Ab buildup in a mPrPAPPswe/PS1dE9AD transgenic mouse model when administered at 25 mg/kg/day for 3 months (Lv et al. 2015).

The useful components of phytochemical extraction and identification aided the growth of analytical chemistry. The majority of the simple-to-synthesize compounds have been incorporated into the screening bank, making them insufficient for additional high-throughput screening. As a result, the screen library in the modern pharmaceutical business needs to be updated with increased structural complexity. The derivatives of phytochemicals could be a valuable source for this revolution. The benefits of phytochemical derivatives also stem from a variety of factors, including their endogeneity and inexpensive cost of biosynthesis. They also have a relatively high level of biological reactivity. De novo chemical synthesis for medicinal phytochemicals, which aims for reduced costs and improved purity, is emerging as a significant synthetic chemistry study in recent years. Consequently, it is expected that phytochemicals and their derivatives will continue to provide valuable insights for pharmaceutical sciences in the future.

## 14.8 Nanodelivery Systems Provide Solutions for Phytochemical Brain Therapies

The BBB-permeable or -impermeable phytochemicals, which are titled with a number of therapeutic potentials, have been mentioned in the discussions above. Their bioavailability across the BBB still limits the in vivo or future clinical application in brain treatment, nevertheless. Solubility, stability, target specificity, effective accumulation, and passage through the BBB are only a few of the challenges that nanoparticles could resolve in relation to the clinical applications of phytochemicals. Materials or structures that are smaller than a nanometer are referred to as nanoparticles. The nanoparticles' size makes it possible for them to potentially penetrate cell barriers (Liu et al. 2022). Liposomes, polymers, dendrimers, micelles, synthetic biomaterials, and inorganic nanocarriers are general categories for therapeutic nanoparticles (Zheng et al. 2018). When it comes to acting as drug carriers, different nanoparticle formulations each have their own advantages and disadvantages; however, it is the straightforward physical and chemical characterization that can be altered as needed that satisfies the optimization of phytochemicals in vivo or even clinical application toward brain diseases (Liu et al. 2022). Additionally, the targeted delivery and release capability of nanoparticles could possibly significantly reduce the therapeutic dosage when administered intravenously or orally. The intrinsic qualities of the targeted drug formula, such as electrical charges or water solubility, are typically used in the loading of its cargo. For a nanomedicine to be successful, release-at-site is also necessary after administration (Tu et al. 2021).

### 14.8.1 Neuroprotective Signaling Pathways

Dietary phytochemicals influence the neurological system in a positive way by altering cellular signaling pathways involved in the stress response (Lee et al. 2014). Numerous phytochemicals are known to increase neuronal stress resistance at low dosages; however, many phytochemicals can be hazardous at high concentrations. This is an illustration of a dose–response relationship that exhibits biphasic effects, with stimulatory or advantageous effects at low levels and inhibiting or undesirable effects at high concentrations (Mattson et al. 2007). Thus, dietary phytochemicals function as minor stressors to encourage the adaptive expression of stress-protective genes in neuronal cells and to increase resistance to the processes that cause brain aging.

Against O &NS stress and neuroinflammation, the transcription factor nuclear factor E2-related factor 2 (Nrf2) has become a crucial cytoprotective regulator (Sandberg et al. 2014). The cap'n'collar (CNC) family of transcription factors for stress sensing includes Nrf2. A simple leucine zipper protein called Nrf2 binds to certain DNA locations known as antioxidant response elements (AREs) or electrophile response elements (EpRE) after heterodimerizing with the tiny Maf protein in the nucleus (Sykiotis and Bohmann 2010). The Kelch-like ECH-associated protein 1 (Keap1), Nrf2's repressor protein, prevents Nrf2 from being transcriptionally active and sequestered in the cytoplasm under normal circumstances. The turnover of Nrf2 is provided by this binding through proteasomal degradation. Sulfhydryl-rich protein Keap1 serves as a specialized sensor to measure stress when electrophilic xenobiotics and oxidative stressors are present.

### 14.8.2 The Ability of Phytochemicals to Protect Against Neuro-Inflammaging

Dietary phytochemicals may have a substantial effect on several areas of neuro-inflammaging. Numerous studies have shown that phytochemicals' neuroprotective effects often take place in a dose- and time-dependent way (Lee et al. 2014). The neuroprotective effects of phytochemicals appear to be mediated by a number of different pathways. Although research has primarily concentrated on the antioxidant properties of dietary phytochemicals, it has also shown that certain phytochemicals specifically target important stress-activated signaling pathways involved in neuroprotection. These compounds have the power to neutralize the effects of free radicals in the brain. Numerous genes producing prosurvival proteins, such as antioxidant enzymes, neurotrophic, or anti-apoptotic factors, have been shown to be influenced by a variety of phytochemicals, according to strong in vitro and in vivo evidence. The major categories of dietary phytochemicals and their growing function as hormetic inducers of neuroprotective pathways pertinent to brain aging will be discussed in the following sections.

### 14.8.3 Neuroprotective Herbs

The general public's perception of naturally derived compounds and extracts as being intrinsically safer and more desired than synthetic chemical products has altered as a result of the "Green" movement in Western civilization, which has led to an increase in sales of herbal treatments (Capasso et al. 2000; Plotkin 1992). About 80% of people in underdeveloped nations rely on phytomedicine for basic human and animal healthcare (Harvey 1999). Numerous plants have been employed in traditional medical practices to treat cognitive disorders, including memory-related disorders and neurodegenerative diseases like Alzheimer's disease (AD). Complex mixes of organic substances, including fatty acids, sterols, alkaloids, flavonoids, glycosides, saponins, tannins, terpenes, and others, can be found in herbal products. In contrast to the individual chemicals of conventional medicines, which are isolated by pharmacologists, supporters of herbal medicines claim that a plant's therapeutic value comes from the synergistic effects of its various components. As a result, it is thought that traditional medicines are effective and have few to no side effects. About 25 to 50 percent of today's medications were first sourced from plants. The primary and growing scientific interest in recent years has been the identification and characterization of new medicinal plants to treat neurodegenerative illnesses and brain damage brought on by a stroke. In Asian nations, there are more than 120 traditional medicines used in the treatment of illnesses of the central nervous system (CNS) (Kumar 2006).

### 14.8.4 Antioxidants Derived from Phytochemicals Used in Neuroregeneration

The brain has a high potential for oxidation but a low capacity to reduce oxidative stress (Weiss and Fintelmann 2000). The mechanisms causing neuronal cell injury in a variety of disease states of the brain, including neurodegenerative illnesses, have been linked to oxidative stress. Although it makes up less than 2% of body weight, the brain uses 20% of the oxygen produced by respiration. Thus, the brain is the organ most vulnerable to oxidative injury due to its high oxygen requirement (Weiss and Fintelmann 2000). Due to their therapeutic potential, phytopharmaceuticals are becoming more and more important in both traditional and modern medicine. Novel antioxidants might provide a powerful and secure way to increase the body's resistance to free radicals. Utilizing a limited number of resources, such as vitamins, bioactive compounds, lipoic acid, antioxidant enzymes, and redox-sensitive protein transcriptional factors, central nervous system cells can battle oxidative stress. However, nutritional antioxidants including polyphenols, flavonoids, terpenoids, fatty acids, etc., can activate or modify this defensive system. Alternative antioxidants from plants (AOX) are thought to be helpful at reducing the consequences of oxidative damage, and as a result, they have influenced what people eat and drink (Viana et al. 1996). Antioxidants have a lot of scientific and practical support for their use in the treatment of neurodegenerative diseases.

Antioxidants made of phytochemicals may play a role in neuroprotection (preventing apoptosis) and neuroregeneration by reducing or reversing cellular damage and delaying the loss of neuronal cells (Moosmann and Behl 2002). AOX is classified as either endogenous or exogenous in nature. The endogenous group consists of proteins like albumin, transferrin, ceruloplasmin, metallothionein, and haptoglobin as well as enzymes like superoxidase dismutase (zinc, manganese, and copper), glutathione peroxide (selenium), and catalase. Dietary phytochemicals including polyphenols, quinones, flavonoids, catechins, coumarins, and terpenoids, as well as smaller molecules like ascorbic acid (Vitamin C), alpha-tocopherol (Vitamin E), beta-carotene vitamin-E, and supplements, constitute the most significant exogenous AOX (Larson 1988). Although their mechanism of action is not yet fully understood and there are few clinical trials using them, AOX represents a promising strategy for halting the progression of neurodegenerative diseases like Alzheimer's, Parkinson's, Huntington's, amyotrophic lateral sclerosis, ischemic, and hemorrhagic stroke (Floyd 1999).

### 14.8.5 Neuroprotection with Phytochemicals

The use of phytoconstituents for neuroprotection or to prevent neurodegenerative disorders has attracted significant public and scientific interest. Numerous phytochemicals have demonstrated neuroprotective effects in animal and cell culture models of neurological diseases. For instance, piceatannol can shield cultured neurons from A-induced death while a chalcone (saffron or yellow B) can shield neurons from ischemic brain damage. Research on neurodegenerative illnesses in animal models and epidemiological studies on human populations have shown that phytochemicals in fruits and vegetables can shield the brain system from disease (Liu 2003). The fact that many of the active compounds have antioxidant activity has been the main focus of studies on the health benefits of phytochemicals. Various phytochemicals' neuroprotective properties are linked to decreased oxidative stress. Reducing oxidative stress and protecting cultured hippocampal neurons from nitric oxide-mediated cell death, for instance, were achieved by resveratrol, quercetin, and catechins (Larson 1988).

## 14.9 Conclusion

Plants are a limitless source of molecules that can be used to enhance human health in the form of foods, spices, and herbs. The chemical diversity that influenced the evolutionary success of plants, favoring their adaptation to a changing environment, is found in the hundreds or thousands of secondary, bioactive metabolites that are present in a single plant. According to this theory, it would be incorrect and unsuitable to attribute the health-promoting properties of a medical herb or plant food to just one molecule or one class of components. Many phytochemicals that have lately been proven to have protective effects against neurological illnesses in a variety of experimental settings have previously been demonstrated to have cytotoxic or cytostatic effects on cancer cells. While a single phytochemical might not hold the key to unlocking all the health benefits of plants, the synergistic interplay between these diverse molecules within a single source offers a powerful approach to promoting well-being. This complexity highlights the limitations of studying isolated components and underscores the importance of a holistic approach when exploring the therapeutic potential of plants. Furthermore, the discovery of neuroprotective properties in phytochemicals with previously known anticancer effects suggests intriguing connections between cellular health and overall disease prevention. As research delves deeper into this fascinating synergy, we can expect the identification of even more targeted and effective interventions for a wide range of age-related health concerns.

**Acknowledgments** Authors acknowledge the SERB for research grant (ref: EEQ/2019/000194) for financial support. The authors are also extremely thankful to the Presidency University, Kolkata and The Bhawanipur Education Society College, Kolkata.

**Author's Contribution** All authors contributed equally.

## References

Abbott NJ, Rönnbäck L, Hansson E (2006) Astrocyte–endothelial interactions at the blood–brain barrier. Nat Rev Neurosci 7(1):41–53

Anderton BH (2002) Ageing of the brain. Mech Ageing Dev 123(7):811–817

Askarizadeh A, Barreto GE, Henney NC, Majeed M, Sahebkar A (2020) Neuroprotection by curcumin: a review on brain delivery strategies. Int J Pharm 585:119476

Barnes C (2003) Long-term potentiation and the ageing. Philos Trans Royal Soc Lond B Biol Sci 358:765–772

Bartzokis G, Cummings J, Sultzer D et al (2003) White matter structural integrity in healthy ageing adults and patients with Alzheimer disease. Arch Neurol 60:393–398

Bellettato CM, Scarpa M (2018) Possible strategies to cross the blood–brain barrier. Ital J Pediatr 44(2):131

Bennett IJ, Madden DJ (2014) Disconnected aging: cerebral white matter integrity and age-related differences in cognition. Neuroscience 276:187–205

Capasso R, Izzo AA, Pinto L, Bifulco T, Vitobello C, Mascolo N (2000) Phytotherapy and quality of herbal medicines. Fitoterapia 71:S58–S65

Chen WF, Wu L, Du ZR, Chen L, Xu AL, Chen XH et al (2017) Neuroprotective properties of Icariin in MPTP-induced mouse model of Parkinson's disease: involvement of PI3K/Akt and MEK/ERK signaling pathways. Phytomedicine 25:93–99

Cole JH, Ritchie SJ, Bastin ME, Valdes Hernandez MC, Munoz Maniega S, Royle N et al (2018) Brain age predicts mortality. Mol Psychiatry 23(5):1385–1392

Cole JH, Marioni RE, Harris SE, Deary IJ (2019) Brain age and other bodily 'ages': implications for neuropsychiatry. Mol Psychiatry 24(2):266–281

Damoiseaux JS (2017) Effects of aging on functional and structural brain connectivity. NeuroImage 160:32–40

Fjell AM, Walhovd KB (2010) Structural brain changes in aging: courses, causes and cognitive consequences. Rev Neurosci 21(3):187–221

Floyd RA (1999) Antioxidants, oxidative stress and degenerative neurological disorders. Proc Soc Exp Biol Med 222:236–245

Gunning-Dixon F, Raz N (2003) Neuroanatomical correlates of selected executive functions in middle aged and older adults: a prospective MRI study. Neuropsychologia 41:1929–1941

Gunning-Dixon FM, Brickman AM, Cheng JC, Alexopoulos GS (2009) Aging of cerebral white matter: a review of MRI findings. Int J Geriatr Psychiatry 24(2):109–117

Guo S, Li K, Hu B, Li C, Zhang M, Hussain A et al (2021) Membrane-destabilizing ionizable lipid empowered imaging-guided siRNA delivery and cancer treatment. Exp Dermatol 1(1):35–49

Harvey AL (1999) Medicines from nature: are natural products still relevant to drug discovery? Trends Pharmacol Sci 20:196–198

Homayun B, Lin X, Choi H-J (2019) Challenges and recent progress in oral drug delivery systems for biopharmaceuticals. Pharmaceutics 11(3):129

Hosseini A, Ghorbani A (2015) Cancer therapy with phytochemicals: evidence from clinical studies. Avicenna J Phytomed 5(2):84–97

Khan AR, Liu M, Khan MW, Zhai G (2017) Progress in brain targeting drug delivery system by nasal route. J Control Release 268:364–389

Kirkwood TB (2005) Understanding the odd science of aging. Cell 120(4):437–447

Kumar V (2006) Potential medicinal plants for CNS disorders: an overview. Phytother Res 20: 1023–1035

Larson RA (1988) The antioxidants of higher plants. Phytochemistry 27:969–978

Lee J, Jo DG, Park D, Chung HY, Mattson MP (2014) Adaptive cellular stress pathways as therapeutic targets of dietary phytochemicals: focus on the nervous system. Pharmacol Rev 66:815–868

Levine B, Cabeza R, McIntosh A et al (2002) Functional reorganisation of memory after traumatic brain injury: a study with H2 15O positron emission tomography. J Neurol Neurosurg Psychiatry 73:173–181

Liu RH (2003) Health benefits of fruit and vegetables are from additive and synergistic combinations of phytochemicals. Am J Clin Nutr 78:517S–520S

Liu ZJ, Li ZH, Liu L, Tang WX, Wang Y, Dong MR et al (2016) Curcumin attenuates beta-amyloid-induced neuroinflammation via activation of peroxisome proliferator-activated receptor-gamma function in a rat model of Alzheimer's disease. Front Pharmacol 7:261

Liu Y, Chen Z, Li A, Liu R, Yang H, Xia X (2022) The phytochemical potential for brain disease therapy and the possible Nanodelivery solutions for brain access. Front Oncol 12:936054

Lopez-Otin C, Blasco MA, Partridge L, Serrano M, Kroemer G (2013) The hallmarks of aging. Cell 153(6):1194–1217

Lv C, Wang L, Liu X, Yan S, Yan SS, Wang Y et al (2015) Multi-faced neuroprotective effects of geniposide depending on the RAGE-mediated signaling in an Alzheimer mouse model. Neuropharmacology 89:175–184

Mattson M, Maudsley S, Martin B (2004) BDNF and 5-HT: a dynamic duo in agerelated neuronal plasticity and neurodegenerative disorders. Trends Neurosci 27:589–594

Mattson MP, Son TG, Camandola S (2007) Viewpoint: mechanisms of action and therapeutic potential of neurohormetic phytochemicals. Dose Response 5:174–186

Moosmann B, Behl C (2002) Antioxidants as treatment for neurodegenerative disorders. Expert Opin Investig Drugs 11:1407–1435

Muldoon LL, Alvarez JI, Begley DJ, Boado RJ, Del Zoppo GJ, Doolittle ND et al (2013) Immunologic privilege in the central nervous system and the blood–brain barrier. J Cereb Blood Flow Metab 33(1):13–21

Murphy D, DeCarli C, McIntosh A et al (1996) Sex differences in human brain morphometry and metabolism: an in vivo quantitative magnetic resonance imaging and positron emission tomography study on the effect of ageing. Arch Gen Psychiatry 53:585–594

Nyberg L, Bäckman L (2004) Cognitive ageing: a view from brain imaging. In: Dixon R, Bäckman L, Nilsson L (eds) New frontiers in cognitive ageing. Oxford University Press, Oxford, pp 135–160

Peters R (2006) Ageing and the brain. Postgrad Med J 82:84–88

Plotkin MJ (1992) Ethnomedicine: past, present and future. In: Baba S, Akerele O, Kawaguchi Y (eds) Natural resources and human health: plants of medicinal and nutritional value. Elsevier, Amsterdam, pp 79–86

Ransohoff RM, Engelhardt B (2012) The anatomical and cellular basis of immune surveillance in the central nervous system. Nat Rev Immunol 12(9):623–635

Raz N (2004) The ageing brain: structural changes and their implications for cognitive ageing. In: Dixon R, Bäckman L, Nilsson L (eds) New frontiers in cognitive ageing. Oxford University Press, Oxford, pp 115–134

Ronan L, Alexander-Bloch AF, Wagstyl K, Farooqi S, Brayne C, Tyler LK et al (2016) Obesity associated with increased brain age from midlife. Neurobiol Aging 47:63–70

Sandberg M, Patil J, D'Angelo B, Weber SG, Mallard C (2014) NRF2-regulation in brain health and disease: implication of cerebral inflammation. Neuropharmacology 79:298–306

Scahill R, Frost C, Jenkins R et al (2003) A longitudinal study of brain volume changes in normal ageing using serial registered magnetic resonance imaging. Arch Neurol 60:989–994

Svennerholm L, Boström K, Jungbjer B (1997) Changes in weight and compositions of major membrane components of human brain during the span of adult human life of swedes. Acta Neuropathol 94:345–352

Sweeney MD, Sagare AP, Zlokovic BV (2018) Blood–brain barrier breakdown in Alzheimer disease and other neurodegenerative disorders. Nat Rev Neurol 14(3):133–150

Sykiotis GP, Bohmann D (2010) Stress-activated cap'n'collar transcription factors in aging and human disease. Sci Signal 3(112):re3

Trollor J (2001) Valenzuela M Brain ageing in the new millennium. Aust N Z J Psychiatry 35(6): 788–805

Tu L, Liao Z, Luo Z, Wu Y-L, Herrmann A, Huo S (2021) Ultrasound-controlled drug release and drug activation for cancer therapy. Exp Dermatol 1(3):20210023

Viana M, Barbas C, Banet B, Bonet MV, Castro M, Fraile MU et al (1996) In vitro effect of aflavonoid-rich extract on LDL oxidation. Athelosclerosis 123:83–91

Weiss RF, Fintelmann V (2000) Herbal medicine. Thieme, Stuttgart

Zheng M, Tao W, Zou Y, Farokhzad OC, Shi B (2018) Nanotechnology-based strategies for siRNA brain delivery for disease therapy. Trends Biotechnol 36(5):562–575

Zhu L, Liu Z, Ren Y, Wu X, Liu Y, Wang T et al (2021) Neuroprotective effects of salidroside on ageing hippocampal neurons and naturally ageing mice via the PI3K/Akt/TERT pathway. Phytother Res 35(10):5767–5780

# Repurposing of Drugs in Aging

15

Dhivya Sundaram, Hemamalini Vedagiri,
Gowtham Kumar Subbaraj, and Premkumar Kumpati

**Abstract**

Aging is a complex biological process characterized by a gradual decline in physiological functions and an increased susceptibility to age-related diseases. With the global population aging rapidly, there is a growing need to develop effective interventions to promote healthy aging and extend lifespan. Traditional drug discovery approaches are time-consuming and costly, leading researchers to explore alternative strategies. Repurposing existing drugs has emerged as a promising avenue to accelerate the development of anti-aging therapies. This abstract reviews the current state of knowledge on repurposed drugs for anti-aging purposes. It highlights the advantages of repurposing over traditional drug discovery, such as reduced cost, shorter development timelines, and known safety profiles. The discussion focuses on specific repurposed drugs that have shown potential in preclinical and clinical studies for their anti-aging effects. The article also delves into the underlying mechanisms through which repurposed drugs exert their anti-aging properties. These mechanisms include the modulation of cellular senescence, inflammation, oxidative stress, mitochondrial dysfunction, and DNA damage repair pathways. By targeting these key hallmarks of aging, repurposed drugs hold promise for mitigating age-related decline and extending health span. Furthermore, the abstract explores the challenges and opportunities

D. Sundaram · P. Kumpati (✉)
Department of Biomedical Science, Bharathidasan University, Tiruchirappalli, Tamil Nadu, India
e-mail: prems@bdu.ac.in

H. Vedagiri
Department of Bioinformatics, Bharathiar University, Coimbatore, Tamil Nadu, India

G. K. Subbaraj
Faculty of Allied Health Sciences, Chettinad Hospital and Research Institute I, Chettinad Academy of Research and Education, Kelambakkam, Tamil Nadu, India



S. Pathak, A. Banerjee (eds.), *Neuroprotective Effects of Phytochemicals in Brain Ageing*, https://doi.org/10.1007/978-981-99-7269-2_15

associated with repurposing drugs for anti-aging interventions. It discusses the importance of rigorous clinical trials to evaluate the safety and efficacy of repurposed drugs and the need for personalized approaches to account for individual variations in drug response. Phytochemicals, a vast array of natural compounds found in plants, are emerging as potential allies in combating the various aspects of aging. Their antioxidant, anti-inflammatory, and other potent properties can target processes linked to age-related decline. These natural compounds have the potential to protect organs like the brain and skin, and promote overall health and wellbeing throughout the aging process. While the research continues to evolve, phytochemicals hold exciting promise in the pursuit of healthier, longer lives. In conclusion, repurposing existing drugs for anti-aging purposes represents a cost-effective and time-efficient strategy to identify potential interventions. This chapter highlights the potential of phytochemicals as repurposed drugs to protect age-related decline and promote healthy aging. Continued research in this field has the potential to unlock new phytochemical-based therapeutic options and improve the quality of life for aging populations worldwide.

**Keywords**

Anti-aging · Repurposed drugs · Metformin · Rampamycin · Aspirin

## 15.1 Introduction

The increasing average lifespan of human beings has not been accompanied by a corresponding improvement in overall health and well-being. The rising global population of older individuals poses a substantial challenge to modern society. The process of getting older plays a major role in the development of various serious health conditions, including Parkinson's disease, Alzheimer's disease, cardiovascular disorders, inflammatory and metabolic ailments, and cancer. These diseases associated with aging are the leading causes of illness and death rates in developed nations. While studying and treating these diseases individually has provided valuable knowledge and improved patients' quality of life, it is crucial to recognize that they often coexist. Dealing with multiple age-related diseases simultaneously can introduce significant complexities when it comes to implementing effective treatment interventions. Treating each individual age-related co-morbidity can be a challenging task. However, the concept of targeting the aging process itself as a means to address the underlying causes of multiple diseases concurrently holds great appeal. By focusing on slowing down or reversing the aging process, it becomes possible to tackle the root causes of various age-related diseases simultaneously. This approach offers the potential for a more comprehensive and holistic approach to combatting these conditions. Phytochemicals like curcumin and resveratrol have shown promise in protecting brain cells from oxidative stress and inflammation, key contributors to cognitive decline. They may help maintain memory function and reduce the risk of neurodegenerative diseases. Phytochemicals hold promise for slowing or mitigating the effects of brain aging. These naturally occurring plant compounds exhibit diverse properties, including antioxidant and anti-inflammatory actions, which can protect neurons and support healthier brain function. Researchers are exploring the repurposing of existing drugs, originally developed for other conditions, that have mechanisms of action potentially beneficial for brain aging. This strategy offers a faster and more cost-effective path than developing entirely new drugs. Phytochemicals with antioxidant and anti-inflammatory properties can protects free radical damage, a major culprit behind wrinkles and loss of skin elasticity. Green tea, rich in polyphenols, and berries packed with anthocyanins are examples of such sources. Phytochemicals can help maintain healthy blood pressure and cholesterol levels. Garlic, rich in allicin, and cocoa, containing flavanols, are examples. These can reduce the risk of heart disease, a common concern with aging. Certain phytochemicals, like isoflavones found in soy, may help improve bone mineral density and reduce the risk of osteoporosis, a condition more prevalent with age. This chapter aims to assess the current progress in translating fundamental aging research into practical treatments. It will examine ongoing efforts to repurpose existing drugs and explore potentially phytochemical based repurposable drugs on recent findings in pre-clinical models. Additionally, it will discuss emerging therapeutics designed specifically to target aging. Moreover, as reliable biomarkers of aging play a crucial role in assessing interventions aimed at slowing down the aging process, the article provides a brief overview of the quest for effective indicators of physiological age. To concentrate on approaches that hold the greatest promise for influencing human aging, only drugs that have exhibited effects in mammals will be taken into account. Previous analyses have explored treatments that enhance longevity in non-mammalian models.

Two decades ago, Richard Doll, a renowned epidemiologist, argued against the existence of aging as a unified phenomenon. According to Doll, based on the available evidence at the time, it was more reasonable to view each age-related disease as distinct and unrelated, requiring separate treatment approaches. In the field of gerontology, there have been notable advancements that highlight the existence of common underlying causes among many age-related diseases. This discovery opens up the possibility of shared interventions aimed at reversing or preventing these diseases. This shift in scientific consensus is significant, as it challenges the notion that aging is an inevitable and irreversible process. Instead, it suggests that aging is influenced by specific and modifiable biochemical pathways. Early studies conducted on nematodes demonstrated the potential of genetic manipulation to extend lifespan. Subsequent research has focused on unraveling the genetic factors that contribute to human longevity. These studies provide further evidence that aging is not solely determined by random processes but can be influenced by specific genes and molecular mechanisms. This growing understanding of the genetic and biochemical aspects of aging paves the way for the development of targeted interventions that can potentially slow down or even reverse the aging process, thereby addressing multiple age-related diseases simultaneously. It represents an exciting frontier in medical research and offers hope for the development of novel approaches to promote healthy aging and extend the human lifespan.

Aging is believed to be a result of developmental processes that initially evolved to ensure successful reproductive maturation but continue beyond that stage. This phenomenon, known as antagonistic pleiotropy (Williams 1957), refers to the concept that certain genes or traits can have both positive and negative effects on an organism's fitness. In the context of aging, antagonistic pleiotropy gives rise to molecular and physiological irregularities that render organisms fragile and less adept at handling stress. These irregularities manifest across various levels of biological organization, encompassing molecular, cellular, and tissue biology. In the case of aged individuals, several cellular abnormalities become evident, such as protein aggregation, dysfunctional mitochondria, genetic and epigenetic mutations, and shortened telomeres. These factors collectively contribute to cellular senescence and the decline of active stem cell populations, consequently compromising regenerative capabilities and accelerating tissue aging. Moreover, at a systemic level, aging is correlated with immunodegeneration, disrupted circadian rhythms, impaired nutrient sensing, and dysregulated autocrine and paracrine signaling processes. The culmination of these factors fosters the development of an "aged" extracellular environment.

This environment can cause or exacerbate organ failure, resulting in a significant increase in morbidity and mortality among aging populations. Public health efforts promote healthy aging through nutrition and physical activity, while pharmacological therapies, such as natural compounds like curcumin and polyamines, show the potential to prolong health span. Repurposing existing medications approved by regulatory agencies for specific diseases is also being explored. To explore these mechanisms in detail, several reviews have been conducted. This chapter provide comprehensive insights into the molecular and physiological changes associated with aging, shedding light on the complex processes that underlie age-related diseases and decline.

## 15.2 Druggability of Aging: Has the Time Come?

The Food and Drug Administration and other regulatory bodies do not currently see aging as a viable target for pharmaceutical intervention due to a lack of awareness of aging as a medical issue. However, in 2018, the WHO included the word "old age" in the International Classification of Diseases (ICD) (Konopka and Miller 2019). Some authors have proposed that identifying aging as a medical condition could spur the development of novel study strategies and business models that specifically target aging as a curable illness, leading to improved health outcomes and financial advantages for society in general (Zhavoronkov and Bhullar 2015). The FDA-approved Targeting Aging with MEtformin (TAME) clinical study is an excellent illustration of this principle in action since it investigates if the use of metformin, an antidiabetic medicine, may prevent or slow the progression of various types of age-related diseases (Konopka and Miller 2019; Barzilai et al. 2016).

Advancements in biomedicine are rapidly progressing, particularly in the development of medications that target aging-related processes. Gain- or loss-of-function techniques are commonly used to identify genetic pathways suitable for anti-aging interventions. Various pharmacological classes, such as caloric restriction mimetics, anti-inflammatory medications, autophagy inducers, epigenetic modulators, and senolytics, are frequently employed in anti-aging research. Recent studies indicate that these drugs have multiple effects, targeting various molecular and cellular processes. However, there is inconsistency and variation in the results of observational and clinical research regarding the effectiveness and long-term safety of these drugs. To address these challenges and facilitate their implementation in clinical practice, innovative research ideas and additional translational steps are necessary.

### 15.2.1 Aging as a Clinical Indication

The effects of aging on people, including the heightened likelihood of serious illnesses that undermine autonomy and self-respect, establish significant ethical and financial reasons to tackle and avert these age-associated maladies. Recognizing the complexities of aging itself, rather than treating its associated diseases separately, has led to the emergence of various companies and government-funded initiatives dedicated to tackling aging as a distinct problem. Notable examples include Calico (California Life Sciences Company), a venture funded by Alphabet (formerly Google), which focuses on aging research. The National Institute on Aging (NIA) administers the Interventions Testing Program (ITP) to assess the potential of various drugs in prolonging life. J. Craig Venter's co-founded organization, Human Longevity Inc., is committed to studying and managing the fundamental genetic and epigenetic factors leading to diseases associated with aging. The SENS Research Foundation carries out its research while backing other institutions to exploit regenerative medicine, gene therapy, and pharmacology to reverse the aging phenomenon. Unity Biotechnology is working on creating "senolytics" following research that suggests eliminating senescent cells enhances health in aging mice (Baker et al. 2016). Chronos Therapeutics, based in Oxford, UK, is concentrated on creating drugs to battle diseases linked to aging. These examples highlight the increasing recognition of aging as an actionable and potentially profitable field, attracting both new startups and established companies, rather than being viewed solely as an obscure academic matter.

### 15.2.2 Repurposing Drugs to Combat Aging

Drug repurposing is a widely used strategy in which existing therapeutics are explored for new applications. This approach aims to expand the market reach of these drugs and benefit a larger number of patients (Dudley et al. 2011). The strategy of repurposing drugs that have already received approval for certain applications presents several benefits. The process of developing a new drug is considerably costly and lengthy, requiring billions of dollars and about 10–17 years from the onset of clinical trials to its availability in the market (Adams and Brantner 2006). Drug repurposing offers a valuable advantage by bypassing the need for extensive preliminary investigations into mechanisms of action, drug formulation, manufacturing processes, and pharmacokinetics. Furthermore, repurposed drugs have already undergone clinical trials, establishing their safety and efficacy for their original intended use. As a result, repurposed drugs can be developed in a significantly shorter timeframe, ranging from 3 to 12 years. Drug repurposing can be pursued via two principal strategies: drug-centered and disease-centered, as elaborated by Dudley et al. (2011). In drug-centered repurposing, attention is given to the drug's physicochemical characteristics, such as its structural resemblances to other drugs, or its biological impacts, such as analogous reactions noted at the transcriptome level among various drugs. Shared side effects among drugs approved for distinct diseases can also be taken into account. In addition, computational modeling approaches, like molecular docking simulations, have emerged as a means to perform high-throughput screening to pinpoint potential interactions between drugs and proteins.

The disease-focused approach to drug repurposing capitalizes on the similarities between different diseases to identify potential drugs for repurposing. If two diseases share common molecular foundations, therapies used for one disease may hold promise for the other. These strategies are not mutually exclusive, and depending on the available data regarding the drug or disease in question, either or both methods can be employed. To aid in drug repurposing efforts targeting aging specifically, resources like geroprotectors.org provide a database with over 250 compounds known to extend lifespan in various model organisms (Moskalev et al. 2016). Among the 103 compounds approved for human use listed on geroprotectors.org, there is considerable interest in repurposing at least two: the immunosuppressant drug rapamycin and the anti-diabetic drug metformin. These compounds have garnered attention due to their potential in interventions aimed at mitigating the aging process.

### 15.2.3 Rampamycin

Rapamycin has garnered significant attention in the field of gerontology due to its remarkable properties. Various studies have shown that it can prolong the lifespan and improve the health of mice, even when administered later in their lives (Wilkinson et al. 2012; Harrison et al. 2009). The way rapamycin works is by binding to a protein called FK506 binding protein and inhibiting the mTOR complex I (mTORC1) kinase (Banaszynski et al. 2005). This inhibition of mTORC1 has multiple effects that are believed to enhance the balance of proteins and cellular components. It accomplishes this by alleviating the inhibition of autophagy, the process of cellular recycling, and enhancing proteasome function. At the same time, it suppresses protein translation, contributing to the overall improvement of cellular homeostasis (Qian et al. 2010; Yu et al. 2010). Furthermore, mTOR is also involved in regulating processes like the creation of new mitochondria and the maintenance of lipid balance, achieved through interactions with PPARγ and PGC1α (Laplante and Sabatini 2009). Rapamycin has displayed promising results in various disease models. In experiments involving mice with Alzheimer's disease, rapamycin treatment has been proven to reverse the clumping of amyloid-β, enhance brain blood flow, and improve cognitive function. These effects have been observed in both disease models and healthy mice. Similarly, in toxin-induced models of Parkinson's disease, rapamycin has been shown to prevent the death of nerve cells. While rapamycin is primarily approved as an immunosuppressant for post-organ transplantation, a related compound called RAD001 has shown potential for improving the response to influenza vaccination in elderly patients. It has been suggested that RAD001 may target chronic inflammatory conditions rather than general immune function by adjusting mTORC1 activity to levels seen in younger individuals. However, further research is necessary to confirm this hypothesis. It is important to compare the increased vaccination response with young controls for a more accurate assessment. Moreover, the effects of rapamycin on extending lifespan in mice are specific to sex and depend on specific genetic backgrounds. It is worth noting that rapamycin has demonstrated some adverse effects in certain situations. For example, in diabetic mice, it was found to increase mortality. Furthermore, the administration of high doses of rapamycin to a diverse group of mice resulted in an increased occurrence of cataracts and shrinkage of the testicles (Wilkinson et al. 2012). These studies underscore the significance of personalized medicine in identifying individuals who would derive the greatest benefit from specific treatments. To investigate the influence of genetic and environmental factors on the lifespan-extending effects of rapamycin, the University of Washington is currently conducting trials on pet dogs as part of the Dog Aging Project (DAP) (dogagingproject.com). The rationale behind this project is that domesticated dogs, as a result of selective breeding, exhibit considerable genetic variation and are exposed to similar environmental conditions as their human owners. If rapamycin proves to have a significant positive impact on the longevity and health span of dogs, despite genetic and environmental differences, it would hold promising implications for potential trials involving human populations.

### 15.2.4 Metformin

Metformin, a medication used to treat diabetes, is well-known for its ability to activate adenosine monophosphate-activated kinase (AMPK), which plays a crucial role in regulating glucose balance (Shaw et al. 2005). However, recent studies indicate that metformin has broader effects and can impact various biological processes associated with aging (Foretz et al. 2010). Animal studies suggest that metformin may possess systemic anti-aging properties when administered early in life, potentially unrelated to its ability to lower blood sugar levels (Anisimov et al. 2011). Building on these findings, an innovative clinical trial, funded by the American Federation of Aging Research, is currently underway to investigate whether metformin can truly prevent aging, marking the first study of its kind. The TAME

(Targeting Aging with Metformin) study is specifically designed to explore whether metformin possesses longevity-enhancing properties that are distinct from its anti-diabetic effects. In addition to studying metformin, this trial aims to establish a standard in clinical trial design that can be used to evaluate other compounds with potential anti-aging effects (Barzilai et al. 2016). Assessing the rate of aging in human populations poses methodological challenges. In model organisms, the conventional approach to determining aging rates involves constructing Kaplan-Meier survival curves and subsequently employing Gompertz analysis, a mathematical model proposed by Benjamin Gompertz in 1825. The traditional method of constructing survival curves and performing Gompertz analysis is not practical for clinical trials involving human subjects due to the lengthy timescales required. In the TAME study, an alternative approach is being used, which involves using the time before the appearance of the next co-morbidity as a proxy for the rate of aging. As populations age, overall health tends to decline exponentially, resulting in a higher incidence of co-morbidities and a subsequent drop in survival observed in survival curves. By examining the time until the next co-morbidity, researchers can gain insights into the rate of aging within the studied population. However, this approach still requires a significant amount of time to observe meaningful results. In addition to this approach, there is ongoing research focused on identifying biomarkers that can serve as indicators of biological age and age-related processes. These biomarker-based approaches hold the potential to assess aging more efficiently and quickly. By identifying specific biomarkers associated with aging, researchers can potentially evaluate the effects of interventions on aging in a more expedited manner.

## 15.3 Nonsteroidal Anti-inflammatory Drugs

Immunosenescence, one of the primary indications of aging, is marked by a gradual decline in both innate and adaptive immune responses (Neves and Sousa-Victor 2020). As per Fulop et al. (2018), chronic, low-grade inflammation, commonly referred to as "inflammaging," contributes to the onset and progression of age-related diseases (Franceschi et al. 2018). NSAIDs, or nonsteroidal anti-inflammatory drugs, are considered the most effective treatment approach for aging. However, their utilization in anti-aging therapy poses a challenge due to known adverse effects, including gastrointestinal bleeding, cardiovascular complications, and nephrotoxicity (Wongrakpanich et al. 2018). We will discuss the research findings for the most extensively investigated anti-aging agents.

### 15.3.1 Aspirin

Acetylsalicylic acid, or aspirin, is an NSAID derived from white willow tree (Salix alba) bark extract. This includes: (Montinari et al. 2019) works by blocking cyclo-oxygenase (COX) to stop the production of inflammatory prostaglandins (Botting 2010). It is commonly used to treat cancer and other life-threatening conditions, including heart disease and stroke. Data from Hua et al. (2019) in animal studies have been shown to delay the onset of age-related decreases in function and increase longevity. Research from several sources (Strong et al. 2008) supports this notion. Aspirin's most prevalent use in humans is in the prevention of arterial thrombotic events in those with preexisting atherosclerotic diseases. Mekaj et al. (2015), in spite of its widespread use, aspirin may not be the best choice for primary prevention of cardiovascular disease, according to several recent research. Although aspirin's use in preventing heart attacks and strokes is widespread, the drug's effectiveness is constrained by major side effects. According to a recent study (Huang et al. 2011), the ambiguity of the benefit/risk ratio is reflected in the divergence of expert opinion over the use of aspirin for primary prevention. There is now widespread agreement that each patient's aspirin prescription should be made with careful consideration of the potential benefits and risks involved.

### 15.3.2 Ibuprofen

Ibuprofen, the initial non-aspirin NSAID to be licensed for over-the-counter usage, is thought to be more effective than paracetamol and at least as effective as aspirin for pain management (Moore 2007). It has been proven to lengthen longevity by up to 15% in animal mice, with minimal risk of gastrointestinal problems and cardiovascular side effects (Varrassi et al. 2020). No positive results have been found in clinical studies of ibuprofen for treating pathological diseases related to aging. Recent research has shown that ibuprofen has no positive effect on cognition and can even make Alzheimer's sufferers sick. Ibuprofen is associated with an increased risk of acute myocardial infarction and stroke, as well as renal and hepatic adverse effects, endocrine-disrupting effects, and gastrointestinal side effects. The severity of these side effects also varies with the dosage, other drugs being used, and the patient group (Varrassi et al. 2020).

## 15.4 Statins

Statins are a group of medications used to treat high cholesterol and decrease the risk of cardiovascular disease and death (Ward et al. 2019). They have been demonstrated to lessen the chances of having a nonfatal heart attack, a nonfatal stroke, or dying from cardiovascular disease or any cause. However, research shows that the pleiotropic effects of these medications may be mostly responsible for these advantages. Statins have been shown to have life-extending qualities in animal and human research. As stated by (Spindler et al. 2012), Statin users in the US veteran community have been shown to have a 2-year greater mean age at death compared to nonstatin users while being at a higher risk of mortality overall. This includes those older than 74 years without Type 2 diabetes; using a statin is linked to a lower risk of atherosclerotic cardiovascular disease and overall mortality (Mehta et al. 2006). According to research (Ramos et al. 2018), New-onset diabetes, cognitive impairment, and hemorrhagic stroke are only a few of the possible side effects. Recent large-scale investigations have shown that the occurrence of these unfavorable consequences is minimal, providing more evidence that the benefits much exceed the dangers. Post-stroke individuals who take statins have a lower chance of developing complications, including dementia, cataracts, peripheral neuropathy, tendinitis, sexual dysfunction, and cancer. However, it is still unclear if statin usage increases the risk of dementia, cataracts, peripheral neuropathy, tendinitis, erectile dysfunction, or cancer. Statins should be prescribed to older people with prudence after carefully weighing the potential benefits against the potential risks.

### 15.4.1 NAD+ Supplementation

Despite the practical and commercial advantages of repurposing existing therapeutics, there are still numerous untapped targets for anti-aging interventions. A specific objective in the field of aging research is the use of nicotinamide adenine dinucleotide (NAD+) supplementation. Mitochondrial dysfunction and metabolic disruptions are widely recognized as characteristic signs of aging. Gomes et al. (2013) conducted a study demonstrating that NAD+ levels decline in the nucleus of skeletal muscle as individuals age, causing an imbalance between the expression of genes in the nucleus and mitochondria. They also discovered that replenishing NAD+ levels could reverse various biochemical changes associated with skeletal muscle aging.

Studies have demonstrated that feeding mice with nicotinamide mononucleotide (NMN) can increase circulating levels of NAD+ when included in their diet despite NAD+ itself being poorly absorbed. This discovery suggests that NMN, along with another precursor called nicotinamide riboside (NR), which is currently available as a supplement, could potentially be marketed as an anti-aging drug. However, it is important to note that while NMN feeding has shown promise in reversing certain biochemical features associated with muscle aging, it remains uncertain whether this translates into improved muscle function or overall health span. It is also crucial to acknowledge that the studies conducted on NMN, NR, or NAD+ supplementation have been limited to mice, and their impact on humans remains largely untested. Moreover, considering that NR is naturally present in small amounts in milk, it raises questions about the necessity of costly supplementation when dietary adjustments could suffice.

### 15.4.2 Senolytics

A different approach to address the aging process involves the utilization of senolytics. In a study conducted by Baker et al. (2016), it was demonstrated that the removal of senescent cells from elderly mice resulted in a prolongation of their lifespan and a reversal of organ deterioration, such as that observed in the heart and kidney during the aging process. The presence of senescent cells serves two primary purposes: first, to prevent the development of cancer, and second, to attract macrophages for the elimination of senescent cells. Nevertheless, as individuals age, a combination of a decline in immune function and abnormal communication between cells can result in senescent cells becoming resistant to natural cell death processes. Consequently, these senescent cells persist in tissues. To address this issue, a potential solution is the utilization of a small molecule called ABT-737. Research has demonstrated that ABT-737 can selectively eliminate senescent cells in living organisms by inhibiting the specific anti-apoptotic proteins that these cells rely on to survive (Yosef et al. 2016). By removing senescent cells, it is possible to prevent further tissue remodeling caused by the senescence-associated secretory phenotype (SASP) and promote the deposition and remodeling of the extracellular matrix, thereby improving the regenerative capacity (Calhoun et al. 2016). It is indeed important to acknowledge that senescent cells and the senescence-associated secretory proteome (SASP) can have diverse functions in different biological contexts. For instance, in certain species of salamanders, the SASP-induced recruitment of immune cells has been identified as a crucial mechanism contributing to their impressive regenerative capabilities (Yun et al. 2015). Additionally, the senescence of pancreatic β-cells has been shown to enhance insulin secretion and improve glucose regulation in a mouse model of diabetes (Helman et al. 2016). Considering the complex and multifaceted characteristics of the SASP, a more refined approach could involve manipulating the SASP itself to address long-term negative consequences, such as chronic inflammation and impaired tissue regeneration, while still preserving short-term benefits like cancer prevention and wound healing. A study conducted by Malaquin et al. (2016) discussed this strategy, highlighting the importance of striking a balance between modulating the SASP for long-term benefits and maintaining its positive contributions in the short term. This approach would aim to harness the beneficial aspects of the SASP while mitigating its potential detrimental effects, thus offering a more nuanced strategy for addressing aging-related issues. This approach aims to find a middle ground by dampening harmful aspects of the SASP while preserving its advantageous functions. Through deeper comprehension of the specific elements and signaling pathways involved in the SASP, it may be feasible to develop interventions that selectively target the detrimental aspects while safeguarding the beneficial effects associated with senescence.

## 15.5 Antioxidant Vitamins

Antioxidants have been established to be useful in maintaining homeostasis under chronic oxidative stress, as predicted by Denham Harman's free radical theory of aging from the 1950s, although no pharmaceutical antioxidants have been authorized as of yet (Sadowska-Bartosz and Bartosz 2014; Margaritelis et al. 2018). Although dietary antioxidants have been touted as a means of delaying the onset of age-related diseases, clinical trials and observational research have shown that they are linked to unfavorable health outcomes (Poprac et al. 2017). The "antioxidant paradox" describes the contradictions between laboratory and human research on antioxidants' lifespan-extending effects. It is possible that exogenous antioxidants' dual function in ROS generation and incapacity to address oxidative stress and inflammation are to blame for this seeming contradiction. Manipulating endogenous antioxidants may be more suitable than focusing on dietary antioxidants to combat ROS-induced diseases. There is an immediate need to converge on a universal scale of oxidative stress and define "normal values" for each bodily fluid that account for differences in age and gender. To wit: (Poljsak et al. 2013).

### 15.5.1 Biomarkers for Detecting Drug Efficacy

Accurately assessing the effectiveness of potential therapies for human aging requires suitable biomarkers that are closely associated with and likely contribute to increased physiological age. A biomarker for aging can be defined as a biological characteristic of an organism that, either alone or in combination with other factors, predicts functional capacity at a later age more accurately than just the individual's chronological age without the presence of disease (Baker 1988). It is crucial for these ideal biomarkers to be easy to measure, non-invasive, and not cause significant discomfort or harm to individuals. Molecular biomarkers indicating physiological age have been a subject of debate, often showing limited predictive ability for subsequent illness or mortality. It is important to note that discrepancies in sampling methods for the same parameter can present challenges when making comparisons (Cooper et al. 2011). Proposed molecular biomarkers of aging tend to be specific to certain cell types and may not fully represent the physiological age of the entire organism. Therefore, the identification and validation of reliable biomarkers that accurately reflect the overall physiological age of an organism, are minimally invasive and have strong predictive power for the functional decline are essential for assessing the efficacy of potential anti-aging interventions. Further investigation in this field is essential to create reliable biomarkers that can adequately assess the effects of treatments on the aging process in humans. A comprehensive analysis performed by Chen et al. (2016) has demonstrated a significant association between changes in methylation patterns of particular CpG dinucleotides and overall mortality among different ethnic groups, indicating that modifications in DNA methylation may potentially function as a biomarker for aging.

Another suggested biomarker for the process of aging is the length of telomeres, which are protective structures located at the ends of chromosomes. Telomeres play a crucial role in safeguarding genetic information during DNA replication cycles and shielding chromosome ends from being perceived as DNA damage. Nevertheless, recent studies have indicated that telomere length might not be a more reliable indicator of age-related health issues when compared to a person's chronological age (Glei et al. 2016). The measurement of telomere length is typically performed on leukocytes obtained from blood samples. It is important to note that leukocytes are specialized cells that have reached the end stage of their life cycle and are unlikely to divide further. As a result, the telomere length in leukocytes may not accurately reflect an individual's capacity for regeneration or their ability to cope with stress.

These findings highlight the complexity of identifying reliable biomarkers for aging. While DNA methylation patterns show promise, further research is needed to validate their predictive power. Similarly, the role of telomere length as a biomarker for overall physiological age remains debated, as its association with age-related morbidity is not consistently observed. Developing biomarkers that accurately represent the aging process and reflect an individual's functional decline and health span remains an ongoing challenge.

### 15.5.2 Finding Anti-aging Medicines Using Computational Biology

The field of aging research has seen significant advancements in databases dedicated to the annotation and curation of genes, proteins, their functions, pathways, and the interactions between drugs and these biological entities. These databases provide valuable resources for studying aging and identifying potential interventions. One category of methods used in drug repurposing for targeting aging involves the use of structural information to predict drugs that may interact with specific proteins. Structural databases, such as the Protein Data Bank (PDB), contain three-dimensional structural information of proteins, which can be used to screen for small molecules or drugs that could potentially interact with target proteins involved in aging processes. Another category of methods focuses on similarities in molecular structures, interactions, pathways, or networks between aging-related drugs or genes. These approaches leverage computational techniques to identify drugs that may have similar mechanisms of action or therapeutic effects as known aging-related drugs. By exploring similarities in molecular features, drug-gene interactions, or pathway associations, researchers can identify promising candidates for repurposing existing drugs to target aging.

### 15.5.3 Anti-aging Gene Virtual Screening

Two studies used techniques to anticipate drug-target interactions to identify therapeutics already known to target aging-related genes. Three temperature-sensing proteins (TRP7, S6P, and FhBC) have been linked to aging in the rotifer Brachionus manjavacas, and the first study set out to find new treatments that target these proteins. Only 5 of the 12 chemicals studied showed any meaningful prolongation of rotifer life. The FINDSITE comb program was used to conduct an in-computer screening of 94 targets, with the top 1% binding molecules rated by cumulative lifetime extension. Seven medicines were tested in whole-life survival assessments, two of which resulted in a median lifetime extension of 13–42%. AMP-activated protein kinase (AMPK) was the only gene tested in another silico screening investigation.

### 15.5.4 Methods Based on Similarity

Similarity-based techniques are used to identify potential anti-aging medications based on age-related genes, prolongevity treatments, or gene expression patterns.

### 15.5.5 Identifying Drugs That Affect Known Genes Associated with Aging

Fernandes et al. looked for medications that target orthologs of human genes linked to longevity in animal models. The total number of medications obtained was 376, with 20 having statistically meaningful results. Thirteen histone deacetylases were targeted, three of which have been associated with increased longevity in animal models. Fuentealba et al. also utilized a composite group of age-related genes with compelling evidence for their impact on human aging. Among the top ten compounds, three were already recognized for their ability to extend lifespan in animal models, while the remaining seven were hypothesized to promote longevity through the repurposing of alternative medications.

### 15.5.6 Discovering New Drugs with Similar Effects on Aging

Liu et al., used machine learning to foresee potential novel C. elegans lifespan medicines. They used a semi-supervised algorithm trained using high-confidence prolongevity medications and genes involved with their aging processes. Using a lifetime test, they ranked 785 medicines for their ability to extend worm longevity. Chemical descriptors of existing lifespan medicines and functional annotation of their targets were used to train a distinct machine-learning technique. Drugs expected to increase longevity were ranked using a supervised method (random forest), but the results were not validated.

### 15.5.7 Comparing Transcriptome Signatures from Aging and Drugs

According to a study by Yang et al., the Connectivity Map Resource provides drug-induced expression profiles for 1309 compounds. By comparing these profiles with gene expression signatures associated with aging, potential drugs that induce similar expression changes to those observed in aging can be identified using gene-set enrichment analysis. This approach eliminates the need for a predefined list of 'aging genes' and has the potential to discover novel targets based solely on expression profile similarities. The researchers employed a network-based methodology called ANDRU (aging network-based drug discovery) to identify differentially expressed genes within aging-related networks and drugs capable of reversing these changes. The study revealed three drugs, ranking among the top five, which had not been previously reported as lifespan modulators. These drugs target pathways such as metabolic enzymes and lipid metabolism, which exhibit age-related changes in expression.

### 15.5.8 The Comparison In-Situ Studies

The research was conducted with a unique set of objectives, techniques, and data. To compare the studies, a summary of the pharmaceuticals found the genes that these drugs target and all of the KEGG biological pathways were included. The findings were also compared to humanly curated databases of medications and genes associated with aging (GenAge and DrugAge).

### 15.5.9 Drugs

The information regarding the prioritization of drugs in the DrugAge database and the variations in medications among research studies, as well as the experimental validation of the effects of trichostatin, geldanamycin, tanespimycin, and vorinostat on longevity in animal models, was reported (Author(s), Year). Furthermore, it was found that 66% of the drugs known to target proteins involved in aging were prioritized by the computational studies reviewed. In terms of gene discovery, 34% of the human genes in GenAge (103 genes) and 10% of the model organism genes in GenAge (94 genes) were discovered in at least one research, which may partially reflect the differences in dataset sizes. The most important details in this text are the computational approaches used to select candidate pharmaceuticals that are projected to control aging. These include genes such as DDIT3 and ERBB2, which were the focus of eight trials where medications were aimed. Nine research also predicted medications regulating ABCB1, which has not been previously linked to human aging, and 10 studies indicated pharmaceuticals targeting BIRC5 and KRAS. However, the discovery of genes in several studies does not always indicate a greater significance to aging; it may represent research bias. Additionally, 12 computational investigations suggested candidates for geroprotective medications, 80% of which target the same genes as the known prolongevity treatments.

### 15.5.10 Pathways

The 92% of druggable KEGG pathways have at least one gene targeted by medicines. GenAge genes (83% Model GenAge and 74% Human GenAge) and DrugAge drug targets (88% of all genes) follow this trend. This suggests that aging impacts every system equally, but it is possible that genes are continually identified due to their fundamental function in illnesses and regulatory processes.

## 15.6 Drug Repositioning for Personalized Medicine

The process of developing medications involves optimizing their affinity and selectivity for a specific target. When considering repurposing efforts for old pharmaceuticals to target a different indication, the potency values for the new target are likely to be lower than the primary target. This means that repurposed medications may have reduced effectiveness for the new indication. Additionally, if there are already approved medications available for the intended indication, it may be challenging to demonstrate the superiority of the repurposed medication in terms of efficacy and safety during phases III and IV of clinical trials. Drug repurposing is achievable by collecting and sharing target profile data; however, patent filings may limit commercial applicability. Due to side effect tolerance, some therapeutic applications may be more repurposable. Drug repurposing prospects should also evaluate illness relevance and intellectual property. These initiatives go beyond finding novel targets for established medications.

Drug repurposing is achievable by collecting and sharing target profile data; however, patent filings may limit commercial applicability. Due to side effect tolerance, some therapeutic applications may be more repurposable. Drug repurposing prospects should also evaluate illness relevance and intellectual property. These initiatives go beyond finding novel targets for established medications.

## 15.7 Disease-Mechanism Matching

Knowing the mechanism of action of each medicine might allow fixed dosage combinations (FDCs) to harness possible synergies and increase efficacy and/or decrease adverse effects using any of the aforementioned methods. This method might increase efficacy and postpone or avoid medication resistance in HIV, TB, and cancer by regulating two biological targets. After being designed to suppress HIV protease, ritonavir was shown to inhibit CYP450-3A4 and PgP-mediated drug efflux pathways. Ritonavir is used alongside others.

## 15.8 Databases and Resources Available that Aid in Drug Repositioning Efforts

Drug repositioning, also known as drug repurposing, is the process of identifying new therapeutic uses for existing drugs. This approach is gaining popularity in the field of medical and life sciences as it offers several advantages, such as reduced development time and costs compared to developing new drugs from scratch. There are several databases and resources available that aid in drug repositioning efforts.

**DrugBank:** DrugBank is a comprehensive database that provides information on drugs, their targets, pharmacology, and therapeutic indications. It can be used to explore potential repositioning opportunities by identifying drugs with known mechanisms of action that could be effective for different diseases.

**PubChem:** PubChem is a free chemical database maintained by the National Center for Biotechnology Information (NCBI). It contains information on the biological activities of small molecules, including drugs. Researchers can use PubChem to identify compounds with similar chemical structures or properties to known drugs, which may suggest potential repositioning candidates.

**ClinicalTrials.gov:** ClinicalTrials.gov is a database of clinical trials conducted worldwide. Researchers can search for ongoing or completed trials to identify drugs being tested for new indications. This information can guide drug repositioning efforts by providing insights into the safety and efficacy of drugs in different disease contexts.

**ChEMBL:** ChEMBL is a database of bioactive molecules and their targets. It provides information on the activities of compounds against specific protein targets, including drugs. Researchers can utilize ChEMBL to identify drugs that have demonstrated activity against targets relevant to a particular disease, indicating potential repositioning opportunities.

**LINCS (Library of Integrated Network-Based Cellular Signatures):** LINCS is a large-scale project that aims to create a comprehensive map of cellular responses to various perturbations, including drugs. It provides gene expression profiles and other molecular data for different drugs across various cell lines. This resource can help researchers identify drugs that produce similar molecular signatures to those associated with specific diseases, suggesting potential repositioning candidates.

**Connectivity Map (CMap):** CMap is a database that links gene expression signatures of drugs and diseases. It can be used to identify drugs that induce gene expression patterns opposite to those observed in a particular disease. Such drugs may have potential therapeutic value for repositioning.

## 15.9 Future of Drug Repurposing in Aging

The potential of medication repurposing in the aging process offers tremendous promise for the maintenance and enhancement of the health and well-being of older people in the future. The process of discovering new therapeutic applications for existing medications that were first created for other indications is referred to as drug repurposing, drug repositioning, or drug reprofiling. In recent years, this strategy has garnered a lot of interest due to the fact that it has the potential to speed up the process of drug development, cut costs, and make better use of the enormous knowledge base that is currently accessible for pharmaceuticals that have been authorized. The repurposing of drugs has a number of potential benefits in the context of becoming older. To begin, it enables researchers to avoid the time-consuming and financially burdensome process of creating new therapies from the ground up, which may take anywhere from 10 to 15 years and cost billions of dollars. If we repurpose medications that are already on the market, we can move our attention to determining whether or not they are effective and safe for treating illnesses that are associated with aging, which might possibly lead to more rapid clinical translation.

In spite of the progress that has been made in medical research, there are a great many disorders throughout the world that do not have any available treatment options. This presents a problem for researchers to identify adequate therapeutic alternatives. In this context, the idea of repurposing an existing medicine provides a wealth of opportunities for researching the latent potential of the drug in question and recycling the drug itself. In spite of the fact that researchers in this day and age have the good fortune to have access to such computational tools, there is still a certain hole that has to be filled in order to address the post-prediction deficiencies in areas such as optimizing the dosage regimen, formulation, and selection of target populations, etc. Even if there were some snags along the way, the value of medication repurposing utilizing in silico techniques has gained prominence in the field of drug development in today's pharma businesses.

An online tool (http://drugrepo.org/) provides further help for computational analysis by allowing users to pick a disease and then receive repurposed pharmaceuticals or rescued chemicals along with other parameters. DrugRepo's graphical user interface (GUI) allows users to choose an illness, view the authorized pharmaceuticals and targets connected with those treatments, and then download the drugs that can be repurposed for that disease. Both the TC and DrugRepo score cutoffs are customizable by users.

## 15.10 Challenges

Lack of Financial Incentives: Repurposing drugs typically involves using existing medications, which often have generic formulations. This lack of exclusive patent protection limits the financial incentives for pharmaceutical companies to invest in repurposing efforts, as the potential return on investment may be lower compared to developing a new drug. Limited Clinical Evidence: Repurposing drugs for new indications requires robust clinical evidence to support their efficacy and safety. However, since these drugs were not initially developed for the repurposed indication, there may be a lack of well-designed clinical trials and specific regulatory pathways to approve their use in new therapeutic areas.

Complex Regulatory Landscape: The regulatory framework for drug repurposing can be challenging, as it often involves navigating through multiple regulatory agencies and their specific requirements. The absence of clear guidelines and streamlined processes for repurposed drugs adds to the complexity and can hinder the efficient development and approval of such treatments.

## 15.11 Possible Solutions

Public-Private Partnerships: Collaboration between academia, pharmaceutical companies, and government agencies can help overcome the financial barriers associated with drug repurposing. By sharing resources, expertise, and funding, these partnerships can support the exploration and development of repurposed drugs. Real-World Evidence: The use of real-world evidence, such as patient data from electronic health records and observational studies, can provide valuable insights into the effectiveness and safety of repurposed drugs. This data can complement traditional clinical trials and help generate the necessary evidence for regulatory approval. Regulatory Reforms: Regulatory agencies can play a pivotal role in facilitating drug repurposing by establishing dedicated pathways and streamlined processes for evaluating and approving repurposed drugs. Clear guidelines and regulatory incentives can encourage pharmaceutical companies to invest in repurposing efforts. Despite the challenges of individual variation and the need for rigorous clinical trials, repurposing existing drugs offers a promising, cost-effective approach to protecting against age-related disease. By targeting the fundamental mechanisms of aging, these interventions have the potential to tackle multiple age-related diseases simultaneously, promoting healthy aging and improving the quality of life for our growing elderly population. Moreover, the field of phytochemicals holds immense potential for advancing brain aging research. Ongoing research aims to identify specific phytochemicals with the greatest neuroprotective potential and to develop effective delivery methods for maximizing their bioavailability within the brain. As these studies progress, we can anticipate the development of novel phytochemical-based therapies that could significantly mitigate cognitive decline and preserve brain health throughout the aging process. The combination of repurposed drugs and cutting-edge phytochemical interventions could add to our approach to promoting longevity and protecting the debilitating effects of brain aging. Phytochemical-based drug repurposing presents a particularly exciting avenue for combating brain aging. Researchers are analyzing the molecular mechanisms of potent phytochemicals like curcumin or resveratrol, identifying existing drugs that target similar pathways. This strategy offers the potential to obtain the known safety profiles of these drugs while harnessing the neuroprotective benefits of the phytochemicals that inspired their repurposing. This approach could lead to the expedited development of novel therapies to mitigate brain aging and protect age-related neurodegeneration.

## References

Adams CP, Brantner VV (2006) Estimating the cost of new drug development: is it really $802 million? Health Aff 25(2):420–428

Anisimov VN, Berstein LM, Popovich IG, Zabezhinski MA, Egormin PA, Piskunova TS, Semenchenko AV, Tyndyk ML, Yurova MN, Kovalenko IG, Poroshina TE (2011) If started early in life, metformin treatment increases life span and postpones tumors in female SHR mice. Aging (Albany, NY) 3(2):148

Baker GT (1988) I11 and SPROTT, RL biomarkers of aging. Exp Gerontol 23:223–239

Baker DJ, Childs BG, Durik M, Wijers ME, Sieben CJ, Zhong JA, Saltness R, Jeganathan KB, Verzosa GC, Pezeshki A, Khazaie K (2016) Naturally occurring p16Ink4a-positive cells shorten healthy lifespan. Nature 530(7589):184–189

Banaszynski LA, Liu CW, Wandless TJ (2005) Characterization of the FKBP.rapamycin.FRB ternary complex. J Am Chem Soc 127(13):4715–4721

Barzilai N, Crandall JP, Kritchevsky SB, Espeland MA (2016) Metformin as a tool to target aging. Cell Metab 23(6):1060–1065
Botting RM (2010) Vane's discovery of the mechanism of action of aspirin changed our understanding of its clinical pharmacology. Pharmacol Rep 62(3):518–525
Calhoun C, Shivshankar P, Saker M, Sloane LB, Livi CB, Sharp ZD, Orihuela CJ, Adnot S, White ES, Richardson A, Jourdan Le Saux C (2016) Senescent cells contribute to the physiological remodeling of aged lungs. J Gerontol A Biomed Sci Med Sci 71(2):153–160
Chen BH, Marioni RE, Colicino E, Peters MJ, Ward-Caviness CK, Tsai PC, Roetker NS, Just AC, Demerath EW, Guan W, Bressler J (2016) DNA methylation-based measures of biological age: meta-analysis predicting time to death. Aging (Albany, NY) 8(9):1844
Cooper R, Hardy R, Aihie Sayer A, Ben-Shlomo Y, Birnie K, Cooper C, Craig L, Deary IJ, Demakakos P, Gallacher J, McNeill G (2011) Age and gender differences in physical capability levels from mid-life onwards: the harmonisation and meta-analysis of data from eight UK cohort studies. PLoS One 6(11):e27899
Dudley JT, Deshpande T, Butte AJ (2011) Exploiting drug–disease relationships for computational drug repositioning. Brief Bioinform 12(4):303–311
Foretz M, Hébrard S, Leclerc J, Zarrinpashneh E, Soty M, Mithieux G, Sakamoto K, Andreelli F, Viollet B (2010) Metformin inhibits hepatic gluconeogenesis in mice independently of the LKB1/AMPK pathway via a decrease in hepatic energy state. J Clin Invest 120(7):2355–2369
Franceschi C, Garagnani P, Parini P, Giuliani C, Santoro A (2018) Inflammaging: a new immune–metabolic viewpoint for age-related diseases. Nat Rev Endocrinol 14(10):576–590
Fulop T, Witkowski JM, Olivieri F, Larbi A (2018) The integration of inflammaging in age-related diseases. Semin Immunol 40:17–35
Glei DA, Goldman N, Risques RA, Rehkopf DH, Dow WH, Rosero-Bixby L, Weinstein M (2016) Predicting survival from telomere length versus conventional predictors: a multinational population-based cohort study. PLoS One 11(4):e0152486
Gomes AP, Price NL, Ling AJ, Moslehi JJ, Montgomery MK, Rajman L, White JP, Teodoro JS, Wrann CD, Hubbard BP, Mercken EM (2013) Declining NAD+ induces a pseudohypoxic state disrupting nuclear-mitochondrial communication during aging. Cell 155(7):1624–1638
Harrison DE, Strong R, Sharp ZD, Nelson JF, Astle CM, Flurkey K, Nadon NL, Wilkinson JE, Frenkel K, Carter CS, Pahor M (2009) Rapamycin fed late in life extends lifespan in genetically heterogeneous mice. Nature 460(7253):392–395
Helman A, Klochendler A, Azazmeh N, Gabai Y, Horwitz E, Anzi S, Swisa A, Condiotti R, Granit RZ, Nevo Y, Fixler Y (2016) p16Ink4a-induced senescence of pancreatic beta cells enhances insulin secretion. Nat Med 22(4):412–420
Hua H, Zhang H, Kong Q, Wang J, Jiang Y (2019) Complex roles of the old drug aspirin in cancer chemoprevention and therapy. Med Res Rev 39(1):114–145
Huang ES, Strate LL, Ho WW, Lee SS, Chan AT (2011) Long-term use of aspirin and the risk of gastrointestinal bleeding. Am J Med 124(5):426–433
Konopka AR, Miller BF (2019) Taming expectations of metformin as a treatment to extend healthspan. Geroscience 41(2):101–108
Laplante M, Sabatini DM (2009) mTOR signaling at a glance. J Cell Sci 122(20):3589–3594
Malaquin N, Martinez A, Rodier F (2016) Keeping the senescence secretome under control: molecular reins on the senescence-associated secretory phenotype. Exp Gerontol 82:39–49
Margaritelis NV, Paschalis V, Theodorou AA, Kyparos A, Nikolaidis MG (2018) Antioxidants in personalized nutrition and exercise. Adv Nutr 9(6):813–823
Mehta JL, Bursac Z, Hauer-Jensen M, Fort C, Fink LM (2006) Comparison of mortality rates in statin users versus nonstatin users in a United States veteran population. Am J Cardiol 98(7): 923–928
Mekaj YH, Daci FT, Mekaj AY (2015) New insights into the mechanisms of action of aspirin and its use in the prevention and treatment of arterial and venous thromboembolism. Ther Clin Risk Manag 11:1449–1456

Montinari MR, Minelli S, De Caterina R (2019) The first 3500 years of aspirin history from its roots–a concise summary. Vasc Pharmacol 113:1–8

Moore N (2007) Ibuprofen: a journey from prescription to over-the-counter use. J Royal Soc Med 100(48_Suppl):2–6

Moskalev A, Chernyagina E, Tsvetkov V, Fedintsev A, Shaposhnikov M, Krut'ko V, Zhavoronkov A, Kennedy BK (2016) Developing criteria for evaluation of geroprotectors as a key stage toward translation to the clinic. Aging Cell 15(3):407–415

Neves J, Sousa-Victor P (2020) Regulation of inflammation as an anti-aging intervention. FEBS J 287(1):43–52

Poljsak B, Dahmane R, Godic A (2013) Skin and antioxidants. J Cosmet Laser Ther 15(2):107–113

Poprac P, Jomova K, Simunkova M, Kollar V, Rhodes CJ, Valko M (2017) Targeting free radicals in oxidative stress-related human diseases. Trends Pharmacol Sci 38(7):592–607

Qian SB, Zhang X, Sun J, Bennink JR, Yewdell JW, Patterson C (2010) mTORC1 links protein quality and quantity control by sensing chaperone availability. J Biol Chem 285(35): 27385–27395

Ramos R, Comas-Cufí M, Martí-Lluch R, Balló E, Ponjoan A, Alves-Cabratosa L, Blanch J, Marrugat J, Elosua R, Grau M, Elosua-Bayes M (2018) Statins for primary prevention of cardiovascular events and mortality in old and very old adults with and without type 2 diabetes: retrospective cohort study. BMJ 362:k3359

Sadowska-Bartosz I, Bartosz G (2014) Effect of antioxidants supplementation on aging and longevity. Biomed Res Int 2014:404680

Shaw RJ, Lamia KA, Vasquez D, Koo SH, Bardeesy N, DePinho RA, Montminy M, Cantley LC (2005) The kinase LKB1 mediates glucose homeostasis in liver and therapeutic effects of metformin. Science 310(5754):1642–1646

Spindler SR, Li R, Dhahbi JM, Yamakawa A, Mote P, Bodmer R, Ocorr K, Williams RT, Wang Y, Ablao KP (2012) Statin treatment increases lifespan and improves cardiac health in Drosophila by decreasing specific protein prenylation. PLoS One 7(6):e39581

Strong R, Miller RA, Astle CM, Floyd RA, Flurkey K, Hensley KL, Javors MA, Leeuwenburgh C, Nelson JF, Ongini E, Nadon NL (2008) Nordihydroguaiaretic acid and aspirin increase lifespan of genetically heterogeneous male mice. Aging Cell 7(5):641–650

Varrassi G, Pergolizzi JV, Dowling P, Paladini A (2020) Ibuprofen safety at the golden anniversary: are all NSAIDs the same? A narrative review. Adv Ther 37:61–82

Ward NC, Watts GF, Eckel RH (2019) Statin toxicity: mechanistic insights and clinical implications. Circ Res 124(2):328–350

Wilkinson JE, Burmeister L, Brooks SV, Chan CC, Friedline S, Harrison DE, Hejtmancik JF, Nadon N, Strong R, Wood LK, Woodward MA (2012) Rapamycin slows aging in mice. Aging Cell 11(4):675–682

Williams GC (1957) Pleiotropy, natural selection, and the evolution of senescence. Evolution 11: 398–411

Wongrakpanich S, Wongrakpanich A, Melhado K, Rangaswami J (2018) A comprehensive review of non-steroidal anti-inflammatory drug use in the elderly. Aging Dis 9(1):143

Yosef R, Pilpel N, Tokarsky-Amiel R, Biran A, Ovadya Y, Cohen S, Vadai E, Dassa L, Shahar E, Condiotti R, Ben-Porath I (2016) Directed elimination of senescent cells by inhibition of BCL-W and BCL-XL. Nat Commun 7(1):11190

Yu L, McPhee CK, Zheng L, Mardones GA, Rong Y, Peng J, Mi N, Zhao Y, Liu Z, Wan F, Hailey DW (2010) Autophagy termination and lysosome reformation regulated by mTOR. Nature 465(7300):942

Yun MH, Davaapil H, Brockes JP (2015) Recurrent turnover of senescent cells during regeneration of a complex structure. elife 4:e05505

Zhavoronkov A, Bhullar B (2015) Classifying aging as a disease in the context of ICD-11. Front Genet 6:326

# 16 Advantages and Disadvantages of Current Therapeutics and Phytochemicals for Age-Related Brain Disorders

Ahamed Basha Abdul Bari, Madhan Krishnan, and Shyamaladevi Babu

**Abstract**

Age-related brain disorders are a group of conditions characterised by the progressive loss of structure and function of neurons in the brain. These disorders primarily affect older individuals and include conditions such as Alzheimer's disease, Parkinson's disease, Huntington's disease, etc. Age-related brain disorders present significant challenges to individuals and healthcare professionals worldwide. Recent advancements in current therapeutics and phytochemicals play a crucial role in managing these ageing disorders of the brain. Current therapeutics have several advantages, like symptom management, modifying the underlying disease progression, established efficacy, and targeted treatment. On the other hand, it has several disadvantages, which include side effects, cost and accessibility, treatment complexity, etc. Similarly, the advantages of phytochemical usage in age-related brain disorders include that phytochemicals are derived from natural sources, they have significant antioxidant and anti-inflammatory properties, they exhibit potential neuroprotective effects, and they serve as a potential adjunct therapy. However, there is no universal standardisation or quality control, and they can interact with other medications, potentially affecting their efficacy or safety. Thus, considering the higher advantage when compared to the disadvantages regarding the current therapeutics and phytochemicals, it can be concluded that recent advancements in the treatment of age-related brain disorders provide promise for improved

A. B. A. Bari (✉)
Department of Physiology, Chettinad Hospital and Research Institute, Chettinad Academy of Research and Education, Kelambakkam, Tamil Nadu, India

M. Krishnan · S. Babu
Research, Chettinad Hospital and Research Institute, Chettinad Academy of Research and Education, Kelambakkam, India, Tamil Nadu



S. Pathak, A. Banerjee (eds.), *Neuroprotective Effects of Phytochemicals in Brain Ageing*, https://doi.org/10.1007/978-981-99-7269-2_16

health care in the future. Despite these limitations, recent advancements in both current therapeutics and phytochemical research offer promise for improved management of age-related brain disorders in the future. Further studies are crucial to fully understand the mechanisms of action of phytochemicals, establish optimal dosages, and develop standardized extracts to ensure consistent clinical outcomes.

**Keywords**

Ageing · Brain disorders · Phytochemicals

## 16.1 Introduction

Ageing causes several structural and physiochemical changes in living organisms, and these changes lead to a decrease in the functional capacity of almost all the organ systems in the body, with the brain being no exception (Li et al. 2021). The changes in the brain lead to the development of age-related brain disorders. It includes a variety of conditions that impair normal brain morphology and function. The primary causes of age-related brain disorders are due to alterations in brain neuronal structure, neurotransmitters, fluids, and other neuropsychological factors (Hou et al. 2019). Further, genetic predisposition exhibits a significant impact on the cognition, behaviour, emotions, and quality of life of the elderly (Brouwer et al. 2022). In the current decade, the incidence of age-related brain disorders is rising at an alarming rate.

Numerous studies have shown that beyond the age of 40, the volume and/or weight of the brain begin to drop at a rate of roughly 5% each decade, with the actual rate of loss probably rising with age, particularly after the age of 70 (Hedman et al. 2012). The shrinkage of the grey matter is commonly attributed to the loss of neuronal cells. The neurotransmitters that are most frequently mentioned in connection with ageing are dopamine and serotonin (Morgan et al. 1987). Dopamine levels start to decrease by about 10% every ten years after early adulthood, which has been associated with deficits in cognitive and motor abilities (Morgan 1987). It is probable that as individuals age, the frontal cortex-striatum dopaminergic connections lose some of their effectiveness. Additionally, receptor binding or synaptic density may decline in addition to the decrease in dopamine levels. Synaptic plasticity and neuronal damage in the adult brain may be connected to age-related reductions in serotonin and brain-derived neurotrophic factor levels (Bathina and Das 2015).

Additionally, age-related increases in monoamine oxidase, a chemical linked to neurotransmitter levels, have the potential to cause reactions that produce free radicals in excess and cause an imbalance in the body's antioxidant defence mechanism (Santin et al. 2021). The malfunctioning of the mitochondria, the generation of reactive oxygen species, and the dysregulation of calcium are further factors associated with brain ageing (Singh et al. 2019). The ageing brain may have altered glucose metabolism or a decreased input of glucose or oxygen as cerebrovascular efficiency falls, although the decrease in glucose may be caused more by atrophy than by any changes in glucose metabolism (Kapogiannis and Avgerinos 2020). White matter lesions and stroke are two additional frequent findings in aged brains linked with ischaemia, along with major changes in the vasculature (Mattson and Arumugam 2018). Further, several neurodegenerative diseases of the brain, such as Alzheimer's disease, Parkinson's disease, Huntington's disease, and frontotemporal lobar dementia, have been linked to ageing, and the elderly are more likely to develop these neurodegenerative diseases than other age groups (Hou et al. 2019). Over the years, research and medical progress have concentrated on developing therapeutics to slow down, halt, or perhaps even reverse the progression of these disorders. However, the current therapeutic management has its own pros and cons. In this chapter, recent therapeutic developments and their advancements in the area of common age-related brain disorders, their advantages and disadvantages, and their patient-related implications are discussed below.

## 16.2 Alzheimer's Disease

Alzheimer's disease, the most common kind of dementia, is distinguished by memory loss, cognitive deterioration, and poor reasoning (Soria Lopez et al. 2019). Available literature indicates that there is no known cure for Alzheimer's disease. However, there are several therapeutic approaches that aim to manage the symptoms, slow down the progression of the disease, and improve the quality of life for individuals with Alzheimer's (Cummings et al. 2019; Breijyeh and Karaman 2020). Promising current therapeutic options include cholinesterase inhibitors (Sharma 2019); medications such as donepezil, rivastigmine, and galantamine are commonly prescribed to manage cognitive symptoms by increasing levels of acetylcholine, which is a chemical involved in learning and memory, in the brain (Hansen et al. 2008). However, the disadvantages of their usage include that it may cause side effects like reduced appetite, nausea, gastric upset, headache, and insomnia (Marucci et al. 2021). The above medications are only used as medical management, not as a permanent cure.

Next, the N-methyl-D-aspartate receptor antagonist, memantine, is an FDA-approved medication that regulates the activity of glutamate, a neurotransmitter involved in learning and memory. It is often used in moderate to severe stages of Alzheimer's disease (Olivares et al. 2012; Wang and Reddy 2017; Grossberg et al. 2019). Thus, memantine may, according to studies, slightly postpone the decline in cognitive (mental) function. It is anticipated to be linked to higher quality-adjusted life years, a longer period before reliance and institutionalisation, and cheaper care costs. Further, the safety and tolerability of memantine are proven to be very effective. Adding to this, the development of amyloid-beta plaques in the brain is one of the underlying reasons that more recent treatment strategies aim to address (Busche and Hyman 2020). The therapeutics include anti-amyloid treatments, where monoclonal antibodies like aducanumab have demonstrated promise for removing amyloid-beta plaques from the brain (Tolar et al. 2020). These medications target and aid in the removal of harmful protein aggregates, potentially delaying the onset of disease. However, there are still numerous safety concerns about the effectiveness of the anti-amyloid in patients that need to be documented (van Dyck 2018).

Next, the tau proteins, which are linked to neuronal damage in Alzheimer's disease, are being investigated by researchers as potential therapeutics. Clinical trials for a number of tau-targeting medications are ongoing at various stages (Breijyeh and Karaman 2020). They cause microtubule stabilisation, i.e. one of the major jobs of tau proteins is to maintain the stability of microtubules, which are critical for maintaining the shape and movement of neurons. Microtubule organisation and integrity are maintained by tau proteins, which are essential for normal neuronal activity. Further, tau proteins support the establishment of healthy neural connections and aid in directing axon growth throughout brain development. They play a crucial role in the development and maintenance of neural networks. It also regulates signal transduction pathways inside neurons, and they can influence the activity of many signalling molecules via interactions, which may affect cellular functions, including synaptic plasticity, neuronal survival, and neurotransmitter release. It also transports several cellular components to ensure optimal distribution inside the cell; they aid in the movement of vesicles, organelles, and other necessary molecules along the microtubule pathways. Thus, tau proteins are thought to have a role in synaptic plasticity, which is important for learning and memory processes. They support cognitive processes by preserving the form and functionality of synapses and the connections between neurons (Avila et al. 2004; Pîrşcoveanu et al. 2017; Gao et al. 2018).

Adding to this, supportive therapies include non-drug approaches such as cognitive stimulation, occupational therapy (Piersol et al. 2018), and psychosocial interventions that can help individuals with Alzheimer's disease maintain their cognitive abilities, enhance communication, and manage behavioural and psychological symptoms (Khalsa 2015; Gómez Gallego and Gómez García 2017). Thus, it is crucial for individuals with Alzheimer's disease to work closely with healthcare professionals to develop a personalised treatment plan that suits their needs. Additionally, staying mentally and physically active, eating a healthy diet, and maintaining social connections may also contribute to overall well-being and potentially help manage the symptoms of the disease.

## 16.3 Parkinson's Disease

Parkinson's disease is a neurodegenerative disorder that manifests as tremors, bradykinesia, and rigidity in the muscles (Reich and Savitt 2019; Cabreira and Massano 2019). New therapeutic developments aim to protect brain neurons and improve dopaminergic function (Elsworth 2020). The primary medical management is levodopa, and dopamine agonists are the most prevalent drugs (Bloem et al. 2021). However, prolonged levodopa usage may cause motor fluctuations or wearing-off or on-off events (Hansen et al. 2022). Each dosage may provide shorter, less predictable symptom relief over time. This may cause periods of excellent (on) and poor (off) symptom management. Further, long-term levodopa medication may cause uncontrollable movements (Hansen et al. 2022). Choreic or dystonic movements may affect the face, torso, or limbs. Moreover, levodopa may produce nausea, vomiting, and appetite loss (Pfeiffer et al. 2020). These adverse effects may worsen early in therapy but improve afterwards. Adjusting the dose or taking the drug with meals might reduce these effects. Levodopa may also cause mood swings, anxiety, hallucinations, and disorientation. Higher dosages and a psychiatric history are associated with these outcomes. Further, levodopa-treated Parkinson's patients may develop compulsive gambling, shopping, or hypersexuality. These behaviours may necessitate dosage modifications or extra drugs due to brain dopamine alterations (Lenka et al. 2022). Further, monoamine oxidase B inhibitors, catechol-O-methyltransferase inhibitors, and anticholinergics may also be administered to increase dopamine levels or relieve symptoms (Connolly and Lang 2014; Gray et al. 2022). They found them beneficial in alleviating the majority of the symptoms of Parkinson's disease. On the other hand, the evidence supporting the above for motor and nonmotor aspects is not nearly as well established.

The next therapeutic option is gene therapy. Scientists are investigating gene therapies to deliver particular genes into affected brain regions in an effort to protect or restore dopaminergic neurons (Merola et al. 2020). Preclinical research and early-stage clinical trials have demonstrated the potential of this strategy. But again, gene therapy alters cell function by inserting genetic material. Unintended repercussions are always possible. Long-term gene therapy safety requires substantial preclinical and clinical trials (Abeliovich et al. 2021). Further, gene therapy is now being investigated for advanced Parkinson's patients with motor fluctuations and dyskinesias. It may not benefit all Parkinson's patients, especially those with fewer symptoms. Gene therapy involves the efficient and focused delivery of genetic material to brain regions. The brain is complex and protective, making this difficult. Research is focused on developing delivery technologies that accurately and consistently target damaged brain areas. Additionally, gene therapy delivery techniques like viral vectors may cause immunological reactions (Björklund and Davidsson 2021). The vectors may be recognised as foreign by the immune system and cause unpleasant effects or reduce therapeutic efficacy. To add to this, early gene therapy clinical trials showed encouraging results, but its long-term efficacy and durability need to be assessed. Gene therapy may wear off over time, requiring recurrent treatment.

The other advancement in management includes deep brain stimulation, which modifies abnormal neural activity by implanting electrodes in particular brain regions and delivering electrical stimulation (França et al. 2022). Recent improvements in programming and electrode design have enhanced treatment results and minimised side effects (Malek 2019; Kogan et al. 2019; Krauss et al. 2021). It also reduces the dose and frequency of levodopa. This reduces drug adverse effects such as dyskinesias and motor irregularities. Its advantages are maximised with long-term follow-up and supervision. However, it involves implanting brain electrodes surgically. Infection, haemorrhage, and anaesthesia side effects are dangers of deep brain stimulation, as seen in every operation (Ward et al. 2021).

Electrode implantation may also harm adjacent brain regions. Further, it needs precise programming and adjustments to regulate symptoms. Trial and error, many programming sessions, and follow-up visits are involved. Some individuals need continual programming tweaks as their condition evolves. Mild speech or language impairments, dystonia, and tingling or soreness at the stimulation site are common adverse effects. Infection, cognitive abnormalities, and increasing symptoms are uncommon but possible. Thus, it is not appropriate for all Parkinson's disease patients; a neurologist or movement disorder expert must evaluate them, and it should be chosen depending on personal choices, requirements, risks, and advantages.

## 16.4 Age-Associated Stroke

The term age-related stroke is used to characterise how the risk of stroke increases with age. Interruption or reduction of blood flow to the brain can result in a stroke, which can damage or destroy brain cells. The incidence of stroke increases with age, making ageing one of the most significant risk factors for stroke (Beishon et al. 2021). Age-related stroke treatment varies based on the type of stroke, severity of symptoms, location of the cerebral injury, and general health (Kalaria et al. 2016; Baltan et al. 2019). Early treatment for a stroke may reduce brain injury and enhance outcomes. Intravenous plasminogen activator is the drug of preference for removing the blockage and resuming blood flow in cases of acute age-related ischemic stroke (Alberts 2017; Kim et al. 2017). However, it may not be suitable for all elderly people. Due to the risk of bleeding complications, plasminogen activator therapy may not be appropriate for patients with recent significant surgery, bleeding disorders, uncontrolled high blood pressure, or severe trauma. In specific cases, mechanical thrombectomy (Blanc et al. 2020), in which the clot is physically removed using a specialised catheter, is the next treatment option. Mechanical thrombectomy is safe, but complications may arise. These include vascular injury, bleeding, and damage to cerebral tissue. Anticoagulants may also increase the risk of haemorrhaging during surgery. In addition, after mechanical thrombectomy, patients need continuous monitoring and treatment for complications related to stroke. This may necessitate interdisciplinary care coordination to enhance outcomes.

## 16.5 Brain Tumours in the Elderly

Brain tumours can affect individuals of any age, including the elderly. Nevertheless, elderly individuals are more likely to be diagnosed with specific types of brain tumours. There are some crucial aspects of brain tumours in older people. Older people are more likely to be diagnosed with glioblastoma meningioma and primary central nervous system lymphoma (Nayak and Iwamoto 2010). Age, general health, and the type of tumour all play a role in determining how to treat brain tumours in geriatric patients.

Radiation therapy is one of the current choices of treatment (Minniti et al. 2017), which entails the use of high-energy radiation to kill or inhibit the development of cancer cells. However, radiation therapy can cause a variety of side effects. Typical side effects include fatigue, skin changes such as erythema, dehydration, or irritation, hair loss in the treated area, vertigo, and digestive issues. The elderly patient's quality of life during treatment may be substantially impacted by these adverse effects, even though they are typically transient and improve over time after treatment is discontinued.

Chemotherapy drugs like temozolomide, carmustine, lomustine, procarbazine, methotrexate, cisplatin, and carboplatin are the next choices (Omuro and Abrey 2006). This is especially advantageous for tumours that have migrated to the brain from other regions of the body (brain metastases) or for tumours that have diffusely infiltrated the brain. However, it has numerous adverse effects, some of which can be extremely severe. The exact medications used, the dosage, and the individual's response all influence these adverse effects differently. Typical adverse effects include fatigue, nausea, vomiting, hair loss, decreased appetite, mouth ulcers, diarrhoea or constipation, and increased susceptibility to infections (Omuro and Abrey 2006).

The other therapeutic option is brain tumour surgery (Vogelbaum et al. 2021); however, risks associated with this procedure in the elderly population include fitness for anaesthesia, bleeding, infection, and unfavourable medication reactions. The surrounding brain tissue, blood vessels, or nerves may be harmed as a result of brain surgery, among other negative effects. There might be post-operative bleeding, blood clots, oedema, infection, or cerebrospinal fluid leakage, among other concerns. Neurological problems may also show up as changes in speech, motor control, or cognition, depending on the location and extent of the surgery.

Another cutting-edge method for treating brain tumours in older adults is target therapy. Epidermal growth factor receptor inhibitors, for instance, have been studied in glioblastoma, the most aggressive brain tumour (He et al. 2021). Studies using its single-agent inhibitor, however, have not shown any survival improvements. Results could be enhanced by drug combinations or patient subgroups with specific genetic disorders. Additionally, v-raf murine sarcoma viral oncogene homolog B1 gene alterations may be present in primary brain tumours. In these circumstances, medications like vemurafenib and dabrafenib have been successfully identified as inhibitors (Szklener et al. 2022). However, in non-mutated forms of the tumours, it may result in treatment resistance and little effectiveness. Additionally, bevacizumab, an anti-angiogenic drug, improves symptoms and progression-free survival in certain elderly people with recurrent glioblastoma. Age-related factors are thus crucial in the treatment of brain tumours. In comparison with younger patients, older patients may need a different treatment strategy for brain cancer, taking into consideration aspects including general health, functional level, and possible treatment-related hazards.

## 16.6 Age-Related Cognitive Decline and Other Neuropsychological Changes

Cognitive impairment brought on by ageing is known as age-related cognitive impairment (Jafari et al. 2021). The advancement of brain health and the enhancement of cognitive function are the major objectives of therapeutic treatments. New developments include brain training software, and computerised cognitive exercises are being created for cognitive training in order to improve memory, attention, and executive skills. Studies have shown that these programmes may be useful in reducing cognitive ageing, and they employ adaptive learning algorithms to customise training to each participant's requirements (Chen et al. 2022). Additionally, a number of lifestyle factors, such as physical exercise, a healthy diet, social engagement, and mental stimulation, have been linked to enhanced cognitive performance and brain health (Erickson et al. 2022). In order to promote healthy ageing and perhaps reduce the risk of cognitive decline as individuals age, it is essential to adopt a holistic strategy, according to recent research (Mintzer et al. 2019). But instead of addressing the underlying reasons, the present approach to treating cognitive decline focuses on symptom management. They may not be able to stop cognitive deterioration entirely. A loss in processing speed, which is the amount of time it takes to detect and react to information, is a common occurrence in older people in addition to cognitive impairment. For older people, it could take longer to complete tasks that call for fast thinking, decision-making, and response times. In addition, older individuals could be more prone to interference, which makes it difficult for them to block out extraneous information or distractions while concentrating on a job. This might have an impact on how well people pay attention and remember things. Additionally, as individuals age, their ability to hold knowledge in working memory tends to decline, making it harder to retain and process new information at the same time. In order to lessen the effects of other age-related neuropsychological changes, one should maintain a healthy lifestyle, take part in cognitively challenging activities, and maintain social engagement. A better understanding of a person's cognitive health and any potential issues may be gained by regular cognitive evaluations and talks with healthcare specialists.

## 16.7 Amyotrophic Lateral Sclerosis

Amyotrophic lateral sclerosis, or Lou Gehrig's disease, is a rare, progressive neurodegenerative condition that destroys nerve cells in the ageing brain and spinal cord (Dandaba et al. 2017). Any age may experience it, although older people are more susceptible. The motor neurons in charge of regulating voluntary muscular movements are its main targets. Although the earliest signs might be different, they often involve muscular twitching, weakness, and trouble breathing, speaking, or swallowing. People may have gradual muscular atrophy, loss of motor function, and eventually paralysis as the condition worsens. Despite the fact that there is no cure, current therapeutic care mainly focuses on controlling symptoms, halting the course of the illness, and enhancing quality of life. FDA-approved medications like riluzole and edaravone may help halt the spread of the condition (Jaiswal 2019). Riluzole and edaravone, however, have shown advantages in patients by reducing the course of illness (Oskarsson et al. 2018; Chiò et al. 2020; Witzel et al. 2022). Effects might vary, and not every patient will have access to effective treatment alternatives. Additionally, a number of treatments, including respiratory support, occupational therapy, speech therapy, and physical therapy, may aid with symptom management, mobility improvement, and quality of life improvement. Wheelchairs, communication aids, and respiratory support equipment are examples of assistive technologies that may be used to help with everyday living. However, they solely concentrate on alleviating symptoms, delaying the onset of illness, and enhancing quality of life. These therapies do not provide a permanent cure, but they may help reduce certain symptoms and lengthen survival.

## 16.8 Huntington's Disease

Huntington's disease is an inherited genetic disorder that causes the progressive degeneration of nerve cells in the brain. Movement, cognition, and behavioural skills all deteriorate along with physical and mental ability. To control some of its symptoms, a variety of drugs are advised. Tetrabenazine (Caron et al. 1993), for instance, is a dopamine-depleting medication that may help lessen chorea or uncontrollable movements. Tetrabenazine has various drawbacks and possible side effects, including drowsiness, exhaustion, sadness, and suicidal thoughts, even though it may be useful in controlling the symptoms of Huntington's disease. While antipsychotic drugs may be used to treat mental symptoms like agitation, hostility, or hallucinations, antidepressants can be used to treat mood problems brought on by the illness. The need for palliative care, support, physical therapy, and occupational therapy must be included. In order to create prospective disease-modifying therapeutics for Huntington's disease, continuing research is being carried out. A more thorough analysis is required to establish the efficacy and safety of novel experimental therapies that try to address the disease's underlying genetic aetiology.

## 16.9 Diet and Life Style Modification in Age-Related Brain Disorders

Studies showing that a diet high in energy and low in antioxidants is a risk factor, as well as studies showing that energy restriction may lengthen life, reduce oxidative damage, or protect against cognitive decline, suggest that diet may play a role in biological ageing and the emergence of cognitive decline (McGrattan et al. 2019). Increased consumption of fish and shellfish, even if only once a month, may be advantageous and reduce the risk of stroke. Exercise has other benefits as well; studies have shown that it enhances executive function and even delays the usual reduction in the density of white and grey matter that occurs as we age. The aforementioned research shows that biological and chronological ages do not always correspond, particularly when it comes to cardiovascular risk factors. Reduced cardiovascular risk may be linked to younger biological ages and vice versa (Caron et al. 1993).

## 16.10 Phytochemicals in Age-Related Brain Disorders

Conventional treatment options, as discussed above, have limitations and often focus on symptom management to a greater extent than completely addressing the underlying disease processes. In order to combat the onset and development of age-related brain disorders, there is growing interest in discovering natural substances (phytochemicals) that can target a variety of pathways.

Phytochemicals have gained attention for their potential neuroprotective effects and their ability to modulate various pathological processes. These substances have demonstrated a wide range of biological potential activities, such as anti-inflammatory, antioxidant, and anti-amyloid properties. Age-related brain disorders may be prevented or delayed in their progression by phytochemicals by targeting oxidative stress, neuroinflammation, protein misfolding, and other pathological features and alleviating their associated symptoms. Natural compounds derived from plants, such as polyphenols, flavonoids, non-flavonoids, isoprenoid derivatives, alkaloids, and steroidal lactones, have gained attention for their potential roles in combating age-related brain disorders. Hence, this chapter additionally examines the therapeutic roles of the major bioactive compounds from plant sources and their advantages and disadvantages in age-related brain disorders.

Following are the major group of phytochemicals, and how they affect various ailments of age-related brain disorders is discussed below.

## 16.11 Polyphenols

Polyphenols are a diverse group of bioactive compounds found abundantly in fruits, vegetables, tea, cocoa, and other plant-based sources. They exhibit antioxidant and anti-inflammatory properties, which contribute to their neuroprotective effects (Scapagnini et al. 2010). Polyphenols, such as resveratrol, curcumin, and catechins, have shown promising results in reducing oxidative stress, combating neuroinflammation, and modulating signalling pathways involved in neuroprotection (Rajaram et al. 2019; Stacchiotti and Corsetti 2020). These compounds also demonstrate the potential in reducing protein aggregation, promoting synaptic plasticity, and enhancing cognitive function (Kim et al. 2010; Sawikr et al. 2017).

## 16.12 Flavonoids

Flavonoids, a subclass of polyphenols, are widely distributed in plants and have been extensively studied for their neuroprotective effects. Quercetin, hesperidin, and epicatechin are examples of flavonoids that exhibit antioxidant, anti-inflammatory, and neuroprotective properties. They scavenge free radicals, reduce neuroinflammation, and modulate various signalling pathways in the brain (Graf et al. 2005). Flavonoids have also shown potential in enhancing synaptic plasticity, improving cognitive function, and inhibiting the formation of beta-amyloid plaques and tau protein tangles associated with Alzheimer's disease (Airoldi et al. 2018).

## 16.13 Non-flavonoids

Non-flavonoids, another group of polyphenolic compounds, have demonstrated neuroprotective effects in age-related brain disorders. Stilbenes, including resveratrol, exhibit antioxidant and anti-inflammatory properties, protecting against oxidative stress and neuroinflammation. They also enhance mitochondrial function and promote neuronal survival (Doré 2005; Varesi et al. 2023). Other non-flavonoids, such as ellagic acid and tannins, have shown potential in reducing neurodegenerative processes and improving cognitive function (Bui and Nguyen 2017; Gupta et al. 2021).

## 16.14 Isoprenoid Derivatives

Isoprenoid derivatives, including terpenoids and their derivatives, are bioactive compounds found in various plants. Ginkgolides and bilobalide from Ginkgo biloba extract exhibit antioxidant and anti-inflammatory effects, protecting neurons from oxidative damage and reducing neuroinflammation. Isoprenoid derivatives also enhance cerebral blood flow, improve cognitive function, and modulate neurotransmitter systems, offering potential therapeutic benefits in age-related brain disorders (Natarajan et al. 2013; Pelleieux et al. 2018).

## 16.15 Alkaloids

Alkaloids, naturally occurring nitrogen-containing compounds found in plants, have shown promise for age-related brain disorders. Vincamine, derived from Vinca minor, enhances cerebral blood flow and cognitive function (Moglia et al. 1984). Vinpocetine, a derivative of vincamine, exhibits antioxidant and anti-inflammatory effects, protects against oxidative stress, and enhances memory and cognitive function (Gulyás et al. 2011). Alkaloids such as huperzine A, derived from Huperzia serrata, demonstrate acetylcholinesterase inhibitory activity, leading to increased acetylcholine levels and improved cognitive function (Yang et al. 2013; Friedli and Inestrosa 2021).

## 16.16 Steroidal Lactones

Steroidal lactones, also known as withanolides, are bioactive compounds predominantly found in Ashwagandha (Withania somnifera) and other plants. Withanolides exhibit antioxidant and anti-inflammatory properties, protecting neurons from oxidative damage and reducing neuroinflammation. They also modulate signalling pathways involved in neuroprotection, synaptic plasticity, and cognitive function. Withanolides have shown potential in reducing beta-amyloid accumulation and improving memory and cognitive function in age-related brain disorders (Das et al. 2021; Shukla et al. 2023).

## 16.17 Disadvantages of Phytochemicals in Age-Related Brain Disorders

Phytochemicals may be beneficial for age-related brain diseases, but it's vital to weigh any possible risks as well. Here are a few things to think about: One important disadvantage is limited research. Our knowledge of the processes behind and the efficacy of phytochemicals in the treatment of age-related brain diseases is still developing. Numerous studies use animal or cell models, and the conclusions drawn from these studies may not necessarily apply to humans. Next, bioavailability and absorption: Because phytochemicals may not be fully absorbed or used by the body, they may have a restricted bioavailability. The medicinal potential of several phytochemicals may be hampered by their quick metabolism or poor absorption rates. Further, phytochemicals may interact negatively with pharmaceuticals or cause adverse effects. For instance, some phytochemicals may suppress or stimulate the activity of drug-metabolising enzymes, which might have an impact on the effectiveness or safety of drugs. Next, due to individual variations due to differences in metabolism, heredity, and general health state, humans might react to phytochemicals differently. It's possible that what works for one person won't work the same way for another, and some people may react negatively or have allergies to certain phytochemicals. Additionally, regarding standardisation and quality control, there are several variations in phytochemical supplement manufacture and composition. To guarantee constant quantities of active chemicals and reduce impurities, phytochemical products must be standardised and subject to quality control. Further, the disease complexity of Alzheimer's disease and other age-related brain illnesses is multiple ailments with intricate underpinning processes. While phytochemicals may focus on certain symptoms of various illnesses, they may not cover the whole range of associated pathological processes. While short-term studies may suggest benefits, long-term safety data on the use of phytochemical supplements is often limited. There's a need to investigate whether prolonged, high-dose use presents any unforeseen risks in the context of brain aging. Some phytochemicals can act as pro-oxidants in certain circumstances, potentially causing cellular damage rather than offering protection. A thorough understanding of a phytochemical's context-dependent actions is crucial for safe application. Plant foods contain complex mixtures of phytochemicals. Isolating individual compounds in supplement form might eliminate beneficial synergies or create unintended imbalances in their effects. Phytochemical supplements often fall into a less strictly regulated category compared to pharmaceuticals. Variations in quality control and potential contamination raise concerns about safety and efficacy across different products. The optimal dosage of phytochemicals for brain health is still under investigation. While low doses might have beneficial effects, higher doses could potentially become toxic. The concept of hormesis, where a compound is beneficial at low doses but harmful at high doses, might play a role in understanding the effects of phytochemicals on the brain. The potential benefits and risks of phytochemicals may vary depending on the specific age-related brain disorder. For example, a phytochemical that shows promise in reducing inflammation associated with Alzheimer's disease might not be as relevant for Parkinson's disease, where different pathways are involved. Phytochemicals should be viewed as one potential part of a multifaceted approach to promoting brain health. Combining phytochemicals with a balanced diet, regular exercise, cognitive stimulation, stress management, and sufficient sleep creates a more robust foundation for maintaining a healthy brain as we age. While the potential of phytochemicals is exciting, it's crucial to have realistic expectations. Instead of a "miracle cure," it's helpful to think of phytochemicals as tools that can contribute to overall brain health when used responsibly in conjunction with other healthy lifestyle practices. Thus, it is crucial to proceed cautiously when using phytochemicals to treat age-related brain diseases, and for optimal brain health, it is often advised to combine current therapeutic management with a balanced diet full of a range of whole foods high in phytochemicals and with other lifestyle choices, including consistent physical activity, mental stimulation, and social interaction.

## 16.18 Conclusion

Recent developments in the treatment of age-related brain disorders give hope for better outcomes. It is obvious that the brain changes with increasing chronological age; however, the rate of change, the biological age of the brain, and the processes involved are less clear. Brain changes that can affect cognition and behaviour occur at the molecular, intercellular, and intracellular levels, as well as at the tissue and organ levels. Many areas of research are being investigated in order to understand the mechanisms of ageing and to try to alleviate age-related disorders, particularly dementias, which have the greatest impact on the population. In terms of personal brain ageing, studies indicate that a healthy lifestyle that lowers cardiovascular risk benefits the brain as well. Researchers are making great strides towards comprehending and treating these challenging conditions, from focusing on the underlying disease mechanisms to investigating gene therapies and using non-invasive methods like deep brain stimulation. Although there are still difficulties, the potential for advances in the coming years holds promise for better management and possibly slowing the progression of age-related brain disorders, ultimately improving the quality of life for affected individuals. Future research must fully account for these elements, and "cross-sequential" studies, which combine cross-sectional and longitudinal research, have been suggested. It is evident that while our knowledge of the ageing brain continues to advance, much more research is still needed, which is crucial given the prevalence of the elderly in society and the potential severity of their cognitive decline. In the future, randomised controlled trials of therapeutic interventions may, where appropriate, pave the way for greater comprehension.

Additionally. polyphenols, flavonoids, non-flavonoids, isoprenoid derivatives, alkaloids, and steroidal lactones derived from plants offer potential benefits for age-related brain disorders. Their antioxidant, anti-inflammatory, and neuroprotective properties, along with their effects on signalling pathways, synaptic plasticity, and cognitive function, make them attractive candidates for therapeutic interventions. However, further research is needed to optimise dosages, evaluate safety profiles, and explore potential synergistic effects with conventional treatments, ultimately facilitating their translation into clinical practice.

## References

Abeliovich A, Hefti F, Sevigny J (2021) Gene therapy for Parkinson's disease associated with GBA1 mutations. J Parkinsons Dis 11:S183–S188. https://doi.org/10.3233/JPD-212739

Airoldi C, La Ferla B, Orazio DG et al (2018) Flavonoids in the treatment of Alzheimer's and other neurodegenerative diseases. Curr Med Chem 25:3228–3246. https://doi.org/10.2174/0929867325666180209132125

Alberts MJ (2017) Stroke treatment with intravenous tissue-type plasminogen activator: more proof that time is brain. Circulation 135:140–142

Avila J, Lucas JJ, Perez M, Hernandez F (2004) Role of tau protein in both physiological and pathological conditions. Physiol Rev 84:361–384. https://doi.org/10.1152/physrev.00024.2003

Baltan S, Shi Y, Keep RF, Chen J (2019) The effect of aging on brain injury and recovery after stroke. Neurobiol Dis 126:1–2

Bathina S, Das UN (2015) Brain-derived neurotrophic factor and its clinical implications. Arch Med Sci 11:1164–1178. https://doi.org/10.5114/aoms.2015.56342

Beishon L, Clough RH, Kadicheeni M et al (2021) Vascular and haemodynamic issues of brain ageing. Pflugers Arch 473:735–751. https://doi.org/10.1007/s00424-020-02508-9

Björklund T, Davidsson M (2021) Next-generation gene therapy for Parkinson's disease using engineered viral vectors. J Parkinsons Dis 11:S209–S217. https://doi.org/10.3233/JPD-212674

Blanc R, Escalard S, Baharvadhat H et al (2020) Recent advances in devices for mechanical thrombectomy. Expert Rev Med Devices 17:697–706. https://doi.org/10.1080/17434440.2020.1784004

Bloem BR, Okun MS, Klein C (2021) Parkinson's disease. Lancet (London, England) 397:2284–2303. https://doi.org/10.1016/S0140-6736(21)00218-X

Breijyeh Z, Karaman R (2020) Comprehensive review on Alzheimer's disease: causes and treatment. Molecules 25. https://doi.org/10.3390/molecules25245789

Brouwer RM, Klein M, Grasby KL et al (2022) Genetic variants associated with longitudinal changes in brain structure across the lifespan. Nat Neurosci 25:421–432. https://doi.org/10.1038/s41593-022-01042-4

Bui TT, Nguyen TH (2017) Natural product for the treatment of Alzheimer's disease. J Basic Clin Physiol Pharmacol 28:413–423. https://doi.org/10.1515/jbcpp-2016-0147

Busche MA, Hyman BT (2020) Synergy between amyloid-β and tau in Alzheimer's disease. Nat Neurosci 23:1183–1193. https://doi.org/10.1038/s41593-020-0687-6

Cabreira V, Massano J (2019) Parkinson's disease: clinical review and update. Acta Medica Port 32:661–670. https://doi.org/10.20344/amp.11978

Caron NS, Wright GEB, Hayden MR (1993) Huntington disease. 1998 Oct 23 [updated 2020 Jun 11]. In: Adam MP, Feldman J, Mirzaa GM, Pagon RA, Wallace SE, Bean LJH, Gripp KW, Amemiya A (eds) GeneReviews® [Internet]. University of Washington, Seattle, WA. https://pubmed.ncbi.nlm.nih.gov/20301482/

Chen L, Jiao J, Zhang Y (2022) Therapeutic approaches for improving cognitive function in the aging brain. Front Neurosci 16:1060556. https://doi.org/10.3389/fnins.2022.1060556

Chiò A, Mazzini L, Mora G (2020) Disease-modifying therapies in amyotrophic lateral sclerosis. Neuropharmacology 167:107986. https://doi.org/10.1016/j.neuropharm.2020.107986

Connolly BS, Lang AE (2014) Pharmacological treatment of Parkinson disease: a review. JAMA 311:1670–1683. https://doi.org/10.1001/jama.2014.3654

Cummings JL, Tong G, Ballard C (2019) Treatment combinations for Alzheimer's disease: current and future pharmacotherapy options. J Alzheimers Dis 67:779–794. https://doi.org/10.3233/JAD-180766

Dandaba M, Couratier P, Labrunie A et al (2017) Characteristics and prognosis of oldest old subjects with amyotrophic lateral sclerosis. Neuroepidemiology 49:64–73. https://doi.org/10.1159/000479969

Das R, Rauf A, Akhter S et al (2021) Role of withaferin a and its derivatives in the Management of Alzheimer's disease: recent trends and future perspectives. Molecules 26. https://doi.org/10.3390/molecules26123696

Doré S (2005) Unique properties of polyphenol stilbenes in the brain: more than direct antioxidant actions; gene/protein regulatory activity. Neurosignals 14:61–70. https://doi.org/10.1159/000085386

van Dyck CH (2018) Anti-amyloid-β monoclonal antibodies for Alzheimer's disease: pitfalls and promise. Biol Psychiatry 83:311–319. https://doi.org/10.1016/j.biopsych.2017.08.010

Elsworth JD (2020) Parkinson's disease treatment: past, present, and future. J Neural Transm 127:785–791. https://doi.org/10.1007/s00702-020-02167-1

Erickson KI, Donofry SD, Sewell KR et al (2022) Cognitive aging and the promise of physical activity. Annu Rev Clin Psychol 18:417–442. https://doi.org/10.1146/annurev-clinpsy-072720-014213

França C, Carra RB, Diniz JM et al (2022) Deep brain stimulation in Parkinson's disease: state of the art and future perspectives. Arq Neuropsiquiatr 80:105–115. https://doi.org/10.1590/0004-282X-ANP-2022-S133

Friedli MJ, Inestrosa NC (2021) Huperzine a and its neuroprotective molecular Signaling in Alzheimer's disease. Molecules 26. https://doi.org/10.3390/molecules26216531

Gao Y, Tan L, Yu J-T, Tan L (2018) Tau in Alzheimer's disease: mechanisms and therapeutic strategies. Curr Alzheimer Res 15:283–300. https://doi.org/10.2174/1567205014666170417111859

Gómez Gallego M, Gómez García J (2017) Music therapy and Alzheimer's disease: cognitive, psychological, and behavioural effects. Neurologia 32:300–308. https://doi.org/10.1016/j.nrl.2015.12.003

Graf BA, Milbury PE, Blumberg JB (2005) Flavonols, flavones, flavanones, and human health: epidemiological evidence. J Med Food 8:281–290. https://doi.org/10.1089/jmf.2005.8.281

Gray R, Patel S, Ives N et al (2022) Long-term effectiveness of adjuvant treatment with catechol-O-methyltransferase or monoamine oxidase B inhibitors compared with dopamine agonists among patients with Parkinson disease uncontrolled by levodopa therapy: the PD MED randomized clinical trial. JAMA Neurol 79:131–140. https://doi.org/10.1001/jamaneurol.2021.4736

Grossberg GT, Tong G, Burke AD, Tariot PN (2019) Present algorithms and future treatments for Alzheimer's disease. J Alzheimers Dis 67:1157–1171. https://doi.org/10.3233/JAD-180903

Gulyás B, Vas A, Tóth M et al (2011) Age and disease related changes in the translocator protein (TSPO) system in the human brain: positron emission tomography measurements with [11C] vinpocetine. NeuroImage 56:1111–1121. https://doi.org/10.1016/j.neuroimage.2011.02.020

Gupta A, Singh AK, Kumar R et al (2021) Neuroprotective potential of ellagic acid: a critical review. Adv Nutr 12:1211–1238. https://doi.org/10.1093/advances/nmab007

Hansen CA, Miller DR, Annarumma S et al (2022) Levodopa-induced dyskinesia: a historical review of Parkinson's disease, dopamine, and modern advancements in research and treatment. J Neurol 269:2892–2909. https://doi.org/10.1007/s00415-022-10963-w

Hansen RA, Gartlehner G, Webb AP et al (2008) Efficacy and safety of donepezil, galantamine, and rivastigmine for the treatment of Alzheimer's disease: a systematic review and meta-analysis. Clin Interv Aging 3:211–225

He Y, Li J, Koga T et al (2021) Epidermal growth factor receptor as a molecular determinant of glioblastoma response to dopamine receptor D2 inhibitors. Neuro-Oncology 23:400–411. https://doi.org/10.1093/neuonc/noaa188

Hedman AM, van Haren NEM, Schnack HG et al (2012) Human brain changes across the life span: a review of 56 longitudinal magnetic resonance imaging studies. Hum Brain Mapp 33:1987–2002. https://doi.org/10.1002/hbm.21334

Hou Y, Dan X, Babbar M et al (2019) Ageing as a risk factor for neurodegenerative disease. Nat Rev Neurol 15:565–581. https://doi.org/10.1038/s41582-019-0244-7

Jafari Z, Kolb BE, Mohajerani MH (2021) Age-related hearing loss and cognitive decline: MRI and cellular evidence. Ann N Y Acad Sci 1500:17–33. https://doi.org/10.1111/nyas.14617

Jaiswal MK (2019) Riluzole and edaravone: a tale of two amyotrophic lateral sclerosis drugs. Med Res Rev 39:733–748. https://doi.org/10.1002/med.21528

Kalaria RN, Akinyemi R, Ihara M (2016) Stroke injury, cognitive impairment and vascular dementia. Biochim Biophys Acta 1862:915–925. https://doi.org/10.1016/j.bbadis.2016.01.015

Kapogiannis D, Avgerinos KI (2020) Brain glucose and ketone utilization in brain aging and neurodegenerative diseases. Int Rev Neurobiol 154:79–110. https://doi.org/10.1016/bs.irn.2020.03.015

Khalsa DS (2015) Stress, meditation, and Alzheimer's disease prevention: where the evidence stands. J Alzheimers Dis 48:1–12. https://doi.org/10.3233/JAD-142766

Kim J, Lee HJ, Lee KW (2010) Naturally occurring phytochemicals for the prevention of Alzheimer's disease. J Neurochem 112:1415–1430. https://doi.org/10.1111/j.1471-4159.2009.06562.x

Kim J-T, Fonarow GC, Smith EE et al (2017) Treatment with tissue plasminogen activator in the Golden hour and the shape of the 4.5-hour time-benefit curve in the National United States get with the Guidelines-Stroke Population. Circulation 135:128–139. https://doi.org/10.1161/CIRCULATIONAHA.116.023336

Kogan M, McGuire M, Riley J (2019) Deep brain stimulation for Parkinson disease. Neurosurg Clin N Am 30:137–146. https://doi.org/10.1016/j.nec.2019.01.001

Krauss JK, Lipsman N, Aziz T et al (2021) Technology of deep brain stimulation: current status and future directions. Nat Rev Neurol 17:75–87. https://doi.org/10.1038/s41582-020-00426-z

Lenka A, Di Maria G, Lamotte G et al (2022) Practical pearls to improve the efficacy and tolerability of levodopa in Parkinson's disease. Expert Rev Neurother 22:489–498. https://doi.org/10.1080/14737175.2022.2091436

Li Z, Zhang Z, Ren Y et al (2021) Aging and age-related diseases: from mechanisms to therapeutic strategies. Biogerontology 22:165–187. https://doi.org/10.1007/s10522-021-09910-5

Malek N (2019) Deep brain stimulation in Parkinson's disease. Neurol India 67:968–978. https://doi.org/10.4103/0028-3886.266268

Marucci G, Buccioni M, Ben DD et al (2021) Efficacy of acetylcholinesterase inhibitors in Alzheimer's disease. Neuropharmacology 190:108352. https://doi.org/10.1016/j.neuropharm.2020.108352

Mattson MP, Arumugam TV (2018) Hallmarks of brain aging: adaptive and pathological modification by metabolic states. Cell Metab 27:1176–1199. https://doi.org/10.1016/j.cmet.2018.05.011

McGrattan AM, McGuinness B, McKinley MC et al (2019) Diet and inflammation in cognitive ageing and Alzheimer's disease. Curr Nutr Rep 8:53–65. https://doi.org/10.1007/s13668-019-0271-4

Merola A, Van Laar A, Lonser R, Bankiewicz K (2020) Gene therapy for Parkinson's disease: contemporary practice and emerging concepts. Expert Rev Neurother 20:577–590. https://doi.org/10.1080/14737175.2020.1763794

Minniti G, Filippi AR, Osti MF, Ricardi U (2017) Radiation therapy for older patients with brain tumors. Radiat Oncol 12:101. https://doi.org/10.1186/s13014-017-0841-9

Mintzer J, Donovan KA, Kindy AZ et al (2019) Lifestyle choices and brain health. Front Med 6:204. https://doi.org/10.3389/fmed.2019.00204

Moglia A, Alfonsi E, Zandrini C et al (1984) Acute i.v. vincamine teprosilate administration: quantified investigation in elderly subjects. Int J Clin Pharmacol Res 4:303–306

Morgan DG (1987) The dopamine and serotonin systems during aging in human and rodent brain. A brief review. Prog Neuro-Psychopharmacol Biol Psychiatry 11:153–157. https://doi.org/10.1016/0278-5846(87)90053-4

Morgan DG, May PC, Finch CE (1987) Dopamine and serotonin systems in human and rodent brain: effects of age and neurodegenerative disease. J Am Geriatr Soc 35:334–345. https://doi.org/10.1111/j.1532-5415.1987.tb04641.x

Natarajan S, Shunmugiah KP, Kasi PD (2013) Plants traditionally used in age-related brain disorders (dementia): an ethanopharmacological survey. Pharm Biol 51:492–523. https://doi.org/10.3109/13880209.2012.738423

Nayak L, Iwamoto FM (2010) Primary brain tumors in the elderly. Curr Neurol Neurosci Rep 10:252–258. https://doi.org/10.1007/s11910-010-0110-x

Olivares D, Deshpande VK, Shi Y et al (2012) N-methyl D-aspartate (NMDA) receptor antagonists and memantine treatment for Alzheimer's disease, vascular dementia and Parkinson's disease. Curr Alzheimer Res 9:746–758. https://doi.org/10.2174/156720512801322564

Omuro AMP, Abrey LE (2006) Chemotherapy for primary central nervous system lymphoma. Neurosurg Focus 21:E12. https://doi.org/10.3171/foc.2006.21.5.13

Oskarsson B, Gendron TF, Staff NP (2018) Amyotrophic lateral sclerosis: an update for 2018. Mayo Clin Proc 93:1617–1628. https://doi.org/10.1016/j.mayocp.2018.04.007

Pelleieux S, Picard C, Lamarre-Théroux L et al (2018) Isoprenoids and tau pathology in sporadic Alzheimer's disease. Neurobiol Aging 65:132–139. https://doi.org/10.1016/j.neurobiolaging.2018.01.012

Pfeiffer RF, Isaacson SH, Pahwa R (2020) Clinical implications of gastric complications on levodopa treatment in Parkinson's disease. Parkinsonism Relat Disord 76:63–71. https://doi.org/10.1016/j.parkreldis.2020.05.001

Piersol CV, Jensen L, Lieberman D, Arbesman M (2018) Occupational therapy interventions for people with Alzheimer's disease. Am J Occup Ther 72:7201390010p1–7201390010p6. https://doi.org/10.5014/ajot.2018.721001

Pîrşcoveanu DFV, Pirici I, Tudorică V et al (2017) Tau protein in neurodegenerative diseases—a review. Rom J Morphol Embryol = Rev Roum Morphol Embryol 58:1141–1150

Rajaram S, Jones J, Lee GJ (2019) Plant-based dietary patterns, plant foods, and age-related cognitive decline. Adv Nutr 10:S422–S436. https://doi.org/10.1093/advances/nmz081

Reich SG, Savitt JM (2019) Parkinson's disease. Med Clin North Am 103:337–350. https://doi.org/10.1016/j.mcna.2018.10.014

Santin Y, Resta J, Parini A, Mialet-Perez J (2021) Monoamine oxidases in age-associated diseases: new perspectives for old enzymes. Ageing Res Rev 66:101256. https://doi.org/10.1016/j.arr.2021.101256

Sawikr Y, Yarla NS, Peluso I et al (2017) Neuroinflammation in Alzheimer's disease: the preventive and therapeutic potential of polyphenolic nutraceuticals. Adv Protein Chem Struct Biol 108: 33–57. https://doi.org/10.1016/bs.apcsb.2017.02.001

Scapagnini G, Caruso C, Calabrese V (2010) Therapeutic potential of dietary polyphenols against brain ageing and neurodegenerative disorders. Adv Exp Med Biol 698:27–35. https://doi.org/10.1007/978-1-4419-7347-4_3

Sharma K (2019) Cholinesterase inhibitors as Alzheimer's therapeutics (review). Mol Med Rep 20: 1479–1487. https://doi.org/10.3892/mmr.2019.10374

Shukla R, Pahal S, Gupta A et al (2023) Modulation of GPCR receptors common to gut inflammatory diseases and neuronal disorders, Alzheimer's and Parkinson's diseases as druggable targets through Withania somnifera bioactives: an in silico study. J Biomol Struct Dyn 41:4485–4503. https://doi.org/10.1080/07391102.2022.2068072

Singh A, Kukreti R, Saso L, Kukreti S (2019) Oxidative stress: a key modulator in neurodegenerative diseases. Molecules 24. https://doi.org/10.3390/molecules24081583

Soria Lopez JA, González HM, Léger GC (2019) Alzheimer's disease. Handb Clin Neurol 167: 231–255. https://doi.org/10.1016/B978-0-12-804766-8.00013-3

Stacchiotti A, Corsetti G (2020) Natural compounds and autophagy: allies against neurodegeneration. Front Cell Dev Biol 8:555409. https://doi.org/10.3389/fcell.2020.555409

Szklener K, Mazurek M, Wieteska M et al (2022) New directions in the therapy of glioblastoma. Cancers (Basel) 14. https://doi.org/10.3390/cancers14215377

Tolar M, Abushakra S, Hey JA et al (2020) Aducanumab, gantenerumab, BAN2401, and ALZ-801-the first wave of amyloid-targeting drugs for Alzheimer's disease with potential for near term approval. Alzheimers Res Ther 12:95. https://doi.org/10.1186/s13195-020-00663-w

Varesi A, Campagnoli LIM, Carrara A et al (2023) Non-enzymatic antioxidants against Alzheimer's disease: prevention, diagnosis and therapy. Antioxidants (Basel, Switzerland) 12. https://doi.org/10.3390/antiox12010180

Vogelbaum MA, Parney IF, Elder JB, Cahill D (2021) Neurosurgical involvement in clinical trials for CNS tumors. J Neuro-Oncol 151:367–373. https://doi.org/10.1007/s11060-020-03438-3

Wang R, Reddy PH (2017) Role of glutamate and NMDA receptors in Alzheimer's disease. J Alzheimers Dis 57:1041–1048. https://doi.org/10.3233/JAD-160763

Ward M, Ahmed M, Markosian C et al (2021) Complications associated with deep brain stimulation for Parkinson's disease: a MAUDE study. Br J Neurosurg 35:625–628. https://doi.org/10.1080/02688697.2021.1935727

Witzel S, Maier A, Steinbach R et al (2022) Safety and effectiveness of long-term intravenous administration of Edaravone for treatment of patients with amyotrophic lateral sclerosis. JAMA Neurol 79:121–130. https://doi.org/10.1001/jamaneurol.2021.4893

Yang G, Wang Y, Tian J, Liu J-P (2013) Huperzine a for Alzheimer's disease: a systematic review and meta-analysis of randomized clinical trials. PLoS One 8:e74916. https://doi.org/10.1371/journal.pone.0074916

# The Controversy Surrounding Drugs Against Neurodegenerative Disorders: Benefit or Harm?

17

Priyadarshini Roy, Dibyajyoti Ghosh, Rupa Sanyal, Nithar Ranjan Madhu, and Abhijit Dey

**Abstract**

The global healthcare system and public health are burdened by Alzheimer's, Parkinson's, and Huntington's diseases. Due to their increasing prevalence and incapacity, extensive research has been done to develop appropriate treatments. Pharmacological interventions to slow or stop neurodegeneration have garnered interest, but their benefits and risks are still debated. This academic review examines the consequences of neurodegenerative disease treatments from many perspectives. Drug treatments and drug use are the main topics of discussion. Many feel that neurodegenerative drugs can slow disease progression, improve cognitive function, and improve quality of life. Clinical investigations suggest these medicines reduce symptoms and delay functional decline. Neuroprotective medications can also change disease processes. These perspectives emphasise the importance of early diagnosis and pharmaceutical treatment. Sceptics doubt the efficacy and safety of neurodegenerative disease medicines. Some contend that clinical research's small results may not necessarily benefit patients. These drugs also have side effects. These adverse events may be modest or severe. Due to financial strain on healthcare systems, excessive treatment costs may prohibit many from receiving it. The treatment for neurodegenerative diseases is outlined in this chapter. Tetrabenazine, L-DOPA/Levodopa, donepezil, and dopamine

P. Roy · D. Ghosh · A. Dey (✉)
Department of Life Sciences, Presidency University, Kolkata, West Bengal, India
e-mail: abhijit.dbs@presiuniv.ac.in

R. Sanyal
Department of Botany, Bhairab Ganguly College, Belgharia, West Bengal, India

N. R. Madhu
Department of Zoology, Acharya Prafulla Chandra College, New Barrackpore, Kolkata, West Bengal, India



S. Pathak, A. Banerjee (eds.), *Neuroprotective Effects of Phytochemicals in Brain Ageing*, https://doi.org/10.1007/978-981-99-7269-2_17

modulators are examples. It investigates the processes by which medications achieve their effects, as well as the molecular basis of action, most prescribed drug, therapeutic role, clinical data, and common side effects.

**Keywords**

Neurodegeneration · Parkinson's Disease · Alzheimer's Disease

## 17.1 Introduction

A collection of chronic, progressive diseases known as neurodegenerative disorders impact the nervous system and eventually result in the loss of cognitive and/or motor function (Fig. 17.1). Normal brain function depends on a sufficient blood supply and the anatomical and functional integrity of blood vessels. On the other hand, abnormalities in the blood-brain barrier and cerebral blood flow are early signs of neurodegenerative diseases in both human subjects and animal models (Sweeney et al. 2018). Neurodegenerative disorders are a growing public health concern worldwide. It is estimated that over 50 million people worldwide are affected by Alzheimer's disease alone, and this number is expected to triple by 2050. Other neurodegenerative disorders such as Huntington's disease, amyotrophic lateral sclerosis, and frontotemporal dementia also have significant impacts on individuals and their families. Despite ongoing research efforts, there are currently no cures for these diseases, making early detection and intervention critical for managing symptoms and improving quality of life. It is essential to continue advancing our understanding of the causes and mechanisms underlying neurodegenerative disorders to develop effective treatments and improve outcomes for those affected.

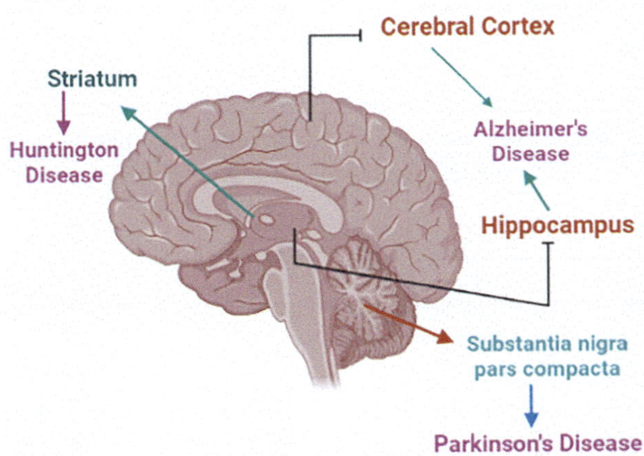


**Fig. 17.1** The brain parts affected by indifferent neurodegenerative disorders are shown in a schematic representation (adapted from Kumar et al. 2018)

## 17.2 Types of Neurodegenerative Disorders

The precise molecular pathways underlying neurodegenerative disorders are intricate and multifaceted, involving a variety of genetic, lifestyle, and environmental factors, with some known to cause the accumulation of aberrant protein aggregates in the brain, subsequently leading to the death of neurons. Numerous molecular pathways, including oxidative stress, neuroinflammation, mitochondrial dysfunction, and decreased protein clearance, have been hypothesised to underlie these processes. In this review, we aim to discuss primarily four types of neurodegenerative disorders, namely Alzheimer's disease (AD), Parkinson's disease (PD), and Huntington's disease (HD). In Table 17.1, we present a summary of the most commonly prescribed drugs, their molecular mechanisms of action, therapeutic purposes, and potential side effects for these neurodegenerative diseases.

### 17.2.1 Alzheimer's Disease

More than 45 million people around the world are afflicted with Alzheimer's disease, as stated by Alzheimer's Disease International (2015) reports. Despite the efforts of researchers, the disease's primary molecular pathways are still missing from our understanding. In AD, the accumulation of amyloid beta (Aβ) and tau proteins disrupts neuronal function and triggers a cascade of events that lead to neurodegeneration. These amyloid beta-tau protein aggregates form plaques that interfere with the communication between neurons, eventually causing neuronal apoptosis. Aβ is a heterogeneous mix of small peptides with a length of 37–43 amino acids, generated by β- and γ-secretase sequentially cleaving the amyloid precursor protein (APP) (Giorgi et al. 2017). Both autophagy and ubiquitin-proteasome system (UPS) serve a role in Aβ and tau degradation, the failure to do which results in Aβ and tau accumulation and toxicity in AD experimental models and humans (Uddin et al. 2018).

### 17.2.2 Parkinson's Disease

The Parkinson's Foundation claims that millions of individuals worldwide are afflicted by Parkinson's disease (PD), a neurological condition. Although the precise causation of PD is unknown, it is thought to be a result of both genetic and environmental factors. The loss of dopaminergic neurons in the substantia nigra, a part of the brain involved in movement control, is one well-established mechanism for PD. Dopamine levels fall as a result of the loss of these neurons, causing the tremors, stiffness, and bradykinesia that are hallmarks of PD. The molecular pathways underlying PD have been better understood thanks to numerous investigations. Parkinson's disease has been associated with mutations in genes, including SNCA, LRRK2, and PARKIN, for instance (Nuytemans et al. 2010). These genes are involved in processes that are linked to the development of PD,

**Table 17.1** The most frequently prescribed drugs for a particular neurodegenerative disease, along with its molecular mechanism of action, therapeutic role, potential side effects, and controversies associated with its treatment

| S. no. | Disease | Most prescribed drug | Molecular basis of action | Therapeutic Role | Side Effects and Controversy Regarding Treatment | References |
|---|---|---|---|---|---|---|
| 1 | Huntington's disease (HD) | Tetrabenazine | • By decreasing VMAT2, which decreases dopamine absorption and storage in the vesicles, dopamine is selectively depleted in the Striatum of the brain<br>• The amount of dopamine available for signalling in the brain is decreased as a result of a decrease in the release of dopamine into the synapse | • Restores neurotransmitter equilibrium to reduce excessive neuronal signalling and hyperactivity that contribute to chorea | • Sedation<br>• Depression<br>• Parkinsonism-like symptoms<br>• Akathisia<br>• Increased risk of suicidal thoughts or behaviours Individuals respond differently to tetrabenazine, and although some may benefit significantly and experience few adverse effects, others may respond poorly or experience unfavourable side effects. | • Poon et al. (2010)<br>• Kenney et al. (2007)<br>• Travessa et al. (2017) |
| 2 | Parkinson's disease (PD) | L-DOPA/ Levodopa | • Dopamine precursor can pass through the blood-brain barrier and enter the brain, where it is processed into the neurotransmitter by the enzyme aromatic L-amino acid decarboxylase (AADC) with the aid of vitamin B6 as a cofactor | • Levodopa reduces the motor symptoms of Parkinson's disease such as tremors, rigidity, and bradykinesia (slowness of movement) by restoring dopamine levels.<br>• Movement control is enhanced, and normal signalling in the brain's motor circuits is restored thanks to the elevated dopamine levels. | • The onset of motor problems such as dyskinesias (involuntary movements) and variations in motor responsiveness might be brought on by prolonged use | • Cilia et al. (2020)<br>• Cabreira et al. (2019) |

| | | | | | | |
|---|---|---|---|---|---|---|
| 3 | Alzheimer's disease (AD) | Donepezil | • Acts by attaching to and blocking the function of acetylcholinesterase (AChE), an enzyme that breaks down acetylcholine<br>• This inhibition of AChE enables acetylcholine to stay in the synapses for an extended duration, leading to improved cholinergic neurotransmission | • The elevated acetylcholine levels in the brain resulting from the action of donepezil help counterbalance the cholinergic deficiency observed in Alzheimer's disease<br>• This can potentially enhance cognitive abilities like memory, thinking, and behaviour | • Bradycardia<br>• Insomnia<br>• Loss of appetite | • Vecchio et al. (2021)<br>• Gao et al. (2023) |

including protein aggregation, mitochondrial dysfunction, and defective autophagy. Parkinson's disease susceptibility is impacted by both genetic and environmental variables. While only a small portion of Parkinson's cases are caused by genetic abnormalities, environmental variables such as pesticides and chemicals have been linked to an elevated risk of the condition. However, there is still much to learn about how genetic predisposition and environmental factors interact specifically in PD.

### 17.2.3 Huntington's Disease

Millions of people around the world are afflicted by the inherited neurological ailment known as Huntington's disease (HD). The Huntingtin (HTT) gene mutation, which is indicated by an unnatural amplification of CAG repeats in the gene, is the cause of the condition. The basal ganglia in particular experience gradual degeneration as a result of the build-up of mutant Huntingtin protein (mHTT) in neurons. The molecular pathways causing HD have been clarified by many investigations. It has been demonstrated by DiFiglia et al. (1997) and Sathasivam et al. (2013) that the mHTT is subjected to an abnormal process that leads to the formation of damaging fragments and aggregates. These aggregates hinder cellular processes and obstruct vital intracellular pathways involved in the regulation of transcription, mitochondrial activity, and protein quality. The mutant HTT gene is inherited from affected parents and is directly associated with the development of HD. Each kid of a person with the disease has a 50% chance of inheriting the faulty gene because the condition is autosomal dominant. Bigger repeat size is typically associated with an earlier onset and a more severe condition, and the number of CAG repeats in the HTT gene also affects the age at which symptoms first appear and how severe they are. The development of treatment approaches that address the processes of mHTT toxicity has been the main focus of research. Tabrizi et al. (2019) state that some of the techniques include decreasing the expression of mutant HTT, increasing the clearance of mutant HTT, and modifying downstream pathways that are affected by the toxicity of mutant HTT.

## 17.3 Alzheimer's Disease—The Controversy

To add to the expanding list of unsuccessful attempts to identify viable treatments, two highly publicised clinical studies looking into therapeutics for Alzheimer's disease failed in the early months of 2018. The amyloid hypothesis, which has dominated Alzheimer's research for more than 25 years, served as the foundation for these experiments. According to the amyloid hypothesis, the disease is primarily brought on by the build-up of amyloid peptide. The theory has not, however, ever received widespread acceptance, and the negative results of pharmacological studies have stoked scepticism and sparked discussions among scientists.

The validity of the amyloid hypothesis, according to its detractors, is called into question by the discovery of amyloid plaques in the brains of elderly people who are cognitively healthy.

The theory's proponents contend that not all plaques are equally hazardous and that these plaques may signify an early stage of Alzheimer's disease. According to studies, soluble oligomers rather than deposited plaques may have a higher effect on memory problems and synaptic impairment. Concerns concerning the veracity of the amyloid hypothesis have been raised in the light of the failure of medication studies that target Aβ. Numerous treatment strategies have been investigated, such as immunotherapy and inhibition of β- or γ-secretase, but they have had mixed results or possibly accelerated cognitive deterioration.

Given that Alzheimer's-related processes can have advanced significantly by the time individuals enrol in clinical trials, the failure of these trials has led to the realisation that the timing of drug administration may be important. The amyloid hypothesis has faced difficulties and disagreements, but researchers are still hopeful about the prospect of new treatments. Early studies of the amyloid plaque-clearing antibody aducanumab have yielded promising results. Its effectiveness in preventing cognitive deterioration in Alzheimer's patients will be determined by ongoing phase III trials. Additionally, studies looking into ways to completely prevent AD are of considerable interest to both patients and researchers since they may shed light on the amyloid hypothesis.

Finally, it should be noted that the amyloid hypothesis, which holds that Aβ build-up is the primary cause of Alzheimer's disease, has come under fire and has been tested in recent years. Scepticism has been encouraged by the unimpressive results of medication studies and the existence of plaques in people who are cognitively healthy. To develop efficient treatments and preventive measures for Alzheimer's in the future, current studies and clinical trials continue to examine the intricate mechanisms underlying the illness.

There is hope for people with Alzheimer's disease as we enter a new age of medical research and technology. Novel approaches to diagnosis, prevention, and treatment have emerged as a result of the dogged quest to decipher the complex mechanisms underpinning this fatal illness.

The capacity to identify tiny changes in the brain is being revolutionized by advancements in neuroimaging technology, such as positron emission tomography (PET) and functional magnetic resonance imaging (fMRI). This enables earlier and more accurate diagnosis. These discoveries shed light on the early phases of Alzheimer's disease, possibly paving the way for therapies while their therapeutic impact is at its peak.

Another promising new area of research is the understanding of Alzheimer's disease and its complicated genetic and molecular components. Customized therapies that take into account each person's unique genetic composition may soon be a reality, thanks to the development of precision medicine. There is promise for more successful and less intrusive treatments for Alzheimer's if we can identify and target certain genetic markers. This could lead to individualized therapy methods.

New medications and treatments are being developed at a rapid pace in the pharmaceutical industry. Immunotherapies that target beta-amyloid plaques and medications that modulate tau protein levels are only two of the novel approaches being investigated. We are getting closer to a future when Alzheimer's disease-modifying medicines are a clinical reality because of collaborative efforts between academics, industry, and advocacy groups. This atmosphere is conducive to ground-breaking discoveries.

People with Alzheimer's also benefit greatly from technological developments in care and assistance. Both patients and caregivers are benefiting from the increased quality of life brought about by the utilization of assistive technologies, artificial intelligence, and smart home solutions. These technologies are designed to empower individuals and make life easier for those living with the disease. They range from apps that train the brain to smart devices that track daily activities.

There will be obstacles along the way, but the progress we have made so far in Alzheimer's research gives us hope. The search for a cure for Alzheimer's is gaining momentum because of international cooperation, more financing, and a greater understanding of the disease's social consequences. The story of Alzheimer's disease may soon be rewritten from one of hopelessness to one of resilience if a multi-pronged strategy integrating early detection, individualized medicine, and novel treatments is implemented. The fight against Alzheimer's is far from over, but recent advances point to a future where the disease is seen as a problem that can be overcome by empathy, creativity, and a determination to improve people's lives.

## 17.4 Parkinson's Disease—The Controversy

Alpha-synuclein-based Lewy bodies, which are a hallmark of PD, are aberrant protein deposits that accumulate. Alpha-synuclein's precise contribution to the onset of the disease is still up for dispute. The precise mechanisms causing neuronal degeneration and disease progression are still poorly understood, despite the fact that protein aggregation is a common hallmark in many neurodegenerative illnesses. The major toxic trigger for neuronal susceptibility in Parkinson's disease and the precise configuration in which alpha-synuclein exerts its effects are still being debated. Furthermore, it is unclear whether Lewy bodies play a protective role by trapping harmful forms of alpha-synuclein or if they cause neuronal death.

There is also a current study into the spread and templating of alpha-synuclein as a significant element in the development of the disease as well as the potential for therapies involving antibodies that scavenge extracellular alpha-synuclein.

In this situation, several crucial inquiries come up: What alpha-synuclein hazardous conformations lead to the selective death of dopaminergic neurons? Is it feasible for alpha-synuclein to be transferred from one cell to another, and if so, which species are involved? Is there a seeding and templating mechanism that leads to the diseased forms of physiologically normal alpha-synuclein? Despite the growing body of research evidence, there are still ongoing debates in the scientific community, and no one hypothesis is supported by a strong consensus. Findings can be interpreted in a variety of ways, frequently leaving an opportunity for alternative mechanisms, and it is still unclear how certain factors cause other factors.

The use of bimolecular fluorescence complementation (BiFC) technology, which relies on fluorescent proteins that have been genetically modified, is discouraged by Tiago Outeiro (2020) in response to a study conducted by Frey et al. (2021). Outeiro claims that some of the findings from the study are similar to earlier findings from his own team's team in 2008. In his answer, Hilal Lashuel highlights some drawbacks of the BiFC method for detecting alpha-synuclein oligomers, notably with regard to estimating the proportion of fragments with monomeric versus oligomeric status. BiFC is nevertheless a useful method for identifying alpha-synuclein oligomers in cellular models, despite these drawbacks.

Researchers at the University of California San Diego School of Medicine conducted a retrospective study to investigate the mortality risk related to pimavanserin (Nuplazid), a medication used to treat Parkinson's disease psychosis (PDP), a symptom characterised by hallucinations and delusions. The study examined the medical records of Parkinson's patients treated with pimavanserin or quetiapine (Seroquel) in the UC San Diego Health System between April 2016 and April 2018. Concerns about pimavanserin's limited clinical studies and observed adverse effects have been raised since the FDA approved it as a breakthrough therapy for PDP in 2016.

In 2023, there have been enormous developments in Parkinson's disease research and innovation. The complexity of the disease has been largely unveiled by emerging technology, such as artificial intelligence, genetic investigations, and improved imaging tools. The development of Parkinson's disease can be slowed down by therapies that are now possible because of more accurate early diagnosis technologies.

The identification of critical genetic markers linked to Parkinson's disease through genomic research has opened the door to the development of individualized treatment strategies. The potential for more targeted and efficient interventions, with fewer side effects and better outcomes, is great with personalized therapies derived from an individual's genetic composition.

## 17.5 Huntington's Disease—The Controversy

Since just two substances have been FDA approved to treat HD, developing new medications for this condition is very challenging. Thus, it is essential to assess drug development programmes in order to comprehend the causes of their failure during the initial phases of development.

All trials that were registered in the WHO International Clinical Trials Search Portal up until May 2017 were looked at in order to analyse HD clinical trials. Two separate authors independently selected and extracted the data. By dividing the number of compounds that moved on to the following phase by the total number of compounds in that phase, the success rate of each trial phase was computed. By dividing the total number of compounds by the number of compounds that got regulatory approval, the overall success rate was calculated.

There were 99 trials found, examining 41 substances and 11 non-pharmacological interventions such as devices and cell therapies. Phase 1 trials constituted 24.2% of them, Phase 2 trials 46.5%, Phase 3 trials 20.2%, and Phase 4 trials 2.0%. A total of 60.6% of the trials included industry sponsorship. Creatine, latrepirdine, and pridopidine were the substances that were most frequently examined. From phase 1 to phase 3 studies, the average number of people enrolled was 92.0, and the average length of the treatment was 262.9 days. From Phase 1 to Phase 2, Phase 2 to Phase 3, and Phase 3 to approval, there was a success percentage of 25.0%, 19.4%, and 14.3%. 3.5% was the overall success rate.

Huntington's disease, a degenerative ailment that appears in adults, is a difficult condition that requires genetic testing. In contrast to those who are at risk but choose not to get tested, clinicians who work with HD families spend more time counselling and supporting those who select genetic testing. Only about 12% to 17% of at-risk persons in North America and Europe currently go for testing. The requirement for genetic testing as a criterion for enrolment in recent and continuing HD therapy trials, however, may result in greater uptake. Reach2HD, First-HD, and Roche Htt lowering studies are a few examples of these studies.

In this cohort, the objective of the Prospective Huntington At-Risk Observational Study (PHAROS) was to ascertain if seasoned doctors could reliably recognise early clinical symptoms that indicate the development of HD.

A thorough clinical trial registry found 99 HD trials, despite HD being a rare disorder. A low success rate in the early stages of medication development and an even lower success rate in the latter stages were found by the analysis. The stark difference in success rates between medication discovery and development warrants serious analysis and improvement. Fig. 17.2 depicts the three diseases discussed above and the controversies surrounding them.

As new research on Huntington's disease (HD) emerges, there is hope for the disease's potential future. In order to better diagnose, treat, and eventually cure this terrible neurological disease, researchers are devoted to understanding its intricacies and are investigating novel ways to do so.

Heredity restoration is one of the primary research areas. As our knowledge of the genetic causes of Huntington's disease grows, gene-editing tools like CRISPR-Cas9 present promising new possibilities for individualized treatments. In an effort to slow or reverse the course of HD, researchers are looking into methods to alter or replace the defective gene that causes the disease.

The role of advances in precision medicine is also critical. Improved and more tailored interventions are possible when treatment plans are customized according to each patient's genetic makeup and the unique symptoms of HD. This method has the ability to enhance the results and well-being of individuals impacted by the illness.

Research on biomarkers is also continuing, which could one day allow for the monitoring and early diagnosis of Huntington's disease. The development of a disease can be slowed down if accurate biomarkers can be identified in time for therapeutic interventions to be administered at the earliest possible stage.

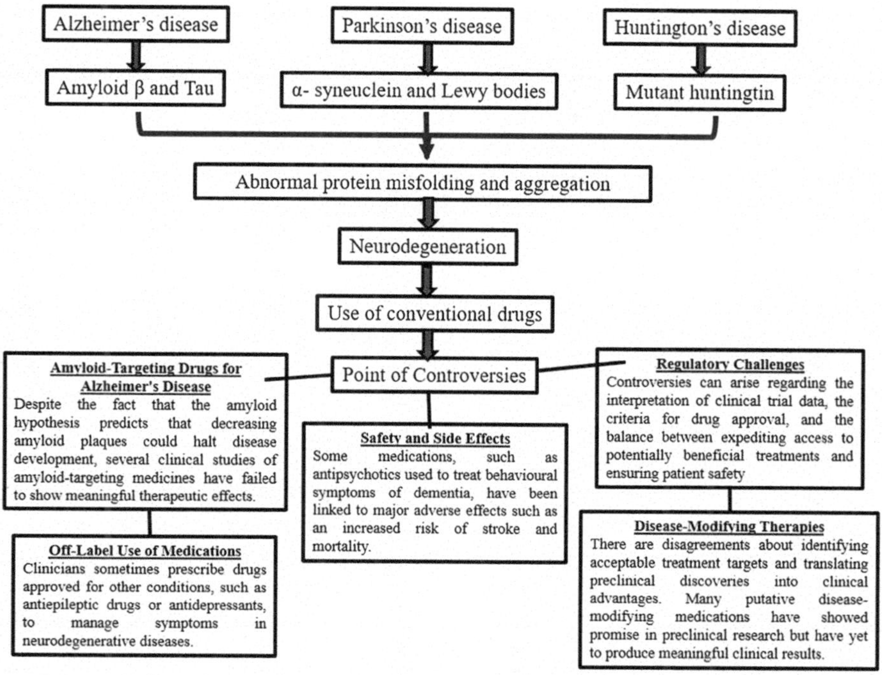


**Fig. 17.2** Schematic diagram representing the controversies surrounding the three diseases discussed above

To make headway, it is essential for researchers, medical professionals, and advocacy groups to work together. We can learn more about Huntington's disease faster and more thoroughly if researchers share their data, resources, and insights. A more inclusive and compassionate approach to handling the issues posed by HD can be achieved through initiatives that aim to raise awareness and develop a supportive community.

Despite ongoing obstacles, the unwavering commitment of researchers and the strength displayed by those impacted by Huntington's disease give us reason to believe in a future where effective therapies and a possible cure will be within reach. New hope for people affected by Huntington's disease is emerging as research shows promise of changing the course of the disease.

## 17.6 Conclusion

Due to the fact that the average age of the world's population is continually rising, more and more people are being diagnosed with neurodegenerative disorders. NDs are age-related disorders that have extremely varied pathophysiologies. Additionally, the origins and mechanisms of these diseases are poorly understood, which results in a paucity of therapeutic options. The prevalence of disorders that impair memory, cognition, and movement has increased in tandem with the ageing of the world's population, which is a trend that is expected to continue. The World Health Organization (WHO) forecasts that in the next 20 years, neurodegenerative diseases (NDs) that mostly affect motor neurons will overtake cancer to rank as the second-most prevalent leading cause of mortality, behind only cardiovascular illnesses. As a result, there is an immediate need for treatments for NDs. As a result of the review works, it has become abundantly evident that repurposing drugs can be an attractive source of candidates for the treatment of diseases different than those for which they were initially employed in therapeutics. The neurodegenerative disorders that are discussed in this article are not one hundred percent treatable and have a severe negative impact on the patient's quality of life in general. These medications, on the other hand, should not be written off as a lost cause, rather; they should be seen as a model starting point for the synthesis of new molecules, which can be accomplished through molecular alterations and ligand-based design. At this time, there is no neurodegenerative disease that can be cured, and the medicines that are available can only control the symptoms or slow down the advancement of the disease. Synthesis, natural products, and current medications are the three primary places from which novel therapeutics can originate. The third and final source is known as drug repurposing, and it offers the greatest number of benefits. This is because the pharmacokinetic and pharmacodynamic profiles of the medicine have already been established, and the amount of investment required for this technique is significantly lower than that required for the traditional creation of new drugs.

In conclusion, the discussion regarding medications that treat neurodegeneration is a contentious topic due to its complexity and varied nature. Concerns have been raised concerning the possibility of these medications causing harm as well as their limits, despite the fact that they may be able to provide major benefits in the treatment and management of neurodegenerative illnesses. On one side, medications that aim to combat neurodegeneration have shown promise in reducing symptoms and enhancing cognitive function associated with illnesses such as Alzheimer's, Huntington's, and Parkinson's diseases. They have the potential to impede the advancement of these diseases and to improve the quality of life of those who are afflicted as well as the lives of those who care for them. Additionally, ongoing research and development activities in this area have the potential to provide treatments that are even more successful in the near future. Nevertheless, there are a number of difficulties and debates that are connected to these medications. Because not all patients react in the same way to treatment, one of the key concerns is that their effectiveness can be reduced in some instances. The improvements that can be attained through the use of these medications are sometimes just transitory and do not deal with the fundamental reasons for neurodegeneration. In addition, there is a possibility of major adverse effects and hazards linked with the use of these medications, including problems with the cardiovascular system and the digestive tract, as well as combinations with other medications. The high price of these medications is another source of contention, since it places a financial burden not just on individuals but also on the healthcare systems that they utilise. There is a possibility that access to these treatments will be restricted, in particular, in areas with poor incomes or nations that lack developed healthcare infrastructure. This raises major ethical considerations about whether or not everyone should have equal access to medicines that could potentially change their lives.

There are many who believe that the emphasis placed on the creation of drugs to treat neurodegenerative illnesses may obscure the significance of preventative measures and treatments that do not include the use of pharmaceuticals. It has been demonstrated that making changes to one's lifestyle, including incorporating cognitive stimulation, regular exercise, a good diet, and social involvement into one's routine, can have a beneficial effect on neurodegenerative disorders. It is possible that a more all-encompassing and holistic strategy for disease management could be achieved by promoting these techniques in conjunction with medication therapy.

In conclusion, anti-neurodegenerative medications have the potential to give major improvements in the treatment and management of these incapacitating illnesses. However, there are concerns over their limited effectiveness, side effects, high prices, and the need for a treatment strategy that is more comprehensive, which has led to the debate that surrounds their use. To resolve these disputes and ensure the development of treatments for neurodegenerative disease that are safe, effective, and easily accessible, it is essential to do additional research, engage in public conversation, and work together with researchers, healthcare professionals, lawmakers, and patient's rights groups.

## References


Cabreira V, Massano J (2019) Doença de Parkinson: Revisão Clínica e Atualização [Parkinson's disease: clinical review and update]. Acta Medica Port 32(10):661–670. https://doi.org/10.20344/amp.11978

Cilia R, Cereda E, Akpalu A, Sarfo FS, Cham M, Laryea R, Obese V, Oppon K, Del Sorbo F, Bonvegna S, Zecchinelli AL, Pezzoli G (2020) Natural history of motor symptoms in Parkinson's disease and the long-duration response to levodopa. Brain 143(8):2490–2501. https://doi.org/10.1093/brain/awaa181

DiFiglia M, Sapp E, Chase KO, Davies SW, Bates GP, Vonsattel JP, Aronin N (1997) Aggregation of huntingtin in neuronal intranuclear inclusions and dystrophic neurites in brain. Science 277 (5334):1990–1993

Frey B, AlOkda A, Jackson MP, Riguet N, Duce JA, Lashuel HA (2021) Monitoring alpha-synuclein oligomerization and aggregation using bimolecular fluorescence complementation assays: What you see is not always what you get. J Neurochem 157(4):872–888

Gao X (2023) Editorial: disease-modifying targets and strategies for Alzheimer's disease and Parkinson's disease. Front Aging Neurosci 15:1247256. https://doi.org/10.3389/fnagi.2023.1247256

Giorgi FS, Ryskalin L, Ruffoli R, Biagioni F, Limanaqi F, Ferrucci M, Busceti CL, Bonuccelli U, Fornai F (2017) The neuroanatomy of the reticular nucleus locus coeruleus in Alzheimer's disease. Front Neuroanat 11:80. https://doi.org/10.3389/fnana.2017.00080

Kenney C, Powell S, Jankovic J (2007) Autopsy-proven Huntington's disease with 29 trinucleotide repeats. Mov Disord 22(1):127–130. https://doi.org/10.1002/mds.21195

Kumar S, Yadav N, Pandey S, Thelma BK (2018) Advances in the discovery of genetic risk factors for complex forms of neurodegenerative disorders: contemporary approaches, success, challenges and prospects. J Genet 97(3):625–648. https://doi.org/10.1007/s12041-018-0953-5

Nuytemans K, Theuns J, Cruts M, Van Broeckhoven C (2010) Genetic etiology of Parkinson disease associated with mutations in the SNCA, PARK2, PINK1, PARK7, and LRRK2 genes: a mutation update. Hum Mutat 31(7):763–780. https://doi.org/10.1002/humu.21277

Poon LH, Kang GA, Lee AJ (2010) Role of tetrabenazine for Huntington's disease-associated chorea. Ann Pharmacother 44(6):1080–1089. https://doi.org/10.1345/aph.1m582

Sathasivam K, Neueder A, Gipson TA, Landles C, Benjamin AC, Bondulich MK, Smith DL, Faull RL, Roos RA, Howland D, Detloff PJ (2013) Aberrant splicing of HTT generates the pathogenic exon 1 protein in Huntington disease. Proc Natl Acad Sci 110(6):2366–2370

Sweeney MD, Kisler K, Montagne A (2018) The role of brain vasculature in neurodegenerative disorders. Nat Neurosci 21:1318–1331. https://doi.org/10.1038/s41593-018-0234-x

Tabrizi SJ, Leavitt BR, Landwehrmeyer GB, Wild EJ, Saft C, Barker RA, Blair NF, Craufurd D, Priller J, Rickards H, Rosser A (2019) Targeting huntingtin expression in patients with Huntington's disease. N Engl J Med 380(24):2307–2316

Travessa AM, Rodrigues FB, Mestre TA, Ferreira JJ (2017) Fifteen years of clinical trials in Huntington's disease: a very low clinical drug development success rate. J Huntingtons Dis 6 (2):157–163. https://doi.org/10.3233/JHD-170245

Uddin MS, Stachowiak A, Mamun AA, Tzvetkov NT, Takeda S, Atanasov AG, Bergantin LB, Abdel-Daim MM, Stankiewicz AM (2018) Autophagy and Alzheimer's disease: from molecular mechanisms to therapeutic implications. Front Aging Neurosci 10:04. https://doi.org/10.3389/fnagi.2018.00004

Vecchio I, Sorrentino L, Paoletti A, Marra R, Arbitrio M (2021) The state of the art on acetylcholinesterase inhibitors in the treatment of Alzheimer's disease. J Cent Nerv Syst Dis 13:11795735211029112. https://doi.org/10.1177/11795735211029113

**SPRINGER NATURE**

# GPSR Compliance

*The European Union's (EU) General Product Safety Regulation (GPSR) is a set of rules that requires consumer products to be safe and our obligations to ensure this.*

*If you have any concerns about our products, you can contact us on ProductSafety@springernature.com*

In case Publisher is established outside the EU, the EU authorized representative is:

Springer Nature Customer Service Center GmbH
Europaplatz 3
69115 Heidelberg, Germany

The manufacturer's authorised representative in the EU is Springer Nature Customer Service Centre GmbH, Europaplatz 3, 69115 Heidelberg, Germany. If you have any concerns regarding our products, please contact ProductSafety@springernature.com

Printed and bound by CPI Group (UK) Ltd, Croydon, CR0 4YY
30/09/2026
02225298-0014